Lecture Notes in Computer Science 7874

Commenced Publication in 1973
Founding and Former Series Editors:
Gerhard Goos, Juris Hartmanis, and Jan van Leeuwen

Carla Gomes   Meinolf Sellmann (Eds.)

Integration of AI and OR Techniques in Constraint Programming for Combinatorial Optimization Problems

10th International Conference, CPAIOR 2013
Yorktown Heights, NY, USA, May 18-22, 2013
Proceedings

 Springer

Volume Editors

Carla Gomes
Cornell University, Department of Computer Science
5133 Upson Hall, 14853 Ithaca, NY, USA
E-mail: gomes@cs.cornell.edu

Meinolf Sellmann
Thomas J. Watson Research Center
10598 Yorktown Heights, NY, USA
E-mail: meinolf@us.ibm.com

ISSN 0302-9743   e-ISSN 1611-3349
ISBN 978-3-642-38170-6   e-ISBN 978-3-642-38171-3
DOI 10.1007/978-3-642-38171-3
Springer Heidelberg Dordrecht London New York

Library of Congress Control Number: 2013936947

CR Subject Classification (1998): G.1.6, G.1, G.2.1, F.2.2, I.2.8, J.1

LNCS Sublibrary: SL 1 – Theoretical Computer Science and General Issues

Typesetting: Camera-ready by author, data conversion by Scientific Publishing Services, Chennai, India

Printed on acid-free paper

Springer is part of Springer Science+Business Media (www.springer.com)

Preface

This volume is a compilation of the research program of the 10th International Conference on the Integration of Artificial Intelligence (AI) and Operations Research (OR) Techniques in Constraint Programming (CPAIOR 2013), held at the IBM worldwide research headquarters, the T.J. Watson Research Center, Yorktown Heights, NY, during May 18–22, 2013. More information about the conference can be found at: http://www.cis.cornell.edu/ics/cpaior2013/

The CPAIOR Conference Series

After a successful series of five CPAIOR international workshops in Ferrara (Italy), Paderborn (Germany), Ashford (UK), Le Croisic (France), and Montreal (Canada), in 2004 CPAIOR evolved into a conference. More than 100 participants attended the first meeting held in Nice (France). In the subsequent years, CPAIOR was held in Prague (Czech Republic), Cork (Ireland), Brussels (Belgium), Paris (France), Pittsburgh (USA), Bologna (Italy), Berlin (Germany), and Nantes (France). This year CPAIOR was held in the USA.

The aim of the CPAIOR conference series is to bring together researchers from constraint programming (CP), artificial intelligence (AI), and operations research (OR) to present new techniques or applications in the intersection of these fields, as well as to provide an opportunity for researchers in one area to learn about techniques in the others. A key objective of the conference is to demonstrate how the integration of techniques from different fields can lead to highly novel and effective new methods for large and complex problems. Therefore, papers that actively combine, integrate, or contrast approaches from more than one of the areas were especially welcome. Application papers showcasing CP/AI/OR techniques on innovative and challenging applications or experience reports on such applications were also strongly encouraged.

Program, Submissions, and Reviewing

The main CPAIOR 2013 program featured invited presentations from Peter van Beek on "Constraint Programming in Compiler Optimization: Lessons Learned," Andreas Krause on "Sequential Decision Making in Experimental Design and Computational Sustainability via Adaptive Submodularity," Vijay Saraswat on "Scalable Concurrent Application Frameworks for Constraint Solving," and an invited tutorial on "Recent Advances in Maximum Satisfiability and Extensions" by Carlos Ansotegui.

Seventy-one full papers were submitted to the conference. Out of these, 20 long papers and, additionally, 11 short papers were selected for presentation in the main technical program of this year's conference. Moreover, more than 15 presentation-only papers were submitted. These latter papers were not formally reviewed and are therefore not part of these proceedings.

As the conference is relatively small, the Program Chairs decided to develop and test a revised reviewing format. In other communities, papers are typically reviewed in depth by three members of the Program Committee (PC). In the Program Chairs' experience, this format works well in terms of filtering papers that are below the conference standards. However, interesting novel contributions that open exciting new research avenues can fall through the cracks more easily than incremental advancements in established fields of research. Clearly, this is not desirable.

To address this issue, the Program Chairs first assembled a large PC of almost 70 distinguished researchers, who we wish to thank whole-heartedly for their voluntary service to our community. Each paper was reviewed in depth by four PC members. The four initial reviews were sent to the authors and they were asked to provide their feedback. The initial reviewers then had a discussion and had the opportunity to adjust their reviews based on the authors' rebuttal. Seven additional PC members were assigned to each paper to simply vote yes or no, without the need to write a review. So in the end there were 11 votes per paper (four initial and seven additional). Those papers with at least six votes in favor were accepted for publication.

This scheme changed the traditional roles of the reviewers. They could no longer decide among themselves, they instead needed to convince seven other PC members. We found that this procedure positively changed the tone of the reviews. Moreover, the voting procedure meant that reviewers did not have to achieve consensus and could provide individual points of view until the very end.

On the other hand, the voting scheme resulted in a relatively large acceptance rate. All 20 long papers in these proceedings turned out to have at least eight votes in favor of publication. Since the Program Chairs did not want to overrule two-third majorities, all of these 20 papers were accepted in the main program.

Masterclass and Workshops

It is a wonderful tradition at CPAIOR that the main conference is preceeded by a weekend full of vibrant talks organized in a masterclass, as well as several workshops. This year's masterclass theme was on "Computational Sustainability: Optimization and Policy-Making." In addition, the program included three workshops on "Parallel Methods for Combinatorial Search and Optimization," on "Algorithm Selection," and on "General Principles in Seeking Feasible Solutions for Combinatorial Problems."

Many thanks to our Masterclass Chair Barry O'Sullivan, our Workshop Chair Horst Samulowitz, all workshop organizers, and all conference presenters. We especially thank our Conference Chair Ashish Sabhrawal, Publicity Chair Bistra Dilkina, Sponsorship Chair Stefan Heinz, and our conference manager Megan McDonald who all did an outstanding job organizing this event. Thanks also once more to all members of the PC and all reviewers.

Sponsors

The cost for holding an event like the CPAIOR conference would be much higher without the help of generous sponsors. We received outstanding support from the Institute for Computational Sustainability at Cornell University, IBM Research, and SAS. We also thank the Association for Constraint Programming (ACP), GAMS, and NICTA, as well as AIMMS, AMPL, Jeppesen, and SICS. Finally, thanks to Springer and EasyChair for their continuing support of the CPAIOR conference series.

February 2013   Carla Gomes
                                                                 Meinolf Sellmann

Table of Contents

Stronger Inference through Implied Literals from Conflicts and Knapsack Covers

Tobias Achterberg[1], Ashish Sabharwal[2], and Horst Samulowitz[2]

[1] IBM, Germany
achterberg@de.ibm.com
[2] IBM Watson Research Center, Yorktown Heights, USA
{ashish.sabharwal,samulowitz}@us.ibm.com

Abstract. Implied literals detection has been shown to improve performance of Boolean satisfiability (SAT) solvers for certain problem classes, in particular when applied in an efficient dynamic manner on learned clauses derived from conflicts during backtracking search. We explore this technique further and extend it to mixed integer linear programs (MIPs) in the context of conflict constraints. This results in stronger inference from clique tables and implication tables already commonly maintained by MIP solvers. Further, we extend the technique to knapsack covers and propose an efficient implementation. Our experiments show that implied literals, in particular through stronger inference from knapsack covers, improve the performance of the MIP engine of IBM ILOG CPLEX Optimization Studio 12.5, especially on harder instances.

1 Introduction

Systematic solvers for combinatorial search and optimization problems such as Boolean satisfiability (SAT), constraint satisfaction (CSP), and mixed integer programming (MIP) are often based on variants of backtracking search performed on an underlying search tree. A key to their effectiveness is the ability to prune large parts of the search tree without explicit exploration. This is done through inference during search, which can take two different forms. First, "probing" techniques [8, 11] are performed in advance to explore the potential effect of making a certain choice. Second, post-conflict analysis [1, 9] is conducted after the search has run into a conflict (i.e., an infeasible or non-improving region of the search space) in an attempt to learn and generalize the "cause" of that conflict. Stronger inference captured in the form of redundant constraints or tighter variable bounds typically leads to stronger propagation of constraints and more pruning of the search space during search. The goal of this work is to develop a technique for stronger inference during search through the concept of *implied literals*, which we apply dynamically (i.e., during search) to enhance the amount of propagation achieved from both the original problem constraints as well as constraints learned from conflicts.

In the context of SAT where the inference mechanism is unit propagation of clauses, one may perform a form of probing [8] that simply applies unit propagation to all individual literals (i.e., variables or their negations) at the root

C. Gomes and M. Sellmann (Eds.): CPAIOR 2013, LNCS 7874, pp. 1–11, 2013.

node of the search in order to detect *failed literals* [5] (i.e., literals setting which to 1 leads to a conflict by unit propagation) or to populate *implication lists* of literals containing implications of the form $x \rightarrow y$. The latter information can then, for instance, be used to shrink clauses by using *hidden literal elimination* (e.g., if $a \rightarrow b$ then $(a \vee b \vee c)$ can be reduced to $(b \vee c)$ [cf. 6]).

Most pertinent to this work, Heule et al. [6], Soos [12], and Matsliah et al. [10] have independently proposed ways to strengthen clause learning performed by SAT solvers by dynamically inferring implied literals, i.e., literals that the newly learned clause entails. A literal l is an *implied literal* for a clause C if all literals of C entail l. For instance, if $a \rightarrow d$, $\neg b \rightarrow d$, and $c \rightarrow d$, then $(a \vee \neg b \vee c)$ entails d. This observation forms the basis of several other techniques such as variations of *hyper binary resolution* and hidden literal elimination briefly mentioned above. To apply the technique during clause learning, one generates and periodically updates implication lists $L(l) = UnitPropagation(l)$ for each literal l. During this computation, one may also add not yet existing binary clauses corresponding to $\neg l \rightarrow \neg p$ for all $l \in L(p)$, and detect failed literals and add and propagate their negations as new unit literals. One may do the same also for all literals in the intersection of $L(p)$ and $L(\neg p)$. This strengthens the inference that unit propagation achieves. Matsliah et al. [10] have shown that by implementing the underlying data structures efficiently and using heuristics to guide choices such as how many implied literals and binary clauses to keep and how often to recompute implication lists, this technique can boost the performance of state-of-the-art SAT solvers such as Glucose 2.0 [3], especially on certain kinds of benchmarks coming from the planning domain.

We explore this concept further and extend it to mixed integer linear programs (MIPs). While implied literals can be derived using probing for failed literals or hyper binary resolution in the case of SAT [4], this turns out not to be the case for MIPs with non-binary variables. Further, rather than imposing the overhead of deriving and maintaining our own implication lists, we capitalize on the fact that MIP solvers often already internally maintain implications in the form of clique tables and implication tables. We propose algorithms to derive implied literals from conflict constraints in MIP solvers as well as from covers of knapsack constraints, while keeping the computational overhead low. Our experimental results on over 3,000 MIP instances show that the application of implied literals improves performance, especially on harder instances. Of the 1,096 benchmark instances that need at least 10 seconds to solve, implied literals detection affects 65% of the instances, speeding up the solution process on these instances by 6% and reducing the number of search tree nodes by 7%.

2   Implied Literals in Mixed Integer Programs

Let $N := \{1, \ldots, n\}$ be an index set for variables $z_j \in \mathbb{R}$, $j \in N$. Further, let $N_X \subseteq N$ be the subset of binary variables and $N_Y = N \setminus N_X$ the subset of non-binary variables. The set $N_I \subseteq N_Y$ denotes the indices of general integer variables. We consider the mixed integer (linear) program

$$\max \quad \sum_{j \in N} c_j z_j$$

$$\text{s.t.} \quad \sum_{j \in N} a_{ij} z_j \leq b_i \text{ for all } i = 1, \ldots, m$$

$$z_j \in \{0, 1\} \qquad \text{for all } j \in N_X$$

$$z_j \in \mathbb{Z} \qquad \text{for all } j \in N_I \subseteq N_Y$$

$$z_j \in \mathbb{R} \qquad \text{for all } j \in N_Y \setminus N_I$$

with objective coefficients $c \in \mathbb{R}^n$, right hand sides $b \in \mathbb{R}^m$, and coefficient matrix $A = (a_{ij})_{ij} \in \mathbb{R}^{m \times n}$. For ease of presentation we use the symbols $x_j := z_j$ for binary variables, $j \in N_X$, and $y_j := z_j$ for non-binary variables, $j \in N_Y$. Moreover, we define $X := \{x_j \mid j \in N_X\}$ and $Y := \{y_j \mid j \in N_Y\}$ to be the sets of binary and non-binary variables, respectively.

For $x \in X$, let $\bar{x} := 1 - x$ denote the "negated" binary variable, and let $\overline{X} := \{\bar{x} \mid x \in X\}$. We define $Z := X \cup \overline{X} \cup Y$. For $z \in Z$, $\mathrm{Dom}(z)$ denotes the domain of z.

Following Achterberg [1], we define *literals* in the context of MIP as bound inequalities of the form $(z \leq u)$ or $(z \geq l)$, where $z \in Z$ and $l, u \in \mathrm{Dom}(z)$. For the sake of simplicity of notation, we will work only with literals of the form $(z \leq u)$ in the rest of this paper. The arguments and constructs naturally generalize to any combination of literals of the form $(z_1 \leq u_1)$ and $(z_2 \geq l_2)$.

Definition 1. *Suppose we are given the implications $(z_i \leq u_i) \rightarrow (z \leq d_i)$ for $i \in \{1, \ldots, k\}$, where $z_i, z \in Z$, $u_i \in \mathrm{Dom}(z_i)$, and $d_i \in \mathrm{Dom}(z)$. Let $d_{max} := \max_{i=1}^{k} d_i$. Let $C := \bigvee_{i=1}^{k} (z_i \leq u_i)$ be a constraint. Then $(z \leq d_{max})$ is called an implied literal derived from C.*

We will sometimes refer to $(z \leq d_{\max})$ simply as an implied literal when C and the implications needed for its derivation are implicit in the context. Note that when $z_i = x$ for $x \in X$, the implication $(z_i \leq u_i) \rightarrow (z \leq d_i)$ in Definition 1 is equivalent to $(x = 0) \rightarrow (z \leq d_i)$. Similarly, when $z_i = \bar{x}$ for $\bar{x} \in \overline{X}$, the implication may equivalently be written as $(x = 1) \rightarrow (z \leq d_i)$.

2.1 Implied Literals from Conflict Constraints

As in the case of SAT, the idea is to efficiently store a number of implications and use them to infer implied literals when, for example, a new constraint is derived or added to the model. Fortunately, current MIP solvers typically already store certain types of implications of literals, which are used in presolving and during the branch and bound process. For example, CPLEX 12.5 internally maintains implications of the following types:

$$(x_i = v_i) \rightarrow (x_j = v_j) \qquad x_i, x_j \in X, v_i, v_j \in \{0, 1\}$$

$$(x = v) \rightarrow (y \leq d) \qquad x \in X, y \in Y, v \in \{0, 1\}, d \in \mathrm{Dom}(y)$$

Implications of the first type are stored in a *clique table* $\mathcal{K}$. For example, the set packing constraint $\sum_{i=1}^{k} x_i \leq 1$ leads to a "clique" of implications ($x_i = 1$) $\rightarrow$ ($x_j = 0$) for $i \neq j$. The clique table stores cliques $K \in \mathcal{K}$, $K \subseteq X \cup \bar{X}$, in aggregated form as set packing constraints $\sum_{x \in K} x \leq 1$, rather than explicitly recording the $|K|(|K|-1)$ implications that are entailed by the clique K. Implications of the second type are stored in an *implication table* $\mathcal{I}$. For example, a "variable upper bound" constraint $-ax + y \leq b$ with $x \in X$ and $y \in Y$ yields the implication ($x = 0$) $\rightarrow$ ($y \leq b$). Cliques and implications can be directly extracted from general linear constraints and computed with techniques such as probing. Note that CPLEX does not store implications between pairs of non-binary variables such as ($y_1 \leq u$) $\rightarrow$ ($y_2 \leq d$) with $y_1, y_2 \in Y$.

Example 1. Consider the constraint $C = (x_1 \geq 1) \vee (x_2 \leq 0) \vee (y_3 \geq 4) \vee (y_4 \leq 8)$ and the following cliques (left) and implications (right):

$$\bar{x}_1 + \bar{x}_2 + \bar{x}_5 \leq 1 \qquad\qquad (x_1 = 0) \rightarrow (y_3 \leq 2)$$
$$x_1 + x_6 \leq 1 \qquad\qquad (x_1 = 0) \rightarrow (y_4 \geq 12)$$
$$\bar{x}_2 + x_6 \leq 1$$

with $x_i \in X$ and $y_j \in Y$. Including the self-implicants, we get:

$$(x_1 \geq 1) \rightarrow (x_1 \geq 1), \qquad (y_3 \geq 4), \qquad\qquad (x_6 \leq 0)$$
$$(x_2 \leq 0) \rightarrow (x_1 \geq 1), (x_2 \leq 0), \qquad\qquad (x_5 \geq 1), (x_6 \leq 0)$$
$$(y_3 \geq 4) \rightarrow (x_1 \geq 1), \qquad (y_3 \geq 4)$$
$$(y_4 \leq 8) \rightarrow (x_1 \geq 1), \qquad\qquad (y_4 \leq 8)$$

Since $x_1 \geq 1$ is implied by all literals of C, it is an implied literal and we can permanently fix $x_1 := 1$.

Since a MIP solver already maintains cliques and implications, we can exploit them "for free" in the implied literals detection. Whenever a conflict constraint

$$C = \bigvee_{i=1}^{k} (z_i \leq u_i)$$

is derived for an infeasible node in the search tree, we apply Algorithm 1 to infer implied literals. The goal is to find new bounds $z \leq d$ that are implied by all of the k literals of C. Thus, for each variable $z \in Z$ we count the number of literals ($z_i \leq u_i$) that imply a bound $z \leq d_i$. If this count reaches $count[z] = k$ at the end of the algorithm, then an implied literal has been identified. In the case of a binary variable $z = x$ (line 18) we have $d_i = 0$ for all i, and the implied literal is ($x = 0$). In the case of a non-binary variable $z = y$ (line 20) the implied literal is ($y \leq maxd[y]$) where $maxd$ is an array that tracks the maximal implied bound d_i for non-binary variables $y \in Y$.

For each of the k literals of the conflict constraint C, the algorithm inspects the implications in the main loop of line 4 to update the *count* and *maxd* arrays. If the literal variable z_i is binary (line 6), then implications on other binary

---

Algorithm 1. Deriving Implied Literals from MIP Conflict Constraints

Input　: variables $Z = X \cup \overline{X} \cup Y$, conflict constraint $C = \bigvee_{i=1}^{k}(z_i \leq u_i)$,
　　　　　clique table $\mathcal{K}$, implication table $\mathcal{I}$
Output: a set L of implied literals derived from C

1　**begin**
2　　initialize $count[z] := 0$ for all $z \in Z$
3　　initialize $maxd[y] := -\infty$ for all $y \in Y$
4　　**forall** $i \in \{1, \ldots, k\}$ **do**
5　　　increment $count[z_i]$
6　　　**if** $z_i \in X \cup \overline{X}$ *(and thus $u_i = 0$)*, **then**
7　　　　**forall** $K \in \mathcal{K}$ with $\bar{z}_i \in K$ **do**
8　　　　　**forall** $x \in K \setminus \{\bar{z}_i\}$ *with* $count[x] = i - 1$ **do**
9　　　　　　increment $count[x]$

10　　　　**forall** $((z_i = 0) \rightarrow (y \leq d)) \in \mathcal{I}$ *with* $count[y] = i - 1$ **do**
11　　　　　increment $count[y]$
12　　　　　$maxd[y] := \max\{maxd[y], d\}$

13　　　**else**
14　　　　$maxd[z_i] := \max\{maxd[z_i], u_i\}$
15　　　　**forall** $((x = 1) \rightarrow (z_i \geq u)) \in \mathcal{I}$ *with* $u > u_i$ and $count[x] = i - 1$ **do**
16　　　　　increment $count[x]$

17　　$L := \emptyset$
18　　**forall** $x \in X$ *with* $count[x] = k$ **do**
19　　　set $L := L \cup \{(x = 0)\}$
20　　**forall** $y \in Y$ *with* $count[y] = k$ **do**
21　　　set $L := L \cup \{(y \leq maxd[y])\}$
22　　**return** L
23　**end**

---

variables x' can be found in the clique table, namely by inspecting the cliques $K \in \mathcal{K}$ with $\bar{z}_i \in K$, see line 7. Implications involving non-binary variables y can be inferred from the implication table, as done in line 10. On the other hand, if z_i is non-binary (line 13), then the implication table yields implications on binary variables by reversing the direction of the implication. Namely, an implication $(x = 1) \rightarrow (y \geq u)$ with binary $x \in X \cup \overline{X}$ can be rewritten as $(y < u) \rightarrow (x = 0)$, resulting in the implication of $(x = 0)$, see line 15.

Note that we need to include the trivial self-implicants $(z_i \leq u_i) \rightarrow (z_i \leq u_i)$ in the counting, see lines 5 and 14, in order to not miss the cases in which one of the literals of C is an implied literal. Note also that in each iteration i we only consider variables for which the *count* value is maximal, i.e., equal to $i - 1$. This makes sure that we do not count variables twice for the same conflict clause literal $(z_i \leq u_i)$ and it prevents unnecessary updates on variables that cannot be implied literals.

Performance improvements. It is easy to see that Algorithm 1 can be enhanced in order to avoid unnecessary work. If there is more than one non-binary variable in C, then we can initialize $count[y] = -1$ for all $y \in Y$, because there are no implications between pairs of non-binary variables and thus no implied literal can be found for non-binary variables. If there is exactly one non-binary variable y_i in C, then we can initialize $count[y] = -1$ for all $y \in Y \setminus \{y_i\}$.

After each iteration $i \in \{1, \ldots, k\}$ of the main loop, let $L_i = \{z \in Z \mid count[z] = i\}$. As soon as $L_i = \emptyset$, the loop can be aborted since no implied literals exist. Moreover, if $L_i \cap Y = \emptyset$, then we no longer need to inspect the implication table $\mathcal{I}$ for binary literal variables $z_j \in X \cup \overline{X}$, $j > i$, because there cannot be any implied literals for non-binary variables. On the other hand, if $L_i \cap (X \cup \overline{X}) = \emptyset$, then we can stop inspecting the clique table $\mathcal{K}$ for literals with binary variable $z_j \in X \cup \overline{X}$, $j > i$, and for non-binary variables $z_j \in Y$, $j > i$, we only need to consider the self-implication by incrementing $count[z_j]$ and updating $maxd[z_j]$.

2.2   Implied Literals from Knapsack Covers

We will now consider the derivation of implied literals based on knapsack constraints of the form $\sum_{i=1}^{k} a_i x_i \leq b$ with $x_i \in X \cup \overline{X}$ and $a_i \geq 0$. A *cover* R is a subset of $\{1, \ldots, k\}$ such that $\sum_{i \in R} a_i > b$. We will be interested in *minimal covers* which are covers whose proper subsets are not covers. Clearly, if $x_i = 1$ for all $i \in R$, the knapsack constraint will be violated. Hence, the cover R entails the implicit constraint $\bigvee_{i \in R}(x_i = 0)$, which we will refer to as the *cover constraint* corresponding to R. Suppose further that we have implications $(x_i = 0) \rightarrow (z \leq d_i)$ for all $i \in R$ and some $z \in Z$. Then, using the cover constraint along with these implications, we can derive the implied literal $z \leq \max\{d_i \mid i \in R\}$.

Remark 1. Suppose we are given an arbitrary linear constraint of the form $\sum_{i=1}^{p} c_i x_i + \sum_{j=1}^{q} e_j y_j \leq r$ where, as before, $x_i \in X$ are binary variables and $y_j \in Y$ are non-binary variables. We can derive implied literals from this constraint by first relaxing it into a knapsack constraint of the form considered above by (i) using global bounds $l_j \leq y_j \leq u_j$, (ii) using the implication table to derive variable bounds of the form $a_{j,l} x_{j,l} + b_{j,l} \leq y_j \leq a_{j,u} x_{j,u} + b_{j,u}$, and (iii) complementing the binary variables if necessary to make their coefficients non-negative.

Since the number of minimal covers of a knapsack constraint can be exponential in the length of the knapsack, it is prohibitive to naïvely enumerate all minimal covers and apply Algorithm 1 to the corresponding cover constraints. Instead, we propose a more efficient method, described as Algorithm 2.

The main idea of the algorithm is to consider the implications $(x_i = 0) \rightarrow (z \leq d_i)$ for a given variable $z \in Z$ in the reverse direction, namely as $(z > d_i) \rightarrow (x_i = 1)$. For $d \in \mathrm{Dom}(z)$ we define the set of *implied knapsack variables*

$$I_{(z>d)} := \left\{ i \in \{1, \ldots, k\} \,\middle|\, \exists((x_i = 0) \rightarrow (z \leq d_i)) \in \mathcal{I} \text{ with } d_i \leq d \right\}$$

Algorithm 2. Deriving Implied Literals from MIP Knapsack Covers

---

Input   : variables $Z = X \cup \overline{X} \cup Y$, knapsack constraint $\sum_{i=1}^{k} a_i x_i \leq b$ with
              $x_i \in X \cup \overline{X}$ for $i = 1, \ldots, k$, clique table $\mathcal{K}$, implication table $\mathcal{I}$
Output: a set L of implied literals derived from the knapsack

```
 1  begin
 2  |    initialize weight[z] := 0 for all z ∈ Z
 3  |    initialize I_y := ∅ for all y ∈ Y
 4  |    forall i ∈ {1, . . . , k} do
 5  |    |    set T := {x_i}
 6  |    |    set weight[x_i] := weight[x_i] + a_i
 7  |    |    forall K ∈ K with x̄_i ∈ K do
 8  |    |    |    forall x ∈ K \ {x̄_i} \ T do
 9  |    |    |    |    set T := T ∪ {x}
10  |    |    |    |    set weight[x] := weight[x] + a_i
11  |    |    forall ((x_i = 0) → (y ≤ d_i)) ∈ I with y ∉ T do
12  |    |    |    set T := T ∪ {y}
13  |    |    |    set weight[y] := weight[y] + a_i
14  |    |    |    set I_y := I_y ∪ {{(x_i = 0) → (y ≤ d_i)}}
15  |    L := ∅
16  |    forall x ∈ X with weight[x] > b do
17  |    |    set L := L ∪ {(x = 0)}
18  |    forall y ∈ Y with weight[y] > b do
19  |    |    I_sorted := I sorted by non-decreasing d_i
20  |    |    set s := 0
21  |    |    forall ((x_i = 0) → (y ≤ d_i)) ∈ I_sorted do
22  |    |    |    set s := s + a_i
23  |    |    |    if s > b then
24  |    |    |    |    L := L ∪ {(y ≤ d_i)}
25  |    |    |    |    break
26  |    return L
27  end
```

---

and call

$$a_{(z>d)} := \sum_{i \in I_{(z>d)}} a_i$$

the *implied weight* of $(z > d)$. If $a_{(z>d)} > b$, then $z > d$ implies that the knapsack constraint is violated, and we can conclude $z \leq d$.

For binary variables $z = x \in X \cup \overline{X}$, all non-trivial implications $(x_i = 0) \to (x \leq d_i)$ have $d_i = 0$, and we can fix $x = 0$ if and only if $a_{(x=1)} > b$. This is done in the algorithm by adding up the implied weights in the *weight* array and updating $weight[x]$ for binary variables x by scanning the clique table $\mathcal{K}$ in line 7. Again, we need to consider the trivial self-implications $(x_i = 0) \to (x_i = 0)$, see line 6. The implied fixings of binary variables are then collected in line 16.

For non-binary variables $z = y \in Y$ we want to find the *smallest* bound d that yields an implied weight $a_{(y>d)} > b$. In order to find this smallest implied bound d we first collect the implications $(x_i = 0) \to (y \le d_i)$ of the implication table $\mathcal{I}$ for all knapsack variables x_i in a set $\mathcal{I}_y$, see line 11. To evaluate these sets, we sort them by non-decreasing d_i in line 19. Then, in the loop of line 21, we consider the non-decreasing sequence of implied weights $a_{(y>d_i)}$ by adding up the knapsack weights a_i in this order until the capacity b is exceeded for the first time at element $i^\star$, that is, $i^\star$ is the first index in the sorted order such that $a_{(y>d_{i^\star})} > b$. If this process succeeds, then $d_{i^\star}$ is the smallest valid implied upper bound for y that can be derived from the knapsack and the implications using the reasoning of the algorithm. If even the total implied weight $weight[y]$ does not exceed the capacity, then we cannot tighten the upper bound of y.

Performance improvements. The additional performance improvements that we applied to Algorithm 2 are a bit more involved than the ones for Algorithm 1. Again, the goal is to avoid unnecessary work and abort as early as possible if no implied literals can be found. This is achieved by tracking the implied knapsack weight $weight[z]$ of the variables $z \in Z$ and the maximal remaining weight $w_i = \sum_{j=i+1}^{k} a_j$ of knapsack items that we did not yet consider at iteration i of the main loop. If $weight[z] + w_i \le b$ at the end of an iteration, it is clear that we will not be able to find an implied literal for z. If this is true for all binary variables, then we can stop looking at the clique table $\mathcal{K}$. If it is the case for all non-binary variables, the implication table $\mathcal{I}$ becomes uninteresting. Finally, we can abort the loop if no variable remains with $weight[z] + w_i > b$.

Now it becomes important in which order we process the knapsack items. To be able to abort as soon as possible, we want to first look at items that have large weight, so that w_i decreases fast. Moreover, items that trigger only few implications should be preferred, which will lead to the $weight[z]$ values staying small. In our implementation, we sort the knapsack items x_i in a non-decreasing order defined by $|\{K \in \mathcal{K} \mid \bar{x}_i \in K\}| - 10\, a_i/b$.

Finally, it is useful to observe that there are no negated self-implications ($x = 1) \to (x = 0)$, $x \in X \cup \overline{X}$, in the clique table; otherwise we would have already fixed $x = 0$ during presolve. As a consequence, we know that nothing can be deduced for x_j, $j = i+1, \ldots, k$, if $weight[x_j] + w_i - a_j \le b$ after iteration i. For set covering knapsacks $\sum_{i=1}^{k} x_i \le k - 1$, this means that we can rule out a variable as soon as there is a knapsack item without an implication to this variable. Hence, Algorithm 2 coincides with Algorithm 1 if applied to set covering constraints.

3   Empirical Evaluation

We evaluated the proposed techniques on a benchmark set containing 3,189 MIP models from public and commercial sources.[1] All experiments were conducted

---

[1] Due to proprietary rights, the commercial benchmarks are not publicly disclosed. However, detailed anonymized data from our experiments may be found at the following URL: http://researcher.watson.ibm.com/researcher/files/us-ashish.sabharwal/CPAIOR2013-impliedLitsMIP.txt

Table 1. Comparison of default CPLEX 12.5 vs. disabling implied literals detection

bracket	models	CPLEX 12.5 tilim	no implied literals detection					affected		
			tilim	faster	slower	time	nodes	models	time	nodes
all	3172	97	101	249	270	**1.01**	**1.01**	1134	**1.03**	**1.03**
[0,10k]	3093	18	22	249	270	**1.01**	**1.01**	1134	**1.03**	**1.03**
[1,10k]	1861	18	22	246	268	**1.02**	**1.02**	1005	**1.04**	**1.04**
[10,10k]	1096	18	22	194	226	**1.04**	**1.04**	713	**1.06**	**1.07**
[100,10k]	579	18	22	128	147	**1.05**	**1.06**	430	**1.06**	**1.08**
[1k,10k]	232	18	22	63	73	**1.06**	**1.06**	191	**1.07**	**1.07**

on a cluster of identical 12 core Intel Xeon E5430 machines running at 2.66 GHz and equipped with 24 GB of memory. A time limit of 10,000 seconds and a tree memory limit of 6 GB was employed for all runs. When the memory limit was hit, we set the solve time to 10,000 seconds and scale the number of nodes processed for the problem instance accordingly.

Implied literals detection was implemented in CPLEX 12.5 (the "default" for the purposes of this section). Algorithm 1 is applied for each conflict constraint that is derived for infeasible search tree nodes, while Algorithm 2 is only applied during the presolving stage since knapsack constraints are not generated on the fly during the tree search of CPLEX.

Table 1 shows a summary of our computational experiments comparing default CPLEX with a modified CPLEX variant where implied literals detection was disabled. Both variants were run with the default parameter settings of CPLEX. We first note that across all models in our MIP test set, Algorithms 1 and 2 never took more than 0.08 and 0.18 seconds, respectively, for each invocation of the algorithm. The times reported for default CPLEX include the overhead of computing and reasoning with implied literals.

Column 1 of the table, "bracket", labels subsets of problem instances with different "hardness", each row representing a different such subset. Subset "all" is the set of all models used for the first row of data. The labels "[n,10k]" represent the subset of "all" models for which at least one of the solvers being compared took at least n seconds to solve, and that were solved to optimality within the time limit by at least one of the solvers.

Column 2, "models", shows the number of problem instances in each subset. Note that only 3,172 rather than 3,189 problem instances are listed in row "all", because we excluded those 17 models for which the two solvers reported different optimal objective values. These inconsistencies result from the inexact floating point calculations employed in CPLEX. For ill-posed problem instances such numerical difficulties cannot be completely ruled out by floating point based MIP solvers, and this does not point to a logical error in any of the two solvers.

Column 3, "tilim", gives the number of models in each subset for which default CPLEX hit the time or memory limit. It is by design that the numbers in the last five rows match, since these models are included in all five of the corresponding subsets. Column 4 gives the corresponding numbers when not using implied literals. Both variants hit the time or memory limit on the same 79 instances of our test set. Default CPLEX hit a limit on 18 additional instances on which modified CPLEX did not, and the converse is true for 22 instances. This marginal difference of four models is likely due to performance variability [cf. 7] across runs and not an indication of the strength or weakness of one variant of CPLEX over the other.

Columns 5, "faster", and 6, "slower", show the number of models in each subset for which disabling implied literals detection resulted in the instance being solved at least 10% faster or slower, respectively, than the baseline solver (which applied implied literals). As with the time limit hits, there is only a marginal difference between the two variants of CPLEX, which is again somewhat in favor of implied literals.

Column 7, "time" displays the shifted geometric mean of the ratios of solution times [2] with a shift of $s = 1$ second.[2] A value $t > 1$ in the table indicates that CPLEX without implied literals detection is a factor of t slower (in shifted geometric mean) than default CPLEX. Column 8, "nodes", is similar to the previous column but shows the shifted geometric mean of the ratios of the number of branch-and-cut nodes needed for the models by each solver, using a shift of $s = 10$ nodes. Note that when a time limit is hit, we use the number of nodes at that point. Recall that when a memory limit is hit, we scale the node count by $10000/t$ with t being the time at which the solving process was aborted.

We observe from the time and nodes ratios that implied literals detection speeds up the solving process on average and reduces the size of the search tree explored. In particular, for models that take more than 100 seconds by at least one of the two solvers, implied literals detection reduces the solving time by 5% and the number of search tree nodes by as much as 6%.

The last three columns, under the heading "affected", report the impact on the subset of models in each bracket for which the use of implied literals had an effect on the *solution path* itself. Here, we assume that the solution path is identical if both the number of nodes and the number of simplex iterations are identical for the two solvers. Column 9, "models", shows that implied literals lead to a path change for about 36% of the models, and this fraction increases as the solving difficulty increases. For instance, for models that take at least 100 seconds to solve, over 74% of the instances are affected and the speed-up on those instances is 6%.

---

[2] The use of arithmetic means also resulted in a similar overall picture as the shifted geometric means we report here. In general, the use of geometric means, as opposed to arithmetic means, prevents situations where a small relative improvement by one solver on one long run overshadows large improvements by the other solver on many shorter runs. Further, a shift of 1 second guarantees that large but practically immaterial relative improvements on extremely short runs (e.g., 0.05 sec improving to 0.01 sec) do not distort the overall geometric mean.

4   Conclusion

We extended the concept of implied literals from SAT literature to the context of conflict constraints and knapsack covers in MIPs. Our empirical results show that while the application of this technique does not significantly improve the number of instances solved within the time limit, it does speed up the solution process. For example, it affected a substantial number of MIP models that need over 10 seconds to solve, where it reduced the solution time by 6% on average and the number of search nodes explored by 7%. We found the technique to generally have a higher impact on the solution path of harder instances and provide larger performance improvements on them.

We close with a contrast to the SAT domain. While implied literals detection can be very beneficial on certain SAT benchmark families [10], its general application can be prohibitive due to the significant computational overhead. Consequently, implied literals detection is currently not a standard technique in state-of-the-art SAT solvers. However, as our results on over 3,000 MIP instances demonstrate, implied literals detection can be employed in a way that serves as a useful generic additional inference technique in the context of MIP solvers.

References

[1] Achterberg, T.: Conflict analysis in mixed integer programming. Discrete Optimization 4(1), 4–20 (2007); Special issue: Mixed Integer Programming
[2] Achterberg, T.: Constraint Integer Programming. PhD thesis, Technische Universität Berlin (July 2007)
[3] Audemard, G., Simon, L.: Predicting learnt clauses quality in modern SAT solvers. In: 21st IJCAI, Pasadena, CA, pp. 399–404 (July 2009)
[4] Bacchus, F., Biere, A., Heule, M.: Personal Communication (2012)
[5] Freeman, J.: Improvements to Propositional Satisfiability Search Algorithms. PhD thesis, University of Pennsylvania (1995)
[6] Heule, M.J.H., Järvisalo, M., Biere, A.: Efficient CNF simplification based on binary implication graphs. In: Sakallah, K.A., Simon, L. (eds.) SAT 2011. LNCS, vol. 6695, pp. 201–215. Springer, Heidelberg (2011)
[7] Koch, T., Achterberg, T., Andersen, E., Bastert, O., Berthold, T., Bixby, R.E., Danna, E., Gamrath, G., Gleixner, A.M., Heinz, S., Lodi, A., Mittelmann, H., Ralphs, T., Salvagnin, D., Steffy, D.E., Wolter, K.: MIPLIB 2010. Mathematical Programming Computation 3, 103–163 (2011)
[8] Lynce, I., Marques-Silva, J.: Probing-based preprocessing techniques for propositional satisfiability. In: 15th ICTAI, Sacramento, CA, pp. 105–111 (November 2003)
[9] Marques-Silva, J.P., Sakallah, K.A.: GRASP: A search algorithm for propositional satisfiability. IEEE Transactions of Computers 48, 506–521 (1999)
[10] Matsliah, A., Sabharwal, A., Samulowitz, H.: Augmenting clause learning with implied literals. In: Cimatti, A., Sebastiani, R. (eds.) SAT 2012. LNCS, vol. 7317, pp. 500–501. Springer, Heidelberg (2012)
[11] Savelsbergh, M.W.P.: Preprocessing and probing techniques for mixed integer programming problems. ORSA Journal on Computing 6, 445–454 (1994)
[12] Soos, M.: CryptoMiniSat 2.9.x (2011), http://www.msoos.org/cryptominisat2

Recent Improvements Using Constraint Integer Programming for Resource Allocation and Scheduling

Stefan Heinz[1,*], Wen-Yang Ku[2], and J. Christopher Beck[2]

[1] Zuse Institute Berlin, Takustr. 7, 14195 Berlin, Germany
heinz@zib.de
[2] Department of Mechanical & Industrial Engineering
University of Toronto, Toronto, Ontario M5S 3G8, Canada
{wku,jcb}@mie.utoronto.ca

Abstract. Recently, we compared the performance of mixed-integer programming (MIP), constraint programming (CP), and constraint integer programming (CIP) to a state-of-the-art logic-based Benders manual decomposition (LBBD) for a resource allocation/scheduling problem. For a simple linear relaxation, the LBBD and CIP models deliver comparable performance with MIP also performing well. Here we show that algorithmic developments in CIP plus the use of an existing tighter relaxation substantially improve one of the CIP approaches. Furthermore, the use of the same relaxation in LBBD and MIP models significantly improves their performance. While such a result is known for LBBD, to the best of our knowledge, the other results are novel. Our experiments show that both CIP and MIP approaches are competitive with LBBD in terms of the number of problems solved to proven optimality, though MIP is about three times slower on average. Further, unlike the LBBD and CIP approaches, the MIP model is able to obtain provably high-quality solutions for all problem instances.

1 Introduction

In previous work, we provided empirical evidence showing that models based on mixed-integer programming (MIP) and constraint integer program (CIP) were competitive with logic-based Benders decomposition (LBBD) for a class of resource allocation and scheduling problems [1]. A weakness in this work was that we were not able to achieve the same performance with LBBD as in previous work (e.g., [2,3]), which made our conclusions necessarily conservative. In this paper we show that equivalent performance of LBBD to that in the literature can be obtained by using a stronger sub-problem relaxation, strengthening the Benders cuts, and employing a commercial CP solver for the sub-problems. None of these results come as a surprise as they already exist in the literature [2], but

---

* Supported by the DFG Research Center MATHEON *Mathematics for key technologies* in Berlin.

C. Gomes and M. Sellmann (Eds.): CPAIOR 2013, LNCS 7874, pp. 12–27, 2013.

(re)establishing these results replicates the literature, as well as allowing our conclusions with respect to other approaches to be placed on a firmer foundation.
More interestingly, we demonstrate that:

- The further integration of global constraint-based reasoning within the CIP framework, combined with the same stronger relaxation, results in a CIP model with equivalent performance to the LBBD model.
- The existing MIP model can itself be augmented with the stronger relaxation and this extension leads to substantially improved performance to the point MIP is competitive with the improved LBBD and CIP model.

Our experimental investigations show that the instances in one problem set (with unary resources) are now all easily solved by all models in a few seconds. For the more challenging set of instances with non-unary resources, LBBD and one of our CIP models (CIP[CP]) achieve essentially equivalent performance in terms of solving problems to optimality and finding the best known solutions, while CIP[CP] is able to find provably high quality solutions on more problem instances. The extended MIP model and the CIP model based on the MIP formulation (CIP[MIP]) find slightly fewer optimal solutions and require more run-time to do so. However, unlike both LBBD and the CIP[CP] model, the MIP and CIP[MIP] models are able to find solutions with a small optimality gap for *all* instances. Furthermore, the time for the MIP model to find a first feasible solution is twenty times faster than LBBD and twice as fast as CIP[CP] in geometric mean.

Based on our results, declaring a single winner among these three approaches is therefore fraught and perhaps not of fundamental interest (see [4]). However, our results with extended models reinforce our previous conclusions [1]: both CIP and MIP are competitive with LBBD for these scheduling problems and should be considered as core technologies for more general scheduling problems.

The rest of the paper is organized as follows. In Section 2, we formally define our problem. Section 3 presents the necessary background, including a discussion of logic-based Benders decomposition and a short summary of the results from our previous paper. In Section 4, we present the models used in this paper and we discuss detailed results of our experiments in Section 5. Section 6 provides a discussion of our results and we then conclude in Section 7.

2   Problem Definition

We study two scheduling problems referred to as UNARY and MULTI [5,3,1], which are defined by a set of jobs $\mathcal{J}$ and a set of resources $\mathcal{K}$. Each job j must be assigned to a resource k and scheduled to start at or after its release date, $\mathcal{R}_j$, end at or before its deadline, $\mathcal{D}_j$, and execute for p_{jk} consecutive time units. Each job also has a resource assignment cost c_{jk} and a resource requirement r_{jk}. We denote $\mathcal{R}$ as the set of all release dates and $\mathcal{D}$ as the set of all deadlines. Each resource $k \in \mathcal{K}$ has a capacity C_k and a corresponding constraint which states that the resource capacity must not be exceeded at any time. In the UNARY problems, each resource has unary capacity and all jobs require only one unit of resource. For the MULTI problems, capacities and requirements may be non-unary. A feasible solution is an assignment where each job is placed on exactly one resource and a start time is assigned to each job such that no resource exceeds its capacity at any time point. The goal is to find an optimal solution, that is, a feasible solution which minimizes the total resource assignment cost.

3   Background

In this section we give the necessary background w.r.t. the logic-based Benders decomposition and revisit our previous results.

3.1   Logic-Based Benders Decomposition

Logic-based Benders decomposition (LBBD) is a problem decomposition technique that generalizes Benders decomposition [6,7]. Conceptually, some of the variables and constraints of a global problem model are removed, creating a master problem (MP) whose solution (in the case of minimization) forms a lower-bound on the globally optimal solution. The extracted problem components form one or more sub-problems (SPs) where each SP is an inference dual [6]. Based on an MP solution, each sub-problem is solved, deriving the tightest bound on the MP cost function that can be inferred from the current MP solution and the constraints and variables of the SP. If a bound produced by an SP is not satisfied by the MP solution, a *Benders cut* is introduced to the MP. For global convergence, the cut must remove the current MP solution from the feasibility space of the MP without removing any globally optimal solutions.

The standard solution procedure for an LBBD model is to iteratively solve the MP to optimality, solve each sub-problem, add the Benders cuts, and re-solve the MP. Iterations continue until all SPs are satisfied by the MP solution, which has thereby been proved to be globally optimal.

For the resource allocation and scheduling problem, since the MP assigns each job to a resource and there are no inter-job constraints, the SPs are independent, single-machine feasibility scheduling problems. If all SPs are feasible, then the assignment found by the MP is valid for all resources and the corresponding cost is the global minimum. Otherwise, each infeasible SP generate a Benders cut involving a set of jobs that cannot be feasibly scheduled.

Experience with LBBD models has shown that two aspects of the formulation are critical for achieving good performance: the inclusion of a relaxation of each SP in the MP and a strong, but easily calculated Benders cut [8].

The Sub-Problem Relaxation. For the problem studied here two relaxations have been proposed. Here we label them as the *single* relaxation [9] and the *interval* relaxation [2]. The former consists of one linear constraint per SP representing to total area (i.e., time by capacity) available on that resource. Formally, the relaxation can be formulated as follows:

$$\sum_{j \in \mathcal{J}} p_{jk} r_{jk} \, x_{jk} \leq C_k \cdot (\max_{j \in \mathcal{J}}\{\mathcal{D}_j\} - \min_{j \in \mathcal{J}}\{\mathcal{R}_j\}) \qquad \forall k \in \mathcal{K} \qquad (1)$$

where x_{jk} is the binary resource choice variable equal to 1 if and only if job j is assigned to resource k.

The interval relaxation consists of $O(|\mathcal{J}|^2)$ linear constraints per SP representing the total area of a number of overlapping intervals and sets of jobs. The interval relaxation is formulated as follows:

$$\sum_{j \in \mathcal{J}(t_1, t_2)} p_{jk} r_{jk} \, x_{jk} \leq C_k \cdot (t_2 - t_1) \qquad \forall k \in \mathcal{K}, \ \forall (t_1, t_2) \in \mathcal{E} \qquad (2)$$

where $\mathcal{E} = \{(t_1, t_2) \mid t_1 \in \mathcal{R}, t_2 \in \mathcal{D}, t_1 < t_2\}$. We denote with $\mathcal{J}(t_1, t_2)$ the set of jobs that execute between t_1 and t_2: $\mathcal{J}(t_1, t_2) = \{j \in \mathcal{J} \mid t_1 \leq \mathcal{R}_j, t_2 \geq \mathcal{D}_j\}$.

If all jobs have the same time window the interval relaxation collapses to the single relaxation.

The Benders Cut. Given that the SPs are feasibility problems without any visibility to the global optimization function, the only possible Benders cut is a no-good constraint preventing the same set of jobs from being assigned to the resource again. Therefore, the cut will take the form of Constraint (11) in Model 3. Note that $\mathcal{J}_{hk}$ is a set of jobs that cannot be feasibly scheduled together on resource k. A strengthened cut can be produced by finding a subset of $\mathcal{J}_{hk}$ that also cannot be feasibly scheduled on resource k. Hooker [2] suggests a greedy procedure to find a minimal infeasible set by removing each job, one by one, from $\mathcal{J}_{hk}$ and resolving the SP. If the SP is still infeasible the corresponding job can be removed from the infeasible set, otherwise it is replaced in the set and the greedy procedure continues.

3.2   Previous Results

Our previous work compared five models: constraint programming (CP), mixed-integer programming (MIP), logic-based Benders decomposition (LBBD) and two constraint integer programming models (CIP[CP] and CIP[MIP]) [1]. The LBBD model used the single relaxation and the non-strengthened cut. While the need for a cut is unique to LBBD, it may be possible and useful to include the problem relaxation in any model that makes use of an linear programming relaxation. In particular, to be consistent with the LBBD formulation, in our previous work we used the single relaxation in CIP[CP] tested. The MIP and CIP[MIP] models were based on a different formulation so it was not obvious how to incorporate this type of relaxation.[1]

Table 1a (Section 5) reproduces the summary of our previous results [1], omitting the CP results as they are not extended here. Based on these results,

---

[1] Below we show how to do this.

we concluded that both CIP and MIP technologies are at the least competitive with LBBD as a state-of-the-art technique for these problems. One caveat to this conclusion (noted in Heinz & Beck [1]) was that previous work had achieved stronger results for LBBD [2,3]. However, even taking into account those stronger results, both MIP and CIP models found provably high-quality feasible solutions for all instances while LBBD does not.

4   Model and Solver Extensions

The primary contributions of this paper are:

- Implementation of new presolving, propagation, and primal heuristics in the SCIP solver and their application to the resource allocation/scheduling problems.
- Extension of all models to include the interval relaxation. As noted, this extension is not new for LBBD.
- Replication of previous results using LBBD with the interval relaxation and strengthened cuts.

In this section, we present the extensions to the CIP solving techniques and the mathematical models used in our experiments.

4.1   Constraint Integer Programming

In our previous work [1], we presented the first integration of the optcumulative global constraint, a cumulative resource constraint with optional activities, into the paradigm of CIP [10,11]. We focused mainly on its linear relaxation and incorporated a straightforward propagation via the cumulative constraints.

Here we continue the integration of the optcumulative global constraint into a CIP framework, including the addition of presolving techniques, general purpose primal heuristics focusing in the clique structure, and the interval relaxation. All of these techniques have been previously presented in the literature, separately, and not all in the context of CIP. Therefore the techniques, in themselves, do not represent a contribution of this paper. Rather, our contributions here are the integration of these techniques in CIP and the demonstration that their combination leads to state-of-the-art performance.

The Integration of optcumulative. In this section, we discuss the integration of the optcumulative into the CIP framework via presolving, propagation, conflict analysis, linear relaxation, and primal heuristics.

Presolving. Before the tree search starts, presolving detects and removes redundant constraints and variables. In case of the optcumulative, one shrink the time windows of each job and remove *irrelevant* jobs from the scope of the constraint since this leads to potentially tighter linear relaxation (see Equation (2)). In particular, we have developed *dual reduction techniques* that are able to remove

$$\min \sum_{k \in \mathcal{K}} \sum_{j \in \mathcal{J}} c_{jk}\, x_{jk}$$

$$\text{s.t.} \sum_{k \in \mathcal{K}} x_{jk} = 1 \qquad\qquad \forall j \in \mathcal{J} \qquad\qquad (3)$$

$$\texttt{optcumulative}(\boldsymbol{S}_{\cdot k}, \boldsymbol{x}_{\cdot k}, \boldsymbol{p}_{\cdot k}, \boldsymbol{r}_{\cdot k}, C_k) \qquad \forall k \in \mathcal{K}$$

$$\sum_{j \in \mathcal{J}(t_1, t_2)} p_{jk} r_{jk}\, x_{jk} \le C_k \cdot (t_2 - t_1) \qquad \forall k \in \mathcal{K},\ \forall (t_1, t_2) \in \mathcal{E} \qquad (4)$$

$$\mathcal{R}_j \le S_{jk} \le \mathcal{D}_j - p_{jk} \qquad\qquad \forall j \in \mathcal{J},\ \forall k \in \mathcal{K}$$

$$x_{jk} \in \{0, 1\} \qquad\qquad \forall j \in \mathcal{J},\ \forall k \in \mathcal{K}$$

$$S_{jk} \in \mathbb{Z} \qquad\qquad \forall j \in \mathcal{J},\ \forall k \in \mathcal{K}$$

Model 1. A CIP model extending CIP[CP] [1]

redundant jobs [12] from the cumulative constraints. We apply these reductions to the `optcumulative` constraint by assuming that all potentially scheduled jobs are assigned to a resource. Due to the monotonicity of the inference performed, any redundant jobs detected under the all-jobs assumption remain redundant when a subset of jobs is assigned to a resource. Additionally, we can detect a redundant `optcumulative` constraint by assuming all possible jobs are assigned to the resource and checking if the resultant cumulative constraint has a feasible solution. If so, the corresponding constraint can be removed from the problem formulation because a feasible schedule exists with all possible jobs. These inferences are specializations of the existing general dual inference techniques [12].

Propagation. During the tree search, we collect jobs which are assigned to a resource and apply the cumulative propagator [13,14]. For the remaining jobs, we run singleton arc consistency to detect jobs which can no longer be feasibly scheduled and fix the corresponding binary choice variable to zero. The extension here is that if all resource assignment variables for a given resource are fixed, we try to solve the remaining individual cumulative constraint by itself, triggering a backtrack if no such solution exists. The same data structure used in presolving [12], can be used to perform this detection in a sound and general manner. In contrast to LBBD, these (indirect) sub-problems do not need to be solved. If a solution is found or the problem is proved infeasible, the global search space is reduced. However, if they are not solved, the main search continues.

Conflict analysis. We use the explanation algorithms corresponding to the cumulative propagator [15,16] and extend the generated explanations to include only the binary resource choice variables for those start time variables which are part of the explanation. This is different from our previous implementation where we included all binary variables in the conflict. Adding only the binary variables which are part of the cumulative explanation is analogous to the strengthening techniques of the Benders cuts described above.

$$\min \sum_{k \in \mathcal{K}} \sum_{j \in \mathcal{J}} c_{jk} \, x_{jk}$$

$$\text{s.t.} \sum_{k \in \mathcal{K}} x_{jk} = 1 \qquad\qquad \forall j \in \mathcal{J} \tag{5}$$

$$\sum_{t = \mathcal{R}_j}^{\mathcal{D}_j - p_{jk}} y_{kjt} = x_{jk} \qquad\qquad \forall j \in \mathcal{J}, \ \forall k \in \mathcal{K} \tag{6}$$

$$\sum_{j \in \mathcal{J}} \sum_{t' \in T_{jkt}} r_{jk} \, y_{jkt'} \leq C_k \qquad\qquad \forall k \in \mathcal{K}, \ \forall t \tag{7}$$

$$\mathcal{R}_j + \sum_{t = \mathcal{R}_j}^{\mathcal{D}_j - p_j} (t - \mathcal{R}_j) \cdot y_{jkt} = S_{jk} \qquad\qquad \forall j \in \mathcal{J}, \ \forall k \in \mathcal{K} \tag{8}$$

$$\texttt{optcumulative}(\boldsymbol{S}_{\cdot k}, \boldsymbol{x}_{\cdot k}, \boldsymbol{p}_{\cdot k}, \boldsymbol{r}_{\cdot k}, C_k) \qquad \forall k \in \mathcal{K}$$

$$\sum_{j \in \mathcal{J}(t_1, t_2)} p_{jk} r_{jk} \, x_{jk} \leq C_k \cdot (t_2 - t_1) \qquad \forall k \in \mathcal{K}, \ \forall (t_1, t_2) \in \mathcal{E} \tag{9}$$

$$\mathcal{R}_j \leq S_{jk} \leq \mathcal{D}_j - p_{jk} \qquad\qquad \forall j \in \mathcal{J}, \ \forall k \in \mathcal{K}$$

$$x_{jk} \in \{0, 1\} \qquad\qquad \forall j \in \mathcal{J}, \ \forall k \in \mathcal{K}$$

$$S_{jk} \in \mathbb{Z} \qquad\qquad \forall j \in \mathcal{J}, \ \forall k \in \mathcal{K}$$

$$y_{jkt} \in \{0, 1\} \qquad\qquad \forall j \in \mathcal{J}, \ \forall k \in \mathcal{K}, \ \forall t \in \{\mathcal{R}_j, \ldots, \mathcal{D}_j - p_{jk}\}$$

Model 2. CIP[MIP]: A CIP model based on the MIP model with channeling Constraints (8). $T_{jkt} = \{t - p_{jk}, \ldots, t\}$.

Linear relaxation. As discussed above, we use the interval relaxation (Equation (2)) instead of the single relaxation (Equation (1)) in our CIP models. See below for the details of the CIP[CP] and CIP[MIP] models. To generate the relaxation we use the algorithm presented by Hooker [2] to impose only non-redundant constraints.

Primal heuristic. Inspired by the clique structure of the problem (i.e., each job has to be assigned to one resource), we implemented a general purpose primal heuristic that assigns jobs to resources and solves the resulting decomposed scheduling problems. In MIP and CIP, a clique structure refers to a sets of binary variables that must sum to at most one. This structure is easily detectable within a model and can be used within a diving heuristic.

Extended Models. In this section, we present the full CIP models, one based on the CP formulation (CIP[CP]) and the other based on the MIP formulation (CIP[MIP]). Both are extensions of correspondingly named existing models [1].

Model 1 presents the CIP[CP] model with the resource choice variable x_{jk} equal to 1 if and only if job j is assigned to resource k. The objective function is defined in terms of the resource choice variables. Constraints (3) ensure that each job is assigned to exactly one resource, where the resource capacities are enforced by the global `optcumulative` constraints. Constraints (4) state the interval relaxation. This model is equivalent to the existing CIP[CP] model [1]

$$(\text{MP}) \quad \min \sum_{k \in \mathcal{K}} \sum_{j \in \mathcal{J}} c_{jk} \, x_{jk}$$

$$\text{s.t.} \sum_{k \in \mathcal{K}} x_{jk} = 1 \qquad\qquad\qquad \forall j \in \mathcal{J}$$

$$\sum_{j \in \mathcal{J}(t_1, t_2)} p_{jk} r_{jk} \, x_{jk} \le C_k \cdot (t_2 - t_1) \quad \forall k \in \mathcal{K}, \; \forall (t_1, t_2) \in \mathcal{E} \qquad (10)$$

$$\sum_{j \in \mathcal{J}_{hk}} (1 - x_{jk}) \ge 1 \qquad\qquad \forall k \in \mathcal{K}, \; \forall h \in \{1, \dots, H-1\} \quad (11)$$

$$x_{kj} \in \{0, 1\} \qquad\qquad\qquad \forall j \in \mathcal{J}, \; \forall k \in \mathcal{K}$$

$$(\text{SP}) \quad \texttt{cumulative}(\boldsymbol{S}, \boldsymbol{p}_{\cdot k}, \boldsymbol{r}_{\cdot k}, C_k)$$

$$\mathcal{R}_j \le S_j \le \mathcal{D}_j - p_{jk} \qquad\qquad \forall j \in \mathcal{J}_k$$

$$S_j \in \mathbb{Z} \qquad\qquad\qquad\qquad \forall j \in \mathcal{J}_k$$

Model 3. Logic-based Benders decomposition: master problem (MP) on top and subproblem (SP) for resource k below

except for Constraints (4). The interval relaxation is added in the same way as it is included in the LBBD model (see Hooker [2] and Model 3).

For the CIP model based on the MIP formulation (CIP[MIP]–Model 2), we incorporate the interval relaxation by Constraint (9). The primary decision variables of the time-indexed formulation, y_{kjt}, are equal to 1 if and only if job j starts on resource k at time point t. We add an auxiliary set of binary decision variables, x_{jk}, which are assigned to 1 if and only if job j is assigned to resource k.

The objective function is defined with the new set of binary decision variables. Constraints (5) ensure that each job is assigned to exactly one resource. The two sets of decision variables are linked via Constraints (6). The resource capacities are enforced by the knapsack constraints (7) which are given for each time point. The global `optcumulative` constraints are added to achieve additional propagation. Finally, the Constraints (9) state the (redundant) interval relaxation which potentially strengthens the linear programming relaxation.

4.2 Logic-Based Benders Decomposition

We use the LBBD model from Hooker [2] which uses both the interval relaxation and the strengthened cuts. For completeness, we present the model in Model 3.

4.3 Mixed Integer Programming

Given the presence of cumulative constraints in the MULTI version of the problem, the standard MIP model uses a *time-indexed* formulation [8,9,1] employing a set of binary decision variables, y_{jkt}, which are equal to 1 if and only if job j starts

$$\min \sum_{k \in \mathcal{K}} \sum_{j \in \mathcal{J}} c_{jk}\, x_{jk}$$

$$\text{s.t.} \sum_{k \in \mathcal{K}} x_{jk} = 1 \qquad\qquad\qquad \forall j \in \mathcal{J} \tag{12}$$

$$\sum_{t=\mathcal{R}_j}^{\mathcal{D}_j - p_{jk}} y_{kjt} = x_{jk} \qquad\qquad \forall j \in \mathcal{J},\ \forall k \in \mathcal{K} \tag{13}$$

$$\sum_{j \in \mathcal{J}} \sum_{t' \in T_{jkt}} r_{jk}\, y_{jkt'} \le C_k \qquad\qquad \forall k \in \mathcal{K},\ \forall t \tag{14}$$

$$\sum_{j \in \mathcal{J}(t_1, t_2)} p_{jk} r_{jk}\, x_{jk} \le C_k \cdot (t_2 - t_1) \quad \forall k \in \mathcal{K},\ \forall(t_1, t_2) \in \mathcal{E} \tag{15}$$

$$x_{jk} \in \{0, 1\} \qquad\qquad\qquad \forall j \in \mathcal{J},\ \forall k \in \mathcal{K}$$

$$y_{jkt} \in \{0, 1\} \qquad\qquad\qquad \forall j \in \mathcal{J},\ \forall k \in \mathcal{K},\ \forall t \in \{\mathcal{R}_j, \dots, \mathcal{D}_j - p_{jk}\}$$

Model 4. Mixed integer programming model with $T_{jkt} = \{t - p_{jk}, \dots, t\}$

at time t on resource k. As with the CIP[MIP] model above, we extend the MIP model to include a second set of binary variables, x_{kj}, which are equal to 1 if and only if job j is assigned to resource k. This second set of variables introduces the decomposition aspect of the problem into the MIP model since the cost for an assignment is determined only by this set of variables. In addition, these variables make it natural to express the interval relaxation.

Our MIP model is stated in Model 4. The constraints are almost identical to those presented above in the CIP[MIP] model, with the exception that the start time variables, the global `optcumulative` constraints, and the necessary channeling constraints are absent.

Note that the second set of decision variables is redundant. Our preliminary experiments showed that the solver achieves much higher performance with the redundant formulation. The interval relaxation itself is also redundant given Constraints (14). However, they introduce a connection between the capacity constraints of each resource, strengthening the linear programming relaxation.

5   Computational Results

In this section, we compare the performance of the LBBD, the MIP, and the two CIP models. We use the same test sets and the same computational environment as our previous work [1] to allow direct comparison of the results.

5.1   Experimental Setup

Test Sets. The problem instances were introduced by Hooker [5]. Each set contains 195 problem instances with the number of resources ranging from two to four and the number of jobs from 10 to 38 in steps of two. The maximum number of jobs for the instances with three and four resources is 32 while for

Table 1. Summary of the results presented in Heinz & Beck [1] and the results for the extended model of this paper

(a) Results of Heinz & Beck [1], omitting the CP model.

	UNARY				MULTI			
	MIP	LBBD	CIP[CP]	CIP[MIP]	MIP	LBBD	CIP[CP]	CIP[MIP]
feasible	**195**	175	**195**	**195**	**195**	119	125	**195**
optimal found	**195**	175	194	**195**	**148**	119	124	142
optimal proved	191	175	194	**195**	109	119	123	**133**
best known found	**195**	175	194	**195**	**155**	119	124	146
total time	12	28	**10**	19	442	228	**212**	395
time to best	**7**	28	9	17	209	228	**200**	217

(b) Results for the extended models.

	UNARY				MULTI			
	MIP	LBBD	CIP[CP]	CIP[MIP]	MIP	LBBD	CIP[CP]	CIP[MIP]
feasible	**195**	**195**	**195**	**195**	**195**	174	190	**195**
optimal found	**195**	**195**	**195**	**195**	167	**174**	167	142
optimal proved	**195**	**195**	**195**	**195**	155	**174**	163	126
best known found	**195**	**195**	**195**	**195**	172	**174**	168	146
total time	1.7	**1.0**	1.3	10.4	159.6	**37.8**	54.3	383.3
time to best	1.6	**1.0**	1.3	9.8	121.5	37.8	**6.3**	198.4
time to first	1.3	**1.0**	**1.0**	1.8	**2.4**	37.8	5.0	18.7

two resources the number of maximum number of jobs is 38. In addition, there are five instances for each problem size. For the MULTI problems, the resource capacity is 10 and the job demands are generated with uniform probability on the integer interval $[1, 9]$. See Hooker [5] for further details w.r.t. the generation of instances, and the appendix of [17] for further problem instance characteristics.

Computational Environment. All experiments are performed on Intel Xeon E5420 2.50 GHz computers (in 64 bit mode) with 6 MB cache and 6 GB of main memory, running Linux. For solving the MIP models we used IBM ILOG CPLEX 12.4 in its default setting. The master problem of the LBBD approach is solved with SCIP 3.0.0 [11] using SoPlex [18] version 1.7.0.1 as the linear programming solver. For the sub-problems we used IBM ILOG CP Optimizer 12.4 using the default settings plus extended filtering and depth first search. For the CIP models we used the same solver as for the master problem of the LBBD. For each instance we enforced a time limit of 2 hours and allow for a single thread.

5.2   Previous Results

Table 1a presents the summary of the previous results [1], omitting the pure CP model which we do not extend here. For each test set (UNARY and MULTI) and each model, Table 1a states the number of instances for which (i) a feasible solution was found, (ii) an optimal solution was found, (iii) an optimal was found and proved, and (iv) the best known solution was found. Secondly we present the shifted geometric mean[2] for the total running time and the time until the best solution was found. The time to the best solution is only an upper bound in case of CPLEX since the output log does not display this time point explicitly.

5.3   Results

In the same fashion as in Table 1a, we summarize the results for the extended models in Table 1b. Additionally, we included the shifted geometric mean for the time when the first feasible solution was found.

The UNARY Test Set. The results for the UNARY test set show that all models improved drastically w.r.t. our previous results. All approaches are able to solve all instances in a few seconds or less. While this was expected for LBBD and the CIP models, it comes as a bit of a surprise for the MIP model. Analyzing the results for LBBD, we see that it needs only one iteration for each instance: the first optimal master solution is always proved feasible for all sub-problems. This result indicates that the interval relaxation tightens the master problem significantly. Since the MIP and CIP models have basically the same linear relaxation as LBBD, we believe that the tightness of the interval relaxation explains the improved results for these models as well.

The MULTI Test Set. For the MULTI test set, we observe a substantial improvement for all models except CIP[MIP]. The MIP models solves 155 instances of 195 compared to 109, the LBBD approach proves optimality for 174 instances compared to 119, and the CIP[CP] formulation handles 163 instances compared to 123 before. Similar observations can be made for the running times: LBBD, MIP, and CIP[CP] are now a factor six, three, and four faster than before, respectively. In terms of relative speed, the close-to-uniform speed-ups results in basically the same ratios among the different models as in Beck & Heinz. Only the CIP[MIP] performance remained consistent, while all other models improved in a similar way.

As in our previous results, the MIP and CIP[MIP] models are able to find a feasible solution for all instances. The CIP[CP] does that for 190 instances compared to 125 instances before, while LBBD is only able to find feasible solutions for the problems that it solves to optimality. Comparing the quality of the solutions among the solvers, we observe that the MIP model finds the optimal or best known solutions for 172 instances.[3] By the same metric LBBD and CIP[CP] find best known solutions for 174 and 168 instances, respectively. If the best of the four models is chosen for each instance (resulting in the *virtual best solver*), 187 instances can be solved to proven optimality with a total running time in shifted geometric mean of 19.9 seconds.

For a more detailed indication of the results for the MULTI test set, Table 2 presents results for each problem size for the CIP, MIP, and LBBD models. The

---

[2] The shifted geometric mean of values $t_1, \ldots, t_n$ is $\left(\prod (t_i + s) \right)^{1/n} - s$, with shift $s = 10$.

[3] For the MULTI test set the optimal solution value is known for 189 instances.

Table 2. Detailed results for the MULTI test set. For each resource job combination consists of 5 instances (for a total of 195) we display on line.

		MIP				LBBD				CIP[CP]				CIP[MIP]			
$|\mathcal{K}|$	$|\mathcal{J}|$	opt	feas	arith	geom	opt	feas	arith	geom	opt	feas	arith	geom	opt	feas	arith	geom
2	10	5	5	1.0	**1.0**	5	5	1.0	**1.0**	5	5	1.0	**1.0**	5	5	1.9	1.8
	12	5	5	1.1	1.1	5	5	1.0	**1.0**	5	5	1.0	**1.0**	5	5	3.7	3.5
	14	5	5	1.0	**1.0**	5	5	1.0	**1.0**	5	5	1.0	1.0	5	5	4.9	4.7
	16	5	5	13.1	8.0	5	5	1.0	**1.0**	5	5	8.3	4.7	5	5	40.3	29.9
	18	5	5	36.2	16.9	5	5	1.3	1.3	5	5	1.8	**1.7**	5	5	75.4	66.2
	20	5	5	89.6	29.0	5	5	4.1	3.7	5	5	1.6	**1.5**	5	5	120.2	70.4
	22	4	5	2983.3	812.4	3	5	796.8	**51.4**	3	5	3090.8	382.5	4	5	3036.0	1293.9
	24	3	5	3026.7	883.0	4	4	1733.8	**214.8**	2	5	4321.5	573.4	2	5	4399.3	1508.8
	26	4	5	3013.5	1069.2	5	5	912.1	**209.0**	4	4	2122.4	464.9	3	5	3414.9	1746.9
	28	4	5	2394.7	378.9	5	5	993.7	536.5	4	5	1444.4	**42.0**	3	5	3822.6	1910.5
	30	3	5	3788.2	861.2	3	3	2930.3	**401.2**	2	5	4321.8	587.6	1	5	5802.8	3590.5
	32	3	5	3054.7	**792.1**	0	0	–	–	2	4	4400.1	1140.5	0	5	–	–
	34	3	5	3444.0	**879.7**	2	2	4400.3	1745.1	1	5	5760.4	1995.3	1	5	5843.7	4089.2
	36	2	5	4386.6	1534.1	1	1	5942.7	4770.2	3	4	3476.7	**548.4**	2	5	4709.5	2319.1
	38	2	5	5590.6	4980.2	1	1	6268.8	5848.7	2	5	4360.6	**1334.0**	0	5	–	–
3	10	5	5	1.0	**1.0**	5	5	1.0	**1.0**	5	5	1.0	**1.0**	5	5	1.4	1.3
	12	5	5	1.0	**1.0**	5	5	1.0	**1.0**	5	5	1.0	**1.0**	5	5	4.5	4.3
	14	5	5	1.0	**1.0**	5	5	1.0	**1.0**	5	5	1.2	1.2	5	5	11.3	10.3
	16	5	5	4.5	4.2	5	5	3.3	**3.0**	5	5	3.4	3.3	5	5	43.6	33.1
	18	5	5	76.4	46.0	5	5	7.1	5.8	5	5	5.0	**4.8**	5	5	594.9	244.0
	20	4	5	1470.9	98.5	5	5	1.5	**1.5**	5	5	7.8	6.9	4	5	1622.8	355.1
	22	4	5	1832.6	554.6	5	5	2.4	**2.3**	5	5	6.9	6.6	4	5	2543.7	1008.3
	24	5	5	1703.2	304.5	5	5	9.3	**6.7**	5	5	346.6	78.6	4	5	2225.2	967.5
	26	3	5	3826.9	1652.8	5	5	31.8	**19.8**	5	5	98.4	40.2	1	5	5942.0	4766.2
	28	3	5	3901.1	987.6	5	5	85.3	**35.4**	3	5	2885.5	194.9	2	5	4782.6	3030.9
	30	3	5	4028.1	3100.2	4	4	1523.6	**178.3**	4	5	1911.1	520.9	0	5	–	–
	32	2	5	4840.8	3601.3	4	4	2882.6	1951.8	3	5	2969.5	**559.0**	1	5	6125.7	5475.9
4	10	5	5	1.0	**1.0**	5	5	1.0	**1.0**	5	5	1.0	**1.0**	5	5	1.0	**1.0**
	12	5	5	1.0	**1.0**	5	5	1.0	**1.0**	5	5	1.0	**1.0**	5	5	2.4	2.3
	14	5	5	1.0	**1.0**	5	5	1.6	1.6	5	5	1.7	1.7	5	5	12.8	11.7
	16	5	5	1.2	1.2	5	5	1.0	**1.0**	5	5	1.7	1.7	5	5	22.3	20.4
	18	5	5	3.3	3.1	5	5	2.5	**2.5**	5	5	4.5	4.4	5	5	96.5	76.6
	20	5	5	43.8	25.3	5	5	1.8	**1.8**	5	5	4.4	4.3	5	5	159.3	116.0
	22	5	5	128.2	60.0	5	5	4.3	**3.7**	5	5	20.7	15.0	5	5	1244.4	482.9
	24	4	5	2695.6	1399.0	5	5	16.0	**12.1**	5	5	59.3	42.9	1	5	6236.7	5773.1
	26	3	5	3825.4	2787.8	5	5	15.7	**14.9**	5	5	293.0	112.7	1	5	6227.4	5750.3
	28	3	5	5361.9	2124.2	5	5	9.9	**9.6**	4	5	1562.9	200.0	2	5	5010.5	3646.9
	30	2	5	5035.9	3253.6	5	5	112.6	**31.7**	4	5	2243.0	581.0	0	5	–	–
	32	1	5	5927.9	4691.0	5	5	343.6	**118.3**	2	5	4412.3	1519.1	0	5	–	–
		155	195	1962.5	159.6	174	174	929.5	**37.8**	163	190	1286.1	54.3	126	195	2825.3	383.3

first two columns define the instance size in terms of the number of resources $|\mathcal{K}|$ and the number of jobs $|\mathcal{J}|$. For each model, we report the number of instances solved to proven optimality "opt" and the number instances for which a feasible solution was found, "feas", including the instances which are solved to optimality. For the total running time we report the arithmetic mean ("arith") and the shifted geometric mean ("geom") with shift $s = 10$. All running times that are less than 1.0 second are set to 1.0. For each resource-job combination, the best time is shown in bold. For clarity, when a model did not solve any instances of a given size, we use '–' instead of 7200 for the running time.

The table indicates that all models appear to scale exponentially with the number of jobs. The results for LBBD and CIP[CP] show the increase at a lower rate than for MIP and CIP[MIP]. Nonetheless, LBBD and CIP[CP] both fail to find and prove optimal solutions on some of the larger instances. It is interesting to note that for the instances with two resources, all models suddenly start to

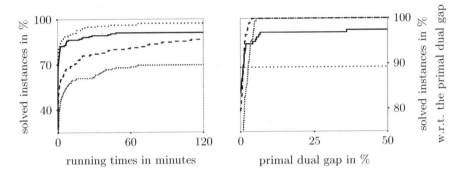

Fig. 5. Performance diagrams for the MULTI test set. The MIP model is dashed (‑ ‑ ‑), the LBBD model dotted (......), the CIP[CP] model solid (——), and the CIP[MIP] model is densely dotted (........).

struggle with 22 or more jobs: the shifted geometric means of the run-time for all models increase one or two orders of magnitude in moving from 20 to 22 jobs. We return to this observation below.

Since the models do not solve or fail to solve exactly the same instances, we depict two performance diagrams for the MULTI test set in Figure 5. The left-hand graph shows the evolution of the number of problems solved to optimality over time. It can be observed that LBBD and CIP behave very similarly while MIP performs worse in the beginning but increases its success with more run-time. CIP[MIP] performs consistently worse than all other models. The right-hand graph displays the percentage of instances for which a solution with given optimality gap (primal bound minus dual bound divided by the primal bound) or better was found. On this basis, both MIP and CIP[MIP] models outperform the other two models by finding solutions with an optimality gap of less than 5% for all problem instances. CIP finds solutions with a gap of 10% or better on about 97% of the instances while LBBD finds the optimal solution on 89% of the instances and is, of course, unable to find any sub-optimal feasible solutions.

6   Discussion

The results of the experiments presented above support and reinforce the conclusions of Heinz & Beck [1]: both CIP[CP] and MIP should be considered to be state-of-art models, along with LBBD, for the tested resource allocation and scheduling problems. On the basis of the number of problem instances solved to optimality LBBD has a marginal advantage over CIP[CP] which itself is marginally better then MIP. However, on other measures of solution quality (number of instances with feasible solutions and the quality of those solutions), the ranking is reversed.

An examination of the sub-problems in the two-resource instances that LBBD and CIP[CP] fail to solve reveals that most of the cumulative constraint/sub-problems which have to be proven to be feasible or infeasible have a very small slack and the jobs have a wide and often identical time windows. Slack is the difference between the rectangular area available on the resource (time by capacity) and the sum of the areas (processing time by resource requirement) of the jobs. Alternatively, slack can be understood to be the tightness of the single relaxation (Equation (1)). The small slack results from the fact that one resource is consistently less costly than the others and so it appears promising to assign many jobs as possible within the limits of the interval relaxation.

All approaches suffer from not being able to handle small slack and wide time window problems efficiently on cumulative resources. This is the underlying reason for the main disadvantage of LBBD which gets stuck at such a sub-problems and fails completely to find a feasible solution. All other approaches have the same issue of not been able to solve these implicit sub-problems, but are able to provide high quality primal solutions. To overcome this issue, stronger cumulative inference techniques [19] may be worth consideration.

As we are comparing the CIP approach against a start-of-the-art LBBD implementation, we should also compare with a state-of-the-art commercial MIP solver when solving MIP models. It has, however, been standard for the past few years for commercial MIP solvers to use multiple cores. If we run IBM ILOG CPLEX with its default settings (using all available cores, eight in our case) on the MULTI instances we can solve 171 instances to proven optimality with a shifted geometric mean of 71.4 seconds. The fact that this performance is only marginally better than what we observe for CIP and similar to LBBD results, strengthens their claims to state-of-the-art status.

Finally, the results of the single-core virtual best solver, solving 187 instances with a shifted geometric mean of 19.9 second, indicate that none of the models is dominant. One of the arguments for pursuing CIP is that it is a framework that strives to combine the advantages of the other approaches in order to overcome the individual disadvantages.

Future Work. There are a number of areas of future work both on extending these approaches to related scheduling problems and in developing the technology of CIP for scheduling.

Continued development of CIP. We have demonstrated through the integration of the `optcumulative` constraint that global constraint-based presolving, inferences, and relaxations can lead to state-of-the-art performance. We intend to further pursue the integration of global constraint reasoning into a CIP framework for scheduling and other optimization problems.

Other scheduling problems. Hooker [2] has presented LBBD models for extensions of the problem studied here with a number of different optimization functions. For such problems, LBBD is able to produce feasible sub-optimal solutions without necessarily finding an optimal solution. Therefore, one of the main advantage of the MIP and CIP techniques compared to LBBD does not appear. It will be valuable to understand how adaptations of the MIP and CIP models presented here perform on such problems. Another important class of scheduling problems has temporal constraints among jobs on different resources. Such constraints destroy the independent sub-problem structure that LBBD and, to a lesser extent, the other models exploit. However, exact techniques currently struggle on such problems including flexible job shop scheduling [20].

Scaling. As shown in Table 2, all models are unable to find optimal solutions as the number of jobs increases. With even more jobs, the only achievable performance measure will be the quality of feasible solutions that are found. We expect LBBD to perform poorly given that it cannot find sub-optimal solutions. However, as the problem size increases the time-indexed formulation on the MIP model will also fail due to model size. CIP[CP] and the pure CP model [1] would appear to be the only exact techniques likely to continue to deliver feasible solutions. Confirming this conjecture, as well as comparing the model performance to incomplete techniques (i.e., heuristics and metaheuristics) is therefore another area for future work.

7   Conclusions

The primary conclusions of these experiments with more sophisticated problem models are consistent with and reinforce those of Heinz & Beck [1]: CIP is a promising scheduling technology that is comparable to the state-of-the-art manual decomposition approach on resource allocation/scheduling problems and MIP approaches, though often discounted by constraint programming researchers, deserve consideration as a core technology for scheduling.

In arriving at these conclusions, we used two primary measures of model performance: the number of problem instances solved to proven optimality and the proven quality of solutions found, given that not all instances were solved to optimality. CIP comes second to LBBD by the former measure and to MIP by the latter. Depending on the importance placed on these measures any of the three algorithms could be declared the "winner". For practical purposes, we believe that the importance of proven solution quality should not be underestimated: in an industrial context it is typically better to consistently produce proven good solutions than to often find optimal solutions but sometimes fail to find any feasible solution at all.

References

1. Heinz, S., Beck, J.C.: Reconsidering mixed integer programming and MIP-based hybrids for scheduling. In: Beldiceanu, N., Jussien, N., Pinson, É. (eds.) CPAIOR 2012. LNCS, vol. 7298, pp. 211–227. Springer, Heidelberg (2012)
2. Hooker, J.N.: Integrated Methods for Optimization. Springer (2007)
3. Beck, J.C.: Checking-up on branch-and-check. In: Cohen, D. (ed.) CP 2010. LNCS, vol. 6308, pp. 84–98. Springer, Heidelberg (2010)
4. Hooker, J.N.: Testing heuristics: We have it all wrong. Journal of Heuristics 1, 33–42 (1995)

5. Hooker, J.N.: Planning and scheduling to minimize tardiness. In: van Beek, P. (ed.) CP 2005. LNCS, vol. 3709, pp. 314–327. Springer, Heidelberg (2005)
6. Hooker, J.N., Ottosson, G.: Logic-based Benders decomposition. Mathematical Programming 96, 33–60 (2003)
7. Benders, J.: Partitioning procedures for solving mixed-variables programming problems. Numerische Mathematik 4, 238–252 (1962)
8. Hooker, J.N.: Planning and scheduling by logic-based Benders decomposition. Operations Research 55, 588–602 (2007)
9. Yunes, T.H., Aron, I.D., Hooker, J.N.: An integrated solver for optimization problems. Operations Research 58(2), 342–356 (2010)
10. Achterberg, T.: Constraint Integer Programming. PhD thesis, Technische Universität Berlin (2007)
11. Achterberg, T.: SCIP: Solving Constraint Integer Programs. Mathematical Programming Computation 1(1), 1–41 (2009)
12. Heinz, S., Schulz, J., Beck, J.C.: Using dual presolving reductions to reformulate cumulative constraints. ZIB-Report 12-37, Zuse Institute Berlin (2012)
13. Baptiste, P., Pape, C.L., Nuijten, W.: Constraint-Based Scheduling. Kluwer Academic Publishers (2001)
14. Beck, J.C., Fox, M.S.: Constraint directed techniques for scheduling with alternative activities. Artificial Intelligence 121(1-2), 211–250 (2000)
15. Heinz, S., Schulz, J.: Explanations for the cumulative constraint: An experimental study. In: Pardalos, P.M., Rebennack, S. (eds.) SEA 2011. LNCS, vol. 6630, pp. 400–409. Springer, Heidelberg (2011)
16. Schutt, A., Feydy, T., Stuckey, P.J., Wallace, M.G.: Explaining the cumulative propagator. Constraints 16(3), 250–282 (2011)
17. Heinz, S., Beck, J.C.: Reconsidering mixed integer programming and MIP-based hybrids for scheduling. ZIB-Report 12-05, Zuse Institute Berlin (2012)
18. Wunderling, R.: Paralleler und objektorientierter Simplex-Algorithmus. PhD thesis, Technische Universität Berlin (1996)
19. Beldiceanu, N., Carlsson, M., Poder, E.: New filtering for the *cumulative* constraint in the context of non-overlapping rectangles. In: Trick, M.A. (ed.) CPAIOR 2008. LNCS, vol. 5015, pp. 21–35. Springer, Heidelberg (2008)
20. Fattahi, P., Saidi Mehrabad, M., Jolai, F.: Mathematical modeling and heuristic approaches to flexible job shop scheduling problems. Journal of Intelligent Manufacturing 18(3), 331–342 (2007)

Cloud Branching

Timo Berthold[1,*] and Domenico Salvagnin[2]

[1] Zuse Institute Berlin, Takustr. 7, 14195 Berlin, Germany
berthold@zib.de
[2] DEI, Via Gradenigo, 6/B, 35131 Padova, Italy
salvagni@dei.unipd.it

Abstract. Branch-and-bound methods for mixed-integer programming (MIP) are traditionally based on solving a linear programming (LP) relaxation and branching on a variable which takes a fractional value in the (single) computed relaxation optimum. In this paper we study branching strategies for mixed-integer programs that exploit the knowledge of *multiple* alternative optimal solutions (a *cloud*) of the current LP relaxation. These strategies naturally extend state-of-the-art methods like strong branching, pseudocost branching, and their hybrids.

We show that by exploiting dual degeneracy, and thus multiple alternative optimal solutions, it is possible to enhance traditional methods. We present preliminary computational results, applying the newly proposed strategy to full strong branching, which is known to be the MIP branching rule leading to the fewest number of search nodes. It turns out that cloud branching can reduce the mean running time by up to 30% on standard test sets.

1 Introduction

In this paper we address branching strategies for the exact solution of a generic mixed-integer program (MIP) of the form (w.l.o.g.):

$$\min\{cx : Ax \leq b \quad x_j \in \mathbb{Z} \quad \forall j \in J\}$$

where $x \in \mathbb{R}^n$ and $J \subseteq N = \{1, \dots, n\}$.

Good branching strategies are crucial for any branch-and-bound based MIP solver. Unsurprisingly, the topic has been subject of constant and active research since the very beginning of computational mixed-integer programming, see, e.g., [1]. We refer to [2, 3, 4] for some comprehensive studies on branching strategies.

In mixed-integer programming, the most common methodology for branching is to split the domain of a single variable into two disjoint intervals. In this paper we will address the key problem of how to select such a variable. Let $x^\star$ be an optimal solution of the linear programming (LP) relaxation at the current

---

* Timo Berthold is supported by the DFG Research Center MATHEON *Mathematics for key technologies* in Berlin.

C. Gomes and M. Sellmann (Eds.): CPAIOR 2013, LNCS 7874, pp. 28–43, 2013.

node of the branch-and-bound tree and let $F = \{j \in J : x_j^\star \notin \mathbb{Z}\}$ denote the set of fractional variables. A general scheme for branching strategies consists in computing a score s_j for each fractional variable $j \in F$, and then picking the variable with maximum score. Different branching rules then correspond to different ways of computing this score.

Several branching criteria have been studied in the literature. The simplest one (*most-fractional branching*) is to branch on the variable whose fractional part is as close as possible to 0.5; however, this is well known to perform poorly in practice [5]. A more sophisticated branching strategy is *pseudocost branching* [1], which consists in keeping a history of how much the *dual bound* (the LP relaxation) improved when branching on a given variable in previous nodes, and then using these statistics to estimate how the dual bound will improve when branching on that variable at the current node. Pseudocost branching is computationally cheap since no additional LPs need to be solved and performs reasonably well in practice. Yet at the very beginning, when the most crucial branching decisions are taken, there is no reliable historic information to build upon.

Another effective branching rule is *strong branching* [6, 7]. The basic idea consists in simulating branching on the variables in F and then choosing the actual branching variable as the one that gives the best progress in the dual bound. Interestingly, this greedy local method is currently the best w.r.t. the number of nodes of the resulting branch-and-bound tree, but introduces quite a large overhead in terms of computation time, since $2 \cdot |F|$ auxiliary LPs need to be solved at every node. Many techniques have been studied to speedup the computational burden of strong branching, in particular by heuristically restricting the list of branching candidates and imposing simplex iteration limits on the strong branching LPs [2] or by ruling out inferior candidates during the strong branching process [8]. However, according to computational studies, a pure strong branching rule is still too slow for practical purposes. Branching rules such as *reliability branching* [3] or *hybrid branching* [9], that combine ideas from pseudocost branching and strong branching, are considered today's state of the art.

Other approaches to branching include the *active constraint* method [10], which is based on the impact of variables on the set of active constraints, branching on general disjunctions [11], *inference branching* and *VSIDS* [12, 13, 4] based on SAT-like domain reductions and conflict learning techniques. Finally, information collected through restarts is at the heart of the methods in [14, 15].

All branching strategies described so far are naturally designed to deal with only one optimal fractional solution. History-based rules use the statistics collected in the process to compute the score of a variable starting from the current fractional solution. Even with strong branching, the list of branching candidates is defined according to the current fractional solution $x^\star$.

However, LP relaxations of MIP instances are well-known for often being massively degenerate; multiple equivalent optimal solutions are the rule rather than the exception. Thus branching rules that consider only one optimal solution risk taking arbitrary branching decisions (thus contributing to performance variability, see [16]), or being unnecessarily inefficient. In the present paper we study the extension of some branching strategies to exploit the knowledge of multiple optimal solutions of the current LP relaxations.

The contribution of the present paper is twofold. First, we introduce for the first time, to the best of our knowledge, a branching strategy that makes use of multiple relaxation solutions and show how it can be naturally integrated into existing branching rules. Second, we evaluate one particular implementation of it in the context of full strong branching, the branching rule commonly known to be most efficient w.r.t. the number of branch-and-bound nodes [4, 17]. We show that it leads to significant savings in computation time while not increasing the number of nodes.

The remainder of the paper is organized as follows. In Section 2 we discuss how to generate alternative optimal solutions (a *cloud* of solutions) and how to exploit this information to enhance some standard branching rules such as pseudocost branching and strong branching. In Section 3 we give more details on the technique applied to full strong branching, while in Section 4 we report a preliminary computational evaluation of the proposed method. Some conclusions are finally drawn in Section 5.

2   A Cloud of Solutions

In order to extend standard branching strategies to deal with multiple LP optima at the same time, we need to solve two problems:

1. How to generate efficiently multiple optimal solutions of the current LP relaxation?
2. How to make use of the additional information provided by these solutions?

The first problem can be effectively solved by restricting the search to the optimal face of the LP relaxation polyhedron. On this face, an auxiliary objective function can be used to move to different bases. From the computational point of view, fixing to the optimal face can be easily and safely implemented by fixing all variables (structural and artificial) whose reduced costs are non-zero, using the reduced costs associated to the starting optimal basis. As far as the choice of the second level objective function(s) is concerned, different strategies can be used. One option is to try to minimize and maximize each variable which is not yet fixed: this is what optimality-based bound tightening techniques do (see, e.g., [18, 19]), with the additional constraint of staying on the optimal face. Another option is to use a feasibility pump [20] like objective function, in which the current LP point is rounded and a Hamming distance function is generated to move to a different point (more details will be given in the next section): this is related to the pump-reduce procedure that Cplex performs to achieve more integral LP optima [21]. Finally, a random objective function might be used.

Suppose now that we have constructed, in one way or another, a *cloud* $C = \{x^1, \ldots, x^k\}$ of alternative optimal solutions to the current LP relaxation. We assume that the initial fractional solution $x^\star \in C$. Given C, we can define our initial set of branching candidates $F(C)$ as

$$F(C) = \{j \in J \mid \exists x^i \in C : x_j^i \notin \mathbb{Z}\}$$

i.e., $F(C)$ contains all the variables that are fractional in at least one solution of the cloud. For each variable in $F(C)$ it is then possible to calculate its *cloud interval* $I_j = [l_j, u_j]$, where:

$$l_j = \min\{x_j^i \mid x^i \in C\}$$

$$u_j = \max\{x_j^i \mid x^i \in C\}$$

Given the cloud interval for each branching candidate, we partition the set $F(C)$ into three subsets, depending on the relative intersection between each interval I_j and the *branching interval* $B_j = [\lfloor x_j^\star \rfloor, \lceil x_j^\star \rceil]$. In particular, we define:

$$F_2 = \{j \in F(C) \mid \lfloor x_j^\star \rfloor < l_j \wedge u_j < \lceil x_j^\star \rceil\}$$

$$F_0 = \{j \in F(C) \mid l_j \leq \lfloor x_j^\star \rfloor \wedge \lceil x_j^\star \rceil \leq u_j\}$$

$$F_1 = F(C) \setminus (F_2 \cup F_0)$$

In particular for binary variables, F_2 contains exactly those variables which are fractional for all $x^i \in C$, or differently spoken: $F(C)$ is the union (taken over C) of all branching candidates, F_2 is the intersection. If C contained all vertices of the optimal face, then F_2 would be exactly the set of variables that are guaranteed to improve the dual bound in both child nodes. The hope is that also with a limited set of sample point in C, F_2 is still a good approximation to that set.

A variable being contained in the set F_0 is a certificate that branching on it will not improve the dual bound on either side since alternative optima exist which respect the bounds after branching. For the same reasoning, variables in F_1 are those for which the objective function will stay constant for one child, but hopefully not for the other.

The details about how branching rules can be extended to deal with this additional information, namely this three-way partition of the branching candidates (F_2, F_1, F_0) and the set of cloud intervals I_j of course depends on the particular strategy. For example, a rule based on strong branching can safely skip variables in F_0, thus saving some LPs (more details on how to extend a full strong branching policy to the cloud will be given in the next section). In the remaining part of this section, we will describe how pseudocost branching can be modified to exploit cloud information.

Pseudocost branching consists mainly in two operations: (i) updating the pseudocosts after an actual branching has been performed and the LP relaxations of the child nodes have been solved and (ii) computing the score of a variable using the current pseudocosts when deciding for a branching candidate. When updating the pseudocosts, the objective gains ς_j^+ and ς_j^- per unit change in variable x_j are computed, that is:

$$\varsigma_j^+ = \frac{\Delta^\uparrow}{\lceil x_j^\star \rceil - x_j^\star} \quad \text{and} \quad \varsigma_j^- = \frac{\Delta^\downarrow}{x_j^\star - \lfloor x_j^\star \rfloor} \tag{1}$$

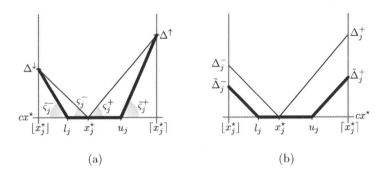

Fig. 1. Graphical representation of pseudocosts update and usage

where $\Delta^{\uparrow}$ and $\Delta^{\downarrow}$ are the differences between the optimal LP objectives of the corresponding child nodes and the current LP value. These gains are then used to update the current pseudocosts Ψ_j^+ and Ψ_j^- which are the averages of the objective gains (per unit step length) that have been observed for that particular variable so far. The thin line in Figure 1(a) illustrates the operation. These estimation formulas are based on the assumption that the objective increases linearly in both directions (hence the resulting triangle). This, however, may be a too crude approximation of the real shape of the projection on the split domain of x_j. In the case of dual degeneracy, there might be many optimal LP solutions with different values for x_j. Which of these values $x_j^\star$ takes is more or less arbitrary, but crucial for the current – and by that also for future – branching decisions.

Using interval I_j on the other hand it is possible to replace this approximation with a more precise model (thick line in Figure 1(a)). The corresponding way to compute gains is then:

$$\tilde{\varsigma}_j^+ = \frac{\Delta^{\uparrow}}{\lceil x_j^\star \rceil - u_j} \quad \text{and} \quad \tilde{\varsigma}_j^- = \frac{\Delta^{\downarrow}}{l_j - \lfloor x_j^\star \rfloor} \tag{2}$$

Where the values for ς^+ and ς^- may vary by chance, $\tilde{\varsigma}^+$ and $\tilde{\varsigma}^-$ will be constant, when the set of all corners of the optimal face is used as a cloud.

As far as the computation of the score s_j is concerned, the standard formulas to predict the objective gains when branching on variable x_j are

$$\Delta_j^+ = \Psi_j^+ (\lceil x_j^\star \rceil - x_j^\star) \quad \text{and} \quad \Delta_j^- = \Psi_j^- (x_j^\star - \lfloor x_j^\star \rfloor) \tag{3}$$

Again, the underlying linear model may give a too optimistic estimate on the dual bound improvements. A more accurate estimate exploiting interval I_j can be obtained as:

$$\tilde{\Delta}_j^+ = \Psi_j^+ (\lceil x_j^\star \rceil - l_j) \quad \text{and} \quad \tilde{\Delta}_j^- = \Psi_j^- (u_j - \lfloor x_j^\star \rfloor) \tag{4}$$

A graphical representation is depicted in Figure 1(b). Furthermore, the following observation holds:

Lemma 2.1. *Let $x^\star$ be an optimal solution of the LP relaxation at a given branch-and-bound node and $\lfloor x_j^\star \rfloor \leq l_j \leq x_j^\star \leq u_j \leq \lceil x_j^\star \rceil$. Then*

1. *for fixed $\Delta^\uparrow$ and $\Delta^\downarrow$, it holds that $\tilde{\varsigma}_j^+ \geq \varsigma_j^+$ and $\tilde{\varsigma}_j^- \geq \varsigma_j^-$, respectively;*
2. *for fixed Ψ_j^+ and Ψ_j^-, it holds that $\tilde{\Delta}_j^+ \leq \Delta_j^+$ and $\tilde{\Delta}_j^- \leq \Delta_j^-$, respectively.*

Proof. Follows directly from Equations (1)–(4).

Thus, under the same preconditions, the standard pseudocosts will be an under-estimation of the pseudocosts based on the cloud intervals, whereas the objective gain, on which the branching decision is made, will be an overestimation. Of course these quantities interact directly which each other: as soon as one of it gets altered, this will have an impact on all upcoming branching decisions and pseudocost computations. The effects of continuous over- and underestimation will amplify each other. The hope is that cloud branching helps to make better, more reliable predictions and thereby leads to better branching decisions.

3   Full Strong Branching with the Cloud

In the present section we detail the extension of a full strong branching strategy to the cloud. The first problem is again how to generate a cloud of optimal LP solutions C. Following some preliminary computational results, we opted for a feasibility pump like objective function, minimizing the distance to the nearest integral point. More precisely, given a fractional solution $x^\star$, we define the objective function coefficient c_j of variable x_j as

$$c_j = \begin{cases} 1 & \text{if } 0 < f_j < 0.5 \\ -1 & \text{if } 0.5 \leq f_j < 1 \\ 0 & \text{otherwise} \end{cases}$$

where $f_j = x_j^\star - \lfloor x_j^\star \rfloor$ is the fractional part of $x_j^\star$. Using the primal simplex, we re-solve the LP (fixed to the optimal face) with this new objective function. We update the interval bound vectors l and u, and iterate, using the new optimum as $x^\star$. If, at a given iteration, the update did not yield a new integral interval bound, we stop.

As far as the three-way partition (F_2, F_1, F_0) is concerned, we perform full strong branching on all variables in the set F_2. If we can find even one variable in this set with a strictly improved dual bound in both child nodes, then we stop and pick the best variable within this set, completely ignoring sets F_1 and F_0. In state-of-the-art solvers such as Cplex or SCIP the score of a variable is computed as the *product* of the objective gains in both directions (maybe using a minimum value of some epsilon close to zero for each factor). By this, the score of all variables in $F_1 \cap F_0$ will be (nearly) zero and therefore none of them will have maximum score.

Note that in this case cloud information is used essentially to filter out variables and solve a smaller number of LPs. If no such variable is found, different strategies can be devised, depending on how we deal with the remaining variables. One option is to proceed with performing strong branching on the variables in set F_1, but solving only one LP per variable (because by definition we already know that in one direction the dual bound change is zero). Note that variables in F_1 are not necessarily a subset of the fractional variables in $x^\star$: as such, while we may still have some speedup because we only solve one LP per variable, the number of variables may indeed be higher than what standard full strong branching would have done. If we can find at least one variable in $F_2 \cap F_1$ with a strictly improved dual bound in one direction, then we can stop and ignore set F_0 for the same reason as before. If this is not the case, then we know that for all variables in $F(C)$ no improvement can be obtained in any child node as far as the dual bound is concerned, and so the branching variable should be chosen with some other criterion.

Another, less computationally expensive, option is to always ignore variables in F_1 and stick to the variables in F_2. Apart from the obvious computational savings, this choice can be justified by the following argument: if there is a variable in F_2 with a strictly improved dual bound in both children, we will not consider $F_1 \cap F_0$ anyway. If there is none, this proves that the global dual bound will not improve independent of the branching decision: at least one of the two children will have the same dual bound as the current node. Therefore, we take the current set of points C as evidence that variables in F_2 are less likely to become integral than variables in F_1, and so should be given precedence as branching candidates.

Note that using additional points to filter out strong branching candidates is similar in spirit to the strategy called *nonchimerical branching* proposed in [8], where the optimal solutions of the strong branching LPs (which might have a different objective function value) were used for this purpose. The two strategies have complementary strengths: nonchimerical branching does not need to solve any additional LP w.r.t. strong branching, but needs the strong branching LPs to be solved to optimality, because of the usage of the dual simplex. Cloud branching on the other hand needs additional LPs, but these are in principle simpler (we are fixed to the optimal face), need not be solved to optimality (primal simplex is used), and do not impose any requirements to the solution of the final strong branching LPs. As such, the two techniques can be easily combined together and might synergize. Moreover, cloud branching can be used independent of strong branching, as argued in Section 2.

4   Computational Experiments

For our computational experiments, we used SCIP 3.0.0.1 [22] compiled with SoPlex 1.7.0 [23] as LP solver. The results were obtained on a cluster of 64bit Intel Xeon X5672 CPUs at 3.20GHz with 12 MB cache and 48 GB main memory, running an openSuse 12.1 with a gcc 4.6.2 compiler. Hyperthreading and Turboboost were disabled. We ran only one job per node to reduce random noise in the measured running time that might be caused by cache-misses if multiple processes share common resources.

Table 1. comparison of Cloud branching and full strong branching on MMM and COR@L instances, averages of success rate, cloud points, saved LPs per node, and rate of saved LPs; shifted geometric means of branch-and-bound nodes and running time in seconds

Test set	cloud statistics				SCIP cloud branch		SCIP strong branch	
	%Succ	Pts	LPs	%Sav	Nodes	Time (s)	Nodes	Time (s)
MMM	12.2	2.19	74.34	21.7	661	68.2	691	72.0
COR@L	40.8	2.71	70.97	51.8	569	118.3	593	157.3

We used two test sets of general, publically available MIP instances: the COR@L test set [24], which mainly contains instances that users worldwide submitted to the NEOS server [25] and the MMM test set which contains all instances from MIPLIB3.0 [26], MIPLIB2003 [17], and MIPLIB2010 [16]. We compare the performance of SCIP when using full strong branching versus a cloud branching version of full strong branching as described in the previous section. In particular, we compare to the cloud branching variant that only considers variables in F_2 as possible branching candidates. Since we want to explicitly measure the impact of using the cloud for variable selection, we did not exploit the alternative LP optima by any other means, e.g. for cutting plane generation, primal heuristics, reduced cost domain propagation, etc. Results by Achterberg [21] indicate that this would be likely to give further improvements on the overall performance. Further, we used the default implementation of full strong branching in SCIP, which does not employ the methods suggested in [8] (yet). We expect that nonchimerical branching and cloud branching will complement each other nicely, however, this is left for future implementation and experiments. We used a time limit of one hour per instance. All other parameters were left at their default values.

For the MMM test set both, SCIP with cloud branching and with full strong branching, both solved the same number of instance; for the COR@L test set, one more instances was solved within the time limit when using cloud branching. Tables 2 and 3 in the Appendix show results for all instances which both variants could solve within the time limit, excluding those which were directly solved at the root node (hence no branching was performed). This leaves 68 instances for MMM and 104 instances for COR@L. Column "%Succ" shows the ratio of nodes on which cloud branching was run successfully, hence at least one additional cloud point was used. Considering those nodes, columns "Pts" and "LPs" depict of how many points the cloud consisted on average and how many strong branching LPs were saved on average per node, i.e., how many integral interval bounds could be found. The Column "%Sav" shows how many percent of all strong branching LPs could be saved for that instance. When the success rate is zero, these three columns show a dash. For both branching variants, "Nodes" and "Time" give the number of branch-and-bound nodes and the computation time needed to prove optimality.

Table 1 shows aggregated results. It gives averages over the corresponding numbers (the success rates, the used points, the saved LPs per node and the percentage of overall saved LPs) from Tables 2 and 3. Shifted geometric means are shown for the number of branch-and-bound nodes and the computation times, which are absolute performance measures. The shifted geometric mean of values $t_1, \ldots, t_n$ with shift s is defined as $\sqrt[n]{\prod(t_i + s)} - s$. We use a shift of $s = 10$ for time and $s = 100$ for nodes in order to reduce the effect of very easy instances in the mean values. Further, using a *geometric* mean prevents hard instances at or close to the time limit from having a huge impact on the measures. Thus, the shifted geometric mean has the advantage that it reduces the influence of outliers in both directions.

The results for the MMM test set show a slight improvement of 6% w.r.t. mean running time and 5% w.r.t. the mean number of nodes when using cloud branching. For COR@L, the mean number of nodes again is slightly larger, about 4%, when using full strong branching instead of cloud branching. The result when comparing computation times is much more explicit: the shifted geometric means differ by about 33%. As can be seen in Table 1, the success rate of cloud branching is much better on the COR@L test set than it is on MMM; and even further, on the successful instances, the average ratio of saved LPs is much larger. Taking these observations together explains why the improvement is much more significant for the COR@L test set.

MIP solvers are known to be prone for an effect called performance variability. Loosely speaking, this term comprises unexpected changes in performance which are triggered by seemingly performance-neutral changes in the environment or the input format. Besides others, peformance variability is caused by imperfect tie breaking [16]. This results in small numerical differences caused by the use of floating point arithmetics which may lead to different decisions being taken during the solution process. A branch-and-bound search often amplifies these effects, which can be similarly observed for all major MIP (and also other optimization) softwares. As a consequence, small changes in performance might in fact be random noise rather than a real improvement or deterioration. This can, e.g., be seen for instance cap6000 from MMM: Although cloud branching was never successful, the number of branch-and-bound nodes alters[1]. Then again, improvements brought by single components of a MIP solver typically lie in the range of 5–10%, see, e.g., [4]. In addition, even if MIP solvers did not exhibit performance variability, we would have the issue of assessing whether the measured difference in performance is statistically significant, a problem common to all empirical studies.

We performed two additional experiments to validate our computational results. First, we ran identical tests on four more copies of the test sets, with perturbed models that were generated by permuting columns and rows of the

---

[1] This can be explained by the intermediate cloud LPs being solved – after this, the original LP basis gets installed again and a resolve without simplex iterations is performed. However, solution values, reduced costs etc. might be slightly different than before.

original problem formulation. This has been introduced in [16] as a good variability generator that affects all types of problems and all components of a typical MIP solver. Another benefit of this experiment is that it counters overtuning since the evaluation testbed is no longer the same as the development test bed.

As can be expected, the results differ in detail from the default permutation run. For MMM, the improvements w.r.t. computation time were 3%, 4%, 4% and 7%, and w.r.t. branch-and-bound nodes -3%, 0%, 1% and 2%. On COR@L, the improvements w.r.t. time were 25%, 29%, 32%, and 42% and w.r.t. branch-and-bound nodes 3%, 5%, 8%, 14%. We conclude the cloud branching was faster in all five times two experiments (including the original ones) and also consistently reduced the number of branch-and-bound nodes on the COR@L test set. For MMM, it can be argued that the changes are performance neutral w.r.t. the number of branch-and-bound nodes.

As far as the statistical significance of these differences is concerned, we performed randomized tests [27] on the detailed results. Randomized tests are standard non-parametric statistics that do not make any assumptions on the underlying population distributions (assumptions are very likely to be violated in our computational settings) but are still as powerful as standard parametric tests. According to these tests, the performance difference, both w.r.t. time and nodes, measured on the MMM is *not* statistically significant. As far as COR@L is concerned, the difference in branch-and-bound nodes is again not significant, while the difference in running times is. Note that on heterogeneous test sets such as MMM and COR@L, it is rather difficult to pass statistical significance tests when testing single MIP solver components, because the improvements are almost always in the single digit range and standard test sets are relatively small. In other words, one method might indeed be better than the other, but not by enough to pass the statistical test. We also applied these randomized tests to the other four copies of the test sets, with consistent results.

Having a closer look at Tables 2 and 3, it can be seen that the success rate of cloud branching is negligible, i.e., close to zero, for a significantly higher ratio of the MMM test set than for the COR@L test set. This is also reflected by the much smaller average success rate shown in Table 1. This partially explains why the differences on COR@L are much more significant than on MMM: there are simply more instances on which degenerate LP solutions are detected in the pump-reduce step of our algorithm. A reason for this might be that MIPLIB instances contain more industry-based models with real, perturbed data whereas COR@L has more combinatorial models which often contain symmetries and are prone for degeneracy.

Our interpretation of the given results therefore is that cloud branching does not hurt a test set where only few degeneracy is detected but is clearly superior on a test set which contains many highly degenerated problems, at least time-wise.

5   Conclusion and Outlook

In this paper, we introduced branching strategies for mixed-integer programs that exploit the knowledge of a cloud of alternative optimal LP solutions. We discussed extensions of full strong branching and pseudocost branching that incorporate this idea. Our computational experiments showed that a version of full strong branching that uses cloud intervals is about 30% faster than default full strong branching on a standard test set with high dual degeneracy. Even the mean number of branch-and-bound nodes could be reduced, though not significantly.

The presented preliminary results are very encouraging for further research on cloud branching. A natural next step is to implement the described modifications on pseudocost branching and a development of hybrid strategies such as reliability branching that make use of the cloud. In this paper, we used multiple optima from a single relaxation as cloud set. In particular in the context of MINLP, employing optima from *multiple, alternative relaxations* seems promising. From the implementation point of view, it could be further exploited that the cloud LPs are solved by the primal simplex algorithm, hence also intermediate, suboptimal solutions will be feasible and could be used as cloud points. Finally, two other improvements of strong branching were suggested recently: nonchimerical branching [8] and a work of Gamrath [28] on using domain propagation in strong branching. It will be interesting to see how these ideas combine and whether it will even be possible to make full strong branching competitive to state-of-the-art hybrid branching rules w.r.t. mean running time.

Acknowledgements. Many thanks to Gerald Gamrath and four anonymous reviewers for their constructive criticism.

References

[1] Benichou, M., Gauthier, J., Girodet, P., Hentges, G., Ribiere, G., Vincent, O.: Experiments in mixed-integer programming. Mathematical Programming 1, 76–94 (1971)

[2] Linderoth, J.T., Savelsbergh, M.W.P.: A computational study of strategies for mixed integer programming. INFORMS Journal on Computing 11, 173–187 (1999)

[3] Achterberg, T., Koch, T., Martin, A.: Branching rules revisited. Operations Research Letters 33, 42–54 (2005)

[4] Achterberg, T.: Constraint Integer Programming. PhD thesis, Technische Universität Berlin (2007), http://opus4.kobv.de/opus4-zib/frontdoor/index/index/docId/1018

[5] Bixby, R., Fenelon, M., Gu, Z., Rothberg, E., Wunderling, R.: MIP: Theory and practice – closing the gap. In: Powell, M., Scholtes, S. (eds.) Systems Modelling and Optimization: Methods, Theory, and Applications, pp. 19–49. Kluwer Academic Publisher (2000)

[6] Applegate, D.L., Bixby, R.E., Chvátal, V., Cook, W.J.: Finding cuts in the TSP (A preliminary report). Technical Report 95-05, DIMACS (1995)

[7] Applegate, D.L., Bixby, R.E., Chvátal, V., Cook, W.J.: The Traveling Salesman Problem: A Computational Study. Princeton University Press, USA (2007)

[8] Fischetti, M., Monaci, M.: Branching on nonchimerical fractionalities. OR Letters 40(3), 159–164 (2012)

[9] Achterberg, T., Berthold, T.: Hybrid branching. In: van Hoeve, W.-J., Hooker, J.N. (eds.) CPAIOR 2009. LNCS, vol. 5547, pp. 309–311. Springer, Heidelberg (2009)

[10] Patel, J., Chinneck, J.W.: Active-constraint variable ordering for faster feasibility of mixed integer linear programs. Mathematical Programming 110, 445–474 (2007)

[11] Karamanov, M., Cornuéjols, G.: Branching on general disjunctions. Mathematical Programming 128(1-2), 403–436 (2011)

[12] Li, C.M., Anbulagan: Look-ahead versus look-back for satisfiability problems. In: Smolka, G. (ed.) CP 1997. LNCS, vol. 1330, pp. 341–355. Springer, Heidelberg (1997)

[13] Moskewicz, M., Madigan, C., Zhao, Y., Zhang, L., Malik, S.: Chaff: Engineering an efficient SAT solver. In: Proceedings of the 38th Annual Design Automation Conference (DAC 2001), pp. 530–535 (2001), doi:10.1145/378239.379017

[14] Kılınç Karzan, F., Nemhauser, G.L., Savelsbergh, M.W.P.: Information-based branching schemes for binary linear mixed-integer programs. Mathematical Programming Computation 1(4), 249–293 (2009)

[15] Fischetti, M., Monaci, M.: Backdoor branching. In: Günlük, O., Woeginger, G.J. (eds.) IPCO 2011. LNCS, vol. 6655, pp. 183–191. Springer, Heidelberg (2011)

[16] Koch, T., Achterberg, T., Andersen, E., Bastert, O., Berthold, T., Bixby, R.E., Danna, E., Gamrath, G., Gleixner, A.M., Heinz, S., Lodi, A., Mittelmann, H., Ralphs, T., Salvagnin, D., Steffy, D.E., Wolter, K.: MIPLIB 2010 - Mixed Integer Programming Library version 5. Mathematical Programming Computation 3, 103–163 (2011), http://miplib.zib.de

[17] Achterberg, T., Koch, T., Martin, A.: MIPLIB 2003. Operations Research Letters 34(4), 1–12 (2006), http://miplib.zib.de/miplib2003/

[18] Zamora, J.M., Grossmann, I.E.: A branch and contract algorithm for problems with concave univariate, bilinear and linear fractional terms. Journal of Global Optimization 14, 217–249 (1999), doi:10.1023/A:1008312714792

[19] Caprara, A., Locatelli, M.: Global optimization problems and domain reduction strategies. Mathematical Programming 125, 123–137 (2010), doi:10.1007/s10107-008-0263-4

[20] Fischetti, M., Glover, F., Lodi, A.: The feasibility pump. Mathematical Programming 104(1), 91–104 (2005), doi:10.1007/s10107-004-0570-3

[21] Achterberg, T.: LP basis selection and cutting planes. Presentation Slides from MIP 2010 Conference in Atlanta (2010), http://www2.isye.gatech.edu/mip2010/program/program.pdf

[22] Achterberg, T.: SCIP: Solving Constraint Integer Programs. Mathematical Programming Computation 1(1), 1–41 (2009), doi:10.1007/s12532-008-0001-1

[23] Wunderling, R.: Paralleler und objektorientierter Simplex-Algorithmus. PhD thesis, Technische Universität Berlin (1996)

[24] COR@L: MIP Instances (2010), http://coral.ie.lehigh.edu/data-sets/mixed-integer-instances/

[25] Czyzyk, J., Mesnier, M., Moré, J.: The NEOS server. IEEE Computational Science & Engineering 5(3), 68–75 (1998), http://www.neos-server.org/neos/

[26] Bixby, R.E., Ceria, S., McZeal, C.M., Savelsbergh, M.W.: An updated mixed integer programming library: MIPLIB 3.0. Optima (58), 12–15 (1998), http://miplib.zib.de/miplib3/miplib.html

[27] Cohen, P.R.: Empirical Methods for Artificial Intelligence. MIT Press (1995)

[28] Gamrath, G.: Improving strong branching by propagation. In: Gomes, C., Sellmann, M. (eds.) CPAIOR 2013. LNCS, vol. 7874, pp. 347–354. Springer, Heidelberg
(2013)

Appendix

Tables 2 and 3 show the detailed results for the computational evaluation given
in Section 4. They report statistics on all instances from our two test sets MMM
and COR@L which SCIP could solve to optimality in less than one hour for either
strong branching variant, but needed more than one node in both cases.

Table 2. comparison of cloud branching and full strong branching on MMM instances,
smaller (better) numbers are bold

instance	cloud statistics %Succ	Pts	LPs	%Sav	SCIP cloud branch Nodes	Time (s)	SCIP strong branch Nodes	Time (s)
10teams	64.3	2.7	50.3	79.3	**129**	**105.3**	348	488.1
aflow30a	0.0	–	–	–	**166**	**19.6**	182	21.7
air04	91.3	2.0	9.1	32.3	**55**	2087.9	57	**2074.6**
air05	46.8	2.0	3.0	3.4	166	1597.0	**153**	**1541.6**
ash608gpia-3col	100.0	4.3	2240.1	86.7	**5**	**1072.1**	9	2406.8
bell3a	0.0	–	–	–	26 588	6.6	26 590	**6.3**
bell5	0.2	2.0	2.0	0.6	**851**	0.7	865	0.7
bienst2	25.5	2.4	6.6	34.1	21 729	**1586.4**	**21 210**	1707.6
binkar10_1	4.4	2.0	4.8	3.3	**45 080**	**1715.7**	48 835	1744.9
blend2	9.3	2.0	2.0	5.4	108	0.8	110	**0.7**
cap6000	0.0	–	–	–	1 601	3.3	**1 545**	**3.1**
dcmulti	0.0	–	–	–	120	**2.3**	120	2.5
dfn-gwin-UUM	0.0	–	–	–	5 897	435.1	5 918	**431.6**
eil33-2	0.0	–	–	–	484	739.8	**480**	**734.2**
enigma	5.2	2.0	9.2	14.6	**27**	**0.5**	249	0.6
fiber	0.0	–	–	–	16	**1.1**	16	1.3
fixnet6	0.0	–	–	–	9	2.3	9	**2.2**
flugpl	0.0	–	–	–	134	0.5	134	0.5
gesa2-o	0.0	–	–	–	5	**1.4**	5	1.5
gesa2	0.0	–	–	–	3	1.0	3	1.0
gesa3	0.0	–	–	–	**11**	**1.4**	15	1.5
gesa3_o	0.0	–	–	–	9	**1.5**	9	1.7
khb05250	0.0	–	–	–	4	0.5	4	0.5
l152lav	3.9	2.0	6.7	3.5	**53**	**4.7**	65	7.1
lseu	15.4	2.1	3.4	12.7	**364**	0.7	382	**0.5**
map18	0.0	–	–	–	103	**1454.7**	**101**	1701.6
map20	0.0	–	–	–	**87**	**1129.0**	91	1384.7
mas74	0.0	–	–	–	574 769	1389.5	574 769	**1321.8**
mas76	0.0	–	–	–	**81 106**	123.7	84 280	**123.0**
mik-250-1-100-1	0.0	–	–	–	**290 018**	1681.4	290 038	**1628.3**
mine-166-5	0.0	–	–	–	2 001	142.6	**1 994**	155.6
misc03	11.7	2.3	10.4	25.4	68	**1.4**	**65**	1.5
misc06	5.9	2.0	4.0	6.7	13	0.8	13	**0.6**
misc07	13.2	2.1	7.1	23.5	**2 300**	62.9	2 365	**57.9**
mod008	0.0	–	–	–	**104**	0.8	111	0.8

Table 2. (*continued*)

instance	cloud statistics %Succ	Pts	LPs	%Sav	SCIP cloud branch Nodes	Time (s)	SCIP strong branch Nodes	Time (s)
mod010	0.0	–	–	–	10	**1.0**	10	1.1
mod011	0.0	–	–	–	321	**989.9**	321	1069.9
modglob	0.0	–	–	–	299	2.9	299	**2.8**
neos-1109824	51.9	2.3	26.0	68.4	1 246	473.0	**1 023**	**390.9**
neos-1396125	69.4	2.2	8.3	55.1	**2 714**	**2355.7**	2 976	2653.2
neos-476283	0.0	–	–	–	445	887.5	**323**	**680.2**
neos-686190	3.7	2.0	9.7	4.8	**1 451**	**540.6**	2 085	774.5
noswot	86.9	2.4	16.5	74.2	337 012	957.6	**210 056**	**869.0**
ns1766074	0.0	2.1	5.5	0.1	**241 641**	492.3	241 801	**470.2**
nw04	0.0	–	–	–	5	54.8	5	**46.4**
p0033	0.0	–	–	–	5	0.5	5	0.5
p0201	47.0	2.3	23.2	64.6	52	**2.6**	**51**	3.0
p0282	0.0	–	–	–	3	0.5	3	0.5
p0548	0.0	–	–	–	5	0.5	5	0.5
p2756	2.6	2.0	4.0	2.5	**82**	**1.9**	146	2.0
pk1	0.1	2.8	16.4	0.6	**76 569**	257.8	77 616	**233.1**
pp08a	0.0	–	–	–	300	3.7	**251**	**3.0**
pp08aCUTS	0.2	2.0	2.0	0.1	**213**	**3.2**	284	4.2
qiu	10.3	2.1	10.3	17.9	**14 858**	**1515.7**	16 290	1895.5
qnet1	17.6	2.0	6.0	4.4	5	3.8	5	**3.4**
qnet1_o	20.0	2.0	3.8	2.8	22	**9.2**	22	10.3
ran16x16	4.4	2.0	2.3	2.6	28 684	1184.3	**27 051**	**964.4**
reblock67	0.0	–	–	–	**28 052**	**1528.8**	33 290	1773.3
rentacar	27.3	2.0	3.3	22.2	**13**	**3.4**	14	3.5
rmatr100-p10	0.1	2.0	2.0	0.0	**163**	952.8	164	**950.2**
rmatr100-p5	0.0	–	–	–	33	1327.1	33	**1321.6**
rout	32.6	2.2	11.8	46.0	**1 561**	**79.3**	1 712	85.8
set1ch	0.0	–	–	–	**16**	**0.9**	17	1.0
sp98ir	2.1	2.0	3.5	1.3	**609**	**404.1**	876	507.4
stein27	29.1	2.3	6.2	22.9	787	2.2	**775**	**2.0**
stein45	20.7	2.1	5.8	11.2	**7 909**	**73.8**	8 446	77.3
tanglegram2	0.0	–	–	–	2	**27.3**	2	34.3
vpm2	9.4	2.0	2.2	3.4	**46**	1.3	48	1.3

Table 3. Comparison of cloud branching and full strong branching on Cor@l instances, smaller (better) numbers are bold

instance	cloud statistics %Succ	Pts	LPs	%Sav	cloud branching Nodes	Time (s)	full strong branching Nodes	Time (s)
22433	0.0	–	–	–	4	**1.1**	4	1.2
23588	0.2	2.0	14.0	0.3	**148**	**6.0**	174	6.3
aligninq	1.8	2.0	9.0	1.1	**24**	**20.6**	32	23.1
bc1	0.0	–	–	–	604	132.6	616	**124.0**
bc	0.0	–	–	–	1 985	2437.7	1 985	**2323.3**
bienst1	31.7	2.2	6.0	39.0	**2 712**	**151.1**	2 737	172.1
bienst2	25.5	2.4	6.6	34.1	21 729	**1587.7**	**21 210**	1703.3
binkar10_1	4.4	2.0	4.8	3.3	**45 080**	**1714.4**	48 835	1744.8
dano3_3	0.0	–	–	–	9	235.5	9	**153.3**
dano3_4	0.0	–	–	–	4	**176.7**	4	177.4
haprp	0.0	–	–	–	20 289	1696.4	**19 844**	**1629.9**
neos-1053591	94.5	2.3	8.7	70.0	**1 794**	**19.2**	46 259	367.4
neos-1109824	51.9	2.3	26.0	68.4	1 246	482.7	**1 023**	**390.7**
neos-1120495	38.7	2.2	19.4	49.7	102	18.7	**75**	**17.6**

Table 3. (*continued*)

instance	cloud statistics				cloud branching		full strong branching	
	%Succ	Pts	LPs	%Sav	Nodes	Time (s)	Nodes	Time (s)
neos-1122047	100.0	3.5	42.0	96.6	3	80.6	2	34.1
neos-1200887	86.2	2.5	13.8	63.5	981	234.4	1465	381.7
neos-1211578	74.9	2.4	10.9	74.6	49619	315.6	32225	323.5
neos-1224597	98.6	6.7	631.1	95.0	70	406.8	80	864.5
neos-1228986	74.8	2.4	11.7	70.2	42072	358.7	39690	400.6
neos-1281048	91.8	5.8	133.8	88.1	59	49.8	80	170.7
neos-1337489	74.9	2.4	10.9	74.6	49619	312.0	32225	320.5
neos-1367061	0.0	–	–	–	16	1601.4	16	1683.2
neos-1396125	69.4	2.2	8.3	55.1	2714	2359.8	2976	2659.6
neos-1413153	81.8	2.9	292.5	88.6	192	219.9	462	2546.4
neos-1415183	90.9	2.6	252.4	85.7	19	6.6	56	43.6
neos-1420205	82.5	2.1	8.0	33.5	10674	46.8	7840	45.5
neos-1437164	74.7	2.2	11.0	47.1	80	1.8	47	1.9
neos-1440225	92.3	4.7	381.7	96.6	6	11.0	134	1387.4
neos-1440447	88.7	2.8	18.4	80.6	7676	207.5	22496	747.9
neos-1441553	78.0	2.3	26.0	58.1	133	16.2	215	86.7
neos-1445743	0.0	–	–	–	2	101.1	2	64.4
neos-1445755	5.0	2.0	4.0	16.7	3	75.0	3	57.6
neos-1445765	1.6	2.0	2.0	0.3	5	374.9	5	236.2
neos-1460265	99.8	4.0	216.2	80.2	6997	1354.8	1125	540.5
neos-1480121	23.0	2.0	3.5	25.0	1288	3.3	1961	4.0
neos-1489999	0.0	–	–	–	21	28.6	21	32.0
neos-476283	0.0	–	–	–	445	885.7	323	687.4
neos-480878	24.2	2.0	3.2	7.3	2803	230.9	3517	279.0
neos-494568	94.2	3.0	181.7	76.8	291	398.2	285	1082.3
neos-501474	48.2	2.0	4.0	46.9	158	1.3	104	0.7
neos-504674	50.5	2.0	5.9	16.6	1256	399.7	1230	426.9
neos-504815	35.6	2.1	6.1	16.5	510	75.4	502	83.3
neos-506422	16.2	2.1	3.7	20.9	1451	540.7	959	337.3
neos-512201	48.7	2.0	6.0	15.5	665	175.7	436	149.2
neos-522351	0.0	–	–	–	3	1.1	3	1.0
neos-525149	55.3	3.0	88.7	45.8	46	17.9	187	193.4
neos-530627	0.0	–	–	–	2	0.5	2	0.5
neos-538867	72.8	3.0	21.4	76.2	6697	318.1	4358	208.9
neos-538916	77.5	3.2	23.9	77.4	4642	371.5	3496	294.6
neos-544324	99.9	2.0	15.2	92.3	7	301.4	7	149.9
neos-547911	90.0	2.1	10.8	85.1	30	244.3	30	184.5
neos-555694	71.3	2.6	98.6	69.9	65	36.5	177	301.7
neos-555771	92.7	2.4	107.7	80.2	32	17.8	70	72.4
neos-570431	71.0	2.0	8.1	59.4	60	290.2	76	314.7
neos-584851	47.0	2.1	20.0	60.3	56	778.1	38	840.5
neos-585192	0.0	–	–	–	333	40.1	345	40.6
neos-585467	1.2	2.0	12.0	1.4	125	10.6	133	10.7
neos-593853	0.0	–	–	–	10157	52.4	12204	56.0
neos-595905	0.0	–	–	–	418	25.2	473	29.5
neos-595925	0.0	–	–	–	1166	51.6	1189	51.8
neos-598183	0.0	–	–	–	488	6.8	486	7.1
neos-611838	0.0	–	–	–	193	89.5	193	94.0
neos-612125	0.0	–	–	–	92	47.1	92	50.3
neos-612143	0.0	–	–	–	130	55.1	128	59.7
neos-612162	0.0	–	–	–	122	74.6	126	80.3
neos-631694	93.9	2.9	56.8	49.5	94	57.3	101	93.6
neos-686190	3.7	2.0	9.7	4.8	1451	537.9	2085	776.4
neos-709469	12.5	2.3	22.4	58.6	1608	3.5	28	1.6
neos-717614	0.0	–	–	–	1059	65.9	1061	65.3
neos-775946	95.4	2.8	93.4	81.3	234	140.6	413	343.6
neos-785899	93.0	2.8	94.2	77.7	179	130.7	266	247.6

Table 3. (*continued*)

instance	cloud statistics				cloud branching		full strong branching	
	%Succ	Pts	LPs	%Sav	Nodes	Time (s)	Nodes	Time (s)
neos-785914	83.8	3.4	135.0	92.0	109	**123.4**	**20**	296.6
neos-801834	0.0	–	–	–	11	841.5	11	**817.6**
neos-803219	0.1	2.0	2.0	0.0	**4 131**	72.8	4 231	**70.9**
neos-803220	0.0	–	–	–	18 713	179.3	**17 175**	**166.5**
neos-806323	28.1	2.0	2.7	11.0	**3 258**	**137.6**	3 645	142.6
neos-807639	4.1	2.0	2.6	3.1	1 130	18.6	**1 120**	**17.2**
neos-807705	20.3	2.0	2.2	6.3	2 373	88.3	**2 241**	**80.0**
neos-808072	72.1	2.3	32.3	51.7	**43**	**379.0**	90	1905.3
neos-810326	34.9	2.0	4.0	6.2	267	**2394.3**	**266**	2431.7
neos-820879	45.1	2.0	3.8	9.6	127	281.1	**114**	**210.3**
neos-825075	84.6	3.9	60.0	80.5	**18**	**3.0**	49	7.8
neos-839859	0.1	2.0	12.0	0.1	**1 084**	**773.1**	1 628	938.6
neos-862348	35.8	2.1	19.8	21.9	99	**33.9**	**70**	38.4
neos-863472	32.8	2.2	15.6	63.2	88 330	2330.9	**68 169**	**2264.0**
neos-880324	62.8	2.4	22.2	78.7	62	1.8	**15**	**1.0**
neos-892255	100.0	2.5	278.6	97.3	8	**720.1**	**5**	1590.8
neos-906865	0.0	–	–	–	7 079	462.4	**7 065**	**453.8**
neos-912015	93.2	5.1	130.8	94.2	791	473.3	**209**	**322.1**
neos-916173	0.0	–	–	–	1 497	**390.2**	**1 478**	392.0
neos-933550	83.3	8.4	638.8	96.6	**5**	**10.1**	25	58.2
neos-933815	47.7	2.0	5.7	32.3	61 797	801.5	**55 797**	**661.4**
neos-934531	99.3	3.4	89.8	96.1	**27**	**293.0**	51	1432.9
neos-941698	97.7	6.1	357.2	95.8	**19**	**14.2**	44	70.5
neos-942323	99.7	4.0	187.8	97.7	**189**	**64.0**	2 205	1240.4
neos-955215	68.4	2.1	9.1	53.0	7 574	61.3	**6 593**	**52.9**
neos-957270	83.1	2.8	159.9	89.0	**14**	471.3	17	**228.4**
nsa	0.0	–	–	–	258	**2.8**	258	3.0
nug08	0.0	–	–	–	3	24.9	3	**23.2**
prod1	0.0	2.0	8.0	0.0	4 053	33.8	**3 820**	**31.4**
prod2	0.0	–	–	–	**25 200**	361.6	25 227	**354.0**
qap10	33.3	2.0	2.0	40.0	2	177.3	2	**157.3**
sp98ir	2.1	2.0	3.5	1.3	**609**	**403.5**	876	503.2
Test3	0.0	–	–	–	10	8.0	10	8.0

Modeling Robustness in CSPs as Weighted CSPs

Laura Climent[1], Richard J. Wallace[2], Miguel A. Salido[1], and Federico Barber[1]

[1] Instituto de Automática e Informática Industrial, Universitat Politècnica de València, Spain
{lcliment,msalido,fbarber}@dsic.upv.es
[2] Cork Constraint Computation Centre, University College Cork, Ireland
r.wallace@4c.ucc.ie

Abstract. Many real life problems come from uncertain and dynamic environments, where the initial constraints and/or domains may undergo changes. Thus, a solution found for the problem may become invalid later. Hence, searching for *robust* solutions for Constraint Satisfaction Problems (CSPs) becomes an important goal. In some cases, no knowledge about the uncertain and dynamic environment exits or it is hard to obtain it. In this paper, we consider CSPs with discrete and ordered domains where only limited assumptions are made commensurate with the structure of these problems. In this context, we model a CSP as a weighted CSP (WCSP) by assigning weights to each valid constraint tuple based on its distance from the edge of the space of valid tuples. This distance is estimated by a new concept introduced in this paper: *coverings*. Thus, the best solution for the modeled WCSP can be considered as a robust solution for the original CSP according to our assumptions.

Keywords: Robustness, Uncertainty, Dynamic CSPs.

1 Introduction

In many real-life situations both the original problem and the corresponding CSP model may evolve because of the environment, or the user or other agents. In such situations, a solution that holds for the original model can become invalid after changes in the original problem. This solution loss can produce negative effects in the situation modeled, which may entail serious economic loss. For example, it could cause the shutdown of the production system, the breakage of machines, delays in transports, or the loss of the material/object in production.

In order to deal with these situations, a number of *proactive approaches* that seek to obtain robust solutions for a given problem, have been proposed. These methods rely on knowledge about possible future changes in order to find solutions with a high probability of remaining valid when faced with these changes (see [13] for a survey). Most of these approaches assume the existence of additional knowledge about the uncertain and dynamic environment (see Section 3), which is often scarce or nonexistent in practice. In this paper, we propose a proactive technique that searches for robust solutions in situations in which only limited (and intuitively reasonable) assumptions are made about possible changes that can occur in CSPs with ordered and discrete domains: namely that changes always take the form of restrictions at the borders of a domain or constraint.

C. Gomes and M. Sellmann (Eds.): CPAIOR 2013, LNCS 7874, pp. 44–60, 2013.

In the simplest case, that of a domain with a single range of values, $[a \ldots b]$, this means increasing the lower bound (a) and/or decreasing the upper bound (b).

Example 1. Figure 1 shows the solution space (continuous lines) of a CSP called P, which is composed of two variables x_0 and x_1. It can be observed that P has 29 solutions (black points).

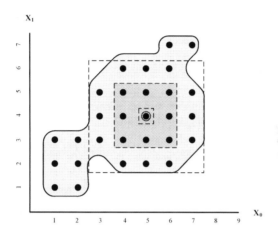

Fig. 1. Solution space of P

If no information is known about unexpected changes in the CSP, it is reasonable to assume that the original constraints and/or domains undergo restrictive or relaxing modifications in the form of range restrictions or expansions, even if this does not cover all possible changes. For instance, in Figure 1, the darkest area represents restrictive modifications. Examples of real life problems that motivate this assumption include scheduling problems. For example, a task may undergo a delay, and therefore, the domains of the subsequent tasks must be reduced.

In this situation, since larger restrictions always include (some) smaller ones, we will assume that values affected by larger restrictions are, in general, less likely to be removed. This is shown in Figure 1, where solutions in the lighter area have a higher probability of becoming invalid. Given these assumptions, *the most robust solution of a CSP with discrete and ordered domains is the solution that is located as far as possible from the bounds of the solution space* and our basic strategy is to search for such solutions. In Example 1, the most robust solution is ($x_0 = 5, x_1 = 4$).

In this paper we present a way of estimating this distance, which is based on the concept of *coverings*. Informally, a covering represents a set of partial or complete solutions that surround another solution and therefore it confers certain level of robustness. These calculations can be incorporated into a WCSP model by assigning weights to the valid tuples based on their coverings. Finally, the modeled WCSP is solved by a general WCSP solver; the best solution is considered to be one of the most robust solutions for the original CSP.

Previously, some work on weighted reformulation have been reported in [2] for Linear CSPs, and [1] for Boolean satisfiability (SAT). These approaches and the approach presented in this paper consider different robustness criterion and/or robustness computation.

Following some technical background (Section 2) and literature review (Section 3), the coverings framework is presented in Section 4. An algorithm for calculating coverings is described in Section 5. In Section 6 our enumeration-based technique is described; an example is given in Section 7. The experimental results obtained with our technique are presented in Section 8. Section 9 gives conclusions.

2 Technical Background

Here, we give some basic definitions that are used in the rest of the paper, following standard notations and definitions in the literature.

Definition 1. *A Constraint Satisfaction Problem (CSP) is represented as a triple $P = \langle \mathcal{X}, \mathcal{D}, \mathcal{C} \rangle$ where $\mathcal{X}$ is a finite set of variables $\mathcal{X} = \{x_1, x_2, ..., x_n\}$, $\mathcal{D}$ is a set of domains $\mathcal{D} = \{D_1, D_2, ..., D_n\}$ such that for each variable $x_i \in \mathcal{X}$ there is a set of values that the variable can take, and $\mathcal{C}$ is a finite set of constraints $\mathcal{C} = \{C_1, C_2, ..., C_m\}$ which restrict the values that the variables can simultaneously take. We denote by $\mathcal{DC}$ the set of unary constraints associated with $\mathcal{D}$.*

Definition 2. *A tuple t is an assignment of values to a subset of variables $\mathcal{X}_t \subseteq \mathcal{X}$.*

For a subset B of $\mathcal{X}_t$, the projection of t over B is denoted as $t \downarrow_B$. For a variable $x_i \in \mathcal{X}_t$, the projection of t over x_i is denoted as t_i. The possible tuples of $C_i \in \mathcal{C}$ are $\prod_{x_j \in var(C_i)} D_j$, where $var(C_i) \subseteq \mathcal{X}$. We denote the set of valid tuples of a constraint $C_i \in (\mathcal{C} \cup \mathcal{DC})$ as $T(C_i)$.

Definition 3. *The tightness of a constraint is the ratio of the number of forbidden tuples to the number of possible tuples. Tightness is defined within the interval [0,1].*

Definition 4. *A Dynamic Constraint Satisfaction Problem (DynCSP) [3] is a sequence of static CSPs $\langle CSP_{(0)}, CSP_{(1)}..., CSP_{(l)} \rangle$, each $CSP_{(i)}$ resulting from a change in the previous one ($CSP_{(i-1)}$) and representing new facts about the dynamic environment being modeled. As a result of such incremental change, the set of solutions of each $CSP_{(i)}$ can potentially decrease (in which case it is considered a restriction) or increase (in which case it is considered a relaxation).*

We only analyze DynCSPs in which the solution space of each $CSP_{(i)}$ decreases over $CSP_{(i-1)}$ (restriction) because relaxations cannot invalidate a solution found previously. As stated, our technique is applied before changes occur. Thus, it is applied to the original CSP ($CSP_{(0)}$) of the DynCSP.

Definition 5. *The most robust solution of a CSP within a set of solutions is the one with the highest likelihood of remaining a solution after changes in the CSP.*

Definition 6. *A Weighted Constraint Satisfaction Problem (WCSP) is a specific sub-class of Valued CSP (see [12]). Here, we consider a variant of WCSP, formalized in [10]. This variant of WCSP is defined as $P = \langle \mathcal{X}, \mathcal{D}, \mathcal{C}, S(W) \rangle$, where:*

- $\mathcal{X}$ *and* $\mathcal{D}$ *are the set of variables and domains, respectively, as in standard CSPs.*
- $S(W) = \langle \{0, 1, ..., W\}, \oplus, > \rangle$ *is the valuation structure, where* $\{0, 1, ..., W\}$ *is the set of costs bounded by the maximum cost* $W \in \mathbb{N}^+$, $\oplus$ *is the sum of costs* $(\forall a, b \in \{0, 1, ..., W\}, a \oplus b = min\{W, a+b\})$ *and* $>$ *is the standard order among natural numbers.*
- $\mathcal{C}$ *is the set of constraints as cost functions* $(C_i : \prod_{x_j \in var(C_i)} D_j \rightarrow \{0, 1, ..., W\})$.

If a tuple t has the maximum cost W for C_i, it means that t is an invalid tuple. The global cost of a tuple t, denoted $\mathcal{V}(t)$, is the sum of all the applicable costs:

$$\mathcal{V}(t) = \bigoplus_{C_i \in \mathcal{C}, var(C_i) \subseteq X_t} C_i(t \downarrow_{var(C_i)}) \tag{1}$$

The tuple t is *consistent* if $\mathcal{V}(t) < W$. The main objective of a WCSP is to find a complete assignment with the minimum cost.

3 Limitations of Earlier Proactive Techniques

The majority of earlier proactive approaches use additional information about the uncertain and dynamic environment and usually involve probabilistic methodologies.

In one example of this type, information is gathered in the form of penalties, in which values that are no longer valid after changes in the problem are penalized [14]. On the other hand, in the Probabilistic CSP model (PCSP) [4], there exists information associated with each constraint, expressing its probability of existence. Other techniques focus on the dynamism of the variables of the CSP. For instance, the Mixed CSP model (MCSP) [5], considers the dynamism of certain *uncontrollable* variables that can take on different values of their uncertain domains. The Uncertain CSP model (UCSP) is an extension of MCSP whose main innovation is that it considers continuous domains [18]. The Stochastic CSP model (SCSP) [15] assumes a probability distribution associated with the uncertain domain of each uncontrollable variable. The Branching CSP model (BCSP) considers the possible addition of variables to the current problem [6]. For each variable, there is a gain associated with an assignment.

In most of these models, a list of the possible changes or the representation of uncertainty is required, often in the form of an associated probability distribution. As a result, these approaches cannot be used if the required information is not known. In many real problems, however, knowledge about possible further changes is either limited or nonexistent.

However, there is one proactive technique that does not consider detailed additional information. In this approach, one searches for *super-solutions* ([8]), which are solutions that can be repaired after changes occur, with minimal changes that can be specified in advance. For CSPs, the focus has been on finding (1,0)-super-solutions.

Definition 7. *A solution is a (1, 0)-super-solution if the loss of the value of one variable at most can be repaired by assigning another value to this variable without changing the value of any other variable ([8]).*

In general, it is unusual to find (1,0)-super-solutions where all variables can be repaired. For this reason, in [8] the author also developed a *branch and bound*-based algorithm for finding solutions that are close to (1,0)-super-solutions, i.e., where the number of repairable variables is maximized (also called maximizing the (1-0)-repairability).

4   Finding Coverings

In order to locate robust solutions, we have developed a technique for calculating *coverings* for valid tuples. The covering of a tuple measures the protection of the tuple against perturbations. It is based on an 'onion topology'. A valid tuple with more layers (its distance to the bounds is higher) is presumed to have a higher probability of remaining valid than a tuple with fewer layers. Here, a layer of a tuple is a convex hull of valid tuples.

To the best of our knowledge, the application of the 'onion structure' to CSPs is a novel idea, although it has been used in robustness for dynamic networks with targeted attacks and random failures. In [9] the authors state: "Our results show that robust networks have a novel "onion-like" topology consisting of a core of highly connected nodes surrounded by rings of nodes with decreasing degree".

We first introduce the concept of *topology* as it provides formulations about proximity relations between the elements that compose it. Subsequently, our own definitions and the associated techniques will be explained.

4.1   Topology of the CSPs

We consider the search space of a CSP with discrete and ordered domains as a set of n-dimensional hyperpolyhedra, where the set of valid tuples of the hyperpolyhedra is denoted by T. There are several distance functions $d(x,y) : T \times T \rightarrow \mathbb{R}^+$ that can be defined over each pair of tuples x and y. We will use the Chebyshev distance, also called L_∞ metric, which measures the maximum absolute differences along any coordinate dimension of two vectors.

$$d(x,y)_{Chebyshev} = \max_i \left(|x_i - y_i| \right) \tag{2}$$

The main reason for selecting this distance metric is that it distinguishes between hypercubes in n-dimensional spaces, which are analogous to squares for $n = 2$ (see Figure 1) and cubes for $n = 3$. In particular, the corners of a cube are at the same distance from the central point as the edges, a feature not obtained with Euclidean distance metric. By checking areas of satisfiability inside these hypercubes, we can ensure minimum distances to the bounds, which is used for the robustness computation (see Section 4.2).

For CSPs with symbolic domains the above definition cannot be applied unless there is an ordering relationship between their values. In this case, a monotonic function has to be applied in order to map the elements by preserving their order. To this end, a monotonic mapping function f is defined over the elements of the CSP domain: $f(x)$: $\mathcal{D} \rightarrow \mathbb{Z}$.

Example 2. We consider a CSP with a symbolic and ordered domain $\mathcal{D}$ which represents clothing sizes: {extra small, small, medium, large, extra large}. In this case, a monotonic function that assigns greater values to the bigger clothing sizes can be defined. For example, $f(extrasmall) = 1$, $f(small) = 2$, $f(medium) = 3$, $f(large) = 4$, etc.

As shown above, a CSP with discrete and ordered domains (both numeric and symbolic) can have a metric function defined over their set of valid tuples T. Therefore, T is a topological space, and the following topological definition presented in [16] can be applied to CSPs:

Definition 8. *A closed ball of a valid tuple $t \in T$ at distance ϵ, is the neighborhood N ($N \subseteq T$) of t composed of $\{y \in T : d(t, y) \leq \epsilon\}$.*

4.2   The Concept of Coverings

As stated earlier, the core of an 'onion structure' is the most robust part of the structure, since the surrounding layers protect it against perturbations. The same is true for the solutions of a CSP: the further away a solution is from the bounds of the solution space, the more robust the solution is. In order to determine how far a valid tuple is from the bounds, we analyze its coverings ('onion layers').

Definition 9. *We define the k-covering(t) of a valid tuple $t \in T$ as its neighborhood $\{y \in T : y \neq t \wedge d(t, y)_{Chebyshev} \leq k\}$, where $k \in \mathbb{N}$.*

From Definition 9 the following property can be deduced: k-covering(t)$\supseteq$(k-1)-covering(t).

Definition 10. *|k-covering(t)| denotes the number of valid tuples inside k-covering(t), without including t.*

Definition 11. *$maxTup(k, |t|)$ denotes the maximum number of tuples that can make up a k-covering(t) (without including t) in a CSP with discrete and ordered domains, where k is the k-covering and |t| represents the arity of t.*

$$maxTup(k, |t|) = (2k + 1)^{|t|} - 1 \qquad (3)$$

Definition 12. *A k-covering(t) is complete if |k-covering(t)| = maxTup(k, |t|).*

If k-covering(t) is complete, it means that t is located at a distance of at least k from the bounds, because inside the k-covering(t) all the tuples are valid. On the other hand, if at least one invalid tuple is inside of the k-covering(t), the unsatisfiability space is not completely outside of k-covering(t) and the minimum distance of t from the bounds of the solution space is the distance to the closest invalid tuple. Note that if at least one of the closest neighbours of t is invalid, it means that t is located on a bound of the solution space.

If there are several tuples with the same number of complete coverings, are they equally robust? The answer is obtained by calculating the number of valid tuples in the minimum incomplete covering (the minimal covering beyond the maximum complete covering). Considering the 'onion topology', if there are holes in an 'onion layer', it is preferable that they be as small as possible.

In Figure 1 (see Example 1) the 1-covering(t) (darkest area) and 2-covering(t) (biggest square) for $t = (x_0 = 5, x_1 = 4)$ are represented with dashed lines. We can see that the 1-covering(t) is complete because $|1\text{-covering}(t)| = \text{maxTup}(1,|t|) = 8$. However, the 2-covering(t) is not complete because $maxTup(2, |t|) = 24$ and $|2\text{-covering}(t)| = 20$. Thus, we can only ensure that $(x_0 = 5, x_1 = 4)$ is located at a distance of at least 1 from the bounds (it has only one completed layer). Note that some bounds of the solution space are located inside the 2-covering(t).

5   Algorithm for Calculating Coverings

To search for a solution located as far as possible from the bounds of the solution space (core of the 'onion'), we must implicitly or explicitly examine the entire solution space of the CSP, which is an NP-complete problem. Our method of search is based on the coverings for each valid tuple of each constraint and domain of the CSP. In this section we present an algorithm for calculating this set of coverings.

5.1   Algorithm Description

Algorithm 1 calculates the coverings of the valid tuples, after first carrying out a global arc-consistency (GAC) process (line 1). In this preliminary step, it searches for a support of each domain value in order to detect tuples that are not globally consistent. For this purpose, we have implemented the well known GAC3 ([11]), but other consistency techniques could be applied. For calculating coverings, Algorithm 1 begins with $k = 1$ (1-covering(t)), increasing this value by one unit in each iteration until the maximum covering of a CSP is reached (or, optionally, a lower bound U). In each iteration, $\forall t \in T(C_i)$, $\forall C_i \in (\mathcal{C} \cup \mathcal{DC})$, k-covering($t$) is computed, iff $k = 1$ or (k-1)-covering(t) is complete (see Definition 12).

Definition 13. *The maximum covering of a CSP is denoted as max-covering(CSP) and it is reached when there is no tuple whose k-covering is complete for some $C_i \in (\mathcal{C} \cup \mathcal{DC})$.*

Furthermore, if the user desires to obtain a lower k-covering(t), she/he can optionally fix a lower bound U. In this case, the algorithm stops after calculating $min(U, \text{max-covering}(CSP))$.

Definition 14. *We define last-covering(t) to be the last k-covering(t) computed by the algorithm for the tuple t. The value of 'last' in last-covering(t) term is equal to $min(U, \text{max-covering}(CSP), (k+1))$, if k-covering(t) is the greatest completed covering of t.*

Algorithm 1 returns the size of last-covering(t) computed for each valid tuple t of each constraint and domain of the CSP. Thus, it gives a measure of robustness for each constraint. In addition, the value of max-covering(CSP) is returned (unless U is provided). The algorithm stores the number of valid tuples whose k-covering(t) is complete for each C_i in $kC[i]$, which is checked in line 11. Thus, it can check whether the current analyzed covering is max-covering(CSP).

Since k-covering(t) $\supseteq$ (k-1)-covering(t), the algorithm only analyzes the new possible neighbors of t, which are the neighbors that belong to it but do not belong to (k-1)-covering(t) (the neighbors that are in the k 'onion layer'). This is due to the fact that the neighbors of the lower coverings have already been calculated in previous iterations and stored in last-covering(t).

Algorithm 1. calculateCoverings (CSP P)

Data: A CSP $P = \langle \mathcal{X}, \mathcal{D}, \mathcal{C} \rangle$ and U (optional)
Result: $|$last-covering(t)$|$ $\forall t \in T(C_i) \forall C_i \in (\mathcal{C} \cup \mathcal{DC})$ and max-covering(P)

1 GAC3(P);
2 **foreach** $C_i \in (\mathcal{C} \cup \mathcal{DC})$ **do**
     $k \leftarrow 1$;
3   max-covering(P) $\leftarrow \infty$;
     $T(C_i) \leftarrow$ Ordered list of valid tuples of C_i;
4   **foreach** $t \in T(C_i)$ **do**
        $|$last-covering(t)$| \leftarrow 0$;

5 **repeat**
6   **foreach** $C_i \in (\mathcal{C} \cup \mathcal{DC})$ **do**
        $kC[i] \leftarrow 0$;
7   **foreach** $t \in T(C_i)$ **do**
8   **foreach** $\{y \in T(C_i): y < t \wedge d(t_1, y_1)_{Chebyshev} \leq k\}$ **do**
9   **if** isNewNeighbor(k,t,y) **then**
                 addNeighbor($k, |$last-covering(t)$|, t$);
                 addNeighbor($k, |$last-covering(y)$|, y$);
10   **if** isComplete($k, |$last-covering(y)$|, |y|$) **then**
                    $kC[i] = kC[i] + 1$;

11   **if** $kC[i] = 0$ **then**
        max-covering(P)$= k$;
    $k \leftarrow k + 1$;
 until $k >$max-covering(P) ;
 return $|$last-coverings$|$, max-covering(P) (unless U is provided)

Firstly, Algorithm 1 initializes some necessary structures (see loop beginning on line 2). Then the sets of valid tuples $T(C_i)$ of each constraint $C_i \in (\mathcal{C} \cup \mathcal{DC})$ are ordered by the value of the first variable of the valid tuples. In this way, a tuple a can only be located in a lower position than a tuple b if $a_1 \leq b_1$ (considering that the subindex 1 indicates the first variable that makes up the tuple). Thus, the tuples whose first variable has the minimum possible value will be placed in the lowest positions. For expressing

---

Procedure isNewNeighbor(k,t,y) : Boolean

$equalDist \leftarrow 0$;

1 **for** $i \leftarrow 1$ **to** $|t|$ **do**

2   **if** $d(t_i, y_i)_{Chebyshev} > k$ **then**
        ∟ **return** *False*

3   **if** $d(t_i, y_i)_{Chebyshev} = k$ **then**
        ∟ $equalDist = equalDist + 1$;

4 **if** $equalDist > 0$ **then**
    ∟ **return** *True*

5 **else**
    ∟ **return** *False*

Procedure isComplete$(k, |\text{last-covering(t)}|, |t|)$: Boolean

---

maxTup$(k, |t|) := (2k + 1)^{|t|}$-1;

1 **if** $|\text{last-covering(t)}| \geq$ maxTup$(k, |t|)$ **then**
    ∟ **return** *True*

2 **else**
    ∟ **return** *False*

---

the order of the tuples, we use the notation $a < b$, which means that the tuple a is located in a lower position than b in the list of ordered tuples.

The implementation of an algorithm for calculating coverings does not strictly require an ordered list of $T(C_i)$, but this ordering allows for a reduction in computation time, because it avoids checking a pair of tuples twice. This temporal benefit is achieved by selecting a reduced subset of possible neighbors of each valid tuple $t \in T(C_i)$ which are located in a lower position of t. If we do not select this reduced set, the algorithm must check all the valid tuples for each valid tuple. The reduced set is a subset composed of the valid tuples that are ordered in a lower position than t in the ordered list of $T(C_i)$ and whose difference between the value of their first variable with respect to t is lower or equal to k. Thus, the reduced set of t is composed of $y \in T(C_i) : y < t \wedge d(t_1, y_1)_{Chebyshev} \leq k$ (line 8).

The procedure isNewNeighbor checks if a valid tuple y is a new neighbor in k-covering(t). This condition is determined by checking that at least one of the variables of y has a value difference of k with respect to t (line 3) and the rest of the variables have a value difference lower or equal to k (line 2). The procedure isComplete determines if a k-covering(t) is complete. Firstly, it calculates the maximum number of tuples of k-covering(t): maxTup$(k, |t|)$ (see Equation 3). Subsequently, it checks if $|\text{last-covering(t)}|$ is equal to maxTup$(k, |t|)$.

The procedure addNeighbor adds 1 to $|\text{last-covering(t)}|$ iff $k = 1$ or (k-1)-covering(t) is complete. Algorithm 1 calls the procedure addNeighbor twice with two different tuples: t and y (y is a valid tuple of the reduced set of t) iff y is a new neighbor of k-covering(t) (line 10). With this process a new neighbor can be added to the last-coverings of two different tuples by doing only one check, which allows further

---

Procedure `addNeighbor` (k,|last-covering(t)|,t)

---

1 **if** $k = 1$ **or** `isComplete`($k - 1$,|last-covering(t)|,|t|) **then**
   ⌞ |last-covering(t)| ← |last-covering(t)| + 1;

---

reduction in computation time. Later, the algorithm checks if k-covering(y) is complete (line 11). Checking the completeness of k-covering(t) is not necessary yet since there are possible new neighbors of k-covering(t) that are ordered in higher positions than k and therefore have not been analyzed yet. They will be analyzed when the tuple t will be an element of their reduced set.

As mentioned previously, Algorithm 1 does not compute k-covering(t) if (k-1)-covering(t) is not complete. However, this restriction could be deleted by skipping the completeness check of the coverings (developed by the `isComplete` procedure). In this case, we cannot use the principle of minimum distance from the constraint (see Section 4.2). In this paper, this principle is necessary because the objective is to find solutions located far away from the constraint boundaries. Nevertheless, the coverings computation does not require the completeness property. Thus, both the covering concepts for CSPs and Algorithm 1 (with the modification discussed above) can be also applied to other areas that do not require this property.

6 Modeling Robustness in a CSP as a WCSP

In this section we introduce an enumeration-based technique for modeling a CSP as a WCSP. The intention is to obtain a CSP model based on the |last-coverings| of the solutions of all the constraints. As argued earlier, |last-covering($s \downarrow_{var(C_i)}$)| for a solution s is a reasonable measure of its robustness for C_i, and it is moreover reasonable to use the sum of |last-covering($s \downarrow_{var(C_i)}$)| for each $C_i \in (\mathcal{C} \cup \mathcal{DC})$ as an approximation of the robustness of s for the CSP. The WCSP model is based on the assumption that the sum of the costs assigned to the tuples of each constraint determines how good a solution is for the WCSP. Thus, we model a CSP as a WCSP after obtaining its |last-coverings| with Algorithm 1.

Although there are other valued CSPs that could conceivably be used to model robustness, these models either involve operations that are questionable (e.g. probabilistic CSP, where valuations based on |last-coverings| would be multiplied) or are insufficiently discriminating (e.g. fuzzy CSP). In addition, the WCSP model adequately incorporates the enumeration aspect of coverings, unlike the other valued CSP models.

The modeling process begins by assigning a cost to each valid tuple t involved in each constraint, which represents its penalty as a function of its |last-covering(t)|. Tuples with the highest last-covering for C_i will have the lowest associated cost, because this value indicates the minimum distance of t from the bounds of C_i. As already noted, the latter can be taken as a measure of its robustness, because it is resistant to more possible changes over the constraint.

Definition 15. *We define* max-|last-covering(C_i)|=*max* {|*last-covering*(t)|, $\forall t \in T(C_i)$}.

The penalty of a valid tuple t for a constraint C_i without considering the rest of the constraints of the CSP is denoted as $p_i(t)$ (see Equation 4).

$$p_i(t) = \text{max-|last-covering}(C_i)| - |\text{last-covering}(t)| \tag{4}$$

However, this is not the final cost assigned to t. We must normalize the penalties of each C_i with respect to the other constraints, because the maximum possible size of the coverings depends on the arity of the tuples (see Equation 3). Otherwise, constraints with higher arities would have higher cost ranges and therefore higher penalties. In this case, we would be assuming that these constraints have a higher likelihood of under-going restrictive modifications, which is not necessarily true according to the limited assumptions we are making for CSPs with discrete and ordered domains. By using a normalization process, we can achieve the same cost range for all the constraints. To obtain normalized scores, we use the maximum penalty assigned to the tuples of each constraint and the maximum penalty assigned to the tuples across all constraints (e is the number of constraints of the CSP according to Definition 1). In this way, the cost function for the tuples of C_i assigns a normalized cost to every tuple of C_i, which is denoted as $C_i(t \downarrow_{var(C_i)})$ (see Equation 5).

$$C_i(t \downarrow_{var(C_i)}) = \begin{cases} 0 & \text{if } t \in T(C_i) \text{ and } p_i(t) = 0 \\ \left\lfloor \frac{p_i(t) * max\{p_j(x), \forall j \in [1...e] \forall x \in T(C_j)\}}{max\{p_i(y), \forall y \in T(C_i)\}} \right\rfloor & \text{if } t \in T(C_i) \text{ and } p_i(t) \neq 0 \\ W, (W \approx \infty) & \text{if } t \notin T(C_i) \end{cases} \tag{5}$$

In line with the version of WCSP that we are using, $C_i(t \downarrow_{var(C_i)})$ assigns a cost of W to each tuple t that does not satisfy the constraint C_i because it is not a partial solution. Note that for valid tuples $C_i(t \downarrow_{var(C_i)}) \in [0, max\{p_j(x), \forall j \in [1 \ldots e] \forall x \in T(C_j)\}]$. A valid tuple t whose |last-covering(t)| = max-|last-covering(C_i)| ($p_i(t) = 0$) has an associated cost of 0. This tuple is not penalized because it has the highest likelihood of remaining valid when faced with future changes in C_i. On the other hand, if |last-covering(t)| = 0 for C_i, which means that t does not have any neighbour in its 1-covering(t) (t is completely non-robust for C_i), t has the maximum possible cost associated: $max\{p_j(x), \forall j \in [1 \ldots e] \forall x \in T(C_j)\}$.

Once the costs have been assigned, the WCSP is generated and solved using a WCSP solver. The solutions obtained for the modeled WCSP are also solutions of the original CSP. In addition, the best solution s with the minimum $\mathcal{V}(s)$ (see Equation 1) is taken to be one of the most robust solutions for the original CSP.

7 Example

In order to clarify the covering concepts, we present an example with three variables (*3-dimensional* hyperpolyhedron CSP). We have modeled this problem as well as the problems presented in Section 8 as WCSPs by following the WCSP file format[1]. In addition, ToolBar2[2] has been used for solving the resultant WCSPs.

Example 3. Figure 2 shows a CSP R, which is composed of variables x_0, x_1 and x_2 with domain $D : \{0, 1, 2\}$. There are three extensional constraints: C_0 (3-ary), C_1 and C_2 (binary constraints) (Figures 2(a), 2(b) and 2(c), respectively). The valid tuples of the constraints and domains are represented with points; the invalid tuples are represented with crosses. Figure 2(d) shows the solutions of R ordered by their relative robustness, as assessed by our technique for calculating coverings.

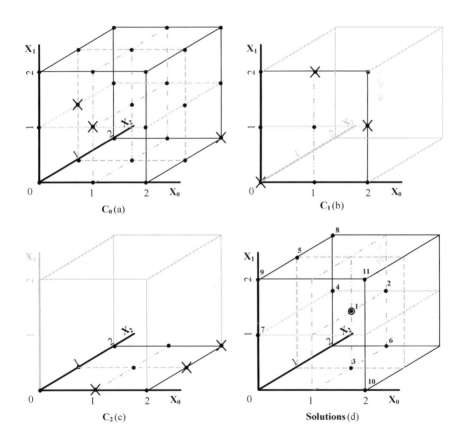

Fig. 2. Example of the 3-dimensional CSP R

Table 1 shows the solutions of the CSP R in decreasing order of robustness (s_i). This order is inversely proportional to the $\mathcal{V}(s)$ obtained after solving the modeled WCSP.

---

[1] http://graphmod.ics.uci.edu/group/WCSP_file_format
[2] http://carlit.toulouse.inra.fr/cgi-bin/awki.cgi/ToolBarIntro

In addition, we present |last-covering(s)| for the solution space of R. Note that its calculation is viable because we are analyzing a toy problem. As expected, the robustness results obtained match with the |last-coverings| order in the solution space: the solutions with higher |last-coverings| in the solution space are identified as more robust by our technique (see the highlighted columns in Table 1).

Table 1 also shows the costs assigned by the constraints (C_i) and domains (X_i), whose sum is $\mathcal{V}(s)$ (see Equation 1). To clarify the process of cost assignment, we explain one case, $C_0(s_2)$, in detail. Taking into account that *max-covering(R)*= 1, the penalty for $s_2 = (1, 1, 2)$ (before the normalization process) is $p_0(s_2) = 23 - 15 = 8$ (see Equation 4), since max-|last-covering(C_0)| $= 23$ (|last-covering((1, 1, 1))| $= 23$ for C_0, which is the maximum for C_0) and |last-covering(s_2)| $= 15$ for C_0. The associated cost (considering the normalization) is: $C_0(s_2) = \lfloor \frac{8*18}{18} \rfloor = 8$ (see Equation 5). Because $max\{p_0(y), \forall y \in T(C_0)\} = 18$ ($p_0((0, 2, 0)) = 23 - 5 = 18$) and this cost is also the maximum penalization for R, since $max\{p_j(x), \forall j \in [1 \ldots e] \forall x \in T(C_j)\} = 18$.

Table 1. Solutions ordered by their robustness

s_i	$s = (x_0, x_1, x_2)$	|*last-covering(s)*|	$\mathcal{V}(s)$	$X_0(s)$	$X_1(s)$	$X_2(s)$	$C_0(s)$	$C_1(s)$	$C_2(s)$
1	(1,1,1)	10	0	0	0	0	0	0	0
2	(1,1,2)	6	35	0	0	18	8	0	9
3	(1,0,1)	6	36	0	18	0	9	9	0
4	(0,1,2)	6	67	18	0	18	13	9	0
5	(0,2,1)	6	67	18	18	0	14	13	4
6	(1,0,2)	4	68	0	18	18	14	9	9
7	(0,1,0)	4	72	18	0	18	14	9	13
8	(0,2,2)	4	93	18	18	18	17	13	9
9	(0,2,0)	3	98	18	18	18	18	13	13
10	(2,0,0)	2	102	18	18	18	17	13	18
11	(2,2,0)	1	107	18	18	18	17	18	18

The best solution found is the solution $s1 = (x_0 = 1, x_1 = 1, x_2 = 1)$ and its |*last-covering(s1)*| $= 10$ in the solution space, which is the highest for R (see Figure 2(d)). As previously mentioned, the solution s with the highest |last-covering(s)| in the solution space, has the highest likelihood of remaining valid faced with future possible restrictive modifications over the bounds of the solution space. Therefore, it is considered the most robust solution for the original CSP. In contrast, the solution $s_{11} = (x_0 = 2, x_1 = 2, x_2 = 0)$ is classified by our technique as the least robust solution. Note that s_{11} only has one neighbor in the solution space (see Figure 2(d)).

8   Experimental Results

In this section, we present results from experiments designed to evaluate the behaviour of our technique for calculating coverings. To the best of our knowledge, there are no benchmarks of DynCSPs (see Definition 4) in the literature, so we generated 500 DynCSPs composed of $l+1$ static CSPs: $\langle CSP_{(0)}, CSP_{(1)}, ..., CSP_{(l)} \rangle$, where $CSP_{(0)}$ is generated randomly by RBGenerator 2.0 [3] and each $CSP_{(i)}$ is derived from $CSP_{(i-1)}$ by making a restrictive modification in some bound (constraint or domain). Thus, an invalid tuple located next to any bound is selected randomly and all the tuples surrounding it to a distance of 1 are invalidated. Thus, only valid tuples located on a bound became invalid. Thus, l indicates the number of restrictive changes that the solution was able to resist, and can be considered a measure of the robustness of the solutions. In order to select equally constraints and domains, we have selected them based on their relative frequency. All tests were executed on a Intel Core i5-650 Processor (3.20 Ghz).

In addition to assessing our technique for max-covering(CSP), we analyzed the robustness of solutions obtained with the proactive technique for maximizing the number of repairable variables for (1,0)-super-solutions by using Branch and Bound and MAC+ algorithms ([8]) and fixing the cutoff to 200 seconds. The main reason for choosing this technique for comparison is that, like our technique, it does not require specific additional information about the future possible changes. In addition, a solution calculated by an ordinary CSP solver has been computed to determine if there are cases in which all solutions have similar robustness. For both techniques, values were selected in lexicographical order. Note that neither of these techniques searches for robust solutions in the same way that our technique does, nor are they based on the same assumptions regarding changes that are inherent in the structure of CSPs with discrete and ordered domains.

Figure 3 shows a robustness analysis when tightness of the constraints is varied. The measure on the left vertical axis is the number of supported changes (both means (continuous lines) and standard deviations are shown in the figure), while the right vertical axis shows the modeling time required by our technique (dashed line). We can observe that with the exception of very highly restricted intances, the mean number of changes before solution breakage for solutions obtained by the k-covering technique is significantly higher than the means obtained with the other two solutions. For instances that are very highly restricted, the number of changes allowed by solutions obtained by the three methods is similar. This is because for this type of instances, the number of solutions is so low that the solutions are scattered within the tuple-space, so the likelihood of a solution being located on the bounds of the solution space is very high. For instance, for the maximum tightness evaluated (0.9), there are only 5 solutions. Even in this case the k-covering algorithm was able to find solutions that remained valid for a few more changes than the solutions found by the other algorithms.

All of the experimental results were evaluated statistically, first with a two-factor Analysis of Variance (ANOVA), followed by the Tukey HSD test for differences between pairs of individual means ([7, 17]). The ANOVA gave F values that were highly significant statistically. Given the small values for HSD (about 0.1 and 0.08), nearly all differences between individual means were statistically significant for $p = 0.01$. Further tests have been done with the other CSP parameters (number of variables, domain size, constraint graph density and constraint arity), and we have obtained the same general results: solutions found with our technique remained valid for a greater number

---

[3] http://www.cril.univ-artois.fr/~lecoutre

of restrictive changes than the other techniques, and the difference is marked when the problems are only slightly constrained CSPs.

The modeling time required by our technique is higher for slight tightness values. This is is related to the number of valid tuples of the constraints, which it has a strong impact in the time spent by Algorithm 1 for finding the coverings.

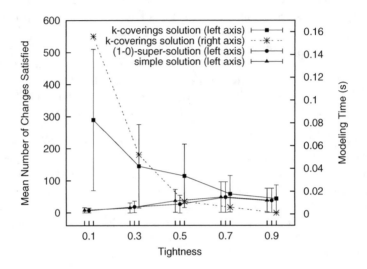

Fig. 3. Robustness analysis ($< arity = 2, |\mathcal{X}| = 40, |D| = 25, |\mathcal{C}| = 120 >$)

9   Conclusions

In this paper we have extended the concept of robustness to CSPs with discrete and ordered domains where only limited assumptions are made about the kinds of changes, commensurate with the structure of these problems. In this context, it is reasonable to assume that the original bounds of the solution space may undergo restrictive modifications. Therefore, the main objective in searching for robust solutions is to find solutions located as far away as possible from these bounds.

In addition, we presented an enumeration-based technique for modeling these CSPs as WCSPs by assigning a cost to each valid tuple of every constraint. The cost of each valid tuple t is obtained by calculating |last-covering(t)| for the corresponding constraint. The obtained solution for the WCSP is a solution for the original CSP that has a higher probability of remaining valid after changes in the original problem.

Another contribution of this paper is the framework introduced in Section 4, which involves the novel concept of an 'onion topology' as applied to CSPs, as well as concepts and definitions built around the idea of coverings. To the best of our knowledge, Algorithm 1 is the first practical method for calculating coverings for CSPs.

In experimental tests we have shown that these techniques can dramatically outperform both ordinary CSP algorithms and algorithms that find (1,0)-super-solutions (or maximize the number of repairable variables in case that there does not exist a (1,0)-super-solution) under a variety of conditions where there are real differences in the robustness of solutions that might be obtained (when the problem is not so tight that there are only a few valid solutions).

Acknowledgements. This work has been partially supported by the research project TIN2010-20976-C02-01 and FPU program fellowship (Min. de Ciencia e Innovación, Spain).

References

[1] Bofill, M., Busquets, D., Villaret, M.: A declarative approach to robust weighted Max-SAT. In: Proceedings of the 12th International ACM SIGPLAN Symposium on Principles and Practice of Declarative Programming (PPDP 2010), pp. 67–76 (2010)

[2] Climent, L., Salido, M., Barber, F.: Reformulating dynamic linear constraint satisfaction problems as weighted CSPs for searching robust solutions. In: Proceedings of the 9th Symposium of Abstraction, Reformulation, and Approximation (SARA 2011), pp. 34–41 (2011)

[3] Dechter, R., Dechter, A.: Belief maintenance in dynamic constraint networks. In: Proceedings of the 7th National Conference on Artificial Intelligence (AAAI 1988), pp. 37–42 (1988)

[4] Fargier, H., Lang, J.: Uncertainty in constraint satisfaction problems: A probabilistic approach. In: Moral, S., Kruse, R., Clarke, E. (eds.) ECSQARU 1993. LNCS, vol. 747, pp. 97–104. Springer, Heidelberg (1993)

[5] Fargier, H., Lang, J., Schiex, T.: Mixed constraint satisfaction: A framework for decision problems under incomplete knowledge. In: Proceedings of the 13th National Conference on Artificial Intelligence (AAAI 1996), pp. 175–180 (1996)

[6] Fowler, D.W., Brown, K.N.: Branching constraint satisfaction problems for solutions robust under likely changes. In: Dechter, R. (ed.) CP 2000. LNCS, vol. 1894, pp. 500–504. Springer, Heidelberg (2000)

[7] Hays, W.: Statistics for the social sciences, 2nd edn., vol. 410. Holt, Rinehart and Winston, New York (1973)

[8] Hebrard, E.: Robust Solutions for Constraint Satisfaction and Optimisation under Uncertainty. PhD thesis, University of New South Wales (2006)

[9] Herrmann, H., Schneider, C., Moreira, A., Andrade Jr., J., Havlin, S.: Onion-like network topology enhances robustness against malicious attacks. Journal of Statistical Mechanics: Theory and Experiment 2011(1), P01027 (2011)

[10] Larrosa, J., Schiex, T.: Solving weighted CSP by maintaining arc consistency. Artificial Intelligence 159, 1–26 (2004)

[11] Mackworth, A.: On reading sketch maps. In: Proceedings of the 5th International Joint Conference on Artificial Intelligence (IJCAI 1977), pp. 598–606 (1977)

[12] Schiex, T., Fargier, H., Verfaillie, G.: Valued constraint satisfaction problems: Hard and easy problems. In: Proceedings of the 14th International Joint Conference on Artificial Intelligence (IJCAI 1995), pp. 631–637 (1995)

[13] Verfaillie, G., Jussien, N.: Constraint solving in uncertain and dynamic environments: A survey. Constraints 10(3), 253–281 (2005)

[14] Wallace, R.J., Freuder, E.C.: Stable solutions for dynamic constraint satisfaction problems. In: Maher, M.J., Puget, J.-F. (eds.) CP 1998. LNCS, vol. 1520, pp. 447–461. Springer, Heidelberg (1998)

[15] Walsh, T.: Stochastic constraint programming. In: Proceedings of the 15th European Conference on Artificial Intelligence (ECAI 2002), pp. 111–115 (2002)

[16] William, F.: Topology and its applications. John Wiley & Sons (2006)

[17] Winer, B.: Statistical principles in experimental design, 2nd edn. McGraw-Hill Book Company (1971)

[18] Yorke-Smith, N., Gervet, C.: Certainty closure: Reliable constraint reasoning with incomplete or erroneous data. Journal of ACM Transactions on Computational Logic (TOCL) 10(1), 3 (2009)

Some New Tractable Classes of CSPs and Their Relations with Backtracking Algorithms[*]

Achref El Mouelhi[1], Philippe Jégou[1], Cyril Terrioux[1], and Bruno Zanuttini[2]

[1] LSIS - UMR CNRS 7296
Aix-Marseille Université
Avenue Escadrille Normandie-Niemen
13397 Marseille Cedex 20, France
{achref.elmouelhi,philippe.jegou,cyril.terrioux}@lsis.org
[2] Normandie Université, France
GREYC, Université de Caen Basse-Normandie, CNRS UMR 6072, ENSICAEN
Campus II, Boulevard du Maréchal Juin
14032 Caen Cedex, France
bruno.zanuttini@unicaen.fr

Abstract. In this paper, we investigate the complexity of algorithms for solving CSPs which are classically implemented in real practical solvers, such as Forward Checking or Bactracking with Arc Consistency (RFL or MAC).. We introduce a new parameter for measuring their complexity and then we derive new complexity bounds. By relating the complexity of CSP algorithms to graph-theoretical parameters, our analysis allows us to define new tractable classes, which can be solved directly by the usual CSP algorithms in polynomial time, and without the need to recognize the classes in advance. So, our approach allows us to propose new tractable classes of CSPs that are naturally exploited by solvers, which indicates new ways to explain in some cases the practical efficiency of classical search algorithms.

1 Introduction

Constraint Satisfaction Problems (CSPs [1]) constitute an important formalism of Artificial Intelligence (AI) for expressing and efficiently solving a wide range of practical problems. A constraint network (or CSP, abusing words) consists of a set of variables X, each of which must be assigned a value in its associated (finite) domain D, so that these assignments together satisfy a finite set C of constraints.

Deciding whether a given CSP has a solution is an NP-complete problem. Hence classical approaches to this problem are based on backtracking algorithms, whose worst-case time complexity is at best of the order $O(e.d^n)$ with n the number of variables, e the number of constraints and d the size of the largest domain. To increase efficiency, such algorithms also rely on filtering techniques during search (among other techniques, such as variable ordering heuristics). With the help of

---

[*] This work was supported by the French National Research Agency under grant TUPLES (ANR-2010-BLAN-0210).

C. Gomes and M. Sellmann (Eds.): CPAIOR 2013, LNCS 7874, pp. 61–76, 2013.

such techniques, despite their theoretical time complexity, algorithms such as Forward Checking [2] (denoted FC), RFL (for Real Full Look-ahead [3]) or MAC (for Maintaining Arc Consistency [4]) for binary CSPs, or nFC$_i$ for non-binary CSPs [5] turn out to be very efficient in practice on a wide range of practical problems.

In a somewhat orthogonal direction, other works have addressed the effectiveness of solving CSPs by defining *tractable classes*. A tractable class is a class of CSPs which can be recognized, and then solved, using polynomial time algorithms. Different kinds of tractable classes have been introduced. Some of them are based on the *structure* of the constraint network, for instance tree-structured networks [6] or more generally, networks of bounded width [7]. This kind of tractable classes has shown its practical interest for benchmarks of small width (e.g. [8]). Other studies have highlighted the interest of certain tractable classes (e.g. [9]) but unfortunately they are very rare. This direction of research has produced and still produces works both numerous and complex and the results are generally difficult to establish (we can find a synthesis in [1]). Unfortunately, most of these results remain only theoretical and thus, the question of their real interest should be raised in the context of constraint programming.

It is not easy to cite one tractable class in the field of CSP that has shown any interest in practice (with the exception of bounded width) and ideally allow it to outperform the efficiency of current solvers. So, we think that it seems necessary today to ask that question to the CPAIOR community, even if this question could be controversial.

In our opinion, the reasons for lack of practical interest of the tractable classes exhibited to date are based on several aspects. Firstly, the identification of a new tractable class requires the development of ad hoc polytime algorithms: one for the recognition of tractable instances, and one for solving them. Secondly, these polytime algorithms are generally neither efficient in practice, and frequently, nor in theory. And most importantly, the proposed tractable classes seem to be artificial in the sense that they do not exist in reality: real benchmarks do not belong to these classes, and even tractable classes only appear in small pieces of real problems, this making them finally completely unusable. And surprisingly, most tractable classes currently exhibited by the community seem to conceal their only interest by their theoretical difficulty. Finally, it seems that these classes have no interest, from a practical point of view, for the CPAIOR community. In addition, to be efficient, solvers need to rely on simple mechanisms that can be efficiently implemented. So, to integrate the use of tractable classes whose treatment would not be in linear time seems almost useless, because their treatment would significantly slow down the efficiency of a solver and thus make it inoperative in practice.

We do not criticize the existence of works on tractable classes, but essentially the direction they take, and we propose to redirect the works in the direction which seems, after several decades of works on the issue, the only one which can be of interest to the CPAIOR community, or at least, offers the best chance of producing useful results. So, we propose here to focus research on the analysis of algorithms such as FC, RFL or nFC$_i$, whose theoretical complexity is exponential, but which are the basis of practical systems for constraint solving, and whose concrete results are often impressive in terms of computational time. Their analysis could lead to identify tractable classes which could then be efficiently exploited in practice, possibly with some slight modifications of solvers. In this respect, our study is very close in spirit to the study by Rauzy about satisfiability and the behaviour of DPLL on known tractable instances [10], and more recently, of the works presented in [11] for CSPs of bounded structural parameters or in [12] for β-acyclic CNFs in SAT.

We do so by reevaluating their time complexity using a new parameter, namely the number of maximal cliques in the *microstructure* [13] of the instance or in the *generalized microstructure* for the non-binary case. For the binary case, writing $\omega_\#(\mu(P))$ for the number of maximal cliques in the microstructure of a CSP P, we show that the complexity of an algorithm such as FC is in $O(n^2 d \cdot \omega_\#(\mu(P)))$. This provides a new perspective on the study of the efficiency of backtracking-like algorithms, by linking it to a well-known graph-theoretical parameter. In particular, reusing known results from graph theory, we propose some tractable classes of CSPs. The salient feature of these classes is that they are solved in polynomial time by *general-purpose, widely used* algorithms, *without the need for the algorithms to recognize the class.*

The paper is organized as follows. We first introduce notations and recall the definitions and basic properties of the microstructure. Then we present our complexity analysis of BT, FC and RFL on binary CSPs, and we introduce the notion of *generalized microstructure* to extend our study to non-binary CSPs and to algorithms of the class nFC$_i$. We then point at new tractable classes issued from graph theory, which can be exploited in the field of CSPs. Finally, we give a discussion and perspectives for future work.

2   Preliminaries

Before reviewing the classical analysis of algorithms, we recall some basic notions about CSPs and their microstructure.

Definition 1 (CSP). *A finite constraint satisfaction problem (CSP) is a triple (X, D, C), where $X = \{x_1, \ldots, x_n\}$ is a set of variables, $D = (D(x_1), \ldots, D(x_n))$ is a list of finite domains of values, one per variable, and $C = \{c_1, \ldots, c_e\}$ is a finite set of constraints. Each constraint c_i is a couple $(S(c_i), R(c_i))$, where $S(c_i) = \{x_{i_1}, \ldots, x_{i_k}\} \subseteq X$ is the scope of c_i, and $R(c_i) \subseteq D(x_{i_1}) \times \cdots \times D(x_{i_k})$ is its relation. The arity of c_i is $|S(c_i)|$.*

We will refer to a binary constraint with scope $\{x_i, x_j\}$ by the notation c_{ij}. A *binary CSP* is one in which all constraints are binary. Otherwise (general case), the CSP is said to be *n-ary*. We assume that all variables appear at least in one scope of constraint and that for a given scope, there is at most one constraint. This is without loss of generality since two constraints over the same set of variables can be merged into one by taking the intersection of their relations (for the purpose of analysis).

Definition 2 (assignment, solution). *Given a CSP (X, D, C), an assignment of values to $Y \subseteq X$ is a set of pairs $t = \{(x_i, v_i) \mid x_i \in Y\}$ (written $t = (v_1, \ldots, v_k)$ when no confusion can arise), with $v_i \in D(x_i)$ for all i. An assignment to $Y \subseteq X$ is said to be* consistent *(or a partial solution) if all constraints $c \in C$ with scope $S(c) \subseteq Y$ are satisfied, i.e., $t[S(c)] \in R(c)$ holds with $t[S(c)]$ the restriction of t to $S(c)$. A* solution *is a consistent assignment to X.*

We consistently write n for the number of variables in a CSP, d for the cardinality of the largest domain, e for the number of constraints, a for the maximum arity over all constraints, and r for the number of tuples of the largest relation.

Given a CSP, the basic question is to decide whether it has a solution, which is well-known to be NP-complete. In order to study CSPs and try to circumvent this difficulty, various points of view can be adopted. As concerns *binary* CSPs, one of them is the *microstructure* of an instance, that is its compatibility graph as we define now. Intuitively, the vertices of this graph code the values, and its edges code their compatibility.

Definition 3 (microstructure). *Given a binary CSP $P = (X, D, C)$, the* microstructure *of P is the undirected graph $\mu(P) = (V, E)$ with:*

- $V = \{(x_i, v_i) : x_i \in X, v_i \in D(x_i)\}$,
- $E = \{ \{(x_i, v_i), (x_j, v_j)\} \mid i \neq j, c_{ij} \notin C \text{ or } (v_i, v_j) \in R(c_{ij})\}$

In words, the microstructure of a binary CSP P contains an edge for all pairs of vertices, except for vertices coming from the same domain and for vertices corresponding to pairs which are forbidden by some constraint. It can easily be seen that the microstructure of a CSP is an n-partite graph, since there is no edge connecting vertices issued from the same domain. In this paper, we will study the complexity of CSP algorithms through cliques in the microstructure.

Definition 4 (clique). *A* complete graph *is a simple graph in which every pair of distinct vertices is connected by an edge. A k-clique in an undirected graph is a subset of k vertices inducing a complete subgraph (all the vertices are pairwise adjacent). A* maximal clique *is a clique which is not a proper subset of another clique. We write $\omega_\#(G)$ for the number of maximal cliques in a graph G.*

The following result follows directly from the fact that in a microstructure, the vertices of a clique correspond to compatible values which are by construction issued from different domains.

Proposition 1. *Given a binary CSP P and its microstructure $\mu(P)$, an assignment $(v_1, ..., v_n)$ to X is a solution of P iff $\{(x_1, v_1), ..., (x_n, v_n)\}$ is an n-clique of $\mu(P)$.*

It can be seen that the transformation of a CSP P to its microstructure $\mu(P)$ can be realized in polynomial time. A polynomial reduction directly follows, from the problem of deciding whether a given CSP has a solution, to the problem of deciding whether a given undirected graph has a clique of a given size (the famous "clique problem"). This transformation has first been exploited by [13], who proposed tractable classes of CSPs based on known tractable classes for the maximal clique problem (*chordal graphs* [14]). A similar approach has been taken for *hybrid* tractable classes [15,16].

Backtracking Algorithms. We now briefly review the complexity of algorithms of interest here: BT, FC, RFL and MAC for both binary and non-binary CSPs. These algorithms essentially cover all the approaches which use backtracking and lookahead (variable and value ordering left apart).

The *Backtracking* algorithm (BT, a.k.a. *Chronological Backtracking*) is a recursive enumeration procedure. It starts with an empty assignment and in the general case, given a current partial solution $(v_1, v_2, \ldots, v_i)$, it chooses a new variable x_{i+1} and tries to assign values of $D(x_{i+1})$ to x_{i+1}. The only check performed while doing so is that the resulting assignment $(v_1, v_2, \ldots, v_i, v_{i+1})$ is consistent. In the affirmative, BT continues with this new partial solution to a new unassigned variable (called a *future variable*). Otherwise (if $(v_1, v_2, \ldots, v_i, v_{i+1})$ is not consistent), BT tries another value from $D(x_{i+1})$. If there is no such unexplored value, BT is in a dead-end, and then it uninstantiates x_i (it performs a *backtrack*). It is easily seen that the search performed by BT corresponds to a depth-first traversal of a semantic tree called the *search tree*, whose root is an empty tuple, while the nodes at the ith level are i-tuples which represent the assignments of the variables along the corresponding path in the tree. Nodes in this tree which correspond to partial solutions are called *consistent nodes*, while other nodes are called *inconsistent nodes*. The number of nodes in the search tree is at most $\Sigma_{0 \leq i \leq n} d^i = \frac{d^{n+1}-1}{d-1}$, hence it is in $O(d^n)$. So, the complexity of BT can be bounded by the number of nodes multiplied by the cost at each node. Assuming that a constraint check can be achieved in $O(a)$, the complexity of BT is $O(e.a.d^n)$.

BT can be considered as a generic algorithm. Algorithms based on BT and used in practice perform some extra work at each node in the search tree, namely, they remove inconsistent values from the domain of future variables (filtering). For *binary* CSPs, FC removes values inconsistent with the current assignment, and RFL moreover enforces full arc-consistency (AC) on the future variables. The complexity of FC can be bounded by $O(nd^n)$. Using an $O(ed^2)$ algorithm for achieving AC, the complexity of RFL is in $O(ed^2 d^{n-1}) = O(ed^{n+1})$. For n-ary CSPs, the algorithms of the class $nFC_i (i = 0, 1 \ldots 5)$ cover the partial and total enforcement of *generalized* arc consistency (GAC) on a subset of constraints involving both assigned variables and future variables. In each case, the filtering is achieved after each variable assignment. So, the complexity of nFC_i depends on the cost of the filtering. Hence, for nFC_5 which achieves the most powerful filtering, the complexity is in $O(eard^n)$. It is the same if we consider the non-binary version of RFL which maintains GAC at each node. In the following, we denote by nBT and nRFL the non-binary versions of BT and RFL.

The algorithm M(G)AC (for Maintaining (Generalized) Arc-Consistency) is slightly different from previous algorithms. Assume that an assignment (x_{i+1}, v_{i+1}) (called a *positive decision*) produces a dead-end. After returning to the current assignment $(v_1, v_2, \ldots v_i)$, and before assigning a new value to x_{i+1}, the value v_{i+1} is deleted from the domain $D(x_{i+1})$, and a (G)AC filtering is realized. All the domains of future variables can be impacted by this filtering. This process is called *refutation* of value x_{i+1}, and can be understood as extending the current partial solution $(v_1, v_2, \ldots, v_i)$ with the "negative" child $(x_{i+1}, \neg v_{i+1})$ (called a *negative decision*), then enforcing arc-consistency, which can end up in a dead end or in further exploration.

Hence the structure and the size of a MAC search tree are different from previous algorithms. First, it is a binary tree verifying particular properties. Each branch of the search tree corresponds to a set of decisions $\Delta = \{\delta_1, \ldots, \delta_i\}$ where each δ_j may be a positive or negative decision. Given an internal node, a negative decision $(x_{i+1}, \neg v_{i+1})$ is produced only after a dead-end has occurred with the positive decision (x_{i+1}, v_{i+1}). Thus, the number of nodes issued from a negative decision is at most the number of nodes issued from standard assignments. Hence for MAC, the number of nodes in the search tree is at most $2 \times \Sigma_{0 \le i \le n-1} d^i = 2\frac{d^n - 1}{d - 1} \in O(d^{n-1})$. Since the cost associated to a node is bounded by the cost of the AC filtering $O(ed^2)$, the worst-case complexity of MAC is the same as for RFL, that is $O(ed^{n+1})$.

We note here that algorithms (n)BT, FC, nFC$_i$, (n)RFL, or M(G)AC may use a *dynamic* variable ordering, that is, which variable (x_{i+1}) to explore next can typically be decided on each assignment.

3   New Complexity Analysis for Binary CSPs

We now come to the heart of our contribution, namely, a complexity analysis of classical algorithms in terms of parameters related to the microstructure. In the following, we say that a node of the search tree is a *maximally deep consistent node* if it is consistent and has no consistent child node (on the next variable in the ordering). Hence, such a node corresponds either to a solution, or to a partial solution which cannot be consistently extended to the next variable. The following result is central to our study.

Proposition 2. *Given a binary CSP $P = (X, D, C)$, there is an injective mapping from the maximally deep consistent nodes explored by BT onto the maximal cliques in $\mu(P)$.*

Proof: Let $(v_1, v_2, \ldots, v_i)$ be a maximally deep consistent node explored by BT. By definition, $(v_1, v_2, \ldots, v_i)$ is a partial solution, hence for all $1 \le j, k \le i$, either there is no constraint c_{jk} in C with scope $\{x_j, x_k\}$, or (v_j, v_k) is in the relation $R(c_{jk})$. In both cases $\{(x_j, v_j), (x_k, v_k)\}$ is an edge in $\mu(P)$. Hence $\{(x_1, v_1), \ldots, (x_i, v_i)\}$ is a clique in $\mu(P)$ and hence, is included in some maximal clique of $\mu(P)$. Write $Cl(v_1, v_2, \ldots, v_i)$ for an arbitrary one.

We now show that Cl forms an injective mapping from maximally deep consistent nodes to maximal cliques. By construction of BT, if $(v_1, v_2, \ldots, v_i)$ and $(v'_1, v'_2, \ldots, v'_{i'})$ are two maximally deep nodes explored, then they must differ on the value of at least one variable. Precisely, they must differ at least at the point where the corresponding paths split in the search tree, corresponding to some variable x_j assigned to some value on one path, and to some other value on the other one[1]. Since there are no edges in $\mu(P)$ connecting two values of the same variable, there cannot be a maximal clique containing both $(v_1, v_2, \ldots, v_i)$ and $(v'_1, v'_2, \ldots, v'_{i'})$, hence Cl is injective. □

Using this proposition, we can easily bound the number of nodes in a search tree induced by a backtracking search, and its time complexity, in terms of the microstructure. As is common, we assume that a constraint check (deciding $(v_i, v_j) \in R(c_{ij})$) requires constant time.

Proposition 3. *The number of nodes $N_{BT}(P)$ in the search tree developed by BT for solving a given binary CSP $P = (X, D, C)$, satisfies $N_{BT}(P) \leq nd \cdot \omega_\#(\mu(P))$. Its time complexity is in $O(n^2 d \cdot \omega_\#(\mu(P)))$.*

Proof: First consider the number of consistent nodes. Because any node in the search tree is at depth at most n and the path from the root to a consistent node contains only consistent nodes, as a direct corollary of Proposition 2 we obtain that the search tree contains at most $n \cdot \omega_\#(\mu(P))$ consistent nodes. Now by definition of BT, a consistent node has at most d children (one per candidate value for the next variable), and inconsistent nodes have none. It follows that the search tree has at most $nd \cdot \omega_\#(\mu(P))$ nodes of any kind.

The time complexity follows directly, since each node corresponds to extending the current partial assignment to one more variable (x_{i+1}), which involves at most one constraint check per other variable (check $c_{j(i+1)}$ for each x_j already assigned). □

It can be seen that in the statement of Proposition 3 and in the forthcoming ones, the number of maximal cliques $\omega_\#(\mu(P))$ could be replaced by the number of maximal cliques *of size at most* $n - 1$. This is because as soon as a path is explored which is contained in an n-clique, that is, in a solution, no backtracking will occur further than this path.

We now turn to forward checking and RFL. Clearly enough, Proposition 2 also holds for both. The number of nodes explored follows from the fact that only consistent nodes are explored. The time complexity follows from the fact that at most n future domains are filtered by FC, and AC is enforced in time $O(ed^2)$ by RFL.

Proposition 4. *The number of nodes $N_{FC}(P)$ in the search tree developed by FC or by RFL for solving a given CSP $P = (X, D, C)$, satisfies $N_{FC}(P) \leq n \cdot \omega_\#(\mu(P))$. The time complexity of FC is in $O(n^2 d \cdot \omega_\#(\mu(P)))$, and that of RFL is in $O(ned^2 \cdot \omega_\#(\mu(P)))$.*

Regarding MAC, unfortunately, the existence of negative decisions in the tree search makes that the proof is not so easy as for other algorithms. So, at present time, we can only claim a conjecture about this result. As a consequence, we

---

[1] We use at this point the assumption that the algorithms explore all the values of a variable before reordering the future variables.

assume too that the time complexity could be a function linear in the number of maximal cliques in the microstructure.

Conjecture 1. *Given a binary CSP $P = (X, D, C)$, there is an injective mapping from the maximally deep consistent nodes explored by MAC onto the maximal cliques in $\mu(P)$.*

4   New Analysis for Non-binary CSPs

We now turn to n-ary CSPs. We first extend the notion of microstructure to this general case, then we analyze the complexity of nFC$_i$ in terms of maximal number of cliques.

4.1   A Generalized Microstructure

Note that a generalization of the notion of microstructure was first proposed in [15]. Nevertheless, this notion is based on hypergraphs and has been little used so far. In contrast, our notion sticks to the simpler framework of graphs. Our *generalized microstructure* is essentially obtained by letting vertices encode the tuples (from relations involved in the CSP) rather than unary assignments (x_i, v_i) of the binary case. Note that other generalizations have been proposed (see [17]) but due to lack of space, we cannot treat them here.

Definition 5 (generalized microstructure). *Given a CSP $P = (X, D, C)$ (not necessarily binary), the* generalized microstructure *of P is the undirected graph $\mu_G(P) = (V, E)$ with:*

- $V = \{(c_i, t_i) : c_i \in C, t_i \in R(c_i)\}$,
- $E = \{ \{(c_i, t_i), (c_j, t_j)\} \mid i \neq j, t_i[S(c_i) \cap S(c_j)] = t_j[S(c_i) \cap S(c_j)]\}$

Like for the microstructure, there is a direct relationship between cliques and solutions of CSPs;

Proposition 5. *A CSP P has a solution iff $\mu_G(P)$ has a clique of size e.*

Proof: By construction, $\mu_G(P)$ is e-partite, and any clique contains at most one vertex (c_i, t_i) per constraint $c_i \in C$. Hence the e-cliques of $\mu_G(P)$ correspond exactly to its cliques with one vertex (c_i, t_i) per constraint $c_i \in C$. Now by construction of $\mu_G(P)$ again, any two vertices (c_i, t_i), (c_j, t_j) joined by an edge (in particular, in some clique) satisfy $t_i[S(c_i) \cap S(c_j)] = t_j[S(c_i) \cap S(c_j)]$. Hence all t_i's in a clique join together, and it follows that the e-cliques of $\mu_G(P)$ correspond exactly to tuples t which are joins of one allowed tuple per constraint, that is, to solutions of P.   □

One can observe that the generalized microstructure corresponds to the microstructure of the dual representation of a CSP [18]. One can also see that our generalization is in fact the *line-graph* (the dual graph) of the hypergraph proposed in [15], in which we add edges for pairs of constraints whose scopes have empty intersections. Hence, in the same spirit, we can propose other generalizations of the microstructure by considering any graph-based representation of a non-binary CSP as soon as the representation has the same set of solutions - wrt. a given bijection - as the original instance (e.g. the hidden variable encoding [19]). However, due to lack of space, we do not deal with these issues here.

4.2   Time Complexity of nBT, nFC and nRFL

We now investigate the complexity of algorithms for solving n-ary CSPs. We first make an assumption about the order in which such algorithms explore variables, then we discuss this restriction.

Definition 6 (compatible with constraints). *Let P be a CSP. A total order $(x_1, x_2, \ldots x_n)$ on X is said to be* compatible with the constraints in C *if there are k constraints $c_{i_1}, c_{i_2}, \ldots c_{i_k}$ in C $(1 \leq k \leq e)$ which satisfy:*

- *$\bigcup_{1 \leq \ell \leq k} S(c_{i_\ell}) = X$*
- *there are k variables $x_{i_1}, x_{i_2}, \ldots x_{i_k}$ such that $\forall \ell \in \{1, \ldots, k\}$, $x_{i_\ell} \in S(c_{i_\ell})$ and $\bigcup_{1 \leq j \leq \ell} S(c_{i_j}) = \{x_i \mid i = 1, \ldots, i_\ell\}$ hold.*

In words, the ordering must be such that the variables in the scope of one distinguished constraint (c_{i_1}) all appear first, then all the variables in the scope of some c_{i_2} (except for those already mentioned by c_{i_1}), etc. The variables $x_{i_1}, \ldots, x_{i_k}$ in the definition are such that x_{i_j} is the last variable assigned in the scope of c_{i_j}. We refer to these variables as *milestones* in the ordering.

For instance, with the notation of the definition we must have $S(c_{i_1}) = \{x_1, x_2, \ldots x_{i_1}\}$ and $S(c_{i_1}) \cup S(c_{i_2}) = \{x_1, x_2, \ldots x_{i_1}, x_{i_1+1}, \ldots x_{i_2}\}$. The variable x_{i_1} is a milestone (last variable assigned in the scope of c_{i_1}).

Under the assumption that such a variable ordering is used, we can give a generalization of Proposition 2.

Proposition 6. *Let P be an n-ary CSP, and assume that nBT explores the variables in some order compatible with the constraints in C. Then there is an injective mapping from the maximally deep consistent nodes (x_i, v_i) in the search tree such that x_i is a milestone, and the maximal cliques in $\mu_G(P)$.*

Proof: Let t be an assignment corresponding to a node as in the statement, and write x_{i_j} for the last variable assigned by t (which is a milestone by assumption). Similarly, let t' be another maximal consistent assignment with the milestone $x_{i_{j'}}$ as its last variable. Write T for the set $\{t[S(c)] \mid c \in C, S(c) \subseteq \{x_1, \ldots, x_{i_j}\}\}$, that is, for the set of all projections of t onto the scopes of constraints fully assigned by t, and similarly for T'. Then T (resp. T') is included in some maximal clique $Cl(t)$ (resp. $Cl(t')$) of $\mu_G(P)$. Now assume $j' \geq j$ (wlog). From $t \neq t'$, the fact that t is maximally consistent, and the fact that t' is consistent, it follows that t' differs from t on at least one variable x_ℓ with $\ell \leq i_j$. Hence this variable is assigned differently by both, and there is some constraint c_{i_ℓ} such that $t[S(c_{i_\ell})]$ is different from $t'[S(c_{i_\ell})]$, and it follows that t, t' cannot be included in a common clique. Hence Cl defines an injective mapping from assignments as in the statement to maximal cliques in $\mu_G(P)$, as desired.   □

Using this property, we can bound the number of nodes in a search tree induced by a backtracking search, and its time complexity, in terms of the generalized microstructure.

Proposition 7. *Let $P = (X, D, C)$ be an n-ary CSP, and assume that nBT uses a variable ordering which is compatible with the constraints in C. Then the number of nodes $N_{nBT}(P)$ in the search tree of nBT on P satisfies $N_{nBT}(P) \leq nd^a \cdot \omega_\#(\mu_G(P))$. Its time complexity is in $O(nea \cdot d^a \cdot \omega_\#(\mu_G(P)))$.*

Proof: From Proposition 6 it follows that the subtree induced by the search tree on milestones contains at most $\omega_\#(\mu_G(P))$ nodes. Now for reaching a milestone from the previous one, that is, for extending an assignment to $x_1, \ldots, x_{i_j}$ onto an assignment to $x_1, \ldots, x_{i_{j+1}}$ (with the notation of Definition 6), nBT explores at most a variables (this is by definition of an ordering compatible with the constraints). Hence it explores at most d^a combinations of values (nBT has no clue for ruling out an assignment before assigning all variables in the scope of a constraint). Since a branch contains at most n milestones, we get the result. The time complexity follows directly since each node requires at most e constraint checks, each one in time $O(a)$ with an appropriate data structure.   □

A similar result holds for nFC_i ($i \geq 2$). Nevertheless, we must consider the additional cost due to applying GAC, that is $O(e \cdot a \cdot r)$ at each node. However, note that contrary to nBT, due to the use of GAC, nFC_i explores only the r tuples allowed by c when exploring the variables in $S(c)$, rather than all d^a combinations of values.

Proposition 8. *Let $P = (X, D, C)$ be an n-ary CSP, and assume that nFC_i ($i \geq 2$) uses a variable ordering which is compatible with the constraints in C. Then the number of nodes $N_{nFC_i}(P)$ in the search tree of nFC_i on P satisfies $N_{nFC_i}(P) \leq nr \cdot \omega_\#(\mu_G(P))$. Its time complexity is in $O(nea \cdot r^2 \cdot \omega_\#(\mu_G(P)))$.*

The same result also holds for nRFL.

As a final note, we have considered a total order in definition 6. A partial order can be used instead, provided the following holds: for each constraint c_{i_j} ($1 < j \leq k$), all the variables of $c_{i_{j-1}}$ have been assigned before the variables of c_{i_j} which do not belong to the scope of a previous constraint. The last assigned variable of each constraint will be the corresponding milestone variable. Moreover, the considered order may be dynamic.

4.3   Time Complexity without Ordering

Arguably, our restriction to variable orderings which are compatible with constraints is not met by all reasonable variable orderings. For instance, there is no reason in general for the well-known dom/deg heuristic to yield such orderings. However, we show here that such restrictions are necessary.

To show this, we build a family of instances which have a linear number of cliques in their generalized microstructure (in its number of vertices), but for which nFC$_5$ explores a search tree of size exponential (in the number of vertices) for some specific variable ordering.

The instances (X, D, C) in this family are built as follows. Writing e for the number of constraints, there are two distinguished variables x_0, x_0' in X (common to all constraints), and $X \setminus \{x_0, x_0'\}$ is partitioned into e sets $X_1, \ldots, X_e$ (X_i is specific to c_i). So each constraint $c_i \in C$ has scope $S(c_i) = \{x_0, x_0'\} \cup X_i$. Now D is $\{v_0, \ldots, v_{e-1}\}$ for all variables ($d = e$). Finally, the tuples allowed by the constraint c_i are precisely those of the form $\{(x_0, v_j), (x_0', v_{i+j}), \ldots\}$, for $j = 1, \ldots, e$ (indices are taken modulo e) and unrestricted assignments to X_i. The point is that for $i \neq i'$, the restrictions of the tuples allowed by c_i and $c_{i'}$ onto $\{x_0, x_0'\}$ never match, so that there are no edges in $\mu_G(P)$. Hence the number of cliques in $\mu_G(P)$ is exactly its number of vertices $|V| = e^{2+(n-2)/e}$.

On the other hand, assume that nFC$_5$ explores all variables in the X_i's and only explores x_0, x_0' after them. Then because all values for x_0 have a support in all constraints, and similarly for x_0', no value will be removed before reaching x_0 or x_0', and hence all $e^{n-2} \sim |V|^e$ combinations of values will be explored, that is, exponentially more than the number of cliques in $\mu_G(P)$.

In the general case, we can bound the time complexity of nFC$_i$ as follows.

Proposition 9. *Let $P = (X, D, C)$ be an n-ary CSP, and assume that nFC$_i$ ($i \geq 2$) uses a variable ordering such that the maximum number of non-milestone variables assigned consecutively is m. Then the number of nodes $N_{nFC_i}(P)$ in the search tree of nFC$_i$ on P satisfies $N_{nFC_i}(P) \leq ndm \cdot \omega_\#(\mu_G(P))$. Its time complexity is in $O(nea \cdot rd^m \cdot \omega_\#(\mu_G(P)))$.*

In our previous example, we have $m = n - 2$.

5   A Few Tractable Classes for Backtracking

The number of cliques in a graph can grow exponentially with the size of the graph [20], and so can the number $\omega_\#(G)$ of *maximal* cliques in a graph G [21]. However, for some classes of graphs, the number of maximal cliques can be bounded by a polynomial in the size of the graph. If the (generalized) microstructure of a (family of) CSP P belongs to one of these classes, then our analysis in the previous sections allows us to conclude that P is solved in polynomial time by classical backtracking algorithms, *without the need to recognize the instance to be in the class*. In this section, we study several such classes of graphs in terms of their relevance to constraint satisfaction problems.

5.1   Triangle-Free and Bipartite Graphs

Recall that a *k-cycle* in a graph $G = (V, E)$ is a sequence $(v_1, v_2, \ldots v_{k+1})$ of distinct vertices satisfying $\forall i, 1 \leq i \leq k, \{v_i, v_{i+1}\} \in E$, and $v_1 = v_{k+1}$.

A *triangle-free* graph is an undirected graph with no 3-cycle. It is easily seen that the number of maximal cliques in a triangle-free graph is exactly its number of edges E. Hence by our analysis, if a class of CSPs has a triangle-free (generalized) microstructure, algorithms (n)BT, (n)FC and (n)RFL correctly solve them in polynomial time. Note however that this is quite a degenerate case, since except for instances having at most two variables (binary case) or two constraints (non-binary case), instances with a triangle-free (generalized) microstructure are inconsistent.

Another degenerate but illustrative case is one of bipartite graphs. A graph is *bipartite* if it does not contain an odd cycle. Again, a bipartite graph cannot contain any clique of more than two vertices, and hence no partial assignment to more than three variables will ever be considered by BT (hence it obviously runs in time $O(d^3)$).

We now turn to more interesting classes, which also essentially contain inconsistent CSPs, but for which our analysis gives a better time complexity than the classical one.

5.2   Planar, Toroidal, and All Embedded Graphs

Definition 7 (planar). *A planar graph is a graph that can be embedded in the plane without crossing edges.*

[20] proved that the number of cliques in a planar graph is at most $8(|V| - 2)$.

Definition 8 (toroidal). *A graph is* toroidal *if it can be embedded on the torus without crossing edges.*

[22] showed that every toroidal graph has at most $8(|V|+9)$ cliques and then that every graph embeddable in some surface has a linear number of cliques ($8(|V| + 27)$ at worst). Since the microstructure $\mu(P)$ (resp. $\mu_G(P)$) of a CSP P contains at most nd vertices (resp. er vertices), it follows that if $\mu(P)$ (resp. $\mu_G(P)$) belongs to one of these classes of graphs, then $\omega_\#(\mu(P))$ (resp. $\omega_\#(\mu_G(P))$) is in $O(nd)$ (resp. $O(er)$). Thanks to Prop. 3, 4, 7 and 8, we immediately get the following.

Theorem 1. *Let Em denote the class of all CSPs whose microstructure is planar, toroidal, or embeddable in a surface. Then instances in Em are solved in time*

- $O(n^2 d \cdot \omega_\#(\mu(P))) = O(n^3 d^2)$ *by BT or FC,*
- $O(ned^2 \cdot \omega_\#(\mu(P))) = O(n^2 ed^3)$ *by RFL,*
- $O(nead^a \cdot \omega_\#(\mu_G(P))) = O(ne^2 ard^a)$ *by nBT,*
- $O(near^2 \cdot \omega_\#(\mu_G(P))) = O(ne^2 ar^3)$ *by nFC_i, nRFL.*

Recall that this family of graphs cannot contain as a minor, an 8-clique (for toroidal graphs), or a 5-clique nor $K_{3,3}$ (for planar graphs). It follows in particular that any binary (resp. non-binary) CSP in Em over at least 8 variables (resp. constraints) is inconsistent. Hence again this class is a little degenerate, however a classical analysis states that, e.g., BT solves these instances in time $O(d^8)$. In case d is large, this is looser that $O(n^3 d^2)$.

5.3 CSG Graphs

We finally turn to the class of *CSG graphs*, which has been introduced by [23] and which generalizes the class of chordal graphs. Given a graph (V, E) and an ordering $v_1, \ldots, v_{|V|}$ of its vertices, we write $N^+(v_i)$ for the *forward neighborhood* of v_i, that is, $N^+(v_i) = \{v_j \in V \mid \{v_i, v_j\} \in E, i < j\}$. For $V' \subseteq V$, we write $G(V')$ for the graph *induced* by E on V', namely, $G(V') = (V', E')$ where $E' = \{\{x, y\} \mid x, y \in V' \text{ and } \{x, y\} \in E\}$.

Definition 9 (CSG graphs). *The class of graphs CSG^k is defined recursively as follows.*

- CSG^0 *is the class of complete graphs.*
- *Given $k > 0$, CSG^k is the class of graphs $G = (V, E)$ such that there exists an ordering $\sigma = (v_1, ..., v_{|V|})$ of V satisfying that for $i = 1, \ldots, |V|$, the graph $G(N^+(v_i))$ is a CSG^{k-1} graph.*

The class of CSG graphs generalizes the class of complete graphs (CSG^0 graphs) and the class of chordal graphs (CSG^1 graphs). Like chordal graphs, CSG graphs have nice properties. For instance, they can be recognized in polynomial time. Moreover, Chmeiss and Jégou have proved that CSG^k graphs have at most $|V|^k$ maximal cliques, and they have proposed an algorithm running in time $O(|V|^{2(k-1)}(|V| + |E|))$ for finding all of them.

   These two algorithms qualify the class of all CSPs which have a CSG^k (generalized) microstructure as a tractable class for any fixed k. We are however able to show that even a *generic* algorithm such as (n)BT, FC, nFC$_i$ or (n)RFL runs in polynomial time on such CSPs, without even the need to recognize membership in this class, nor to compute the microstructure. Again, the result follows from the number of maximal cliques together with Prop. 3, 4, 7 and 8.

Theorem 2. *Given any integer k, the class of all CSPs which have a CSG^k (generalized) microstructure is solved in time*

- $O(n^2 d \cdot \omega_\#(\mu(P))) = O(n^{k+2} d^{k+1})$ *by BT and FC,*
- $O(ned^2 \cdot \omega_\#(\mu(P))) = O(n^{k+1} ed^{k+2})$ *by RFL,*
- $O(nead^a \cdot \omega_\#(\mu_G(P))) = O(ne^{k+1} ar^k d^a)$ *by nBT,*
- $O(near^2 \cdot \omega_\#(\mu_G(P))) = O(ne^{k+1} ar^{k+2})$ *by nFC$_i$ or nRFL.*

Observe that even the time complexities are better than those of the dedicated algorithm. For instance, the latter computes the microstructure of a binary CSP and enumerates all maximal cliques until exhaustion, or an n-clique is found. So it has a time complexity in $O((nd)^{2(k-1)}(nd + n^2 d^2)) = O((nd)^{2k})$. Nevertheless, we can note that this algorithm is defined for general CSG^k graphs while (generalized) microstructures of CSP are very particular graphs.

   CSG graphs are generally speaking less restrictive than the previous classes of graphs. For instance, it is possible to have CSG graphs with n-cliques (resp. e-clique) for any value of n (resp. e), contrary to the case of planar graphs. In particular, there are consistent CSPs with a CSG^k (generalized) microstructure.

It is the case for CSG^0 which are consistent binary CSPs with monovalent domains (one value per domain) or consistent non-binary CSPs with exactly one allowed tuple per relation. Nevertheless, CSPs which have a CSG^1 (generalized) microstructure can be consistent or not and it is easy to build a CSP with several solutions, which corresponds to a collection of cliques of size n (binary case) or e (non-binary case). Furthermore, unlike previous classes, in CSG^k graphs (with $k \geq 1$), there is no restrictions on the values of n, d, e, a or r.

For instance, the *Hanoi* or *Domino* benchmarks used in the CSP solver competitions (e.g. [24]) have a CSG^1 microstructure after applying an AC filtering. Thus, the use of algorithms such as RFL allows them to exploit the tractable classes that we have highlighted here. However, this set of classes of CSPs still has to be studied in detail for assessing its practical interest.

6   Discussion and Perspectives

We have investigated the time complexity of classical, generic algorithms for solving CSPs under a new perspective. Our analysis expresses the complexity in terms of the number of maximal cliques in the (generalized) microstructure of the CSP to be solved. Our analysis reveals that essentially, backtracking and forward checking visit each maximal clique in the (generalized) microstructure at most once. From this analysis we derived tractable classes of CSPs, which can be solved by classical algorithms in polynomial time, *without the need to recognize that the instance at hand is in the class*. So, the results obtained shed a new light on the analysis of CSPs.

Some relationships between tractable classes presented here and other tractable classes of the state of the art are simple to identify. For example, for the classes defined by the structure, that is, CSPs of bounded width, it is easy to see that they are incomparable. It is possible to build a CSP whose constraint network is a complete graph with a polynomial number of maximal cliques. E.g. consider a CSPs defined on n variables with the same domain of size d and with equality constraints between every pair of variables. For this kind of instances, there are d maximal cliques of size n corresponding to the d solutions of the CSP. Conversely, it is easy to see that an acyclic binary CSP can have an exponential number of solutions, and thus, an exponential number of maximal cliques.

For hybrid classes, we can see that the tractable class of binary CSPs satisfying the Broken Triangle Property (BTP) [25] is also incomparable. It is very simple to define a CSP satisfying the BTP property with an exponential number of solutions and thus an exponential number of maximal cliques. In contrast, one can easily define a binary CSP whose microstructure is planar and which does not satisfy the BTP property. For instance, the CSP defined on three variables with domains $D(x_1) = \{a, b\}, D(x_2) = \{c, d\}, D(x_3) = \{e, f\}$, and pairs of values $(a, d), (b, e), (c, f)$ prohibited.

It is thus clear that the tractable classes which we propose here are not necessarily weaker or stronger than the classes of the state of the art. Their main advantage is that they can simply be treated without recourse to ad hoc algorithms, but with state-of-the-art solvers. Despite the lack of a theoretical difficulty in the definition of these tractable classes, they seem to offer an advantage compared to many tractable classes which are for most of them, artificial, if one refers to real benchmarks.

The first perspective of this work is to investigate more classes of graphs with polynomially many maximal cliques. Of particular interest here is the study by [26], who precisely characterize these classes of graphs in terms of intersection graphs. Another important perspective is to relate our analysis to other tractable classes obtained in different manners, in the spirit of the first comparisons given above. An important perspective is also to extend our study to the other possible generalizations of the microstructure for non-binary constraint satisfaction problems. In [27], a close work is presented since it proposes to define a new tractable class which is based on the polynomial size of search trees. The relations with our work should be now clarified.

To complete our study, we must also address the conjecture 1 for the MAC algorithm. Actually, we strongly believe that this conjecture is true, but we have not been able to prove this fact, especially due to the existence of the concept of negative decisions for MAC.

Finally, it would be interesting to analyse a wide set of benchmarks - particularly the ones easily solved by solvers of the state of the art - to investigate cases in which the tractable classes exhibited here are present, in the spirit of the first observations made on benchmarks such as *Hanoi* or *Domino*.

Acknowledgments. Philippe Jégou would like to thank Maria Chudnovsky for their fruitful discussion, about graph theory, perfect graphs and links with classes of graphs related to the clique problem, and the links with the microstructure of CSPs.

References

1. Rossi, F., van Beek, P., Walsh, T.: Handbook of Constraint Programming. Elsevier (2006)
2. Haralick, R., Elliot, G.: Increasing tree search efficiency for constraint satisfaction problems. Artificial Intelligence 14, 263–313 (1980)
3. Nadel, B.: Tree Search and Arc Consistency in Constraint-Satisfaction Algorithms. In: Search in Artificial Intelligence, pp. 287–342. Springer (1988)
4. Sabin, D., Freuder, E.C.: Contradicting Conventional Wisdom in Constraint Satisfaction. In: Borning, A. (ed.) PPCP 1994. LNCS, vol. 874, pp. 10–20. Springer, Heidelberg (1994)
5. Bessière, C., Meseguer, P., Freuder, E.C., Larrosa, J.: On forward checking for non-binary constraint satisfaction. Artificial Intelligence 141, 205–224 (2002)
6. Freuder, E.: A Sufficient Condition for Backtrack-Free Search. JACM 29(1), 24–32 (1982)
7. Gottlob, G., Leone, N., Scarcello, F.: A Comparison of Structural CSP Decomposition Methods. Artificial Intelligence 124, 343–382 (2000)
8. de Givry, S., Schiex, T., Verfaillie, G.: Decomposition arborescente et cohérence locale souple dans les CSP pondérés. In: Proceedings of JFPC 2006 (2006)

9. Purvis, L., Jeavons, P.: Constraint tractability theory and its application to the product development process for a constraint-based scheduler. In: Proceedings of the 1st International Conference on The Practical Application of Constraint Technologies and Logic Programming, pp. 63–79 (1999); This paper was awarded First Prize in the Constraints Technologies area of PACLP 1999

10. Rauzy, A.: Polynomial restrictions of SAT: What can be done with an efficient implementation of the Davis and Putnam's procedure. In: Montanari, U., Rossi, F. (eds.) CP 1995. LNCS, vol. 976, pp. 515–532. Springer, Heidelberg (1995)

11. Jégou, P., Ndiaye, S., Terrioux, C.: A new evaluation of forward checking and its consequences on efficiency of tools for decomposition of CSPs. In: ICTAI (1), pp. 486–490 (2008)

12. Ordyniak, S., Paulusma, D., Szeider, S.: Satisfiability of acyclic and almost acyclic CNF formulas. In: FSTTCS, pp. 84–95 (2010)

13. Jégou, P.: Decomposition of Domains Based on the Micro-Structure of Finite Constraint Satisfaction Problems. In: Proceedings of AAAI 1993, Washington, DC, pp. 731–736 (1993)

14. Golumbic, M.: Algorithmic Graph Theory and Perfect Graphs. Academic Press, New York (1980)

15. Cohen, D.A.: A New Class of Binary CSPs for which Arc-Consistency Is a Decision Procedure. In: Rossi, F. (ed.) CP 2003. LNCS, vol. 2833, pp. 807–811. Springer, Heidelberg (2003)

16. Salamon, A.Z., Jeavons, P.G.: Perfect Constraints Are Tractable. In: Stuckey, P.J. (ed.) CP 2008. LNCS, vol. 5202, pp. 524–528. Springer, Heidelberg (2008)

17. El Mouelhi, A.: Generalized micro-structures for non-binary CSP. In: Doctoral Programme CP, pp. 13–18 (2012), http://zivny.cz/dp12/

18. Dechter, R., Pearl, J.: Tree-Clustering for Constraint Networks. Artificial Intelligence 38, 353–366 (1989)

19. Rossi, F., Petrie, C., Dhar, V.: On the equivalence of constraint satisfaction problems. In: Proceedings of the 9th European Conference on Artificial Intelligence, pp. 550–556 (1990)

20. Wood, D.R.: On the maximum number of cliques in a graph. Graphs and Combinatorics 23, 337–352 (2007)

21. Moon, J.W., Moser, L.: On cliques in graphs. Israel Journal of Mathematics 3, 23–28 (1965)

22. Dujmovic, V., Fijavz, G., Joret, G., Sulanke, T., Wood, D.R.: On the maximum number of cliques in a graph embedded in a surface. European J. Combinatorics 32(8), 1244–1252 (2011)

23. Chmeiss, A., Jégou, P.: A generalization of chordal graphs and the maximum clique problem. Information Processing Letters 62, 111–120 (1997)

24. Third International CSP Solver Competition (2008), http://cpai.ucc.ie/08

25. Cooper, M., Jeavons, P., Salamon, A.: Generalizing constraint satisfaction on trees: hybrid tractability and variable elimination. Artificial Intelligence 174, 570–584 (2010)

26. Rosgen, B., Stewart, L.: Complexity results on graphs with few cliques. Discrete Mathematics and Theoretical Computer Science 9, 127–136 (2007)

27. Cohen, D.A., Cooper, M.C., Green, M.J., Marx, D.: On guaranteeing polynomially bounded search tree size. In: Lee, J. (ed.) CP 2011. LNCS, vol. 6876, pp. 160–171. Springer, Heidelberg (2011)

Revisiting Hyper Binary Resolution[*]

Marijn J.H. Heule[1,3], Matti Järvisalo[2], and Armin Biere[3]

[1] Department of Computer Science, The University of Texas at Austin, United States
[2] HIIT & Department of Computer Science, University of Helsinki, Finland
[3] Institute for Formal Models and Verification, Johannes Kepler University Linz, Austria

Abstract. This paper focuses on developing efficient inference techniques for improving conjunctive normal form (CNF) Boolean satisfiability (SAT) solvers. We analyze a variant of hyper binary resolution from various perspectives: We show that it can simulate the circuit-level technique of structural hashing and how it can be realized efficiently using so called tree-based lookahead. Experiments show that our implementation improves the performance of state-of-the-art CNF-level SAT techniques on combinational equivalent checking instances.

1 Introduction

Boolean satisfiability (SAT) solvers provide the crucial core search engines for solving problem instances arising from various real-world problem domains. This paper focuses on developing efficient inference techniques to improve the robustness of conjunctive normal form (CNF) SAT solving techniques. Especially, our goal is to improve CNF-level techniques on instances of *miter-based combinational equivalence checking* which is an important industrially-relevant problem domain. The main motivation behind this work is to take notable steps towards the ambitious goal of making CNF-level approaches competitive with circuit-level techniques for equivalence checking. This goal is important as it would notably simplify the current state-of-the-art techniques applied in the industry which require alternating between circuit-level techniques and CNF-level SAT solving. To this end, we identify how known CNF-level SAT solving techniques can simulate the circuit-level technique of *structural hashing*—which plays an integral role in solving miter instances—purely on the level of a standard CNF encoding of Boolean circuits. As the main CNF-level approach, we study a variant of hyper binary resolution (HBR), which can be used to learn non-transitive hyper binary resolvents, and analyze this technique from various perspectives. While this variant or HBR has already been studied and implemented previously within the HYPRE [1] and HYPERBINFAST [2] CNF simplifiers, we extend this previous work both from the theoretical and practical perspectives.

Our main theoretical observations include: (i) explanations for how and to what extent the CNF techniques HBR, clause learning, and ternary resolution can simulate structural hashing; (ii) that HBR can be focused in a beneficial way to produce only

---

[*] The first author is supported by DARPA contract number N66001-10-2-4087. The first and third authors are supported by Austrian Science Foundation (FWF) NFN Grant S11408-N23 (RiSE), and the second author by Academy of Finland (grants 132812 and 251170).

C. Gomes and M. Sellmann (Eds.): CPAIOR 2013, LNCS 7874, pp. 77–93, 2013.

non-transitive resolvents that increase transitive reachability of the underlying binary implications; and (iii) providing an explicit quadratic worst-case example on the number of binary clauses added, which applies to all known implementations of HBR, and that has not been explicitly provided before. As the main practical contribution, we show how this variant of HBR can be realized efficiently using so called tree-based lookahead [3]. In fact, the tree-based lookahead algorithm described in this work is a substantially simplified version of the original idea, and is also of independent interest due to its much more general applicability for instance within CDCL SAT solvers. We show experimentally that our *TreeLook* implementation of HBR using tree-based lookahead clearly outperforms state-of-the-art CNF-level SAT techniques on instances encoding on miter-based equivalence checking CNF instances.

The rest of this paper is organized as follows. After preliminaries (Sect. 2), we discuss possibilities of simulating structural hashing on the CNF-level (Sect. 3). Then the considered variant of hyper binary resolution is defined and analyzed (Sect. 4), followed by an in-depth description of tree-based lookahead (Sect. 6) that enables implementing hyper binary resolution efficiently. Before conclusions, experimental results are presented (Sect. 7) and related work is discussed (Sect. 8).

2   Preliminaries

For a Boolean variable x, there are two *literals*, the positive literal x and the negative literal $\neg x$. A *clause* is a disjunction of literals and a CNF formula a conjunction of clauses. A clause can be seen as a finite set of literals and a CNF formula as a finite set of clauses. A (partial) truth assignment for a CNF formula F is a function τ that maps (a subset of) the literals in F to $\{0, 1\}$. If $\tau(x) = v$, then $\tau(\neg x) = 1 - v$. A clause C is satisfied by τ if $\tau(l) = 1$ for some literal $l \in C$. A clause C is falsified by τ if $\tau(l) = 0$ for every literal $l \in C$. An assignment τ satisfies F if it satisfies every clause in F. We denote by $\tau(F)$ the reduced formula for which all satisfied clauses by τ and all falsified literals by τ are removed.

Two formulas are *logically equivalent* if they are satisfied by exactly the same set of assignments. A clause of length one is a *unit clause*, and a clause of length two is a *binary clause*. For a CNF formula F, F_2 denotes the set of binary clauses, and $F_{\geq 3}$ denotes the set of clauses of length three and larger.

Binary Implication Graphs. Given a CNF formula F, the unique *binary implication graph* $\mathsf{BIG}(F)$ of F has for each variable x occurring in F_2 two vertices, x and $\neg x$, and has the edge relation $\{\langle \neg l, l' \rangle, \langle \neg l', l \rangle \mid (l \vee l') \in F_2\}$. In other words, for each binary clause $(l \vee l')$ in F, the two implications $\neg l \to l'$ and $\neg l' \to l$, represented by the binary clause, occur as edges in $\mathsf{BIG}(F)$. A node in $\mathsf{BIG}(F)$ with no incoming arcs is a *root* of $\mathsf{BIG}(F)$ (or, simply, of F_2). In other words, literal l is a root in $\mathsf{BIG}(F)$ if there is no clause of the form $(l \vee l')$ in F_2. The set of roots of $\mathsf{BIG}(F)$ is denoted by $\mathsf{RTS}(F)$.

BCP, Failed Literal Elimination (FLE), and Lookahead. For a CNF formula F, *Boolean constraint propagation* (BCP) (or *unit propagation*) propagates all unit clauses, i.e., repeats the following until fixpoint: if there is a unit clause $(l) \in F$, remove from $F \setminus \{(l)\}$ all clauses that contain the literal l, and remove the literal $\neg l$ from all clauses in F, resulting in the formula $\mathrm{BCP}(F)$. A literal l is a *failed literal* if $\mathrm{BCP}(F \cup \{(l)\})$ contains the empty clause, implying that F is logically equivalent to $\mathrm{BCP}(F \cup \{(\neg l)\})$. FLE removes failed literals from a formula, or, equivalently, adds the complements of failed literals as unit clauses to the formula, until a fixpoint is reached. Failed literal elimination is sometimes also referred to as *lookahead*, and is often applied in non-CDCL DPLL solvers (*lookahead solvers* [4]).

Equivalent Literal Substitution (ELS). The strongly connected components (SCCs) of $\mathrm{BIG}(F)$ represent equivalent classes of literals (or simply *equivalent literals*) in F_2 [5]. *Equivalent literal substitution* refers to substituting in F, for each SCC G of $\mathrm{BIG}(F)$, all occurrences of the literals occurring in G with the representative literal of G. ELS is confluent, i.e., has a unique fixpoint, modulo variable renaming.

Transitive Reduction (TRD). A directed acyclic graph G' is a *transitive reduction* [6] of the directed graph G provided that (i) G' has a directed path from node u to node v if and only if G has a directed path from node u to node v, and (ii) there is no graph with fewer edges than G' satisfying the condition (i). For a CNF formula F, a binary clause $C = (l \vee l')$ is *transitive* in F if l' is reachable from $\neg l$ (equivalently, l is reachable from $\neg l'$) in $\mathrm{BIG}(F \setminus C)$. Applying TRD on $\mathrm{BIG}(F)$ amounts to removing from F all transitive binary clauses in F. TRD is confluent for the class of CNF formulas F for which $\mathrm{BIG}(F)$ is acyclic. This is due to the fact that the transitive reduction of any directed acyclic graph is unique [6]. For directed graphs with cycles, TRD is unique modulo node (literal) equivalence classes.

The main inference rule of interest in this work is the *hyper binary resolution rule*.

Hyper Binary Resolution (HBR). The resolution rule states that, given two clauses $C_1 = \{l, a_1, \ldots, a_n\}$ and $C_2 = \{\neg l, b_1, \ldots, b_m\}$, the clause $C = C_1 \bowtie C_2 = \{a_1, \ldots, a_n, b_1, \ldots, b_m\}$, called the *resolvent* $C_1 \bowtie C_2$ of C_1 and C_2, can be inferred by *resolving* on the literal l. Many different simplification techniques are based on the resolution rule. In this paper of interest is *hyper binary resolution* [7]. Given a clause of the form $(l \vee l_1 \cdots \vee l_k)$ and k binary clauses of the form $(l' \vee \neg l_i)$, where $1 \le i \le k$, the hyper binary resolution rule allows to infer the *hyper binary resolvent* $(l \vee l')$ in one step. HBR is confluent since it only adds clauses to CNF formulas.

3   Simulating Structural Hashing on CNF

In this section we show that hyper binary resolution is surprisingly powerful in that it implicitly—purely on the CNF-level—achieves *structural hashing*, i.e., sharing of equivalent subformula structures, over disjunctive and conjunctive subformulas. This is surprising, as structural hashing is often considered one of the benefits of representing propositional formulas on the higher level of *Boolean circuits* rather than working on the flat CNF form. This result implies that structural hashing can be achieved also during the actual CNF-level solving process by applying HBR on the current CNF formula.

Boolean Circuits are a natural representation form for propositional formulas, offering *subformula sharing* via structural hashing. A Boolean circuit over a finite set $\mathcal{G}$ of *gates* is a set $\mathcal{C}$ of equations of form $g := f(g_1, \ldots, g_n)$, where $g, g_1, \ldots, g_n \in \mathcal{G}$ and $f :$

$\{1,0\}^n \to \{1,0\}$ is a Boolean function, with the additional requirements that (i) each $g \in \mathcal{G}$ appears at most once as the left hand side in the equations in $\mathcal{C}$, and (ii) the underlying directed graph

$$\langle \mathcal{G}, E(\mathcal{C}) = \{\langle g', g\rangle \in \mathcal{G} \times \mathcal{G} \mid g := f(\dots, g', \dots) \in \mathcal{C}\}\rangle$$

is acyclic. Each gate represents a specific subformula in the propositional formula expressed by the set of Boolean equations. If $g := f(g_1, \dots, g_n)$ is in $\mathcal{C}$, then g is an f-gate (or of type f), otherwise it is an *input gate*. The following Boolean functions are some which often occur as gate types: $\text{NOT}(v)$ (1 if and only if v is 0), $\text{OR}(v_1, \dots, v_n)$ (1 if and only if at least one of $v_1, \dots, v_n$ is 1), $\text{AND}(v_1, \dots, v_n)$ (1 if and only if all $v_1, \dots, v_n$ are 1), $\text{XOR}(v_1, v_2)$ (1 if and only if exactly one of v_1, v_2, is 1), and $\text{ITE}(v_1, v_2, v_3)$ (1 if and only if (i) v_1 and v_2 are 1, or (ii) v_1 is 0 and v_3 is 1). The standard "Tseitin" encoding of a Boolean circuit $\mathcal{C}$ into a CNF formula $\text{TST}(\mathcal{C})$ works by introducing a Boolean variable for each gate in $\mathcal{C}$, and representing for each gate $g := f(g_1, \dots g_n)$ in $\mathcal{C}$ the logical equivalence $g \leftrightarrow f(g_1, \dots g_n)$ with clauses.

3.1   Structural Hashing on the CNF-Level via HBR

Structural hashing is a well-known technique for factoring out common sub-expression. It is an integral part of many algorithms for manipulating different data structures representing circuits [8,9,10,11,12].

Given a circuit $\mathcal{C}$ with $g := f(g_1, \dots, g_n), g' := f(g_1, \dots, g_n) \in \mathcal{C}$, *structural hashing* removes $g' := f(g_1, \dots, g_n)$ from $\mathcal{C}$, i.e., detects that g and g' label the same function $f(g_1, \dots, g_n)$ in $\mathcal{C}$. A Boolean circuit $\mathcal{C}$ is structurally hashed if g and g' are the same gate whenever $g := f(g_1, \dots, g_n), g' := f(g_1, \dots, g_n) \in \mathcal{C}$.

Proposition 1. *Let $\mathcal{C}$ be an arbitrary Boolean circuit. Assume that there are two distinct gates $g := f(g_1, \dots, g_n)$ and $g' := f(g_1, \dots, g_n)$ in $\mathcal{C}$, where $f \in \{\text{NOT}, \text{AND}, \text{OR}\}$. Then HBR applied to $\text{TST}(\mathcal{C})$ will produce the clauses $(\neg g \vee g')$ and $(g \vee \neg g')$ representing the fact that g and g' label the same function $f(g_1, \dots, g_n)$ in $\mathcal{C}$.*

Basically the binary clauses in $\text{TST}(\mathcal{C})$ associated with $g := f(g_1, \dots, g_n)$ together with a clause of arity $(n + 1)$ associated with $g' := f(g_1, \dots, g_n)$ always produce the binary clause equivalent to one of the directions of the bi-implication $g \leftrightarrow g'$. The binary clauses in $\text{TST}(\mathcal{C})$ associated with $g := f(g_1, \dots, g_n)$ together with a clause associated with $g' := f(g_1, \dots, g_n)$ will produce the other direction of the bi-implication.

Proof (Proof of Proposition 1). Assume that we have $g := \text{AND}(g_1, \dots, g_n)$ and $g' := \text{AND}(g_1, \dots, g_n)$. On the CNF-level we have the clauses $(\neg g \vee g_i), (g \vee \neg g_1 \vee \dots \vee \neg g_n)$ and $(\neg g' \vee g_i), (g' \vee \neg g_1 \vee \dots \vee \neg g_n)$, where $i = 1..n$. Now the hyper binary resolution rule allows to derive $(\neg g \vee g')$ in one step from $(\neg g \vee g_1), \dots, (\neg g \vee g_n), (g' \vee \neg g_1 \vee \dots \vee \neg g_n)$, and similarly $(\neg g' \vee g)$ in one step from $(\neg g' \vee g_1), \dots, (\neg g \vee g_n), (g \vee \neg g_1 \vee \dots \vee \neg g_n)$. The cases $f \in \{\text{NOT}, \text{OR}\}$ are similar. $\square$

Especially, by Proposition 1 hyper binary resolution can achieve the same effect purely on the CNF-level as circuit-level structural hashing on *And-Inverter Graphs* (AIGs) [9] which are often used for representing circuit-level SAT instances. We say that HBR can hence *simulate* structural hashing of AIGs.

However, HBR is not strong enough to simulate structural hashing for XOR and ITE gates on the standard CNF encoding, simply because the CNF clauses produced by the standard CNF encoding for XOR and ITE gates do not include any binary clauses.

Observation 1. *Given a Boolean circuit C with two gates $g := f(g_1, \ldots, g_n)$ and $g' := f(g_1, \ldots, g_n)$. Assume g and g' label the same function $f(g_1, \ldots, g_n)$. If $f \in \{\text{XOR}, \text{ITE}\}$ then HBR cannot in general derive $g \leftrightarrow g'$ (i.e., establish that g and g' label the same function) from $\text{TST}(C)$.*

3.2 Other Approaches to Structural Hashing on the CNF-Level

Structural Hashing and CDCL. Interestingly, CNF-level *conflict-driven clause learning* (CDCL) SAT solvers can *in principle* simulate structural hashing by learning the bi-implication $g \leftrightarrow g'$. By "in principle" we mean that this requires a CDCL solver to assign the "right" values to the "right" variables in the "right" order, and to restart after each conflict (and possibly to postpone unnecessary unit propagations).

Observation 2. *CDCL can in principle simulate structural hashing of any Boolean circuit C on $\text{TST}(C)$, assuming that the solver assigns variables optimally, restarts after every conflict, and can postpone unit propagation at will.*

The intuition behind this observation is the following. Given any Boolean circuit C containing two gates g and g', where $g := f(g_1, \ldots, g_n)$ and $g' := f(g_1, \ldots, g_n)$. For simplicity, let us assume $g := \text{AND}(g_1, \ldots, g_n)$ and $g' := \text{AND}(g_1, \ldots, g_n)$. Now apply CDCL as follows on $\text{TST}(C)$. First, assign $g = 0$. Notice that unit propagation does not assign values to any g_i based on $g := \text{AND}(g_1, \ldots, g_n)$. Then assign $g' = 1$. Now unit propagation assigns $g_i = 1$ for all $i = 1..n$, resulting in a conflict with $g = 0$. The key observation is that the standard 1-UIP clause learning scheme will now learn the clause $(g \vee \neg g')$, since this is the only 1-UIP conflict clause derivable from the conflict graph restricted to the clauses associated with $g := \text{AND}(g_1, \ldots, g_n)$ and $g' := \text{AND}(g_1, \ldots, g_n)$. Then let the solver restart, and afterward assign similarly first $g' = 0$ and then $g = 1$ in order to learn the clause $(g' \vee \neg g)$.

A similar argument goes through also for XOR and ITE but needs one more decision to learn one auxiliary clause for each of the two implications. Consider for instance $g := \text{XOR}(g_1, g_2)$ and $g' := \text{XOR}(g_1, g_2)$. Assigning $g = 0$, $g' = 1$ and then $g_2 = 0$ allows learning the clause $(g \vee \neg g' \vee g_2)$. After backtracking, unit propagation on this clause assigns $g_2 = 1$ which results in another conflict, from which one of the two implications $(g \vee \neg g')$ is learned. The other implication can be derived in a similar way.

From the practical point of view, however, it is unlikely that CDCL solver implementations would behave in the way just described.

Structural Hashing Using Ternary Resolution. Further we claim that another way of achieving structural hashing of XOR and ITE on the CNF-level is to apply *ternary resolution*, originally suggested in [13] and subsequently applied as an inference technique in the contexts of both complete [14] and local search methods [15] for CNF SAT.

Ternary resolution refers to restricting the resolution rule between two ternary clauses so that only a ternary or binary resolvent are inferred (i.e., added to the CNF).

Proposition 2. *Ternary resolution simulates structural hashing of* ITE *and* XOR.

Proof. Consider the clauses for two ITE gates $x := \text{ITE}(c, t, f)$ and $y := \text{ITE}(c, t, f)$:

$$(\neg x \vee \neg c \vee t) \wedge (\neg x \vee c \vee f) \wedge (x \vee \neg c \vee \neg t) \wedge (x \vee c \vee \neg f)$$
$$(\neg y \vee \neg c \vee t) \wedge (\neg y \vee c \vee f) \wedge (y \vee \neg c \vee \neg t) \wedge (y \vee c \vee \neg f)$$

Using ternary resolution, $(\neg x \vee y \vee \neg c) = (\neg x \vee \neg c \vee t) \bowtie (y \vee \neg c \vee \neg t)$ and $(\neg x \vee y \vee c) = (\neg x \vee c \vee f) \bowtie (y \vee c \vee \neg f)$ can be inferred. These resolvents can be combined to $(\neg x \vee y) = (\neg x \vee y \vee \neg c) \bowtie (\neg x \vee y \vee c)$. In a similar fashion, the other binary clause can be obtained: $(x \vee \neg y \vee \neg c) = (x \vee \neg c \vee \neg t) \bowtie (\neg y \vee \neg c \vee t)$ and $(x \vee \neg y \vee c) = (x \vee c \vee \neg f) \bowtie (\neg y \vee c \vee f)$. Now using these resolvents, we get $(x \vee \neg y) = (x \vee \neg y \vee \neg c) \bowtie (x \vee \neg y \vee c)$. A similar argument applies to XOR. $\square$

4   Capturing Non-transitive HBR

For the following, given a CNF formula F and two literals l and l' that occur in F, we say that l' *dominates* l (or l' is a *dominator* of l) in F if there is a clause $C = (l \vee l_1 \vee \cdots \vee l_k) \in F_{\geq 3}$ such that $(\neg l_1), \ldots, (\neg l_k) \in \text{BCP}(F_2 \cup \{(l')\})$. In other words, l' dominates l in F if there is such a clause C for which each of the literals $\neg l_i$ are reachable from l' in $\text{BIG}(F)$. This implies that by assigning $l' = 1$, unit propagation on F will assign $l = 1$ based on only F_2 and the clause C.

Example 1. Consider the formula $F = (\neg a \vee b) \wedge (\neg a \vee c) \wedge (\neg b \vee d) \wedge (\neg b \vee e) \wedge (\neg c \vee d) \wedge (\neg c \vee e) \wedge (\neg d \vee \neg e \vee f)$. A part of $\text{BIG}(F)$ with a hyperedge on the right showing the ternary clause $(\neg d \vee \neg e \vee f)$ can be illustrated as:

By assigning $a = 1$, unit propagation on F_2 and $(\neg d \vee \neg e \vee f) \in F_{\geq 3}$ will assign $d = 1$ and $e = 1$, and hence also $f = 1$. Thus a dominates f. The literal f has two other dominators: b and c, both of which are implied by a.   ∎

Given a CNF formula F and a literal l in F, the set of *non-transitive hyper binary resolvents* $\text{NHBR}(F, l)$ of F w.r.t. l is the set S of binary clauses arising from the following fixpoint computation. Let $\tau := \{l = 1\}$ and $S := \{\}$. Apply the following (non-deterministic) steps repeatedly until fixpoint:

1. While there is a unit clause $(x) \in \tau(F_2 \cup S)$, let $\tau := \tau \cup \{x = 1\}$.
2. If there is a unit clause $(y) \in \tau(F_{\geq 3})$ and literal l' with $\tau(l') = 1$ that dominates y in $F \cup S$, let $S := S \cup \{(\neg l' \vee y)\}$.

Step 1 corresponds to applying unit propagation under τ on the current set $F_2 \cup S$ of binary clauses. In step 2, it is checked whether a dominator of y has been assigned to true where y is part of a non-binary clause in F that is reduced to the unit (y) under τ. Notice that there is always at least one dominator for each $(y) \in \tau(F_{\geq 3})$, namely l; however, this is not in general the only dominator. Still, only one clause is added per execution of step 2.

Computing $\mathrm{NHBR}(F, l)$ using l as dominator was proposed in [16], while [2] discusses the use of alternative dominators. It should be noted that the above-defined construction algorithm is very similar to the one proposed in [1]. To our best understanding, the main difference is that our definition restricts step 2 to consider only units $in\ \tau(F_{\geq 3})$ in contrast to considering any units inferred by applying BCP on $\tau(F_{\geq 3})$.

In essence, the construction of $\mathrm{NHBR}(F, l)$ consists of applying lookahead on the literal l restricted to F_2, and checking for dominators w.r.t. non-binary clauses in F whenever a BCP fixpoint in reached. Notice that τ may become conflicting (i.e., both $l' = 1$ and $l' = 0$ would be assigned for some literal l') during the computation of $\mathrm{NHBR}(F, l)$. This implies that l is a failed literal, which can in practice be detected on-the-fly during the computation of $\mathrm{NHBR}(F, l)$. For the following analysis, we will always assume that l is not a failed literal.

We call a binary clause C a *non-transitive hyper binary resolvent* w.r.t. a CNF formula F if $C \in \mathrm{NHBR}(F, l)$ for some literal l in F. Given a CNF formula F, the procedure NHBR applies the following until fixpoint: while there is a non-transitive hyper binary resolvent $C \in \mathrm{NHBR}(F, l)$ w.r.t. F for some l, let $F := F \cup \{C\}$. A formula resulting from NHBR is denoted by $\mathrm{NHBR}(F)$. However, this fixpoint is not unique in general, and hence NHBR is not confluent, as will be shown in the following. Among other observations, we will also show that any $C \in \mathrm{NHBR}(F, l)$ for any F, l is indeed *non-transitive* in F, which implies that NHBR can *increase reachability* in the binary implication graph.

4.1 Understanding NHBR

Proposition 3. *For a CNF F and literal l, F is logically equivalent to $F \cup \mathrm{NHBR}(F, l)$.*

Proof. Any assignment that satisfies $F \cup \mathrm{NHBR}(F, l)$ also satisfies F. Now, assume that F is satisfiable, and fix an arbitrary truth assignment τ that satisfies F. Take an arbitrary clause $(\neg l' \vee y) \in \mathrm{NHBR}(F, l)$ with l' being a dominator of y. Notice that $l' \to y$, so $\neg y \to \neg l'$. So either $\tau(y) = 1$ or $\tau(y) = \tau(l') = 0$. Both satisfy $(\neg l' \vee y)$. Thus τ satisfies $\mathrm{NHBR}(F, l)$. □

Example 2. Let $F = (a \vee b) \wedge (a \vee \neg c \vee d) \wedge (\neg b \vee \neg c \vee e) \wedge (\neg b \vee c)$. We have $\mathrm{NHBR}(F, \neg a) = \{(a, d), (\neg b, e)\}$, which means that both of these non-transitive hyper binary resolvents can be added to F while maintaining logical equivalence. ∎

The following proposition shows that all clauses in $\mathrm{NHBR}(F, l)$ for any literal l are indeed non-transitive in F.

Proposition 4. *For any CNF F, literal l, and clause $C \in \mathrm{NHBR}(F, l)$, we have that C is not transitive in F.*

Proof. Consider the first $C = (\neg l' \vee y) \in \text{NHBR}(F, l)$ added to S during the computation of $\text{NHBR}(F, l)$. By definition, l' dominates y in F (recall step 2 of the computation of $\text{NHBR}(F, l)$), S being the empty set. Assume that C is transitive in F. It follows that there is a path from l' to y in $\text{BIG}(F)$. However, by step 1 in the computation of $\text{NHBR}(F, l)$, we would have $\tau(y) = 1$ after step 1, and hence $(y) \notin \tau(F_{\geq 3})$, and thus C would not be added to S.

The claim follows by induction using a similar argument for the $i + 1$ clause added to S assuming that the i clauses added before to S are not transitive in F.   □

This implies that, in case NHBR can add clauses to a CNF formula F, NHBR will increase reachability in the implication graph of F.

Corollary 1. *If* $\text{NHBR}(F) \setminus F \neq \emptyset$, *then it holds that there are two literals* l, l' *such that (i) there is a path in* $\text{BIG}(\text{NHBR}(F))$ *from* l *to* l', *and (ii) there is no path in* $\text{BIG}(F)$ *from* l *to* l'.

However, as an additional observation, we note by adding a clause $C \in \text{NHBR}(F, l)$ to F, some clauses in F_2 may become transitive in the resulting $F \cup \{C\}$.

Example 3. Consider the formula $F := (a \vee b) \wedge (a \vee c) \wedge (a \vee \neg b \vee d) \wedge (c \vee \neg d)$. Notice that $(a \vee d) \in \text{NHBR}(F, \neg a)$. After adding $(a \vee d)$ to F, the clause $(a \vee c)$ is transitive in the resulting formula $F \cup \{(a \vee d)\}$.   ∎

The following clarifies the connection between hyper binary resolvents and non-transitive hyper binary resolvents: in essence, NHBR is a refinement of HBR that focuses on adding the most relevant hyper binary resolvents that improve reachability in the implication graph and hence can contribute to additional unit propagations.

Proposition 5. *Given a CNF formula* F, *and a hyper binary resolvent* C *w.r.t.* F, *it holds that* C *is transitive in* F, *or that* $C \in \text{NHBR}(F, l)$ *for some literal* l.

Proof. Take an arbitrary hyper binary resolvent $C = (l \vee \neg l')$ w.r.t. a CNF formula F and let $D = (l \vee l_1 \vee \cdots \vee l_k)$ be the longest clause used in the hyper binary resolution rule to infer C. Clearly, if D is binary, then C is transitive. Now assume that $D \in F_{\geq 3}$. Because $(l \vee \neg l')$ is a hyper binary resolvent, unit propagation on $F_2 \cup \{(l')\}$ assigns all literals $l_1, \ldots, l_k$ to false. Assume that C is not transitive in F. In this case unit propagation on $F_2 \cup \{(l')\}$ will not assign l to true. Hence, after unit propagation on $F_2 \cup (l')$, $D \in F_{\geq 3}$ becomes the unit clause (l), and hence $(l \vee \neg l') \in \text{NHBR}(F, l')$.   □

As for the number of produced hyper binary resolvents, NHBR does not escape the quadratic worst-case, which, as we show, holds for all known implementations of HBR.

Proposition 6. *For CNF formulas over* n *variables,* NHBR *adds* $\Omega(n^2)$ *hyper binary resolvents in the worst-case. This holds even for formulas with* $\mathcal{O}(n)$ *clauses.*

Proof. There are $2n(n-1)$ different non-tautological binary clauses over n variables. So clearly NHBR adds only $\mathcal{O}(n^2)$ resolvents. As a worst-case example, consider the formula $F = (x_i \vee v) \wedge (x_i \vee w) \wedge (\neg v \vee \neg w \vee y_j)$ with $i, j \in \{1, \ldots, k\}$ having $2k + 2$ variables and $3k$ clauses. Since all $(x_i \vee y_j) \in \text{NHBR}(F, \neg x_i)$, NHBR will add $\Omega(k^2)$ resolvents.   □

We will now address the question of confluence of NHBR.

Proposition 7. NHBR *is not confluent.*

Proof. Consider the formula $F := (a \vee b \vee c) \wedge (\neg b \vee c) \wedge (a \vee \neg d) \wedge (c \vee d \vee e) \wedge (d \vee \neg e)$. Notice that $(a \vee c) \in \text{NHBR}(F, \neg c)$ and $(c \vee d) \in \text{NHBR}(F, \neg d)$. Furthermore, $(c \vee d) \in \text{NHBR}(F \cup \{(a \vee c)\}, \neg d)$, but $(a \vee c) \notin \text{NHBR}(F \cup \{(c \vee d)\}, \neg c)$. Therefore, the resulting formula could only contain $(a \vee c)$ if this resolvent is added before $(c \vee d)$, The reason for the non-confluence in this example is that $(a \vee c)$ is transitive in $F \cup \{(c \vee d)\}$. $\qquad\square$

Example 4. Recall step 2 of the computation of $\text{NHBR}(F, l)$. While l is always guaranteed to be a dominator of $(y) \in \tau(F_{\geq 3})$, there can be other dominators as well (recall Example 1). In case there is a dominator $l' \neq l$, then it is preferable to add $(\neg l' \vee y)$ to S instead of $(\neg l \vee y)$ in the sense that $(\neg l' \vee y)$ is not transitive in $F \cup \{(\neg l \vee y)\}$, while $(\neg l \vee y)$ is transitive in $F \cup \{(\neg l' \vee y)\}$. Recall the formula F in Example 1. The dominators of f are $\neg a$, d, and e (recall Example 1), and b and c are implied by a. Hence $\text{NHBR}(F, a) = \{(\neg a \vee f), (b \vee f), (c \vee f)\}$. Hence, instead of adding $(\neg a \vee f)$, one can add $(b \vee f)$ or $(c \vee f)$. $\qquad\blacksquare$

Although NHBR in itself is not confluent, interestingly, when combining NHBR with ELS and TRD, a unique fixpoint is reached (modulo variable renaming within literal equivalence classes). A similar observation has been previously made in [1, Theorem 1] for the combination of HBR and ELS alone without TRD.

Proposition 8. *For any CNF formula F,* NHBR *followed by the combination of* ELS *and* TRD *until fixpoint is confluent (modulo variable renaming).*

Proof. (sketch) Given any CNF formula F, the implication graph of $\text{ELS}(F)$ is acyclic, and hence $\text{TRD}(\text{ELS}(F))$ is unique (modulo variable renaming). Now assume that there are two literals l, l' and clauses C, C' such that $C \in \text{NHBR}(F, l)$ and $C' \in \text{NHBR}(F, l')$. Assume that C' is transitive in $F \cup \{C\}$ and that C is not transitive in $F \cup \{C'\}$. If NHBR adds the clauses to F in the order C', C, TRD will afterwards remove the transitive C' from $F \cup \{C', C\}$, resulting in $F \cup \{C\}$ to which NHBR would not add C. Finally, since NHBR can only increase reachability in the implication graph, NHBR will not re-introduce any previously added clauses that may have been afterwards removed by TRD. $\qquad\square$

Example 5. As a concrete example, recall that the reason for the non-confluence in the proof of Proposition 7 is that $(a \vee c)$ is transitive in $F \cup \{(c \vee d)\}$. However, a unique result is obtained by applying TRD after NHBR.

5 Realizing Non-transitive HBR

Apart from the classical BCP that removes satisfied clauses and falsified literals, the variant BCP_{NHBR} efficiently adds *non-transitive* hyper binary resolvents by prioritizing binary clauses during propagation. The fact that hyper binary resolution can be achieved through unit propagation is due to [1] and has been extended in [2]. Another extension, called lazy hyper binary resolution (LHBR) [17] is discussed in Sect. 8.

The pseudo-code of BCP_{NHBR} is shown in Fig. 1. Besides a formula F and a literal l, it takes a truth assignment τ (here interpreted as a *stack* of variable-value assignments) as input. For wellformedness, it is required that all assignments in τ are implied by $l = 1$ using only binary clauses. That is, all literals in τ can be reached from l in $\text{BIG}(F)$.

BCP_{NHBR} (formula F, truth assignment τ, literal l)

1      $\tau.push(l = 1)$
2   **while** $\tau(F)$ contains unit clauses **do**
3   **while** $(l') \in \tau(F_2)$ **do** $\tau.push(l' = 1)$
4   **if** $(l'') \in \tau(F_{\geq 3})$ **then** $F := F \cup \{(\neg l \vee l'')\}$
5   **return** $\langle F, \tau \rangle$

Fig. 1. Pseudo-code of the BCP_{NHBR} procedure

First, the input literal l is set to true on the assignment stack τ (line 1). As long as unit clauses exist (line 2), propagation of binary clauses is prioritized (line 3). If there are only unit clauses left originating from $F_{\geq 3}$, then a random one is selected and converted into a non-transitive hyper binary resolvent (line 4). In the end, the resulting formula and extended assignment are returned (line 5).

In practice, implementing BCP_{NHBR} can be expensive. To reduce the computational costs, [2] proposes two optimizations. The first, using alternative dominators is discussed in Sect. 4. The second is restricting computation to $C \in \text{NHBR}(F, l)$ with $l \in \text{RTS}(F)$ (i.e., starting only from literals that are roots in the implication graph). This restriction reduces the costs significantly. For FLE, starting only from literals $l \in \text{RTS}(F)$, will not change the fixpoint [18,19]. Yet, this is not the case for NHBR.

Proposition 9. *By restricting* NHBR *to add only* $C \in \text{NHBR}(F, l)$ *with* $l \in \text{RTS}(F)$, *some non-transitive hyper binary resolvents will not be added.*

Proof. Consider formula $F = (\neg a \vee b) \wedge (\neg a \vee c) \wedge (\neg c \vee d) \wedge (b \vee \neg c \vee \neg d)$. Notice that $(b \vee \neg c) \in \text{NHBR}(F, c)$, while for all $l \in \text{RTS}(F)$ holds that $\text{NHBR}(F, l) = \emptyset$.   □

In the following section, we will discuss an alternative technique, namely, *tree-based lookahead*, that can be used to efficiently compute $\text{NHBR}(F)$ till fixpoint.

6   Tree-Based Lookahead

Tree-based lookahead originates from [3] but has not been properly described in the literature yet. It is a technique to reduce the computational cost to find failed literals and non-transitive hyper binary resolvents by reusing propagations. For some intuition about how this technique works, consider a CNF formula F which contains a binary clause $(\neg a \vee b)$ and several other clauses. Due to the presence of $(\neg a \vee b)$, we know that when propagating $a = 1$, b is forced to 1 as well as all variables that would have been forced by $b = 1$. It is possible to reuse the propagations of $b = 1$ (i.e., without rerunning BCP), by assigning $a = 1$ afterwards *without* unassigning the forced variables. If there is another binary clause $(\neg c \vee b)$, then the effort of propagating $b = 1$ can additionally be shared with the effort of propagating $c = 1$ after backtracking over $a = 1$ and then assigning $c = 1$ without backtracking the assignments implied by $b = 1$.

This concept can be generalized by decomposing $\mathrm{BIG}(F)$ into in-trees: trees in which edges are oriented so that the root is reachable from all nodes (the root has out-degree 0 and other nodes have out-degree 1). For each implication $x \to y$ in the in-trees, y is assigned before x. Note that in-trees in the in-tree decomposition are almost never induced subgraphs, e.g. they are missing some edges; there are edges of $\mathrm{BIG}(F)$ that are not part of any in-tree, and even might connect two different in-trees.

The first step in tree-based lookahead is to create the in-trees, which is realized by the *getQueue* procedure, shown in Fig. 2. First queue Q is initialized and all cycles in $\mathrm{BIG}(F)$ are removed using ELS. Note that applying ELS once to F might produce new binary clauses by shrinking longer clauses, and even introduce new cycles. We thus have to run this process until completion.

Afterwards a random depth-first search is applied starting from the leafs of $\mathrm{BIG}(F)$. Notice that if $\neg l \in \mathrm{RTS}(F)$, l is a leaf. In the *enqueue* procedure, first l is added to Q followed by a recursive call for all literals that imply l and are not in the queue yet. The procedure ends adding the special element $\triangledown$ to Q that denotes that the algorithm should backtrack if that element is dequeued. The resulting Q contains each literal l in F exactly once, and for each literal occurring in Q the special element $\triangledown$ occurs once.

getQueue (F)		enqueue (F, Q, l)	
1	$Q := \{\}$	1	$Q.enqueue(l)$
2	**while** $\mathrm{ELS}(F) \neq F$ **do**	2	**foreach** $(l \vee \neg l') \in F_2$ **do**
3	$\quad F := \mathrm{ELS}(F)$	3	$\quad$ **if** $l' \notin Q$ **then**
4	**foreach** $\neg l \in \mathrm{RTS}(F)$ **do**	4	$\quad\quad Q := enqueue(F, Q, l')$
5	$\quad Q := enqueue(F, Q, l)$	5	$Q.enqueue(\triangledown)$
6	**return** Q	6	**return** Q

Fig. 2.   Left: the *getQueue* procedure.   Right: the *enqueue* sub-procedure.

The *TreeLook* algorithm (Fig. 3) uses the queue Q to compute failed literals and non-transitive hyper binary resolvents efficiently. After initialization (line 1 and 2), it dequeues elements from Q until it is empty (line 3 and 4). In case the current literal l is not $\triangledown$ (line 5), the decision level is increased by pushing $*$ on the assignment stack τ (line 6). If l is assigned to 0 or the current assignment τ falsifies F then the failed literal $(\neg l)$ is found (line 7). Otherwise, if l is still unassigned (line 8), then it is assigned to 1, followed by $\mathrm{BCP}_{\mathrm{NHBR}}$ prioritizing binary clauses, under which unit clauses that originate from non-binary clauses are transformed into a non-transitive hyper binary clause (line 9). If BCP results in a conflict, then a failed literal is found (line 10). Each time the element $\triangledown$ is dequeued, the algorithm backtracks one level, by popping elements from τ until it removes $*$ (line 11). Finally, the resulting F, simplified with failed literals and strengthened by non-transitive hyper binary resolvents (which may be trivially unsatisfiable (line 12)), is returned (line 13).

Example 6. Consider $F = (\neg a \vee \neg b) \wedge (b \vee \neg c \vee e) \wedge (b \vee c) \wedge (c \vee d) \wedge (a \vee \neg d \vee \neg e)$. NHBR can add two clauses to F: $\mathrm{NHBR}(F, b) = \{(b \vee e)\}$ and $\mathrm{NHBR}(F, c) = \{(c \vee \neg e)\}$. The *TreeLook* (F) algorithm can find them as follows. Assume that the result of *getQueue*

TreeLook (formula F)

```
1      τ := {}
2      Q := getQueue(F)
3      while Q is not empty do
4          l := Q.dequeue()
5          if l ≠ ▽ then
6              τ.push(*)
7              if τ(l) = 0 or ∅ ∈ τ(F) then F := BCP(F ∪ {¬l})
8              else if τ(l) ≠ 1 then
9                  ⟨F, τ⟩ := BCP_NHBR(F, τ, l)
10                 if ∅ ∈ τ(F) then F := BCP(F ∪ {¬l})
11         else while τ.pop() ≠ *
12         if ∅ ∈ F then break
13     return F
```

Fig. 3. The *TreeLook* algorithm

(F) is $Q = \{c, \neg b, a, \triangledown, \triangledown, \neg d, \triangledown, \triangledown, d, \neg c, \triangledown, \triangledown, \neg a, b, \triangledown, \triangledown\}$, visiting the leafs in the order c, d, $\neg a$. This Q partitions $\mathrm{BIG}(F)$ (see Fig. 4) in three in-trees by removing the dotted edge $\neg c \dashrightarrow b$. After initialization, τ is extended by pushing $*$ and $c = 1$. This does not result in any units. Now, τ is extended with $*$ and $b = 0$. The clause $(b \vee \neg c \vee e) \in F_{\geq 3}$ becomes unit. Therefore $(b \vee e)$ is added to F, which is unit (e) by construction under τ. Hence τ is extended by $e = 1$. Afterwards, $*$ and $a = 1$ are pushed to τ. No new units exist in $\tau(F)$ and the next element in Q is dequeued which is $\triangledown$. This causes popping $a = 1$ and $*$ from τ as the first backtracking step. The next element is also zero, which pops $e = 1$, $b = 0$, and $*$ from τ. Extending the shrunken τ by pushing $*$ and $d = 0$, does not result in any unit in $\tau(F)$. The first in-tree is now finished and the algorithm will pop all elements from τ due to the double $\triangledown$ element dequeued from Q. The NHBR $(c \vee \neg e)$ is found in the second in-tree: after $d = 1$ and $c = 0$, τ is extended by $b = 1$ and $a = 0$. Now $(a \vee \neg d \vee \neg e) \in F_{\geq 3}$ is unit $(\neg e)$ under τ. The third in-tree does not add any clause to F. Notice that after adding both binaries to F, $(b \vee c)$ becomes redundant (transitive) as well as $(b \vee \neg c \vee e)$ (subsumed). ∎

7   Experiments

The *TreeLook* algorithm (Fig. 3) is implemented in the MARCHRW SAT solver [20]. The SAT Competition version of MARCHRW runs FLE until completion in each node

Fig. 4. $\mathrm{BIG}(F)$ in Example 6 before (left) and after (right) applying NHBR on F. Both graphs have three leafs: $\neg a$, c, and d. The dotted edge $\neg c \dashrightarrow b$ is not in the in-tree decomposition.

of the search-tree. We slightly modified the code such that it runs NHBR until completion. In other words, lookahead runs until no new non-transitive hyper binary resolvent is found, instead of no new failed literal. The resulting version, called MARCHNH (benchmarks, sources and logs at http://fmv.jku.at/treelook), also has the ability to output the formula after preprocessing, so the result is similar to existing implementations of HBR, HYPRE [1] and HYPERBINFAST [2]. The experiments were done on a cluster of computing nodes with Intel Core 2 Duo Quad Q9550 2.8-GHz processors, 8-GB main memory, under Ubuntu Linux. Memory was limited to 7 GB and a timeout of 10 h was enforced for each run.

As benchmarks we used all 818 sequential circuits of the Hardware Model Checking Competition 2010 http://fmv.jku.at/hwmcc10. A miter was constructed from each circuit by connecting the inputs (and latches) of two copies of the same circuit, and by constraining outputs and next state functions to be pairwise equivalent. We used aigmiter for constructing the miters, and translated them to CNF with aigtocnf. Both tools are available from http://fmv.jku.at/aiger. Note that these benchmarks are trivial on the AIG level and can simply be solved by structural hashing. Actually, a non-optimized implementation of structural hashing needs less than 13 seconds for all 818 benchmarks, and less than half a second for the most difficult one (intel048 with 469196 variables and 1300546 clauses).

Running times for the hardest benchmarks using logarithmic scale are shown in Fig. 5: NHBR through tree-based lookahead (MARCHNH with TreeLook) can solve all of

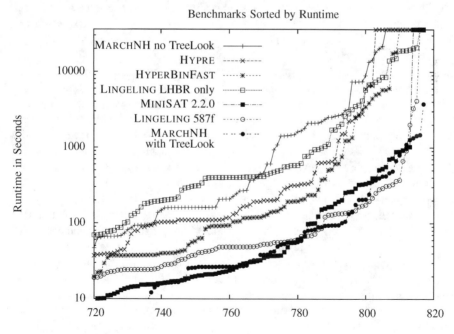

Fig. 5. Runtimes of CNF solving tools on 818 instances generated from HWMCC 2010. The plot starts at 750 because many instances could be solved easily. Notice that only LINGELING and MINISAT perform search. The other tools run NHBR till unsatisfiability is detected.

benchmarks on its own (i.e., without any additional search). Switching off tree-based lookahead (MARCHNH no TreeLook), i.e. always applying $BCP_{NHBR}(F, \tau, l)$ with $\tau = \emptyset$, the timeout is reached on eleven benchmarks and is two orders of magnitude slower. In between are the results of the previous implementations of HBR including LHBR (see next Sect. 8), which take much more time and memory, even though they use ELS. HYPRE hits the memory limit on 15 miters, HYPERBINFAST runs out of memory on eight, and LINGELING (LHBR only) runs out of time on two. Surprisingly state-of-the-art CDCL SAT solvers such as LINGELING 587F and MINISAT 2.2.0 can not solve some of the miters even within 10 hours of search (LINGELING could not solve two miters, MINISAT four).

Although not the main focus here, we also measured the effect of applying NHBR as a preprocessing technique for SAT Competition 2011 application instances. For a clean experiment, we compared plain Lingeling (no pre- and inprocessing) with and without NHBR. With NHBR, Lingeling solved 7 more instances.

8   Related Work and Existing Implementations

A version of the SAT solver PRECOSAT [17] submitted to the SAT Competition 2009 contained an algorithm for cheaply computing hyper binary resolvents *on-the-fly* during BCP in a standard CDCL solver on all decision levels. This method was called *lazy hyper binary resolution* (LHBR), and a preliminary version was implemented in PICOSAT [21] before. It has since then been ported to many other recent SAT solvers, including CIRCUS [22], LINGELING [23], and CRYPTOMINISAT [24]. Extensions of LHBR including a detailed empirical analysis of its benefits, can be found in [22].

The basic idea of LHBR is to restrict the implication graph, made of assigned literals and their forcing antecedents resp. reason clauses, to binary clauses. The implication graph is in general a DAG and the restriction to binary clauses turns it into a forest of trees, which we call *binary implication forest*. This allows us to save for each assigned variable the root of its binary implication tree. If a literal is implied by a non-binary clause, and all its antecedent literals in this clause are in the same tree, or equivalently they have the same root, a binary clause through LHBR is obtained. This can be checked by scanning the forcing non-binary clause, and checking whether all its variables, except the implied one, have the same root. If this is the case, the closest dominator of the antecedents can be computed as least-common ancestor in the tree.

The binary clause derived through LHBR is used as reason instead of the originally forcing non-binary clause, which extends the binary implication tree of the antecedents. It adds an edge from the dominator to the newly forced literal. To avoid adding too many transitive clauses, propagation over binary clauses is run until completion for all assigned literals before non-binary clauses are considered for propagation. This form of LHBR adds a negligible overhead to BCP, because checking for a common root among antecedent literals is cheap and only has to be performed if a non-binary clause becomes forcing. Thus, from the point of view of effectiveness, ease of implementation, and overhead, LHBR is comparable to on-the-fly *self*-subsumption [25]. One difference though is, that the former is implemented as part of BCP and the latter in the analysis algorithm for learning clauses from conflicts.

In practice we observed that the vast majority of binary clauses derived through LHBR are obtained during failed literal probing resp. lookahead at decision level 0 anyhow. Thus a simpler implementation similar to the one used in lookahead solvers *including tree-based lookahead* already gives the largest benefit without the need to store roots of the binary implication forest. The additional advantage of using LHBR even during search is to cheaply learn binary clauses at all decision levels, which are valid globally and can be added permanently. In lookahead solvers binary clauses learned through LHBR have to be removed during backtracking.

The competition version LINGELING 587f used in Sect. 7 uses LHBR during failed literal probing. This time-limited lookahead is one of the many implemented pre- resp. in-processing techniques [26]. We patched LINGELING to run LHBR until completion (http://fmv.jku.at/lingeling/lingeling-587f-lhbrtc.patch) on these instances, but as shown in the experiments the run-times were much worse, even with (full) ELS and (time-limited) TRD.

Recursive Learning [27] and Stålmarck's method [12] work on circuits resp. on data structures (triplets) close to circuits and can easily be combined with structural hashing. This leads to an algorithm similar to congruence closure algorithms used in SMT solvers [28]. There are versions of both Stålmarck's method and Recursive Learning working directly on CNF [29,30]. In both cases only boolean constants are propagated and not equivalences as in the original method of Stålmarck. We conjecture that a combination of these CNF techniques with equivalence reasoning would also simulate structure hashing, but we are not aware of published results along this line.

9 Conclusions

We focused on understanding how non-transitive hyper binary resolvents can be efficiently exploited on the CNF-level. We explained how hyper binary resolution can be implemented through tree-based lookahead, which allows to simulate structural hashing on the CNF-level also in practice much more efficiently than previous CNF-level solutions. As a side-result, we believe our explanation of tree-based lookahead is of independent interest, providing an efficient way of implementing lookahead, which is important for example in the recently proposed *cube & conquer* approach [31].

The motivation for tree-based look-ahead was originally twofold. First, it provides an efficient implementation technique for failed literal probing during pre- and inprocessing [26]. This was the focus of this paper. Second, tree-based look-ahead can also be used to efficiently compute look-ahead heuristics, such as the number of clauses reduced to binary clauses after assuming and propagating a literal. It is unclear at this point whether the second motivation is really important, or whether other cheaper-to-compute metrics could also be used.

While our *TreeLook* implementation significantly improves over existing CNF-level approaches, there is still a large gap between the efficiency of circuit-level structural hashing and of using CNF reasoning alone for identifying equivalences. Future work consists of closing this gap further. As a final remark, as also pointed out by anonymous reviewers, it should be possible to reformulate tree-based lookahead for applying singleton arc consistency in CP and probing in MIP solvers.

Acknowledgements. We thank Donald Knuth for detailed comments and suggestions on a draft version of this paper.

References

1. Bacchus, F., Winter, J.: Effective preprocessing with hyper-resolution and equality reduction. In: Giunchiglia, E., Tacchella, A. (eds.) SAT 2003. LNCS, vol. 2919, pp. 341–355. Springer, Heidelberg (2004)
2. Gershman, R., Strichman, O.: Cost-effective hyper-resolution for preprocessing CNF formulas. In: Bacchus, F., Walsh, T. (eds.) SAT 2005. LNCS, vol. 3569, pp. 423–429. Springer, Heidelberg (2005)
3. Heule, M., Dufour, M., van Zwieten, J., van Maaren, H.: March_eq: Implementing additional reasoning into an efficient look-ahead SAT solver. In: Hoos, H.H., Mitchell, D.G. (eds.) SAT 2004. LNCS, vol. 3542, pp. 345–359. Springer, Heidelberg (2005)
4. Heule, M.J.H., van Maaren, H.: Chapter 5: Look-Ahead Based SAT Solvers. In: Handbook of Satisfiability, pp. 155–184. IOS Press (2009)
5. Van Gelder, A.: Toward leaner binary-clause reasoning in a satisfiability solver. Annals of Mathematics and Artificial Intelligence 43, 239–253 (2005)
6. Aho, A., Garey, M., Ullman, J.: The transitive reduction of a directed graph. SIAM Journal on Computing 1(2), 131–137 (1972)
7. Bacchus, F.: Enhancing Davis Putnam with extended binary clause reasoning. In: Proc. AAAI, pp. 613–619. AAAI Press (2002)
8. Bryant, R.E.: Graph-based algorithms for boolean function manipulation. IEEE Trans. Computers 35(8), 677–691 (1986)
9. Kuehlmann, A., Krohm, F.: Equivalence checking using cuts and heaps. In: Proc. DAC, pp. 263–268. ACM (1997)
10. Williams, P.F., Andersen, H.R., Hulgaard, H.: Satisfiability checking using boolean expression diagrams. STTT 5(1), 4–14 (2003)
11. Abdulla, P.A., Bjesse, P., Eén, N.: Symbolic reachability analysis based on SAT-solvers. In: Graf, S., Schwartzbach, M. (eds.) TACAS/ETAPS 2000. LNCS, vol. 1785, pp. 411–425. Springer, Heidelberg (2000)
12. Sheeran, M., Stålmarck, G.: A tutorial on Stålmarck's proof procedure for propositional logic. Formal Methods in System Design 16(1), 23–58 (2000)
13. Billionnet, A., Sutter, A.: An efficient algorithm for the 3-satisfiability problem. Operations Research Letters 12(1), 29–36 (1992)
14. Li, C.M., Anbulagan: Look-ahead versus look-back for satisfiability problems. In: Smolka, G. (ed.) CP 1997. LNCS, vol. 1330, pp. 341–355. Springer, Heidelberg (1997)
15. Anbulagan, Pham, D.N., Slaney, J.K., Sattar, A.: Old resolution meets modern SLS. In: Proc. AAAI, pp. 354–359 (2005)
16. Heule, M.J.H.: March: Towards a lookahead SAT solver for general purposes, MSc thesis (2004)
17. Biere, A.: P{re,i}coSAT@SC'09. In: SAT 2009 Competitive Event Booklet (2009)
18. Boufkhad, Y.: Aspects probabilistes et algorithmiques du problème de satisfaisabilité, PhD thesis, Univertsité de Paris 6 (1996)
19. Simons, P.: Towards constraint satisfaction through logic programs and the stable model semantics, Report A47, Digital System Laboratory, Helsinki University of Technology (1997)
20. Mijnders, S., de Wilde, B., Heule, M.J.H.: Symbiosis of search and heuristics for random 3-SAT. In: Proc. LaSh (2010)
21. Biere, A.: (Q)CompSAT and (Q)PicoSAT at the SAT'06 Race (2006)

22. Han, H., Jin, H., Somenzi, F.: Clause simplification through dominator analysis. In: Proc. DATE, pp. 143–148. IEEE (2011)
23. Biere, A.: Lingeling, Plingeling, PicoSAT and PrecoSAT at SAT Race 2010. FMV Report Series TR 10/1, JKU, Linz, Austria (2010)
24. Soos, M.: CryptoMiniSat 2.5.0, SAT Race'10 solver description (2010)
25. Han, H., Somenzi, F.: On-the-fly clause improvement. In: Kullmann, O. (ed.) SAT 2009. LNCS, vol. 5584, pp. 209–222. Springer, Heidelberg (2009)
26. Järvisalo, M., Heule, M.J.H., Biere, A.: Inprocessing rules. In: Gramlich, B., Miller, D., Sattler, U. (eds.) IJCAR 2012. LNCS, vol. 7364, pp. 355–370. Springer, Heidelberg (2012)
27. Kunz, W., Pradhan, D.K.: Recursive learning: a new implication technique for efficient solutions to CAD problems-test, verification, and optimization. IEEE T-CAD 13(9) (1994)
28. Barrett, C.W., Sebastiani, R., Seshia, S.A., Tinelli, C.: Chpt. 26: SMT Modulo Theories. In: Handbook of Satisfiability. IOS Press (2009)
29. Marques-Silva, J., Glass, T.: Combinational equivalence checking using satisfiability and recursive learning. In: Proc. DATE (1999)
30. Groote, J.F., Warners, J.P.: The propositional formula checker HeerHugo. J. Autom. Reasoning 24(1/2), 101–125 (2000)
31. Heule, M.J.H., Kullmann, O., Wieringa, S., Biere, A.: Cube and conquer: Guiding CDCL SAT solvers by lookaheads. In: Eder, K., Lourenço, J., Shehory, O. (eds.) HVC 2011. LNCS, vol. 7261, pp. 50–65. Springer, Heidelberg (2012)

Decision Diagrams and Dynamic Programming[*]

John N. Hooker

Carnegie Mellon University, USA
jh38@andrew.cmu.edu

Abstract. Binary and multivalued decision diagrams are closely related to dynamic programming (DP) but differ in some important ways. This paper makes the relationship more precise by interpreting the DP state transition graph as a weighted decision diagram and incorporating the state-dependent costs of DP into the theory of decision diagrams. It generalizes a well-known uniqueness theorem by showing that, for a given optimization problem and variable ordering, there is a unique reduced weighted decision diagram with "canonical" edge costs. This can lead to simplification of DP models by transforming the costs to canonical costs and reducing the diagram, as illustrated by a standard inventory management problem. The paper then extends the relationship between decision diagrams and DP by introducing the concept of nonserial decision diagrams as a counterpart of nonserial dynamic programming.

1 Introduction

Binary and multivalued decision diagrams have long been used for circuit design and verification, but they are also relevant to optimization. A reduced decision diagram can be viewed as a search tree for an optimization problem in which isomorphic subtrees are superimposed, thus removing redundancy.

Dynamic programming (DP) is based on a similar idea. In fact, the state transition graph for a discrete DP can be viewed as a decision diagram, albeit perhaps one in which not all redundancy has been removed. Conversely, the reduced decision diagram for a given problem tends to be more compact when the problem is suitable for solution by DP. This indicates that there may be benefit in clarifying the connection between decision diagrams and DP. In particular, it may be possible to simplify a DP model by regarding its transition graph as a decision graph and reducing it to remove all redundancy.

However, decision diagrams differ from DP in significant ways. Nodes of the DP state transition graph are associated with state variables, whereas there are no explicit state variables in a decision diagram; only decision variables. Furthermore, arcs of a state transition graph are often labeled with costs, and this is not true of a decision diagram. A decision diagram can be given arc costs when the objective function is separable, but costs in a DP transition graph are more complicated because they are *state dependent*: they depend on state variables as well as decision variables.

---

[*] Partial support from NSF grant CMMI-1130012 and AFOSR grant FA-95501110180.

C. Gomes and M. Sellmann (Eds.): CPAIOR 2013, LNCS 7874, pp. 94–110, 2013.

Nonetheless, we show that the elementary theory of decision diagrams can be extended to incorporate state-dependent costs and therefore establish a close parallel with DP. We define a *weighted* decision diagram to be a decision diagram with arbitrary arc costs. Unfortunately, differing arc costs can prevent reduction of a weighted diagram even when the unweighted diagram would reduce. However, we show that costs can often be rearranged on the diagram, without changing the objective function, so as to allow reduction. In fact, we define a unique *canonical* set of arc costs for a given objective function and generalize a well-known uniqueness result for reduced decision diagrams. We show that for a given optimization problem and variable ordering, there is a unique reduced weighted decision diagram with canonical costs that represents the problem.

This opens the possibility of simplifying a DP formulation by converting the transition costs to canonical costs and reducing the state transition diagram that results. In fact, we show this maneuver results in a substantial simplification even for a standard DP formulation of production and inventory management that has appeared in textbooks for decades.

We conclude by extending weighted decision diagram to *nonserial decision diagrams* by exploiting an analogy with nonserial DP.

2   Previous Work

Binary decision diagrams were introduced by [1, 19, 31]. In recent years they have been applied to optimization, initially for cut generation in integer programming [9, 11], post-optimality analysis [25, 26], and 0-1 vertex and facet enumeration [10]. Relaxed decision diagrams were introduced in [3] and further applied in [21, 27, 28] as an alternative to the domain store in constraint programming, and they were used in [14, 15] to obtain bounds for optimization problems. Introductions to decision diagrams can be found in [2, 18].

Dynamic programming is credited to Bellman [12, 13]. A good introductory text is [24], and a more advanced treatment [17]. Nonserial dynamic programming was introduced by [16]. Essentially the same idea has surfaced in a number of contexts, including Bayesian networks [30], belief logics [33, 34], pseudoboolean optimization [22], location theory [20], k-trees [4, 5], and bucket elimination [23].

The identification of equivalent subproblems is known as *caching* in the knowledge representation literature, where it has received a good deal of attention (e.g., [6–8, 29]). However, apparently none of this work deals with state-dependent costs, which are a unique and essential feature of DP, as it is understood in the operations research community.

3   Decision Diagrams

For our purposes, decision diagrams can be viewed as representing the feasible set S of an optimization problem

$$\min_x \{f(x) \mid x \in S\} \tag{1}$$

Table 1. (a) A small set covering problem. The dots indicate which elements belong to each set i. (b) A nonseparable cost function for the problem. Values are shown only for feasible x.

(a)

Set i	1	2	3	4
A	•	•		
B	•		•	•
C		•	•	
D		•		•

(b)

x	$f(x)$
(0,1,0,1)	6
(0,1,1,0)	7
(0,1,1,1)	8
(1,0,1,1)	5
(1,1,0,0)	6
(1,1,0,1)	8
(1,1,1,0)	7
(1,1,1,1)	9

where $f(x)$ is the objective function, and x a tuple $(x_1, \ldots, x_n)$ of discrete variables with finite domains $D_1, \ldots, D_n$, respectively.

An *ordered decision diagram* is a directed, acyclic graph $G = (N, A)$ whose node set N is partitioned into n *layers* $1, \ldots, n$ corresponding to the variables $x_1, \ldots, x_n$, plus a *terminal* layer (layer $n + 1$). Layer 1 contains only a root node r, and the terminal layer contains nodes 0 and 1. For each node u in layer $i \in \{1, \ldots, n\}$ and each value $d_i \in D_i$, there is a directed arc $a(u, d_i)$ in A from u to a node $u(d_i)$ in layer $i+1$, which represents setting $x_i = d_i$. Each path from r to 1 represents a feasible solution of S, and each path from r to 0 represents an infeasible solution. For our purposes, it is convenient to omit the paths to 0 and focus on the feasible solutions.

As an example, consider a set covering problem in which there are four sets as indicated in Table 1(a), collectively containing the elements A, B, C, D. The problem is to select a minimum-cost *cover*, which is a subcollection of sets whose union is $\{A, B, C, D\}$. Let binary variable x_i be 1 when set i is selected, and let $f(x)$ be the cost of subcollection x. The decision diagram in Fig. 1(a) represents the feasible set S. The 9 paths from r to 1 represent the 9 covers.

A decision diagram is a *binary decision diagram* if each domain D_i contains two values, as in the example of Fig. 1. It is a *multivalued decision diagram* if at least one D_i contains three or more values. Decision diagrams can be defined to contain *long arcs* that skip one or more layers, but to simplify notation, we suppose without loss of generality that G contains no long arcs.

A decision diagram is *reduced* when it is a minimal representation of S. To make this more precise, let $G_{uu'}$ be the subgraph of G induced by the set of nodes on paths from u to u'. Subgraphs $G_{uu'}$ and $G_{vv'}$ are *equivalent* when they are isomorphic, corresponding arcs have the same labels, and u, v belong to the same layer. A decision diagram is reduced if it contains no equivalent subgraphs. It is a standard result [19, 35] that there is a unique reduced diagram for any given S and variable order.

The reduced decision diagram can be obtained from a branching tree using a simple procedure. Supposing that the tree contains only the feasible leaf nodes, we first superimpose all the leaf nodes to obtain terminal node 1, and then continue to superimpose equivalent subgraphs until none remain. For example, the branching tree for the set covering problem of Table 1(a) appears in Fig. 2 (ignore the arc labels at the bottom). The tree can be transformed in this manner to the reduced decision diagram of Fig. 1(a).

4 Weighted Decision Diagrams

Given an optimization problem (1), we would like to assign costs to the arcs of a decision diagram to represent the objective function. We will refer to a decision diagram with arc costs as a *weighted* decision diagram. Such a diagram represents (1) if the paths from r to 1 represent precisely the feasible solutions of (1), and length of each path x is $f(x)$. The optimal value of (1) is therefore the shortest path length from r to 1. We will say that two weighted decision diagrams are *isomorphic* if they yield the same unweighted diagram when arc costs are removed.

The assignment of costs to arcs is most straightforward when the objective function is separable. If $f(x) = \sum_i f_i(x_i)$, we simply assign cost $f_i(d_i)$ to each arc $a(u, d_i)$ leaving layer i in the reduced decision diagram representing S. For example, if $f(x) = 3x_1 + 5x_2 + 4x_3 + 6x_4$, the arc costs are as shown in Fig. 1(b), and the shortest path is $x = (0, 1, 0, 1)$ with cost 11.

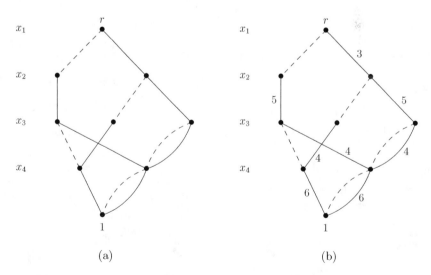

(a)   (b)

Fig. 1. (a) Decision diagram for the set covering problem in Table 1(a). Dashed arcs correspond to setting $x_i = 0$, and solid arcs to setting $x_i = 1$. (b) Decision diagram showing arc costs for a separable objective function. Unlabeled arcs have zero cost.

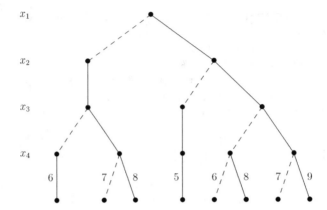

Fig. 2. Branching tree for the set covering problem in Table 1(a). Only feasible leaf nodes are shown.

Arc costs can also be assigned when the objection function is nonseparable. In fact, we will show that the problem is represented by a unique reduced weighted decision diagram, provided arc costs are assigned in a canonical way.

Consider, for example, the nonseparable objective function of Table 1(b). There are many ways to capture the function by putting arc costs on the branching tree in Fig. 2. The most obvious is to assign the value corresponding to each leaf node to the incoming arc, as shown in the figure, and zero cost on all other arcs. However, this assignment tends to result in a large decision diagram.

To find a reduced diagram, we focus on *canonical* arc costs. An assignment of arc costs to a tree or a decision diagram is canonical if for every level $i \geq 2$, the smallest cost on arcs $a(u, d_i)$ leaving any given node u is some predefined value α_i. In the simplest case $\alpha_i = 0$ for all i, but it is convenient in applications to allow other values. We first show that canonical costs are unique.

Lemma 1. *For any given decision diagram or search tree representing a feasible set S, there is at most one canonical assignment of arc costs to G that represents the optimization problem (1).*

Proof. To simplify notation, we assume without loss of generality that each $\alpha_i = 0$. Given node u in layer i, let $c(u, d_i)$ be the cost assigned to arc $a(u, d_i)$, and let $L_u(\bar{d})$ be the length of a path $\bar{d} = (\bar{d}_i, \dots, \bar{d}_n)$ from u to 1 (in a decision diagram) or to a leaf node (in a tree). We show by induction on $i = n, n-1, \dots, 1$ that for any node u, $L_u(\bar{d})$ for any such path $\bar{d}$ is uniquely determined if the arc costs are canonical. It follows that all the arc costs are uniquely determined. First consider any node u on layer n. For any path $d = (d_1, \dots, d_{n-1})$ from r to u, we must have

$$f(d, d_n) - f(d, d'_n) = c(u, d_n) - c(u, d'_n)$$

for any pair of arcs $(u, u(d_n))$, $(u, u(d'_n))$. Because a canonical assignment satisfies $\min_{d_n}\{c(u, d_n)\} = 0$, each of the costs $c(u, d_n)$ is uniquely determined.

For $i = n, n - 1, \ldots, 2$:
  For each node u on layer i:
    Let $c_{\min}(u) = \min_{u' \in U_{\text{out}}} \{c_{uu'}\}$
    For each $u' \in U_{\text{out}}$:
      Let $c_{uu'} \leftarrow c_{uu'} - c_{\min} + \alpha_i$.
    For each $u' \in U_{\text{in}}$:
      Let $c_{u'u} \leftarrow c_{u'u} + c_{\min} - \alpha_i$.

Fig. 3. Algorithm for converting arc costs to canonical arc costs. Here, U_{out} is the set of child nodes of node u, and U_{in} is the set of parent nodes of u.

Now for any node u in layer $i \in \{1, \ldots, n-1\}$, suppose that $L_{u(d_i)}(\bar{d})$ is uniquely determined for any arc $a(u, d_i)$ and any path $\bar{d}$ from $u(d_i)$ to 1 or a leaf node. Then for any path $d = (d_1, \ldots, d_{i-1})$ from r to u and for any pair $(d_i, \bar{d})$, $(d'_i, \bar{d}')$ of paths from u to 1 or a leaf node, we must have

$$f(d, d_i, \bar{d}) - f(d, d'_i, \bar{d}') = L_u(d_i, \bar{d}) - L_u(d'_i, \bar{d}')$$
$$= (c(u, d_i) + L_{u(d_i)}(\bar{d})) - (c(u, d'_i) + L_{u(d'_i)}(\bar{d}'))$$
$$= c(u, d_i) - c(u, d'_i) + \Delta$$

where Δ is uniquely determined, by the induction hypothesis. This and the fact that $\min_{d_i} \{c(u, d_i)\} = 0$ imply that the arc costs $c(u, d_i)$ are uniquely determined. So $L_u(d_i, \bar{d})$ is uniquely determined for any $(d_i, \bar{d})$, as claimed. $\square$

Canonical arc costs can be obtained in a search tree by moving nonzero costs upward in the tree. Beginning at the bottom, we do the following: if $c_{\min}(u)$ is the minimum cost on arcs leaving a given node u in layer i, then reduce the costs on these arcs by $c_{\min}(u) - \alpha_i$, and increase the costs on the arcs entering the node by $c_{\min}(u) - \alpha_i$. The algorithm appears in Fig. 3, and the result for the example appears in Fig. 4 (assuming each $\alpha_i = 0$). If the algorithm is applied to the search tree for the separable objective function $f(x) = 3x_1 + 5x_2 + 4x_3 + 6x_4$, the resulting canonical arc costs are those of Fig. 5(a).

A reduced weighted decision diagram can now be obtained from the branching tree of Fig. 4 much as in the unweighted case. Let subgraphs $G_{uu'}$ and $G_{vv'}$ be equivalent when they are isomorphic, corresponding arcs have the same labels *and costs*, and u, v belong to the same layer. A weighted decision diagram with canonical arc costs is *reduced* if it contains no equivalent subgraphs. A reduced weighted decision diagram can be obtained by superimposing all leaf nodes of the branching tree to obtain node 1, and then continuing to superimpose equivalent subgraphs until none remain. The branching tree of Fig. 4 reduces to the weighted decision diagram of Fig. 5(b). Note that the diagram is larger than the reduced diagram for the separable objective function, which appears in Fig. 5(a).

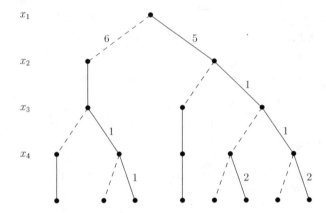

Fig. 4. Branching tree with canonical arc costs. Unlabeled arcs have zero cost.

Theorem 1. *Any given discrete optimization problem (1) is represented by a unique reduced weighted decision diagram with canonical arc costs.*

Proof. We first construct a reduced weighted decision diagram G that represents (1) and has canonical costs, assuming all $\alpha_i = 0$ (the argument is similar for arbitrary α_i). We will then show that G is reduced and unique. Define function g by $g(x) = f(x)$ when $x \in S$ and $g(x) = \infty$ when $x \notin S$. For each $i = 1, \ldots, n$ and each $d = (d_1, \ldots, d_{i-1}) \in D_1 \times \cdots \times D_{i-1}$ define the partial function g_{id} by

$$g_{id}(x_i, \ldots, x_n) = g(d_1, \ldots, d_{i-1}, x_i, \ldots, x_n)$$

Let partial function g_{id} be *finite* if $g_{id}(x_i, \ldots, x_n)$ is finite for some $x_i, \ldots, x_n$. By convention, $g_{n+1,d}() = 0$ for $d \in S$. We say that partial functions g_{id} and $g_{id'}$ are *equivalent* if both are finite and agree on relative values; that is,

$$g_{id}(x_i, \ldots, x_n) - g_{id}(x'_i, \ldots, x'_n) = g_{id'}(x_i, \ldots, x_n) - g_{id'}(x'_i, \ldots, x'_n),$$

for any pair $(x_i, \ldots, x_n)$, $(x'_i, \ldots, x'_n)$.

Now construct G as follows. In each layer $i \in \{1, \ldots, n+1\}$, create a node u in G for each equivalence class of finite partial functions g_{id}. Create outgoing arc $a(u, d_i)$ for each $d_i \in D_i$ such that $g_{i+1,(d,d_i)}$ is finite, where g_{id} is any function in the equivalence class for u. For $i \geq 2$ let arc $a(u, d_i)$ have cost

$$c(u, d_i) = g_{id}(d_i, p_u(d_i)) - c_{\min}(u) \qquad (2)$$

where $p_u(d_i)$ is a shortest path from $u(d_i)$ to 1. Then the arc costs are defined recursively for $i = n, n-1, \ldots, 2$. Arcs $a(r, d_1)$ leaving the root node have cost $c(r, d_1) = g(d_1, p_r(d_1))$.

We will say that G_{u1} represents the finite partial function g_{id} if

$$g_{id}(x_i, \ldots, x_n) - g_{id}(x'_i, \ldots, x'_n) = L_u(x_i, \ldots, x_n) - L_u(x'_i, \ldots, x'_n) \qquad (3)$$

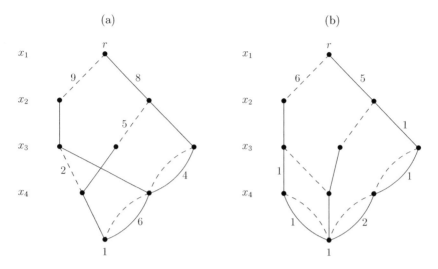

Fig. 5. (a) Weighted decision diagram with canonical arc costs for a separable objective function. (b) Canonical arc costs for a nonseparable objective function.

for all pairs $(x_i, \ldots, x_n)$, $(x'_i, \ldots, x'_n)$. We will show by induction on i that for any node u in layer $i \geq 2$, (i) path $p_u(d_i)$ has length zero for any arc $a(u, d_i)$, and (ii) subgraph G_{u1} represents any function g_{id} in the equivalence class corresponding to u. Given this, for any feasible solution $x = (d_1, d)$, we have

$$
\begin{align}
g(d_1, d) &= g(d_1, p_r(d_1)) + g(d_1, d) - g(d_1, p_r(d_1)) \tag{a}\\
&= g(d_1, p_r(d_1)) + L_r(d_1, d) - L_r(d_1, p_r(d_1)) \tag{b}\\
&= c(r, d_1) + L_r(d_1, d) - L_r(d_1, p_r(d_1)) \tag{c}\\
&= c(r, d_1) + L_r(d_1, d) - [c(r, d_1) + L_{r(d_1)}(p_r(d_1))] \tag{d}\\
&= c(r, d_1) + L_r(d_1, d) - c(r, d_1) = L_r(d_1, d) \tag{e}
\end{align}
$$

where (b) is due to (ii), (c) follows from the definition of $c(r, d_1)$, and (e) is due to (i). This means that for feasible x, $g(x)$ is the length of path x in G. Because the nodes of G represent only finite partial functions g_{id}, G contains no path for infeasible x. Thus, G represents (1).

For the inductive proof, we first observe that for any node u in layer n, path $p_u(d_n)$ is simply node 1 and therefore has length zero, which yields (i). Also for any pair x_n, x'_n, we have for any g_{id} in the equivalence class for u:

$$
\begin{align}
g_{nd}(x_n) - g_{nd}(x'_n) &= (g_{nd}(x_n) - c_{\min}(u)) - (g_{nd}(x'_n) - c_{\min}(u))\\
&= c(u, x_u) - c(u, x'_u) = L_u(x_n) - L_u(x'_n)
\end{align}
$$

This means that G_{u1} represents g_{nd}, and we have (ii).

Supposing that (i) and (ii) hold for layer $i + 1$, we now show that they hold for layer i. To show (i), we note that the length of $p_u(d_i)$ is

$$
\min_{d_{i+1}} \{ c(u(d_i), d_{i+1}) + p_u(d_{i+1})] \} = \min_{d_{i+1}} \{ c(u(d_i), d_{i+1}) \} = 0
$$

where the first equation is due to the induction hypothesis and the second to the definition of $c(u(d_i), d_{i+1})$. To show (ii), it suffices to show (3), which can be written

$$g_{id}(x_i, y) - g_{id}(x'_i, y') = L_u(x_i, y) - L_u(x'_i, y') \tag{4}$$

where $y = (x_{i+1}, \ldots, x_n)$. Note that

$$
\begin{aligned}
g_{id}&(x_i, y) - g_{id}(x'_i, y') \\
&= g_{id}(x_i, p_u(x_i)) - g_{id}(x'_i, p_u(x'_i)) \\
&\qquad + g_{id}(x_i, y) - g_{id}(x_i, p_u(x_i)) - [g_{id}(x'_i, y') - g_{id}(x'_i, p_u(d'_i))] &\text{(a)} \\
&= g_{id}(x_i, p_u(x_i)) - g_{id}(x'_i, p_u(x'_i)) \\
&\qquad + L_{u(x_i)}(y) - L_{u(x_i)}(p_u(x_i)) - [L_{u(x'_i)}(y') - L_{u(x'_i)}(p_u(x'_i))] &\text{(b)} \\
&= g_{id}(x_i, p_u(x_i)) - g_{id}(x'_i, p_u(x'_i)) + L_{u(x_i)}(y) - L_{u(x'_i)}(y') &\text{(c)} \\
&= c(u, x_i) - c(u, x'_i) + L_{u(x_i)}(y) - L_{u(x'_i)}(y') &\text{(d)} \\
&= c(u, x_i) + L_{u(x_i)}(y) - [c(u, x'_i) + L_{u(x'_i)}(y')] &\text{(e)} \\
&= L_u(x_i, y) - L_u(x'_i, y') &\text{(f)}
\end{aligned}
$$

where (b) is due to the induction hypothesis for (ii), (c) is due to the induction hypothesis for (i), and (d) is due to (2). This demonstrates (4).

We now show that G is minimal and unique. Suppose to the contrary that some weighted decision diagram $\bar{G}$ with canonical costs represents (1), is no larger than G, and is different from G. By construction, there is a one-to-one correspondence of nodes on layer i of G and equivalence classes of partial functions g_d. Thus for some node u on layer i if $\bar{G}$, there are two paths d, d' from r to u for which g_d and $g_{d'}$ belong to different equivalence classes. However, $\bar{G}$ represents g, which means that for any path $y = (y_i, \ldots, y_n)$ from u to 1,

$$
\begin{aligned}
g_d(y) &= \bar{L}(d) + \bar{L}_u(y) \\
g_{d'}(y) &= \bar{L}(d') + \bar{L}_u(y)
\end{aligned}
\tag{5}
$$

where $\bar{L}(d)$ is the length of the path d from 1 to u in $\bar{G}$, and $\bar{L}_u(y)$ the length of the path y from u to 1 in $\bar{G}$. This implies that for any two paths y and y' from u to 1 in $\bar{G}$,

$$g_d(y) - g_d(y') = g_{d'}(y) - g_{d'}(y') = \bar{L}_u(y) - \bar{L}_u(y')$$

which contradicts the fact that g_d and $g_{d'}$ belong to different equivalence classes. □

5   Separable Decision Diagrams

A *separable* decision diagram is a weighted decision diagram whose arc costs are directly obtained from a separable objective function. For example, the diagram of Fig. 1(b) is separable. More precisely, a decision diagram is separable if on any layer i, $c(u, d_i) = c(u', d_i) = c_i(d_i)$ for any d_i and any two nodes u, u'.

Separable decision diagrams have the advantage that they can be reduced while ignoring arc costs. That is, the reduced diagram is obtained by removing arc costs, reducing the diagram that results, and then putting cost $c_i(d_i)$ on each arc $a(u, d_i)$.

Converting costs to canonical costs can destroy separability. For example, the separable decision diagram of Fig. 1(b) becomes the nonseparable diagram of Fig. 5(a) when costs are converted to canonical costs. However, the *reduced* diagram remains the same. Thus Fig. 5(a) is a reduced weighted decision diagram.

This means that there is nothing lost by converting to canonical costs when the diagram is separable, and perhaps much to be gained when it is nonseparable, because in the latter case the diagram may reduce further.

Lemma 2. *A separable decision diagram that is reduced when costs are ignored is also reduced when costs are converted to canonical costs.*

Proof. Suppose that G is reduced when costs are ignored, but it is not reduced when costs are converted to canonical costs. Then some two weighted subgraphs G_{u1} and $G_{u'1}$ are equivalent. This means, in particular, that they are isomorphic. But this contradicts the assumption that G without costs is reduced. □

6   Dynamic Programming

A *dynamic programming problem* is one in which the variables x_i are regarded as *controls* that result in transitions from one state to the next. In particular, control x_i takes the system from the current state s_i to the next state

$$s_{i+1} = \phi_i(s_i, x_i), \quad i = 1, \ldots n \tag{6}$$

where the initial state s_1 is given. It is assumed that the objective function $f(x)$ is a separable function of *control/state pairs*, so that

$$f(x) = \sum_{i=1}^{n} c_i(x_i, s_i) \tag{7}$$

The optimization problem is to minimize $f(x)$ subject to (7) and $x_i \in X_i(s_i)$ for each i.

The attraction of a dynamic programming formulation is that it can be solved recursively:

$$g_i(x_i) = \min_{x_i \in X_i(s_i)} \left\{ c_i(s_i, x_i) + g_{i+1}(\phi_i(s_i, x_i)) \right\}, \quad i = 1, \ldots, n \tag{8}$$

where $g_{n+1}(s_{n+1}) = 0$ for all s_{n+1}. The optimal value is $g_1(s_1)$. To simplify discussion, we will suppose that $X_n(s_n)$ is defined so that there is only one final state s_{n+1}.

To make a connection with decision diagrams, we will assume that the control variables x_i are discrete. Then the recursion (8) describes a *state transition*

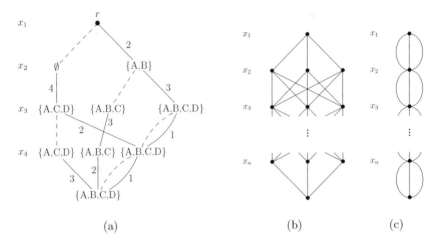

Fig. 6. (a) State transition graph for a set covering instance. (b) State transition graph for a production and inventory management problem. (c) Reduced state transition graph after converting costs to canonical costs.

graph in which each node corresponds to a state, and there is a directed arc $(s_i, \phi_i(s_i, x_i))$ with cost $c_i(s_i, x_i)$ for each control $x_i \in X_i$. The state transition graph is a binary decision diagram in which s_1 is the root node and s_{n+1} the terminal node.

In the set covering example, the state s_i can be defined as the set of elements that have been covered after variables $x_1, \ldots, x_{i-1}$ have been fixed. The resulting state transition graph appears in Fig. 6(a).

We see immediately that the state transition graph, when viewed as a decision diagram, may allow further reduction. Two of the nodes on level 4 of Fig. 6(a) can be merged even though they correspond to different states.

7   Reducing the State Transition Graph

It may be possible to reduce the size of a DP state transition graph by viewing it as a weighted decision diagram. Even when arc costs as given in the graph prevent reduction, conversion of the arc costs to canonical costs may allow significant reduction that simplifies the problem.

For example, this idea can be applied a textbook DP model that has remained essentially unchanged for decades. The objective is to adjust production quantities and inventory levels to meet demand over n periods while minimizing production and holding costs. We will suppose that h_i the unit holding cost in period i, and c_i is the unit production cost. Let x_i be the production quantity in period i, and let the state variable s_i be the stock on hand at the beginning of period i. Then the recursion is

$$g_i(s_i) = \min_{x_i \in X_i(s_i)} \{c_i x_i + h_i s_i + g_{i+1}(s_i + x_i - d_i)\}, \quad i = 1, \ldots, n \qquad (9)$$

where d_i is the demand in period i.

If we suppose that the warehouse has capacity m in each period, the state transition graph has the form shown in Fig. 6(b). Note that the set of arcs leaving any node is essentially identical to the set of arcs leaving any other node in the same stage. The controls x_i and the costs are different, but the controls can be equalized by a simple change of variable, and the costs can be equalized by transforming them to canonical costs.

To equalize the controls, let the control x'_i be the stock level at the beginning of the next stage, so that $x'_i = s_i + x_i - d_i$. Then the controls leaving any node are $x'_i = 0, \ldots, m$. The recursion (9) becomes

$$g_i(s_i) = \min_{x'_i \in \{0,\ldots,m\}} \{c_i(x'_i - s_i + d_i) + h_i s_i + g_{i+1}(x'_i)\}, \quad i = 1, \ldots, n \qquad (10)$$

To transform the costs to canonical costs, we subtract $h_i s_i + (m - s_i)c_i$ from the cost on each arc (s_i, s_{i+1}), and add this amount to each arc coming into s_i. Then for any period i, the arcs leaving any given node s_i have the same set of costs. Specifically, arc (s_i, s_{i+1}) has cost

$$\bar{c}_i(s_{i+1}) = (d_i + s_{i+1} - m)c_i + s_{i+1}h_{i+1} + (m - s_{i+1})c_{i+1}$$

and so depends only on the next state s_{i+1}. These costs are canonical for $\alpha_i = \min_{s_{i+1} \in \{0,\ldots,m\}} \{\bar{c}_i(s_{i+1})\}$.

In any given period i, the subgraphs G_{s_i} are now equivalent, and the decision diagram can be reduced as in Fig. 6(c). There is now one state in each period rather than m, and the recursion is

$$g_i = \min_{x'_i \in \{0,\ldots,n\}} \{c_i(d_i + x'_i - m) + h_{i+1}x'_i + c_{i+1}(m - x'_i) + g_{i+1}\} \qquad (11)$$

If $\bar{x}'_1, \ldots, \bar{x}'_n$ are the optimal controls in (11), the resulting stock levels are given by $s_{i+1} = \bar{x}'_i$ and the production levels by $x_i = \bar{x}'_i - s_i + d_i$.

There is no need for experiments to determine the effect on computation time. The reduction of the state space implies immediately that the time is reduced by a factor of m.

8   Nonserial Dynamic Programming

In serial dynamic programming, the state variables s_i are arranged linearly as a path. In *nonserial dynamic programming*, they are arranged in a tree. Because formal notation for nonserial DP is rather complex, the idea is best introduced by example. To simplify exposition, we will discuss only problems with separable objective functions.

Figure 7(a) shows a small set partitioning problem. The goal is to select a minimum subcollection of the 6 sets that partitions the set $\{A, B, C, D\}$. Thus there are 4 constraints, corresponding to the elements (A, B, C, D), each requiring that only one set containing that element be selected. The 3 feasible solutions are $(x_1, \ldots, x_6) = (0, 0, 0, 1, 1, 1), (0, 1, 1, 0, 0, 0), (1, 0, 0, 0, 0, 1)$, where $x_i = 1$ indicates that set i is selected. The last two solutions are optimal.

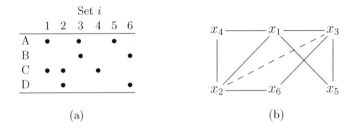

Fig. 7. (a) A small set partitioning problem. The dots indicate which elements belong to each set i. (b) Dependency graph for the problem. The dashed edge is an induced edge.

A nonserial recursion can be constructed by reference to the dependency graph for the problem, shown in Fig. 7(b). The graph connects two variables with an edge when they occur in a common constraint. We arbitrarily select a variable ordering $x_1, \ldots, x_6$ and remove vertices in reverse order, each time connecting vertices adjacent to the vertex removed. Edges added in this fashion are *induced edges*. As in serial DP, we let the state contain the information necessary to determine the feasible choices for x_i. Now, however, the feasible values of x_i depend on the set of variables to which x_i was adjacent when removed. We therefore let x_6 depend on (x_2, x_3), and similarly for the other control variables.

The problem can be solved recursively as follows. Let $C_A(x_1, x_3, x_5)$, $C_B(x_3, x_6)$, $C_C(x_1, x_2, x_4)$, and $C_D(x_2, x_6)$ be the constraints corresponding to elements A, B, C, and D, respectively. The recursion is

$$g_6(x_2, x_3) = \min_{x_6 \in \{0,1\}} \{x_6 \mid C_C(x_3, x_6) \wedge C_D(x_2, x_6)\}$$

$$g_5(x_1, x_3) = \min_{x_5 \in \{0,1\}} \{x_5 \mid C_A(x_1, x_3, x_5)\}$$

$$g_4(x_1, x_2) = \min_{x_4 \in \{0,1\}} \{x_4 \mid C_C(x_1, x_2, x_4)\}$$

$$g_3(x_1, x_2) = \min_{x_3 \in \{0,1\}} \{x_3 + g_6(x_2, x_3) + g_5(x_1, x_3)\}$$

$$g_2(x_1) = \min_{x_2 \in \{0,1\}} \{x_2 + g_4(x_1, x_2) + g_3(x_1, x_2)\}$$

$$g_1() = \min_{x_1 \in \{0,1\}} \{x_1 + g_2(s(1))\}$$

The smallest partition has size $g_1()$, which is infinite if there is no feasible partition. The induced width (treewidth) of the dependency graph is the maximum number of state variables that appear as arguments of a $g_i(\cdot)$, in this case 2.

To write the general recursion for nonserial DP, let $x(J) = \{x_j \mid j \in J\}$. Let each constraint C_i contains the variables in $x(J_i)$, and let each $g_i(\cdot)$ be a function of $x(I_i)$. The recursion is

$$g_i(x(I_i)) = \max_{x_i} \left\{ c_i(x_i) + \sum_{\substack{j > i \\ I_j \subset I_i}} g_j(x(I_j)) \;\middle|\; \bigwedge_{\substack{j \\ J_j \subset I_i}} C_j(x(J_j)) \right\}$$

The complexity of the recursion is $\mathcal{O}(2^{W+1})$, where W is the induced width.

As in serial DP, we can introduce state variables s_i to encode the necessary information for selecting x_i. In the present case, we let s_i be the multiset of elements selected by the variables in $x(I_i)$, where elements selected by two variables are listed twice. For example, s_6 is the multiset of elements selected by x_2 and x_3. In general, we will let $s(I_i)$ denote the multiset of elements selected by the variables in $x(I_i)$, so that $s_i = s(I_i)$. The solution can now be calculated as shown in Table 2.

The state transition graph for the example appears in Fig. 8(a). Here each stage is labeled by the state variable s_i rather than the decision variable x_i. The initial state is associated with state variable s_1. Decision x_i is taken at each value of state variable s_i. Because state variables s_3 and s_4 are identical, they are superimposed in the graph. The choice of x_3 leads to states s_5 and s_6, with two outgoing arcs corresponding to each choice. The choice of x_4 leads to a terminal node.

Feasible solutions correspond to trees that are incident to the initial state and 3 terminal nodes. The tree shown in bold is one of the two optimal solutions. Note that its cost is 2 even though it contains 3 solid arcs, because two of the arcs correspond to the same choice $x_3 = 1$. States that are part of no feasible solution (tree) are omitted from the diagram.

As an illustration, consider state $s_3 = \{A, C\}$. The arcs for $x_3 = 0$ lead to the states $s_5 = \{A,C\}$ and $s_6 = \emptyset$. Arcs for $x_3 = 1$ are not shown because they are not part of a feasible solution. One can check from Table 2 that when $x_3 = 1$,

$$x_3 + g_6(s_6) + g_5(s_5) = 1 + g_6(\{A,B\}) + g_5(\{A,C,C,D\}) = \infty$$

Now consider state $s_4 = \{A, C\}$, which corresponds to the same node. The arc for $x_4 = 0$ leads to a terminal node. The arc for $x_4 = 1$ is not shown because $\{A,C,C\}$ violates constraint C_C.

9 Nonserial Decision Diagrams

The state transition graph for a nonserial DP can be regarded as a decision diagram, as in the case of serial DP. The diagram can also be reduced in a similar fashion. For example, the diagram of Fig. 7(a) reduces to that of Fig. 7(b). Note that two nodes are merged, resulting in a smaller diagram. Feasible solutions correspond to trees that are incident to the root node and the two terminal nodes. The reduction in size can be significant, as in the case of serial decision diagrams.

Table 2. Recursive solution of the set partitioning example

s_6	$g_6(s_6)$	s_5	$g_5(s_5)$	s_4	$g_4(s_4)$	s_3	$g_3(s_3)$	s_2	$g_2(s_2)$
$\emptyset$	1	$\emptyset$	1	$\emptyset$	1	$\emptyset$	2	$\emptyset$	2
$\{A,B\}$	∞	$\{A,C\}$	0	$\{A,C\}$	0	$\{A,C\}$	1	$\{A,B\}$	2
$\{C,D\}$	∞	$\{A,B\}$	0	$\{C,D\}$	0	$\{C,D\}$	1		
$\{A,B,C,D\}$	0	$\{A,A,B,C\}$	∞	$\{A,C,C,D\}$	∞	$\{A,C,C,D\}$	∞	$g_1() = 2$	

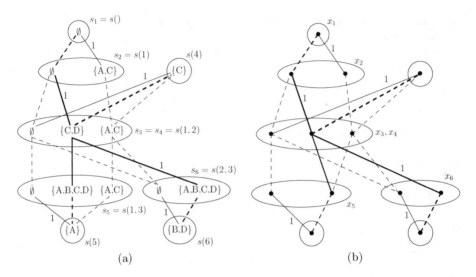

Fig. 8. (a) Nonserial state transition graph for a set partitioning problem. Only nodes and arcs that are part of feasible solutions are shown. Each feasible solution corresponds to a tree incident to the root and both terminal nodes. The boldface tree corresponds to optimal solution $(x_1, \ldots, x_6) = (0, 1, 1, 0, 0, 0)$. (b) Reduced nonserial decision diagram for the same problem.

10   Conclusion

We showed how decision diagrams can be extended to weighted decision diagrams so as to establish a precise parallel with dynamic programming (DP). In particular, we proved that for a given optimization problem and variable ordering, there is a unique reduced decision diagram with canonical arc costs that represents the problem. We also showed how this perspective can allow one to simplify a discrete DP model by transforming arc costs on its state transition graph to canonical arc costs and reducing the diagram that results. Finally, we introduced nonserial decision diagrams as a counterpart to nonserial dynamic programming.

It remains to investigate other possible simplifications of DP models based on the decision diagram perspective, as well as to generalize the uniqueness result to nonserial decision diagrams. Another possible development is to merge relaxed decision diagrams, mentioned in Section 2, with approximate dynamic programming [32]. This may allow algorithms for relaxing a decision diagram to generate an efficient state space relaxation for approximate DP.

References

1. Akers, S.B.: Binary decision diagrams. IEEE Transactions on Computers C-27, 509–516 (1978)
2. Andersen, H.R.: An introduction to binary decision diagrams. Lecture notes, available online, IT University of Copenhagen (1997)

3. Andersen, H.R., Hadzic, T., Hooker, J.N., Tiedemann, P.: A constraint store based on multivalued decision diagrams. In: Bessière, C. (ed.) CP 2007. LNCS, vol. 4741, pp. 118–132. Springer, Heidelberg (2007)
4. Arnborg, S., Corneil, D.G., Proskurowski, A.: Complexity of finding embeddings in a k-tree. SIAM Jorunal on Algebraic and Discrete Mathematics 8, 277–284 (1987)
5. Arnborg, S., Proskurowski, A.: Characterization and recognition of partial k-trees. SIAM Jorunal on Algebraic and Discrete Mathematics 7, 305–314 (1986)
6. Bacchus, F., Dalmao, S., Pitassi, T.: Algorithms and complexity results for #SAT and Bayesian inference. In: Proceedings of the 44th Annual IEEE Symposium on Foundations of Computer Science (FOCS 2003), pp. 340–351 (2003)
7. Bacchus, F., Dalmao, S., Pitassi, T.: Solving #SAT and Bayesian inference with backtracking search. Journal of Artificial Intelligence Research 34, 391–442 (2009)
8. Beame, P., Impagliazzo, R., Pitassi, T., Segerlind, N.: Memoization and DPLL: Formula caching proof systems. In: 18th IEEE Annual Conference on Computational Complexity, pp. 248–259 (2003)
9. Becker, B., Behle, M., Eisenbrand, F., Wimmer, R.: BDDs in a branch and cut framework. In: Nikoletseas, S.E. (ed.) WEA 2005. LNCS, vol. 3503, pp. 452–463. Springer, Heidelberg (2005)
10. Behle, M., Eisenbrand, F.: 0/1 vertex and facet enumeration with BDDs. In: Proceedings of the 9th Workshop on Algorithm Engineering and Experiments and the 4th Workshop on Analytic Algorithms and Combinatorics, pp. 158–165 (2007)
11. Behle, M.: Binary Decision Diagrams and Integer Programming. PhD thesis, Max Planck Institute for Computer Science (2007)
12. Bellman, R.: The theory of dynamic programming. Bulletin of the American Mathematical Society 60, 503–516 (1954)
13. Bellman, R.: Dynamic programming. Priceton University Press, Princeton (1957)
14. Bergman, D., Ciré, A.A., van Hoeve, W.-J., Hooker, J.N.: Optimization bounds from binary decision diagrams. Technical report, Carnegie Mellon University (2012)
15. Bergman, D., van Hoeve, W.-J., Hooker, J.N.: Manipulating MDD relaxations for combinatorial optimization. In: Achterberg, T., Beck, J.C. (eds.) CPAIOR 2011. LNCS, vol. 6697, pp. 20–35. Springer, Heidelberg (2011)
16. Bertele, U., Brioschi, F.: Nonserial Dynamic Programming. Academic Press, New York (1972)
17. Bertsekas, D.P.: Dynamic Programming and Optimal Control, 3rd edn., vol. 1, 2. Athena Scientific, Nashua (2001)
18. Bollig, B., Sauerhoff, M., Sieling, D., Wegener, I.: Binary decision diagrams. In: Crama, Y., Hammer, P.L. (eds.) Boolean Models and Methods in Mathematics, Computer Science, and Engineering, pp. 473–505. Cambridge University Press, Cambridge (2010)
19. Bryant, R.E.: Graph-based algorithms for boolean function manipulation. IEEE Transactions on Computers C-35, 677–691 (1986)
20. Chhajed, D., Lowe, T.J.: Solving structured multifacility location problems efficiently. Transportation Science 28, 104–115 (1994)
21. Ciré, A.A., van Hoeve, W.-J.: MDD propagation for disjunctive scheduling. In: Proceedings of the International Conference on Automated Planning and Scheduling (ICAPS), pp. 11–19. AAAI Press (2012)
22. Crama, Y., Hansen, P., Jaumard, B.: The basic algorithm for pseudoboolean programming revisited. Discrete Applied Mathematics 29, 171–185 (1990)
23. Dechter, R.: Bucket elimination: A unifying framework for several probabilistic inference algorithms. In: Proceedings of the Twelfth Annual Conference on Uncertainty in Artificial Intelligence (UAI 1996), Portland, OR, pp. 211–219 (1996)

24. Denardo, E.V.: Dynamic Programming: Models and Applications. Dover Publications, Mineola (2003)
25. Hadžić, T., Hooker, J.N.: Postoptimality analysis for integer programming using binary decision diagrams. Presented at GICOLAG Workshop (Global Optimization: Integrating Convexity, Optimization, Logic Programming, and Computational Algebraic Geometry), Vienna. Technical report, Carnegie Mellon University (2006)
26. Hadžić, T., Hooker, J.N.: Cost-bounded binary decision diagrams for 0-1 programming. Technical report, Carnegie Mellon University (2007)
27. Hadzic, T., Hooker, J.N., O'Sullivan, B., Tiedemann, P.: Approximate compilation of constraints into multivalued decision diagrams. In: Stuckey, P.J. (ed.) CP 2008. LNCS, vol. 5202, pp. 448–462. Springer, Heidelberg (2008)
28. Hoda, S., van Hoeve, W.-J., Hooker, J.N.: A systematic approach to MDD-based constraint programming. In: Cohen, D. (ed.) CP 2010. LNCS, vol. 6308, pp. 266–280. Springer, Heidelberg (2010)
29. Huang, J., Darwiche, A.: DPLL with a trace: From SAT to knowledge compilation. In: International Joint Conference on Artificial Intelligence (IJCAI 2005), vol. 19, pp. 156–162. Lawrence Erlbaum Associates (2005)
30. Lauritzen, S.L., Spiegelhalter, D.J.: Local computations with probabilities on graphical structures and their application to expert systems. Journal of the Royal Statistical Society B 50, 157–224 (1988)
31. Lee, C.Y.: Representation of switching circuits by binary-decision programs. Bell Systems Technical Journal 38, 985–999 (1959)
32. Powell, W.B.: Approximate Dynamic Programming: Solving the Curses of Diumensionality, 2nd edn. Wiley (2011)
33. Shafer, G., Shenoy, P.P., Mellouli, K.: Propagating belief functions in qualitative markov trees. International Journal of Approximate Reasoning 1, 349–400 (1987)
34. Shenoy, P.P., Shafer, G.: Propagating belief functions with local computation. IEEE Expert 1, 43–52 (1986)
35. Sieling, D., Wegener, I.: NC-algorithms for operations on binary decision diagrams. Parallel Processing Letters 3, 3–12 (1993)

CP Methods for Scheduling and Routing
with Time-Dependent Task Costs

Elena Kelareva[1,2], Kevin Tierney[3], and Philip Kilby[1,2]

[1] Australian National University, Canberra, Australia
[2] NICTA, Canberra, Australia
elena.kelareva@nicta.com.au
[3] IT University of Copenhagen, Copenhagen, Denmark

Abstract. A particularly difficult class of scheduling and routing problems involves an objective that is a sum of time-varying action costs, which increases the size and complexity of the problem. Solve-and-improve approaches, which find an initial solution for a simplified model and improve it using a cost function, and Mixed Integer Programming (MIP) are often used for solving such problems. However, Constraint Programming (CP), particularly with Lazy Clause Generation (LCG), has been found to be faster than MIP for some scheduling problems with time-varying action costs. In this paper, we compare CP and LCG against a solve-and-improve approach for two recently introduced problems in maritime logistics with time-varying action costs: the Liner Shipping Fleet Repositioning Problem (LSFRP) and the Bulk Port Cargo Throughput Optimisation Problem (BPCTOP). We present a novel CP model for the LSFRP, which is faster than all previous methods and outperforms a simplified automated planning model without time-varying costs. We show that a LCG solver is faster for solving the BPCTOP than a standard finite domain CP solver with a simplified model. We find that CP and LCG are effective methods for solving scheduling problems, and are worth investigating for other scheduling and routing problems that are currently being solved using MIP or solve-and-improve approaches.

1 Introduction

Scheduling problems typically aim to select times for a set of tasks so as to optimise some cost or value function, subject to problem-specific constraints. Traditional scheduling problems usually aim to minimise the makespan, or total time, of the resulting schedule. More complex objective functions, such as minimising the total weighted tardiness, may vary with time [30]. Routing problems have many similarities with scheduling – both may have resource constraints and setup time constraints, both have actions that need to be scheduled in time, and both may have complex time-dependent cost functions for actions.

In a number of important, real-world scheduling problems, such as the Liner Shipping Fleet Repositioning Problem (LSFRP) [32, 33] and the Bulk Port Cargo Throughput Optimisation Problem (BPCTOP) [14, 15], the objective is a sum of time-varying costs or values for each task. Additional problems include net present value maximization in project scheduling [26]; satellite imaging scheduling [17, 38]; vehicle routing

C. Gomes and M. Sellmann (Eds.): CPAIOR 2013, LNCS 7874, pp. 111–127, 2013.

with soft time windows [29, 12]; ship routing and scheduling with soft time windows [8, 2]; and ship speed optimisation [10, 20].

Mixed Integer Programming (MIP) is a standard approach used to solve many scheduling and routing problems. Solve-and-improve approaches are also commonly used to solve scheduling and routing problems with complex constraints or complex objective functions, such as objective functions that are the sum of time-dependent task costs. Solve-and-improve approaches initially solve a simplified problem, then improve the solution using the objective function and constraints of the full problem.

For example, in the satellite image scheduling problem, each observation may only be performed for a specified period of time during the satellite's orbit, and the quality of the observation drops off to zero for times before and after the peak quality window [38]. Yao *et al* [39] and Wang *et al* [37] solved this problem by converting the image quality function to hard time windows when a "good enough" image could be obtained. Lin *et al* [17] found solutions within 2% of optimality by first converting the quality function to hard time windows to find feasible solutions, and then heuristically improving the solution quality by minor changes in the schedule.

Solve-and-improve approaches have also often been used in other applications, such as the vehicle routing problem with soft time windows (VRPSTW), which has time-varying penalties for early and late arrival outside each customer's preferred time window. Soft time windows allow better utilisation of vehicles, thus reducing transportation costs compared to hard time windows, while still servicing most customers within their preferred times. However, soft time windows result in a more complex objective function, making the problem significantly more difficult to solve [24]. Some VRPSTW approaches optimise first for the number of vehicles (routes), then for minimal travel time and distance, and finally improving solutions by minimising cost with time window penalties included [12]. Solve-and-improve approaches are a standard technique which have been discussed in a recent review of vehicle routing with time windows [7].

However, for some problems such as [14], Constraint Programming (CP) has been shown to be more effective than MIP. CP is also a very flexible method that can be used to model a wider variety of constraints than MIP, which is limited to linear constraints. A number of recent approaches have combined CP with other techniques such as vehicle routing [16], SAT [21] and MIP [1] in order to combine the flexibility of CP with fast algorithms for specific problems. CP approaches may be worth investigating for other problems that have traditionally been modelled with MIP.

One CP technique in particular which has been found to be effective on a number of scheduling problems is Lazy Clause Generation (LCG) [21] – a method for solving CP problems which allows the solver to learn where the previous search failed. LCG combines a finite domain CP solver with a SAT solver, by lazily adding clauses to the SAT solver as each finite domain propagator is executed. This approach benefits from efficient SAT solving techniques such as nogood learning and backjumping, while maintaining the flexible modelling of a CP solver and enabling efficient propagation of complex constraints [21].

A CP solver that uses LCG was found by Schutt *et al* [27] to be more efficient than traditional finite domain CP solvers for the Resource Constrained Project Scheduling Problem with Net Present Value – another scheduling problem with time-dependent action costs, where each activity has an associated positive or negative cash flow that is discounted over time. This problem has previously been solved by relaxing the problem to remove resource constraints, and using the resource-unconstrained solution as an upper bound in the search [35]; however, a CP solver with LCG was able to find better solutions than any previous state-of-the-art complete method [27].

LCG has also been found to be faster than finite domain solvers for a number of other scheduling problems, including the Concert Hall Scheduling problem [5], project scheduling with generalised precedence constraints [28] and a number of other scheduling problems [11]. LCG solvers can be used for any problem that is modelled as a constraint programming problem, which makes LCG a highly generalisable technique that is worth investigating for other scheduling and routing problems with time-varying task costs as an alternative to MIP or model simplification approaches.

The contribution of this paper is to compare the effectiveness of CP and LCG against traditional MIP and solve-and-improve techniques for two recent scheduling and routing problems with time-varying action costs in the field of maritime transportation. We present a novel CP model for the LSFRP and show that this model is faster than existing approaches including MIP and automated planning, even when a simplified automated planning model with no time-varying costs is used. We show that a CP solver with LCG is more effective at solving the BPCTOP with time-varying draft than a traditional Finite Domain CP solver (CP was found to be faster than MIP for the BPCTOP in an earlier paper [14]). We also show that the LCG solver scales better to large BPCTOP problems than the first step of a solve-and-improve approach that simplifies the time-varying cost function to find an initial solution.

2 Background

2.1 Bulk Port Cargo throughput Optimisation with Time-Varying Draft

Many ports have safety restrictions on the draft (distance between waterline and keel) of ships sailing through the port, which vary with the height of the tide. Most maritime scheduling problems either ignore draft constraints entirely [9], or do not consider time variation in draft restrictions [25, 31]. This simplifies the problem and improves scalability, but may miss solutions which allow ships to sail with higher draft (and thus more cargo) close to high tide.

Introducing time-varying draft restrictions requires a problem to be modelled with a very fine time resolution, as the draft can change every five minutes. This greatly increases the size of the problem, so time-varying draft restrictions have thus far only been applied to the problem of optimising cargo throughput at a single bulk export port – the Bulk Port Cargo Throughput Optimisation Problem (BPCTOP). The objective function for the BPCTOP with time-varying draft restrictions is the sum of maximum drafts for each ship at their scheduled sailing times. The shape of the objective function is similar to satellite image scheduling, as the maximum draft for each ship peaks around high tide, and drops off before and after.

The BPCTOP also includes resource constraints on the availability of tugs, berths or shipping channels, and sequence-dependent setup times between successive ships. Ports may have safety restrictions on the minimum separation times between ships, as ships sailing too close together in a narrow channel may pose a safety risk. Minimum separation times between ships may also depend on which berths the ships sail from.

Kelareva *et al* [14] compared CP and MIP approaches for the BPCTOP, and found that CP with a good choice of search strategy was able to find optimal solutions faster than MIP, and was also able to solve problems with more ships. These approaches also produced significantly better solutions compared to human schedulers [15]. Each extra centimetre of draft can increase profit for the shipper by up to $10,000 [14], so there is a high incentive to find optimal solutions, and non-optimal approaches for this problem have not yet been investigated.

The CP model was able to solve problems with up to 6 ships for even the most tightly constrained problems, where all ships are very large and can continue loading extra cargo right up to the peak of the high tide, and with 10 or more ships for less tightly constrained problems. These are realistic problem sizes – Port Hedland, Australia's largest bulk export port recently set a record of 5 draft-constrained ships sailing on the same high tide [22].

2.2   Liner Shipping Fleet Repositioning

Another recently introduced scheduling and routing problem with time-varying action costs in maritime transportation is the Liner Shipping Fleet Repositioning Problem (LS-FRP) [32]. In liner shipping, vessels are assigned to services that operate like a bus timetable, and vessels regularly need to be repositioned between services in response to changes in the world economy. A vessel begins repositioning when it *phases out* from its original service, and ends repositioning when it *phases in* to its new service.

The total cost of the repositioning depends on fuel usage, and a fixed hourly *hotel cost* paid for the time between the phase-out and phase-in. The cost function for a repositioning action is therefore monotonically increasing with the duration of sailing, as the fuel cost is much lower for vessels sailing at low speed. There may also be opportunities for a vessel to carry empty containers (*sail-equipment* (SE)), or replace another vessel in a regular service (*sail-on-service* (SOS)), which significantly reduces the cost of the repositioning. The LSFRP is not a pure scheduling problem, as there is a choice of actions and action durations.

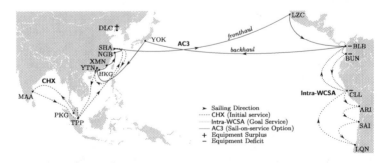

Fig. 1. A subset of a real-world repositioning scenario, from [32]

Figure 1 shows a subset of a repositioning scenario in which the new Intra-WCSA service requires three vessels that must be repositioned from their services in South-East Asia. One of the vessels was originally on the CHX service, and the other two were on other services not shown in this figure. The cost of repositioning in this scenario can be reduced by carrying equipment from China to South America (e.g. DLC to BLB), or using the AC3 service as an SOS opportunity.

There has been little prior research on the LSFRP. In fact, the LSFRP is not even mentioned in the two most comprehensive surveys of maritime research [4, 3]. The LS-FRP was first solved by Tierney *et. al.* [32], who compared three methods for solving the LSFRP. First, the LSFRP was modelled using PDDL 2.1, a modeling language for automated planning problems (see [13]), and solved using the POPF planner [6]. Although POPF found solutions, it could not solve the LSFRP to optimality. This method was therefore compared against a MIP model, and against a Linear Temporal Optimisation Planning (LTOP) approach, which uses temporal planning to build optimisation models, and solves these using an optimisation version of partial-order planning based on branch-and-bound.

Both LTOP and MIP were able to find optimal solutions for problem sizes with up to three ships. These are realistic problem sizes similar to those used by shipping lines. As with the BPCTOP, there is a high motivation to find the optimal solutions, even if solving the problem to optimality is more difficult, since the cost of repositioning a single ship can be hundreds of thousands of dollars [32].

3   CP vs MIP

3.1   CP Model for Liner Shipping Fleet Repositioning

Tierney, *et al* [32] presented MIP and automated planning models for the LSFRP. In this section, we present a novel CP model for the LSFRP, and compare its solution times against the earlier MIP and automated planning models.

Let O_v be the set of possible phase-out actions for the vessel v, and let P be the be the set of possible phase-in ports for the new service. The decision variable $\rho \in P$ is the phase-in port for all vessels. The decision variables $w_v \in \{1, \ldots, W\}$ represent the phase-in week for each vessel v, where W is the number of weeks considered in the problem. For each vessel v we also define a decision variable $q_v \in O_v$ specifying the phase-out action (port and time) used for that vessel.

For each vessel v and phase-out action o, the function $t(v, o)$ specifies the phase-out time for that action. Similarly, $t(p, w)$ specifies the phase-in time for a vessel phasing in at port p in week w. The function $C(v, o, p, w)$ specifies the cost for vessel v using the phase-out action o, and phasing in at port p in week w, with -1 as a flag that indicates vessel v cannot phase in at port p in week w if it phased out using action q (for example, if action q starts too late for vessel v to reach port p in time). The dependent variable c_v specifies the cost for vessel v when the vessel sails directly from the phase-out port to the phase-in port. For each vessel v, $C_H(v)$ specifies its hourly hotel cost, and h_v is the duration of the hotel cost time period (from the phase-out to the phase-in).

We split each SOS opportunity into several *SOS actions*, where each SOS action represents starting the SOS at a different port on the SOS service. SOS opportunities save money by allowing vessels to sail for free between two ports, however a cost for transshipping cargo at each side of the SOS is incurred. Let S be the set of available SOS actions and S' be the set of SOS opportunities. The decision variable $s_v \in S$ specifies the SOS action used for each vessel v, with 0 being a flag indicating that vessel v does not use an SOS action. For each SOS action $s \in S$, the function $y : S \to S'$ specifies which SOS opportunity each SOS action belongs to, with $y(s) = 0$ being a flag that specifies that the vessel is not using any SOS opportunity.

In order to use an SOS opportunity, a vessel must sail to the starting port of the SOS opportunity before a deadline, and after using the SOS, it sails from the end port at a pre-determined time to the phase-in port. The function $C^{to}(v, s, o)$ specifies the cost of vessel v using SOS action s, phasing out at phase-out o going to the SOS action, and $C^{from}(v, s, p, w)$ is the cost of vessel v to sail from SOS action s to phase in port p in week w, with -1 as a flag that indicates that this combination of vessel, SOS action, phase-in port and week is infeasible. The dependent variables σ_v^{to} and σ_v^{from} specify the SOS costs for vessel v for sailing to and from the SOS, respectively. The function $A(v, s)$ specifies the cost savings of vessel v using SOS action s, and the dependent variable σ_v^{dur} specifies the SOS cost savings for vessel v on the SOS.

Let Q be the set of *sail-equipment* (SE) opportunities, which are pairs of ports in which one port has an excess of a type of equipment, e.g. empty containers, and the other port has a deficit. Since we do not include a detailed view of cargo flows in this version of the LSFRP, SE opportunities save money by allowing vessels to sail for free between two ports as long as the vessel sails at its slowest speed. The cost then increases linearly as the vessel sails faster. Let the decision variable $e_v \in E$ be the SE opportunity undertaken by vessel v, with $e_v = 0$ indicating that no SE opportunity is used. Let the decision variables d_v^{to}, d_v^{dur} and d_v^{from} be the duration of vessel v sailing to, during, and from an SE opportunity.

The functions $C^{to}(v, e, o)$, $C^{dur}(v, e)$ and $C^{from}(v, e, p, w)$ specify the fixed costs of sailing to, utilizing, and then sailing from SE opportunity e, where v is the vessel, o is the phase-out port/time, p is the phase-in port and w is the phase-in week. Together with the constant α_v, which is the variable sailing cost per hour of vessel v, the hourly cost of sailing can be computed. This is necessary since SE opportunities are not fixed in time and, thus, must be scheduled. Let the dependent variables $\lambda_v^{to}, \lambda_v^{dur}$ and λ_v^{from} be the fixed costs sailing to, on and from an SE opportunity. Additionally, let $\Delta_{min}^{to}(v, e, o)$, $\Delta_{min}^{dur}(v, e)$ and $\Delta_{min}^{from}(v, e, p, w)$ be the minimum sailing time of v before, during and after the SE opportunity and $\Delta_{max}^{to}(v, e, o)$, $\Delta_{max}^{dur}(v, e)$ and $\Delta_{max}^{from}(v, e, p, w)$ be the maximum sailing time of v before, during and after the SE opportunity.

In this version of the LSFRP, the chaining of SOS and SE opportunities is not allowed, meaning each vessel has the choice of either sailing directly from the phase-out to the phase-in, undertaking an SOS, or performing an SE. The decision variable $r_v \in \{\text{SOS}, \text{SE}, \text{SAIL}\}$ specifies the type of repositioning for each vessel v, where v utilizes an SOS opportunity, SE opportunity, or sails directly from the phase-out to the phase-in, respectively. The CP model is formulated as follows:

$$\min \sum_{v \in V} \left(C_H(v) \left(t(\rho, w_v) - t(v, q_v) \right) + c_v + \sigma_v^{from} + \sigma_v^{dur} + \sigma_v^{to} \right.$$
$$\left. + \lambda_v^{to} + \lambda_v^{dur} + \lambda_v^{from} + \alpha_v (d_v^{to} + d_v^{dur} + d_v^{from}) \right) \tag{1}$$

$$\text{s.t.} \qquad \texttt{alldifferent}(w_v), v \in V \tag{2}$$

$$\max_{v \in V} w_v - \min_{v \in V} w_v = |V| - 1 \tag{3}$$

$$\texttt{alldifferent\_except\_0}(s_v), v \in V \tag{4}$$

$$\texttt{alldifferent\_except\_0}(y(s_v)), v \in V \tag{5}$$

$$r_v = \texttt{SAIL} \rightarrow c_v = C(v, q_v, \rho, w_v), \quad \forall v \in V \tag{6}$$

$$r_v = \texttt{SAIL} \rightarrow \big(s_v = 0 \wedge y(s_v) = 0 \wedge \sigma_v^{dur} = 0 \wedge \sigma_v^{from} = 0 \wedge \sigma_v^{to} = 0$$
$$\wedge e_v = 0 \wedge \lambda_v^{to} = 0 \wedge \lambda_v^{dur} = 0 \wedge \lambda_v^{from} = 0\big), \quad \forall v \in V \tag{7}$$

$$r_v = \texttt{SOS} \rightarrow s_v > 0 \wedge y(s_v) > 0 \wedge \sigma_v^{dur} = -A(v, s_v)$$
$$\wedge \sigma_v^{from} = C^{from}(v, s_v, \rho, w_v) \wedge \sigma_v^{to} = C^{to}(v, s_v, q_v), \quad \forall v \in V \tag{8}$$

$$r_v = \texttt{SOS} \rightarrow c_v = 0 \wedge e_v = 0 \wedge \lambda_v^{to} = 0 \wedge \lambda_v^{dur} = 0 \wedge \lambda_v^{from} = 0, \quad \forall v \in V \tag{9}$$

$$s_v > 0 \vee y(s_v) > 0 \rightarrow r_v = \texttt{SOS}, \quad \forall v \in V \tag{10}$$

$$\texttt{alldifferent\_except\_0}(e_v), \quad \forall v \in V \tag{11}$$

$$r_v = \texttt{SE} \rightarrow e_v > 0 \wedge \lambda_v^{to} = C^{to}(v, e_v, q_v) \wedge \lambda_v^{dur} = C^{dur}(v, e_v)$$
$$\wedge \lambda_v^{from} = C^{from}(v, e_v, \rho, w_v), \forall v \in V \tag{12}$$

$$r_v = \texttt{SE} \rightarrow s_v = 0 \wedge y(s_v) = 0 \wedge \sigma_v^{dur} = 0 \wedge c_v = 0$$
$$\wedge \sigma_v^{from} = 0 \wedge \sigma_v^{to} = 0, \forall v \in V \tag{13}$$

$$e_v > 0 \rightarrow r_v = \texttt{SE} \tag{14}$$

$$\Delta_{min}^{to}(v, e_v, q_v) \leq d_v^{to} \leq \Delta_{max}^{to}(v, e_v, q_v), \quad \forall v \in V \tag{15}$$

$$\Delta_{min}^{dur}(v, e_v) \leq d_v^{dur} \leq \Delta_{max}^{dur}(v, e_v), \quad \forall v \in V \tag{16}$$

$$\Delta_{min}^{from}(v, e_v, \rho, w_v) \leq d_v^{from} \leq \Delta_{max}^{from}(v, e_v, \rho, w_v), \quad \forall v \in V \tag{17}$$

$$\sigma_v^{to}, \sigma_v^{from}, c_v \geq 0, \quad \forall v \in V \tag{18}$$

The objective function (1) minimises the sum of the hotel costs and repositioning action costs minus the cost savings for SOS actions for the set of vessels. Constraints (2) and (3) specify that the vessels must all phase in to the new service on different, successive weeks. Constraints (4) and (5) specify that all vessels using SOS actions must use different actions and action types. `alldifferent_except_0` is a global constraint that requires all elements of an array to be different, except those that have the value 0.

Constraints (6) and (7) set the costs for a vessel if it uses a SAIL repositioning, and ensures that the SOS/SE actions and costs are set to 0, as they are not being used. Constraints (8) and (9) specify that if vessel v uses an SOS (SOS) repositioning action s_v, then its repositioning cost is equal to the costs for sailing to and from that SOS action based on the phase-out action, phase-in port and week, minus the cost savings $A(v, s_v)$ for that SOS action. In addition, the normal repositioning cost c_v and the sail equipment action for that vessel are set to 0. We also add redundant constraints (10) to reinforce that the repositioning type be set correctly when an SOS is chosen.

In constraints (11) we ensure that no two vessels choose the same SE action (unless they choose no SE action), and constraints (12) and (13) bind the costs of the sail equipment action to the dependent variables if an SE is chosen, as well as set the costs of a direct sailing and SOS opportunities for each vessel to 0. The redundant constraints (14) ensure that the repositioning type of vessel v is correctly set if an SE action is chosen. The minimum and maximum durations of the parts of the SE (sailing to the SE from the phase-out, the SE itself, and sailing from the SE to the phase-in) are set in constraints (15), (16) and (17). Constraint (18) requires that all SOS actions and phase-out/phase-in combinations must be valid for each vessel (i.e. transitions with -1 costs must not be used).

3.2 LSFRP CP Results

To compare our CP model for the LSFRP against an earlier MIP model and against the LTOP planner [32], we use 11 problem instances based on a real-world scenario provided by an industrial collaborator. The problem instances contain up to 3 vessels, with varying SOS and sail-with-equipment opportunities that may be used to reduce repositioning costs. Note that this version of the LSFRP lacks the cargo flows present in [33], but is nonetheless a real-world relevant problem.

The LSFRP CP model was formulated in the MiniZinc 1.6 modelling language [19, 18], and solved using the G12 finite domain solver [36]. We compare the CP against a MIP model and the LTOP planner [32], both using CPLEX 12.3. All problems were solved to optimality. Note that in our CP model for MiniZinc we had to add constraints on the maximum duration of SE actions, as well as a constraint on the maximum sum of the objective, in order to prevent integer overflows. These constraints do not cut off any valid solutions from the search tree. Since MiniZinc does not support floating point objective values, the MiniZinc model is a close approximation of the true objective.

We also used several search annotations within MiniZinc to help guide the solver to a solution. The first is to branch on the type of repositioning, r_v, before other variables, attempting at first to find a SOS option for each vessel, then searching through SE options, and finally SAIL options. This search order was the most efficient for the most complex models that include both SOS and SE opportunities, since SE constraints are more complex than SOS constraints, so searching SE options first is more time consuming for models that contain both SOS and SE opportunities.

For instances with SE opportunities, we also add a search annotation to branch on the SE opportunity, e_v, using the "indomain_split" functionality of MiniZinc, which excludes the upper half of a variable's domain. Both annotations use a first failure strategy, meaning the variable the solver branches on is the one with the smallest domain.

Table 1 compares the run times of the CP model against the MIP model and LTOP, all of which solve to optimality[1]. We ran the CP model with search annotations using a search order of SAIL/SOS/SE (CP–A), with search annotations and the SOS/SE/SAIL ordering (CP–AO), as well as with only the redundant constraints and using the solver's default search (CP–R), and using the SOS/SE/SAIL ordering with redundant constraints (CP–AOR). CP–AOR is faster than MIP on all instances, often by an order of magnitude. The main challenge for the CP model are instances with SE, same as for LTOP and the MIP. In fact, the CP model times out on two instances for CP and CP–R, as well as one for CP-A, whereas the MIP and LTOP solve all instances. Of particular note is that the CP model is able to solve AC3_3_3, which is the instance that most closely models

---

[1] Experiments used AMD Opteron 2425 HE processors with a maximum of 4GB each.

Table 1. Computation times in seconds to optimality for the CP model with and with annotations (A), repositioning type order (O), and redundant constraints (R) vs LTOP and the MIP

Problem	LTOP	MIP	CP	CP–A	CP–AO	CP–R	CP–AOR
AC3_1_0	1.1	0.4	0.1	0.2	0.1	0.2	0.1
AC3_2_0	51.0	9.3	0.3	0.3	0.3	0.3	0.3
AC3_3_0	188.3	23.0	0.6	0.6	0.6	0.6	0.6
AC3_1_1e	3.9	3.8	0.7	0.7	0.7	0.8	0.4
AC3_2_2ce	15.2	27.7	-	83.9	25.5	-	11.3
AC3_3_2c	203.2	250.5	6.0	3.1	3.1	6.2	3.0
AC3_3_2e	217.1	228.8	1731.0	15.8	16.2	1742.8	13.2
AC3_3_2ce1	218.2	312.2	32.6	23.0	16.7	31.5	13.9
AC3_3_2ce2	192.4	252.6	64.4	25.4	23.3	63.7	17.1
AC3_3_2ce3	516.9	706.5	-	-	695.4	-	470.3
AC3_3_3	80.0	148.3	18.1	11.0	11.0	18.6	11.2

the actual scenario faced by our collaborator, in only 10 seconds. Such a quick solution time allows for interaction and feedback with a repositioning coordinator within a decision support system. These results show that it is critical to choose a search strategy that quickly finds good quality solutions, as well as constraints that propagate well to cut out infeasible areas of the search. Note that no single improvement alone is enough to give the CP model better performance than both LTOP and the MIP.

Our CP model comes with two limitations. The first limitation is the model's flexibility. A natural extension to this model would be to allow for the chaining of SOS and SE opportunities, which is easy to do in both the LTOP and MIP models, due to automated planning's focus on actions, and our MIP model's focus on flows. However, the CP model is structured around exploiting this piece of the problem. Other natural changes, such as allowing vessels to undergo repairs, would also be difficult to implement. The second limitation is that many of the components of the CP model involve pre-computations that multiply the number of phase-out actions with the number of phase-in ports and weeks. Although the model works well on our real world instance, these pre-computations pose an issue for scaling to larger liner shipping services.

Constraint Programming (CP) is a very flexible method that can model many different types of scheduling and routing problems with complex side constraints. Our CP model for the Liner Shipping Fleet Repositioning Problem (LSFRP) was able to solve all of the problem instances faster than Mixed Integer Programming (MIP) or automated planning, and we also found that CP was much faster than MIP for the Bulk Port Cargo Throughput Optimisation Problem [14]. However, while CP can be very efficient at solving scheduling and routing problems, the solution time is highly dependent on having a good choice of model and search strategy, so applying CP to new applications requires a good understanding of both the application and CP modelling techniques.

4   Lazy Clause Generation

Kelareva *et al* [14] presented CP and MIP models for the Bulk Port Cargo Throughput Optimisation Problem (BPCTOP), and found that CP with a good choice of search strategy was much faster than MIP. However, the CP model solution time was highly dependent on the choice of modelling approach and search strategy used – a number of different modelling approaches and search strategies were investigated in [14, 15]. In this paper, we compare a LCG solver against a traditional finite domain CP solver for the BPCTOP model from [14], with all scalability improvements identified in [14, 15]

4.1   CP Model for Bulk Port Optimisation

Let V be the set of vessels to be scheduled. Let $[1, T_{max}]$ be the range of discretised time indices. Each vessel v has an earliest departure time $E(v)$, a maximum allowable draft $D(v, t)$ at each time t, and a tonnage per centimetre of draft $C(v)$. $ST(v_i, v_j)$ specifies the minimum separation time between the sailing times of every ordered pair of ships $v_i, v_j \in V$. Let B be the set of pairs of incoming and outgoing ships $B_i(b)$ and $B_o(b)$ indexed by b that use the same berth. For every such pair of ships, $d(b)$ is the minimum delay between their sailing times. The binary decision variable $s(v)$ specifies whether ship v is included in the schedule. Let $T(v) \in [1, T_{max}]$ be the decision variable specifying the scheduled sailing time for vessel v. The binary variable $sb(v_i, v_j)$ – SailsBefore(v_i, v_j) – is true iff vessel v_i sails earlier than vessel v_j, defined by Eq. (19).

$$sb(v_i, v_j) = 1 \leftrightarrow (T(v_i) < T(v_j) \wedge sb(v_j, v_i) = 0), \ \forall \ v_i, \ v_j \in V; v_i \neq v_j \quad (19)$$

Let U_{max} be the number of tugs available at the port. Let I and O be the sets of incoming and outgoing ships. $G(v)$ is the number of groups of tugs required for ship v, and $H(v, g)$ is the number of tugs in group g of vessel v. G_{max} is the maximum number of groups of tugs required for any ship. $r(v, g)$ is the "turnaround time" – the time taken for tugs in group g of ship v to become available for another ship in the same direction (incoming vs outgoing). $X(v_i, v_j)$ is the extra time required for tugs from incoming vessel v_i to become available for outgoing vessel v_o. $U(v, t, g)$ is a dependent variable that specifies the number of tugs busy in tug group g of vessel v at time t, assuming the next ship for these tugs is in the same direction. $x(v, t)$ defines the number of extra tugs that are busy at time t for an outgoing vessel v, due to still being in transit from the destination of an earlier incoming ship. Finally, $L(v, t)$ specifies that the extra tug delay time for incoming vessel v overlaps with the sailing time of an outbound vessel at time t. The model is formulated as follows:

$$\text{maximise} \sum_{v \in V} s(v) \cdot C(v) \cdot D(v, T(v)) \quad (20)$$

$$\text{s. t.} \quad s(v) = 1 \Rightarrow T(v) \geq E(v), \ \forall \ v \in V \quad (21)$$

$$s(B_i(b)) = 1 \Rightarrow s(B_o(b)) = 1 \wedge T(B_o(b)) \leq T(B_i(b)) - d(b), \ \forall \ b \in B \quad (22)$$

$$sb(v_i, v_i) = 0, \ \forall \ v_i \in V; v_i \neq v_j \quad (23)$$

$$s(v_i) = 1 \wedge s(v_j) = 1 \Rightarrow \quad (24)$$

$$(sb(v_i, v_j) = 1 \Rightarrow T(v_j) - T(v_i) \geq ST(v_i, v_j)), \ \forall \ v_i, \ v_j \in V$$

$$s(v) = 1 \wedge t \geq T(v) \wedge t < T(v) + r(v, g) \Rightarrow U(v, t, g) = H(v, g), \quad (25)$$

$$\forall \ v \in V, \ t \in [1, T_{max}], \ g \in [1, G_{max}]$$

$$s(v) = 0 \vee t < T(v) \vee t \geq T(v) + r(v, g) \Rightarrow U(v, t, g) = 0, \quad (26)$$

$$\forall \ v \in V, \ t \in [1, T_{max}], \ g \in [1, G_{max}]$$

$$x(v_o, t) = 0, \quad \forall \, v_o \in O, \, t \in [1, T_{max}] \tag{27}$$

$$L(v_i, t) \iff \exists \, v_o \in O \text{ s.t.} \tag{28}$$
$$t = T(v_o) \wedge T(v_i) \leq T(v_o) \wedge T(v_i) + \max_{g \in [1, G(v_i)]} r(v_i, g) + X(v_i, v_o) > T(v_o),$$
$$\forall \, v_i \in I, \, t \in [1, T_{max}]$$

$$x(v_i, t) = \text{bool2int}(L(v_i, t)). \sum_{g \in [1, G(v_i)]} H(v_i, g), \quad \forall \, v_i \in I, \, t \in [1, T_{max}] \tag{29}$$

$$\sum_{v \in I} \sum_{g \in G(v)} U(v, t, g) \leq U_{max}, \quad \forall \, t \in [1, T_{max}] \tag{30}$$

$$\sum_{v_o \in O} \sum_{g \in G(v_o)} U(v_o, t, g) + \sum_{v_i \in I} x(v_i, t) \leq U_{max}, \quad \forall \, t \in [1, T_{max}] \tag{31}$$

The objective function (20) maximises total cargo throughput for the set of ships. Constraint (21) specifies the earliest departure time for each vessel. Equation (22) ensures that the berths for any incoming ships are empty before the ship arrives. Equation (23) specifies that no ship sails before itself. Equation (24) ensures that there is sufficient separation time between successive ships to meet port safety requirements. Equations (25) to (31) specify that the total number of tugs in use at any time must be no greater than the number of tugs available at the port, by splitting the tug constraints into several scenarios that can be considered independently as discussed in [14]. For a more detailed discussion of this model, including scalability issues, see [14, 15].

4.2 Experimental Results for Lazy Clause Generation

We used a CP solver with Lazy Clause Generation (LCG) to solve the above CP model for the BPCTOP. The model was formulated in the MiniZinc modelling language [19], and solved using the CPX solver included in G12 2.0 [36, 11], which uses Lazy Clause Generation. The runtimes were then compared against the G12 2.0 finite domain CP solver, using backtracking search with the fastest variable selection and domain reduction strategies as discussed in [14, 15]. All BPCTOP experiments used an Intel i7-930 quad-core 2.80 GHz processor and 12.0 GB RAM, with a 30-minute cutoff time.

Both solvers were used to solve problems presented in [14], with 4-13 ships sailing on a single tide, based on a fictional but realistic port, similar to the SHIP_SCHEDULING data set used for the 2011 MiniZinc challenge [34]. Table 2 presents calculation times in seconds for CPX and the G12 FD solver, for each problem size that could be solved to optimality within the 30-minute cutoff time, for the problems in the most tightly constrained (and thus most difficult) ONEWAY_NARROW (ON) problem set from [14]. A dash indicates that the optimal solution was not found within the cutoff time. Table 3 shows the largest problem solved for all eight problem types from [14], with runtimes in brackets. Problem types include all combinations of MIXED problems with inbound and outbound ships vs. all ships sailing ONEWAY, NARROW vs WIDE safe sailing windows for each ship, with and without tugs.

Table 2 shows CPX is faster to solve ON problems with tug constraints, and scales better than the FD solver for large problem sizes. However, CPX is slower to solve ON problems without tugs. This may indicate that CPX finds effective nogoods (areas of the search space with no good solutions) for tug constraints, enabling CPX to avoid

Table 2. Runtime (s) for CPX solver with LCG vs. G12 finite domain solver

NShips	No Tugs		With Tug Constraints	
	G12_FD	G12_CPX	G12_FD	G12_CPX
4	0.34	**0.23**	**0.23**	0.33
5	**0.23**	**0.22**	**0.33**	**0.33**
6	0.44	**0.22**	**0.44**	0.67
7	**0.34**	**0.33**	2.95	3.60
8	**0.45**	0.87	47.0	**24.4**
9	**4.70**	23.4	184	**65.7**
10	**99.9**	482	>1800	**817**
11	**609**	>1800	–	>1800
12	>1800	–	–	–

Table 3. CPX solver vs. G12 finite domain solver: largest problem solved for each type, with runtime in seconds in brackets. Bold indicates faster method for this problem type.

Problem	FD	CPX
MIXED_WIDE (MW)	11 (326)	**16 (1040)**
ONEWAY_WIDE (OW)	11 (1480)	**11 (273)**
MIXED_NARROW (MN)	11 (370)	**16 (1100)**
ONEWAY_NARROW (ON)	**11 (609)**	10 (482)
MIXED_WIDE_TUGS (MWT)	10 (11.5)	**13 (1290)**
ONEWAY_WIDE_TUGS (OWT)	9 (81.5)	**11 (1680)**
MIXED_NARROW_TUGS (MNT)	10 (202)	**12 (1350)**
ONEWAY_NARROW_TUGS (ONT)	9 (184)	**10 (817)**

searching large areas of the search space for the problem with tug constraints. The FD solver, on the other hand, cannot eliminate those areas of the search space, leading to excessive backtracking resulting from the highly oversubscribed tug problem.

Table 3 shows that the difference between the CPX and FD solvers is even greater for MIXED problems (less constrained problem with more complex constraints). This indicates that CPX is very fast at dealing with complex constraints, but the speed difference decreases when the problem is tightly constrained. CPX is able to solve larger problems faster for all problem types except for ONEWAY_NARROW without tugs – this is the most constrained problem type, using the simplest constraints (no tugs, and no interaction between incoming and outgoing ships).

The slower performance of CPX on the problem without tugs indicates that there may be room for improvement if better explanations are added for constraints that do not involve tugs, such as the sequence-dependent setup times between ships, and the propagation of the objective function itself. As the tug problem is composed of the No-Tugs problem with additional constraints, speeding up the solution time of the NoTugs problem would likely further improve the solution time of the Tugs problem.

Lazy Clause Generation (LCG) is a very general method that has been successfully used to speed up calculation times for many different types of scheduling problems. However, as it is a recent method, its effectiveness for many other problem types has not yet been investigated. Like other complete methods, LCG does not scale well to large problems. However, like traditional CP solvers, LCG solvers can be combined with decomposition techniques or large neighbourhood search to improve scalability [27]. LCG may be worth investigating as an approach to dealing with other routing and scheduling problems with time-dependent task costs, and we plan to compare LCG against an FD solver for the LSFRP in future work.

5   Solve and Improve

Many scheduling and routing approaches initially solve a simplified model, and then use the constraints and objective function of the full problem to improve it. For routing and scheduling problems with time-dependent action costs, removing the time dependence of the objective function is one way to simplify the problem for the first step of a solve-and-improve approach, as used by Lin *et al* [17].

In this section, we solve simplified models for both problems which ignore time-varying action costs, and compare the improvement in runtime against improvements obtained by the LTOP for the LSFRP and the LCG solver for the BPCTOP. The runtimes for the simplified models are a lower bound on the runtime of a full solve-and-improve approach, as the improvement step would increase the runtime further.

5.1   Results for Bulk Port Cargo throughput Optimisation

We implemented a simplified BPCTOP model by replacing the time-varying objective function by feasible time windows, and solving the problem with the objective of maximising the number of ships scheduled to sail. Table 4 compares the runtimes of the normal and simplified models for the largest problem that was successfully solved by all approaches, as well as for the largest problem solved by any approach. Bold font indicates the fastest runtime, (ie. the approach that results in the largest reduction in runtime).

Table 4 shows that, while the simplified model is faster for small problem sizes, for large problem sizes the CPX solver with Lazy Clause Generation is faster than the simplified model to solve 5 of the 8 problem types, indicating that CPX scales better than the simplified model for 5 of the 8 problem types. As seen earlier in Table 3, CPX is particularly effective on large problems with tugs.

Table 4. Runtime (s) of CPX vs. the FD solver with a simplified BPCTOP model

Problem Type	NShips (small)	FD	CPX	SIMPLIFIED	NShips (large)	FD	CPX	SIMPLIFIED
MW	11	326	**8.19**	188	16	>1800	**1040**	>1800
OW	11	1480	273	**0.56**	12	>1800	>1800	**26.7**
MN	11	370	**7.64**	180	16	>1800	**1100**	>1800
ON	10	99.9	482	**0.34**	12	>1800	>1800	**19.6**
MWT	10	11.5	17.1	**11.4**	13	>1800	**1290**	>1800
OWT	9	81.5	23.9	**4.60**	11	>1800	**1680**	>1800
MNT	10	202	42.8	**8.21**	12	>1800	**1350**	>1800
ONT	9	184	65.7	**0.69**	10	>1800	817	**262**

One interesting observation from Table 4 is that the simplified model gives the largest runtime improvement for the most tightly constrained ON and ONT problems, allowing problems with one more ship to be solved within the 30-minute cutoff time. The least tightly constrained MW and MWT problems show only a small improvement in runtime from relaxing the time window penalties; and the moderately constrained MN, MNT, OW and OWT problems show moderate improvements. This result is similar to the effect of relaxing hard time windows in a vehicle routing problem [24]. Extending the latest delivery time by 10-20 minutes was found to significantly improve schedule costs for tightly constrained problems. For problems with wide time windows, on the other hand, where the time windows did not constrain the problem, relaxing the time windows had little effect on cost and also increased runtime due to increasing the computational complexity of the problem.

5.2   Results for Liner Shipping Fleet Repositioning

We implemented a simplified LSFRP model by fixing all sailing actions to "slow-steaming", i.e. minimum fuel cost with maximum time. This constraint was subsequently loosened for any instances that did not solve to optimality. The problems were then solved using the LTOP planner [32].

Table 5 shows that, while CPU time for most problem instances was reduced, the improved runtimes were still slower than the best CP model (CP-AOR). We can therefore conclude that fixing the length and cost of the sail action is not very effective for the LSFRP, since the problems where the most benefit can be expected from fixing action costs are those in which the optimal answer uses all slow-steaming actions.

A solve-and-improve approach that simplifies away time-varying action costs was found by Lin *et al* [17] to be effective for satellite imaging scheduling. However, they did not compare the calculation speed of the complete problem against the simplified problem, so there is no indication of the improvement in calculation speed produced by simplifying away the time-varying quality function. However, our experiments found that simplifying the LSFRP and BPCTOP models by removing the time-varying cost function did not produce significant speed improvement, and that switching to a Constraint Programming model or a solver with Lazy Clause Generation was more effective.

Table 5. CPU time (s) of CP vs. LTOP vs. simplified LSFRP with optimal windows

Problem	Normal LTOP	Simplified LTOP	CP-AOR
AC3_1_0	1.1	0.82	0.1
AC3_2_0	51.0	38.4	0.3
AC3_3_0	188.3	118	0.6
AC3_1_1e	3.9	3.72	0.4
AC3_2_2ce	15.2	15.9	11.3
AC3_3_2c	203.2	265	3.0
AC3_3_2e	217.1	233	13.2
AC3_3_2ce1	218.2	234	13.9
AC3_3_2ce2	192.4	210	17.1
AC3_3_2ce3	516.9	548	470.3
AC3_3_3	80.0	86.6	11.2

It is possible that simplifying the cost function was more effective for speeding up solution times for Linear Programming problems such as [17], rather than for CP or automated planning. However, more work would need to be done to identify the speed improvement produced by simplifying the quality function in [17].

6   Conclusions and Future Work

While scheduling and routing problems have usually been solved using Mixed Integer Programming (MIP) and solve-and-improve approaches, Constraint Programming (CP) with a good choice of model and search strategy, as well as recent techniques such as Lazy Clause Generation (LCG) have been found to be faster for some problem types.

In this paper, we presented a novel CP model for the Liner Shipping Fleet Repositioning Problem (LSFRP) and compared it against existing MIP and planning models, and found that the CP model was faster than both existing approaches, by an order of magnitude for some instances. CP was also found to be faster than MIP for the Bulk Port Cargo Throughput Optimisation Problem (BPCTOP) in an earlier paper [14]. We also compared a CP solver that uses LCG against a traditional finite domain CP solver for the BPCTOP, and found that LCG was faster for 7 out of 8 problem types. We plan to present a more detailed comparison of different CP models and search strategies, as well as an LCG solver for the LSFRP in a future paper.

Both the problems we investigated in this paper have time-varying action costs, which increase the complexity of the problem. Solve-and-improve approaches have previously used simplified models without time-varying action costs to find initial solutions [17]. We compared the LTOP model of the LSFRP and the full CP model for the BPCTOP against simplified models without time-varying action costs, and found that the speed improvement of removing time-varying costs was less than that obtained by converting the LSFRP to a CP model for all problem instances, and that solving the BPCTOP using an LCG solver scaled better than using a finite domain solver for 5 of 8 problem types, particularly for the most challenging problems with complex tug constraints. While our previous approaches were able to solve realistic-sized problems, the long calculation times limited their usefulness, and the speed improvements presented here may open up new applications such as using the LSFRP as a subproblem of fleet redeployment problems.

In our investigation of both the LSFRP and BPCTOP CP models, we found that the CP model solution time was highly dependent on having a good choice of model and search strategy. However, with this caveat, we find that CP and LCG are efficient and flexible methods that are able to handle complex side constraints, and may therefore be worth investigating for other scheduling and routing problems that are currently being solved using MIP or solve-and-improve approaches.

Acknowledgements. The authors would like to acknowledge the support of ANU and NICTA at which Elena Kelareva is a PhD student. NICTA is funded by the Australian Government as represented by the Department of Broadband, Communications and the Digital Economy and the Australian Research Council through the ICT Centre of Excellence program. We would also like to thank OMC International for their support for the research into the port optimisation problem. The research into the fleet repositioning problem is sponsored in part by the Danish Council for Strategic Research as part of the ENERPLAN research project.

References

[1] Achterberg, T.: SCIP: solving constraint integer programs. Mathematical Programming Computation 1(1), 1–41 (2009)

[2] Christiansen, M., Fagerholt, K.: Robust ship scheduling with multiple time windows. Naval Research Logistics 49(6), 611–625 (2002)

[3] Christiansen, M., Fagerholt, K., Nygreen, B., Ronen, D.: Chapter 4: Maritime transportation. In: Barnhart, C., Laporte, G. (eds.) Transportation. Handbooks in Operations Research and Management Science, vol. 14, pp. 189–284. Elsevier (2007)

[4] Christiansen, M., Fagerholt, K., Ronen, D.: Ship routing and scheduling: status and perspectives. Transportation Science 38(1), 1–18 (2004)

[5] Chu, G., de la Banda, M.G., Mears, C., Stuckey, P.J.: Symmetries and lazy clause generation. In: Proceedings of the 16th International Conference on Principles and Practice of Constraint Programming (CP 2010) Doctoral Programme, pp. 43–48 (September 2010)

[6] Coles, A.J., Coles, A.I., Fox, M., Long, D.: Forward-Chaining Partial-Order Planning. In: Proceedings of the Twentieth International Conference on Automated Planning and Scheduling (ICAPS 2010) (May 2010)

[7] Cordeau, J.-F., Desaulniers, G., Desrosiers, J., Solomon, M.M., Soumis, F.: VRP with time windows. In: Toth, P., Vigo, D. (eds.) The Vehicle Routing Problem. SIAM Monographs on Discrete Mathematics and Applications, vol. 9, ch. 7, pp. 157–194. SIAM (2002)

[8] Fagerholt, K.: Ship scheduling with soft time windows: an optimisation based approach. European Journal of Operational Research 131(3), 559–571 (2001)

[9] Fagerholt, K.: A computer-based decision support system for vessel fleet scheduling - experience and future research. Decision Support Systems 37(1), 35–47 (2004)

[10] Fagerholt, K., Laporte, G., Norstad, I.: Reducing fuel emissions by optimizing speed on shipping routes. Journal of the Operational Research Society 61, 523–529 (2010)

[11] Feydy, T., Stuckey, P.J.: Lazy clause generation reengineered. In: Gent, I.P. (ed.) CP 2009. LNCS, vol. 5732, pp. 352–366. Springer, Heidelberg (2009)

[12] Figliozzi, M.A.: The time dependent vehicle routing problem with time windows: Benchmark problems, an efficient solution algorithm, and solution characteristics. Transportation Research Part E: Logistics and Transportation Review 48(3), 616–636 (2012)

[13] Fox, M., Long, D.: PDDL2.1: An extension to PDDL for expressing temporal planning domains. Journal of Artificial Intelligence Research 20(1), 61–124 (2003)

[14] Kelareva, E., Brand, S., Kilby, P., Thiébaux, S., Wallace, M.: CP and MIP methods for ship scheduling with time-varying draft. In: Proceedings of the 22nd International Conference on Automated Planning and Scheduling (ICAPS 2012), pp. 110–118 (June 2012)

[15] Kelareva, E., Kilby, P., Thiébaux, S., Wallace, M.: Ship scheduling with time-varying draft: Constraint programming and benders decomposition. Transportation Science (2012) (submitted)

[16] Kilby, P., Verden, A.: Flexible routing combing constraint programming, large neighbourhood search, and feature-based insertion. In: Schill, K., Scholz-Reiter, B., Frommberger, L. (eds.) Proceedings 2nd Workshop on Artificial Intelligence and Logistics (AILOG 2011), pp. 43–49 (2011)

[17] Lin, W.C., Liao, D.Y., Liu, C.Y., Lee, Y.Y.: Daily imaging scheduling of an earth observation satellite. IEEE Transactions on Systems, Man and Cybernetics, Part A: Systems and Humans 35(2), 213–223 (2005)

[18] Nethercote, N., Marriott, K., Rafeh, R., Wallace, M., de la Banda, M.G.: Specification of Zinc and MiniZinc (November 2010)

[19] Nethercote, N., Stuckey, P.J., Becket, R., Brand, S., Duck, G.J., Tack, G.: MiniZinc: Towards a standard CP modelling language. In: Bessière, C. (ed.) CP 2007. LNCS, vol. 4741, pp. 529–543. Springer, Heidelberg (2007)

[20] Norstad, I., Fagerholt, K., Laporte, G.: Tramp ship routing and scheduling with speed optimization. Transportation Research 19, 853–865 (2011)

[21] Ohrimenko, O., Stuckey, P.J., Codish, M.: Propagation via lazy clause generation. Constraints 14(3), 357–391 (2009)

[22] OMC International. DUKC helps Port Hedland set ship loading record (2009), http://www.omc-international.com/images/stories/press/omc-20090810-news-in-wa.pdf

[23] Port Hedland Port Authority. 2009/10 cargo statistics and port information (2011), http://www.phpa.com.au/docs/CargoStatisticsReport.pdf

[24] Qureshi, A.G., Taniguchi, E., Yamada, T.: An exact solution approach for vehicle routing and scheduling problems with soft time windows. Transportation Research Part E: Logistics and Transportation Review 45(6), 960–977 (2009)

[25] Rakke, J.G., Christiansen, M., Fagerholt, K., Laporte, G.: The traveling salesman problem with draft limits. Computers & Operations Research 39, 2161–2167 (2011)

[26] Russell, A.H.: Cash flows in networks. Management Science 16(5), 357–373 (1970)

[27] Schutt, A., Chu, G., Stuckey, P.J., Wallace, M.G.: Maximising the net present value for resource-constrained project scheduling. In: Beldiceanu, N., Jussien, N., Pinson, É. (eds.) CPAIOR 2012. LNCS, vol. 7298, pp. 362–378. Springer, Heidelberg (2012)

[28] Schutt, A., Feydy, T., Stuckey, P.J., Wallace, M.G.: Solving the resource constrained project scheduling problem with generalised precedences by lazy clause generation (2010), http://arxiv.org/abs/1009.0347

[29] Sexton, T.R., Choi, Y.: Pickup and delivery of partial loads with soft time windows. American Journal of Mathematical and Management Science 6, 369–398 (1985)

[30] Smith, S.: Is scheduling a solved problem? In: Multidisciplinary Scheduling: Theory and Applications, pp. 3–17 (2005)

[31] Song, J.-H., Furman, K.C.: A maritime inventory routing problem: Practical approach. Computers & Operations Research (2010)

[32] Tierney, K., Coles, A., Coles, A., Kroer, C., Britt, A.M., Jensen, R.M.: Automated planning for liner shipping fleet repositioning. In: Proceedings of the 22nd International Conference on Automated Planning and Scheduling (ICAPS 2012), pp. 279–287 (June 2012)

[33] Tierney, K., Jensen, R.M.: The liner shipping fleet repositioning problem with cargo flows. In: Hu, H., Shi, X., Stahlbock, R., Voß, S. (eds.) ICCL 2012. LNCS, vol. 7555, pp. 1–16. Springer, Heidelberg (2012)

[34] University of Melbourne. MiniZinc Challenge 2011 (2011), http://www.g12.cs.mu.oz.au/minizinc/challenge2011/challenge.html

[35] Vanhoucke, M., Demeulemeester, E.L., Herroelen, W.S.: On maximizing the net present value of a project under renewable resource constraints. Management Science 47, 1113–1121 (2001)

[36] Wallace, M.: G12 - Towards the Separation of Problem Modelling and Problem Solving. In: van Hoeve, W.-J., Hooker, J.N. (eds.) CPAIOR 2009. LNCS, vol. 5547, pp. 8–10. Springer, Heidelberg (2009)

[37] Wang, J., Jing, N., Li, J., Chen, H.: A multi-objective imaging scheduling approach for earth observing satellites. In: Proceedings of the 9th Annual Conference on Genetic and Evolutionary Computation (GECCO 2007), pp. 2211–2218 (2007)

[38] Wolfe, W.J., Sorensen, S.E.: Three scheduling algorithms applied to the earth observing systems domain. Management Science 46(1), 148–168 (2000)

[39] Yao, F., Li, J., Bai, B., He, R.: Earth observation satellites scheduling based on decomposition optimization algorithm. International Journal of Image, Graphics and Signal Processing 1, 10–18 (2010)

An MDD Approach
to Multidimensional Bin Packing[*]

Brian Kell[1] and Willem-Jan van Hoeve[2]

[1] Department of Mathematical Sciences, Carnegie Mellon University
bkell@cmu.edu
[2] Tepper School of Business, Carnegie Mellon University
vanhoeve@andrew.cmu.edu

Abstract. We investigate the application of multivalued decision diagrams (MDDs) to multidimensional bin packing problems. In these problems, each bin has a multidimensional capacity and each item has an associated multidimensional size. We develop several MDD representations for this problem, and explore different MDD construction methods including a new heuristic-driven depth-first compilation scheme. We also derive MDD restrictions and relaxations, using a novel application of a clustering algorithm to identify approximate equivalence classes among MDD nodes. Our experimental results show that these techniques can significantly outperform current CP and MIP solvers.

1   Introduction

Many related problems in combinatorial optimization are collectively referred to as "bin packing problems." In the classical bin packing problem, the input is a list $(s_1, \ldots, s_n)$ of item sizes, each in the interval $(0, 1]$, and the objective is to pack the n items into a minimum number of bins of capacity 1.

In this paper we study a multidimensional variant of the bin packing problem, presented as a satisfaction problem. An instance of this problem consists of a list $(s_1, \ldots, s_n)$ of item sizes and a list $(c_1, \ldots, c_m)$ of bin capacities. Each item size and each bin capacity is a d-tuple of nonnegative integers; e.g., $s_i = (s_{i,1}, \ldots, s_{i,d})$. The objective is to assign each of the n items to one of the m bins in such a way that, for every bin and in every dimension, the total size of the items assigned to the bin does not exceed the bin capacity.

This can be viewed as a constraint satisfaction problem (CSP) with n variables and md constraints. Each variable x_i has domain $\{1, \ldots, m\}$ and denotes the bin into which the ith item is placed. The constraints require that $\sum_{i:x_i=j} s_{i,k} \leq c_{j,k}$ for all $j \in \{1, \ldots, m\}$ and all $k \in \{1, \ldots, d\}$.

Note that the "dimensions" in this problem should not be interpreted as geometric dimensions. In this way the problem studied here differs from the two- and three-dimensional bin packing problems studied, for example, in [10,11],

---

[*] This work was supported by the NSF under grant CMMI-1130012 and a Google Research Grant.

C. Gomes and M. Sellmann (Eds.): CPAIOR 2013, LNCS 7874, pp. 128–143, 2013.

in which the items and bins are geometric rectangles or cuboids. Rather, the dimensions in the problem studied in this paper correspond to independent one-dimensional bin packing constraints that must be satisfied simultaneously.

Multidimensional bin packing (MBP) problems of the kind considered in this paper appear in practice. For example, the Google ROADEF/EURO challenge 2011–2012[1] involves a set of machines with several resources, such as RAM and CPU, running processes which consume those resources. However, these problems have received relatively little attention in the literature. Current CP methods are weak on problems involving simultaneous bin packing constraints. Current MIP methods do better but are still limited in their effectiveness.

In this paper we make the following contributions. We present a new generic exploratory construction algorithm for multivalued decision diagrams (MDDs) and a novel application of the median cut algorithm in the construction of approximate MDDs. We also describe several techniques specific to the use of MDDs for the MBP problem. Our experimental results show that such techniques can yield an improvement on existing methods.

The remainder of the paper is organized as follows. In Section 2 we present several generic approaches to the construction of MDDs. The focus of Section 3 is approximate MDDs, which represent sets of solutions to relaxations or restrictions of problem instances. In Section 4 we discuss techniques that can be used to apply MDDs to the MBP problem. In Section 5 we present experimental results comparing the performance of the techniques described in this paper to that of commercial CP and MIP solvers. We conclude in Section 6.

2   MDD Construction

In this section we present a generic algorithm for the construction of an MDD representing the set of feasible solutions to a CSP. A CSP is specified by a set of constraints $\{C_1, \ldots, C_p\}$ on a set of variables $\{x_1, \ldots, x_n\}$ having domains $D_1, \ldots, D_n$, respectively. A *solution* to a CSP is an n-tuple $(y_1, \ldots, y_n) \in D_1 \times \cdots \times D_n$. A solution is *feasible* if the set of assignments $x_1 = y_1, \ldots, x_n = y_n$ satisfies every constraint C_j.

A *multivalued decision diagram* (MDD) [13] is an edge-labeled acyclic directed multigraph whose nodes are arranged in $n+1$ layers $L_1, \ldots, L_{n+1}$. The layer L_1 consists of a single node, called the *root*. Every edge in the MDD is directed from a node in L_i to a node in L_{i+1}. All of the edges directed out of a node have distinct labels. The nodes in layer L_{n+1} are called *sinks* or *terminals*. In this paper we shall primarily be interested in MDDs having a single sink (which represents feasibility), but the ideas can easily be generalized to MDDs with multiple sinks [13].

Let I be an instance of a CSP. An MDD M can be used to represent a set of solutions to I as follows [1]. The layers $L_1, \ldots, L_n$ correspond respectively to the variables $x_1, \ldots, x_n$ in I. An edge directed from a node in L_i to a node in L_{i+1} and having the label y_i, where $y_i \in D_i$, corresponds to the assignment

---

[1] Online: http://challenge.roadef.org/2012/en/index.php

$x_i = y_i$. Therefore a path from the root to the sink along edges labeled $y_1, \ldots, y_n$ corresponds to the solution $(y_1, \ldots, y_n)$. The MDD M represents the set $\mathcal{M}$ of solutions corresponding to all such paths. Let $\mathcal{F}$ denote the set of feasible solutions to I. If $\mathcal{M} = \mathcal{F}$, $\mathcal{M} \supseteq \mathcal{F}$, or $\mathcal{M} \subseteq \mathcal{F}$, then M is said to be an *exact MDD*, a *relaxation MDD*, or a *restriction MDD* for I, respectively [1,3,4,6]. We shall consider only exact MDDs in this section; relaxation and restriction MDDs will be considered in Section 3.

A path in an MDD from the root to a node in the layer L_{i+1} represents a *partial solution* $y = (y_1, \ldots, y_i) \in D_1 \times \cdots \times D_i$; we shall say that the *level* of this partial solution is i and write $\text{level}(y) = i$. Let $\mathcal{F}(y)$ denote the set of *feasible completions* of this partial solution, that is, $\mathcal{F}(y) = \{ z \in D_{i+1} \times \cdots \times D_n \mid (y, z) \text{ is feasible} \}$. If y and y' are partial solutions with $\mathcal{F}(y) = \mathcal{F}(y')$, then we say that y and y' are *equivalent*. Note that in an exact MDD all paths from the root to a fixed node v represent equivalent partial solutions, and conversely if two partial solutions y and y' are equivalent then the paths in an exact MDD that correspond to y and y' can lead to the same node.

Direct MDD Representation for Multidimensional Bin Packing. Let I be an MBP instance, having n items and m bins. A direct MDD representation of the set of feasible solutions of I has layers $L_1, \ldots, L_n$ corresponding to the variables $x_1, \ldots, x_n$, and also the last layer L_{n+1} which contains the sink. The edge labels are elements of $\{1, \ldots, m\}$. A path from the root to the sink along edges labeled $y_1, \ldots, y_n$ represents the feasible solution $(y_1, \ldots, y_n)$, that is, the feasible solution in which item i is placed into bin y_i.

For example, Figure 1a shows the direct MDD representation for a one-dimensional bin packing instance having two bins, each of capacity 7, and four items, with sizes 5, 3, 2, and 1. There are six paths from the root to the sink, representing the six feasible solutions; for instance, the path following the edges labeled 2, 1, 1, 2 corresponds to the solution in which the item of size 5 is packed in bin 2, the items of size 3 and 2 are packed in bin 1, and the item of size 1 is packed in bin 2. The node labels are states; we discuss these next.

Equivalence of Partial Solutions. Equivalent partial solutions have the same set of feasible completions. Hence, the recognition that two partial solutions are equivalent reduces the size of the MDD, because the corresponding paths can lead to the same node.

In general, determining whether two partial solutions are equivalent is NP-hard for the MBP problem (because it is NP-hard even to determine whether a one-dimensional instance is feasible). However, we can sometimes determine that two partial solutions are equivalent by associating partial solutions with "states." A *state function* for layer i is a map σ_i from the set $Y_i = D_1 \times \cdots \times D_{i-1}$ of partial solutions at layer i into some set S_i of *states*, such that $\sigma_i(y) = \sigma_i(y')$ implies $\mathcal{F}(y) = \mathcal{F}(y')$. In other words, two partial solutions that lead to the same state have the same set of feasible completions. (A "perfect" state function would also allow us to say that two partial solutions that lead to different states have

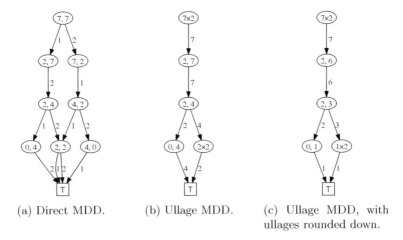

(a) Direct MDD.   (b) Ullage MDD.   (c) Ullage MDD, with
                                                  ullages rounded down.

Fig. 1. MDD representations of a one-dimensional bin packing instance having two
bins, each of capacity 7, and four items, with sizes 5, 3, 2, and 1

different sets of feasible completions, and we strive for this ideal, but for practical
reasons our state function should be easy to compute, so we cannot require this.)

Consider an MBP instance. After items having sizes $s_{i_1}, \ldots, s_{i_k}$ have been
placed into a bin of capacity c_j, the remaining capacity of the bin is $c_j - \sum_{l=1}^{k} s_{i_l}$.
(Recall that the item sizes and bin capacities are d-tuples; here and elsewhere
addition and subtraction of d-tuples is done componentwise.) We shall call this
remaining capacity the *ullage* of the bin; it is a d-tuple. (The word "ullage"
means "the amount by which a container falls short of being full.") Of course,
the ullage of each bin is nonincreasing (componentwise) as the items are placed
one by one into the bins.

A useful state function for the direct MDD representation is the map σ_i
from a partial solution $y = (y_1, \ldots, y_{i-1})$ to the list $(u_1, \ldots, u_m)$ of the ullages u_j of the m bins; in other words, for $j \in \{1, \ldots, m\}$, we take $u_j =$
$c_j - \sum_{k \in \{1, \ldots, i-1\}: y_k = j} s_k$. For example, in Fig. 1a, the path from the root along
the edges labeled 1, 2, 2 represents a partial solution for which the ullages of
the two bins are each 2, so the state of this partial solution is $(2, 2)$. The partial
solution corresponding to the path 2, 1, 1 has the same state. Observe that if
two partial solutions at layer i have the same lists of ullages, then they have the
same set of feasible completions, so this is indeed a state function.

Exact MDD Construction. Behle [2] described a top-down algorithm for
the construction of threshold binary decision diagrams (BDDs), which are exact
representations of solution sets of instances of 0–1 knapsack problems. A general
algorithm for a top-down, layer-by-layer (i.e., breadth-first) construction of an
MDD is presented as Algorithm 1, "Top-down MDD compilation," in Bergman
et al. [4]. The key to the top-down construction of an MDD is the identification
of a *node equivalence test*, which determines when two nodes on the same layer

(each representing one or more partial solutions) have the same set of feasible completions; this is exactly what a state function does.

So far we have spoken of the states of partial solutions. We shall now extend this idea to states of nodes in an MDD. In the MDD that we construct, partial solutions at layer i that lead to the same state will correspond to paths from the root that lead to the same node; we shall associate this state with this node. Now, given a node v in layer L_i in the MDD and its state, which we shall write as state(v), and given a value $y_i \in D_i$, we can determine the state of a child node w of v if the edge (v, w) has label y_i. This is simply the state of the feasible solution (y, y_i), where y is any feasible solution corresponding to the node v. For instance, suppose v is a node in layer L_i of a direct bin packing MDD, and suppose the state of v (i.e., the corresponding list of ullages) is $(u_1, \ldots, u_m)$. Then the child state corresponding to $y_i \in D_i$ is $(u_1, \ldots, u_{y_i-1}, u_{y_i} - s_i, u_{y_i+1}, \ldots, u_m)$.

To be more precise, and to make these ideas applicable to generic CSPs, we make the following definitions. Let $i \in \{1, \ldots, n+1\}$. Let $Y_i = D_1 \times \cdots \times D_{i-1}$ denote the set of partial solutions at level i; take $Y_1 = \{\emptyset\}$, a singleton set having one element representing the empty partial solution. Let S_i be an arbitrary set whose elements are called *states* and which contains a special element $\perp$ indicating infeasibility. Recall that we say that $\sigma_i : Y_i \to S_i$ is a *state function* if $\sigma_i(y) = \sigma_i(y')$ implies $\mathcal{F}(y) = \mathcal{F}(y')$; we also require that $\sigma_i(y) = \perp$ implies $\mathcal{F}(y) = \emptyset$. We assume that we can test the feasibility of a (complete) solution, so for $y \in Y_{n+1}$ we require that $\sigma_{n+1}(y) = \perp$ if $\mathcal{F}(y) = \emptyset$. For $i \in \{1, \ldots, n\}$, we say that $\chi_i : S_i \times D_i \to S_{i+1}$ is a *child state function* if $\chi_i(\sigma_i(y), y_i) = \sigma_{i+1}(y, y_i)$ for all $y \in Y_i$ and all $y_i \in D_i$.

In order to use state information effectively in the construction of an MDD, we must maintain, for each layer L_i, a mapping from states to nodes that have already been constructed in L_i. When we seek a node in L_i having state s, we consult this mapping to see if such a node already exists. Such a mapping can be implemented with a hash table. It is often called the *unique table* because it ensures that the node representing state s in layer L_i is unique [9].

Algorithm 1 constructs an exact MDD. For each $i \in \{1, \ldots, n+1\}$ let σ_i be a state function, and for each $i \in \{1, \ldots, n\}$ let χ_i be a corresponding child state function. Let r be the root node. The algorithm maintains a collection T of nodes to be processed, i.e., nodes whose children need to be constructed. When a node v in layer i is processed, each possible domain value $y \in D_i$ is considered, and the corresponding child state s is computed. The unique table is consulted to see if a node w with state s already exists in layer L_{i+1}; if not, a new node w is constructed and added to T. Then the edge (v, w) is added to the MDD with label y. This is repeated until all nodes have been processed.

Exploratory Construction. The main difference between Algorithm 1 and the top-down exact MDD compilation algorithm of Bergman et al. is the order in which the nodes are processed. Instead of requiring that the nodes be processed layer by layer, we allow the collection T to provide the nodes in any order. This generalization permits exploratory construction of the MDD. For example, if we are constructing the MDD in order to seek a feasible solution, we can build it in

Algorithm 1. Exact MDD Construction

1: $L_1 := \{r\}$
2: $T := \{r\}$
3: **while** T is not empty **do**
4:        select $v \in T$ and remove it from T
5:        $i := \mathrm{layer}(v)$
6:        **for all** $y \in D_i$ **do**
7:                $s := \chi_i(\mathrm{state}(v), y)$
8:                **if** $s \neq \perp$ **then**
9:                        $w := \textit{unique-table}(i + 1, s)$
10:                        **if** $w = $ **nil then**
11:                                $w := $ new node with state s
12:                                add w to L_{i+1}
13:                                add w to T
14:                        add edge (v, w) with label y

---

a depth-first manner by taking T to be a stack. The layer-by-layer behavior of the algorithm of Bergman et al. can be achieved by using a queue for T. Note that if we do construct the MDD layer by layer, we can discard the unique table for each layer as soon as we have finished processing the previous layer.

It is useful to have a heuristic to estimate the "promise" of a partial solution, that is, the likelihood that it has a feasible completion. Such a heuristic can be used to guide the depth-first construction of an MDD in search of a feasible solution. For the MBP problem, we propose the following heuristic. Given a partial solution $(y_1, \ldots, y_i)$ describing the packing of the first i items into bins, we perform a non-backtracking random packing of the remaining items $(i + 1, \ldots, n)$ as well as we can without violating the bin packing constraints. In other words, we iterate through the remaining items in order, and we pack each item into one of the bins that has sufficient ullage, chosen at random; if no such bin exists, we put the item into a trash pile. At the end we count the total size of the items in the trash pile, along all d dimensions, and this number is the score of this packing. This random packing of the remaining items is repeated several times, and the total score of these packings is used as the heuristic value of the partial solution; a low score is better. (Occasionally, while we are computing the heuristic for a partial solution in this way, we may luckily find a feasible completion: the trash pile will be empty. In this case, if we are constructing the MDD merely to seek a feasible solution, we can immediately return the solution thus found.)

With such a heuristic, we can use a priority queue for T to select the most promising nodes to process next. Alternatively, we can use a stack for T and modify Algorithm 1 slightly so that when we process a node we construct all its children, evaluate their heuristics, and then add them to T in reverse order of their promise. This will yield a depth-first algorithm that explores the most promising child of each node first.

This depth-first MDD construction process, especially if it is being used simply to find a feasible solution, is very similar to a backtracking search. It is an improvement, however, because the MDD nodes act as a memoization technique to prevent the exploration of portions of the search tree that can be recognized as equivalent to portions already explored.

3   Approximate MDDs

In general, exact MDDs can be of exponential size, so the use of Algorithm 1 may not be practical because of space limitations. In this case we may be able to use an approximate MDD to get useful results.

The MDDs described in this section are called approximate because their structure approximates the structure of the exact MDD. An approximate MDD represents a set of solutions to a relaxation or a restriction of the problem instance. Hence, if a restriction MDD indicates that an instance is feasible, then every solution it represents (i.e., every path from the root to the sink) is an exact feasible solution to the original instance. Similarly, an indication of infeasibility from a relaxation MDD is a proof that the original instance is infeasible. In this way, relaxation and restriction MDDs can be used together to determine the feasibility or infeasibility of an instance, and to get an exact feasible solution if the instance is feasible. Of course, it is possible for a relaxation MDD to indicate that an instance is feasible while a restriction MDD indicates it is infeasible, in which case nothing is learned. In response, one could construct MDDs representing tighter relaxations or restrictions (probably at the cost of greater time and space requirements), or could embed the MDDs inside a complete search.

Approximation MDDs by Merging. MDDs of limited width were proposed by Andersen et al. [1] to reduce space requirements. In this approach, the MDD is constructed in a top-down, layer-by-layer manner; whenever a layer of the MDD exceeds some preset value W an approximation operation is applied to reduce its size to W before constructing the next layer. For this approximation, Bergman et al. [4] use a relaxation operation $\oplus$ defined on states of nodes so that, given nodes v and v', the state given by $\text{state}(v) \oplus \text{state}(v')$ is a "relaxation" of both $\text{state}(v)$ and $\text{state}(v')$; see also [8].

We can formalize this idea as follows. Let $C_i = D_i \times \cdots \times D_n$ denote the set of completions at level i (independent of any particular partial solution). For a partial solution $y \in Y_i$, the set of feasible completions of y is some subset of C_i, so $\mathcal{F}(y) \in \mathcal{P}(C_i)$, where $\mathcal{P}$ denotes the power set. Recall that a state function $\sigma_i : Y_i \to S_i$ is such that $\sigma_i(y) = \sigma_i(y')$ implies $\mathcal{F}(y) = \mathcal{F}(y')$, and $\sigma_i(y) = \bot$ implies $\mathcal{F}(y) = \emptyset$. The existence of such a function implies the existence of a *completion function* $\tau_i : S_i \to \mathcal{P}(C_i)$ such that $\tau_i(\sigma_i(y)) = \mathcal{F}(y)$ for all $y \in Y_i$. For $i \in \{1, \ldots, n\}$, we say that a binary operation $\vee_i : S_i \times S_i \to S_i$ is a *relaxation merge* if for all $y, y' \in Y_i$ we have $\tau_i(\sigma_i(y) \vee_i \sigma_i(y')) \supseteq \mathcal{F}(y) \cup \mathcal{F}(y')$. In other words, the set of feasible completions implied by the state $\sigma_i(y) \vee_i \sigma_i(y')$ contains all feasible completions implied by the state $\sigma_i(y)$ and all feasible completions implied by the state $\sigma_i(y')$. Similarly, we call $\wedge_i : S_i \times S_i \to S_i$ a *restriction merge* if for all $y, y' \in Y_i$ we have $\tau_i(\sigma_i(y) \wedge_i \sigma_i(y')) \subseteq \mathcal{F}(y) \cap \mathcal{F}(y')$. For simplicity, we

Algorithm 2. Approximate MDD Construction by Merging

```
 1: L₁ := {r}
 2: for i = 1 to n do
 3:     L_{i+1} := ∅
 4:     for all v ∈ L_i do
 5:         for all y ∈ D_i do
 6:             s := χ_i(state(v), y)
 7:             if s ≠ ⊥ then
 8:                 w := unique-table(i + 1, s)
 9:                 if w = nil then
10:                     w := new node with state s
11:                     add w to L_{i+1}
12:                 add edge (v, w) with label y
13:     if |L_{i+1}| > W then
14:         partition L_{i+1} into W clusters A₁, ..., A_W
15:         for j = 1 to W do
16:             w_j := new node with state ⋁ A_j (or ⋀ A_j)
17:             for all v ∈ A_j do
18:                 change every edge (u, v) to (u, w_j) with the same label
19:         L_{i+1} := {w₁, ..., w_W}
```

shall omit the subscript and just write $\vee$ or $\wedge$. These merge operations need not be associative or commutative. However, in a slight abuse of notation, we shall write $\bigvee A$ to denote a combination of all elements $s \in A \subseteq S_i$ using the relaxation merge operation $\vee$, in any order and parenthesized in any way; likewise for $\bigwedge A$.

For the direct MDD representation of an MBP instance, in which node states are lists of ullages $(u_1, \ldots, u_m)$, an appropriate relaxation (respectively, restriction) merge is the componentwise maximum (respectively, minimum).

Bergman et al. give an algorithm to construct a limited-width MDD which iteratively merges pairs of nodes in a layer using a relaxation merge. We propose a refinement of this technique that uses a clustering algorithm to partition the nodes in the layer into W clusters; the nodes in each cluster are then merged into a single node. This is outlined in Algorithm 2.

To perform the clustering of nodes on line 14 of Algorithm 2, we adapted the median cut algorithm of Heckbert [7], which was originally designed for color quantization of images. The median cut algorithm operates on a set of points in q-dimensional Euclidean space (in the original version, $q = 3$, representing the red, green, and blue components of each pixel in the image) and partitions the points into clusters. Initially all of the points are grouped into a single cluster, which is tightly enclosed by a q-dimensional rectangular box. Then the following operation is repeatedly performed: the box having the longest length (among all boxes in all q dimensions) is selected, and it is divided into two boxes along this longest length at the median point, that is, in such a way that each of the two smaller boxes contains approximately half of the points in the original box; the two smaller boxes are then "shrunk" to fit tightly around the points they contain. This process continues until the desired number of clusters (boxes)

Algorithm 3. Restriction MDD Construction by Deletion

1: (lines 1–12 are the same as in Algorithm 2)
13:     **if** $|L_{i+1}| > W$ **then**
14:         use heuristic to select most promising nodes $w_1, \ldots, w_W \in L_{i+1}$
15:         **for all** $w \in L_{i+1} \setminus \{w_1, \ldots, w_W\}$ **do**
16:             delete w from L_{i+1} and delete all edges (u, w)

---

have been generated. The median cut algorithm can be implemented to run in $O\big(K(pq + \log K)\big)$ time, where K is the desired number of clusters, p is the number of points, and q is the number of dimensions.

To apply the median cut algorithm to the nodes in a layer of an MDD, we interpret the state of each node as a point in q-dimensional Euclidean space, for some value of q. For the direct MDD representation of an MBP instance, the state of a node is a list of d-dimensional ullages, one for each of the m bins; so we view this state directly as an md-dimensional point.

If a merged MDD reports that a CSP is feasible, it is desirable to extract a (possibly) feasible solution from it. One way to do this is to maintain a representative partial solution for each node as the MDD is constructed; when two nodes are merged, either of the two corresponding partial solutions can be selected (perhaps in accordance with a heuristic) as the representative partial solution for the merged node. Then the representative (complete) solution at the sink will be a (possibly) feasible solution for the CSP. The representative partial solution can be viewed as auxiliary state information of the node.

Restriction MDDs by Deletion. Algorithm 2 can be used to construct a limited-width MDD by merging nodes when the size of a layer becomes too large. If we are constructing a restriction MDD, however, then another option is simply to delete some of the nodes in the layer [3]. The selection of nodes to keep can be guided by a heuristic. This is described in Algorithm 3.

We note that this deletion algorithm does not use a partitioning algorithm to cluster the nodes in each layer as the merging algorithm does; instead it incurs the cost of computing a heuristic for each node. So the deletion algorithm may be especially beneficial if partitioning the nodes in a layer of the MDD is slower than computing a heuristic for a node.

4   MDD Techniques for Bin Packing

In the previous sections we have presented generic MDD construction algorithms, suitable for any CSP. In this section we specialize some of these techniques to the MBP problem.

Ullage MDD Representation. Let I be an MBP instance, having n items and m bins. One difficulty with the direct MDD representation of I is that it does not take into account the possible symmetry of the bins. For example, suppose that item 1 will fit in any of the m bins. Then the root of the direct MDD will have m outgoing edges labeled 1 through m, indicating the possible bins into which item 1 can be packed. However, if the bins are all identical, these possibilities are essentially equivalent (up to a reordering of the bins). The direct MDD representation cannot recognize this equivalence, because the sets of feasible completions, corresponding to edge-labeled paths in the MDD, are different. For example, in Fig. 1a, the two edges directed out of the root node represent essentially equivalent choices.

To address the possible symmetry of the bins, we can reduce the number of distinct descriptions of feasible solutions by expressing the solutions differently. Rather than assigning items directly to bins, we assign each item to an ullage. For example, instead of saying that item 3 is packed into bin 2, we say that it is packed into a bin with ullage 4. We call this the *ullage description* of the solution; it consists of a list $(u_1, \ldots, u_n)$ of d-tuples, assigning an ullage to each item.

To specify the domains of the variables u_i in the ullage description of a solution, we define the *ullage multiset function* U. If $C = (c_1, \ldots, c_m)$ is the list of bin capacities in I, then $U(C, (u_1, \ldots, u_i))$ denotes the multiset of ullages after the first i items have been placed into bins as described by the list $(u_1, \ldots, u_i)$. This is the same as the multiset of ullages after the first $i - 1$ items have been placed, except that an item of size s_i was placed into a bin having ullage u_i; so an element u_i of the multiset should be removed and replaced by an element $u_i - s_i$. Formally, we can define U recursively as follows:

- $U(C, \emptyset) = C$ (viewing C as a multiset).
- For $i \in \{1, \ldots, n\}$, if $U_{i-1} = U(C, (u_1, \ldots, u_{i-1}))$ is defined and $u_i \in U_{i-1}$, then $U(C, (u_1, \ldots, u_i)) = (U_{i-1} \setminus u_i) \cup \{u_i - s_i\}$.

With this definition of U, the domain of the variable u_i in the ullage description of a solution is $U(C, (u_1, \ldots, u_{i-1}))$. Note that this domain depends on the values that have previously been assigned to $u_1, \ldots, u_{i-1}$.

An *ullage MDD representation* of the set of feasible solutions of I has layers $L_1, \ldots, L_{n+1}$. The label of an edge directed out of a node in layer L_i in an ullage MDD is a d-tuple, representing the ullage of the bin into which item i is to be placed (after items 1 through $i - 1$ have been placed into bins). Therefore the edge labels $u_1, \ldots, u_n$ along a path from the root to the sink in an ullage MDD correspond to an ullage description $(u_1, \ldots, u_n)$ of a feasible solution to I.

Fig. 1b illustrates the ullage MDD representation for the one-dimensional bin packing instance having two bins of capacity 7 and items with sizes $5, 3, 2$, and 1. At the root, the state is $\{7 \times 2\}$, i.e., a multiset containing the element 7 with multiplicity 2. The first item, of size 5, must be placed in a bin having ullage 7; this leads to the state $\{2, 7\}$. Then the second item, of size 3, must be placed in the bin that now has ullage 7, and so forth. Of course, a path from the root to the sink in this ullage MDD can easily be converted into an explicit list of bin assignments if desired.

State Function for the Ullage MDD Representation. For the ullage MDD representation, it is useful to consider the state of a partial solution having ullage description $(u_1, \ldots, u_{i-1})$ to be $U(C, (u_1, \ldots, u_{i-1}))$, that is, the multiset of ullages of the bins.

This idea can be extended to handle side constraints in the CSP. For example, the steel mill slab problem [12] is essentially a (one-dimensional) bin packing problem with the additional constraint that each item has a color and no bin can contain items of more than two colors. To handle a side constraint like this, we can simply augment the state information of a node to include the colors of items that have been packed into it so far.

A few observations can be used to identify further equivalent partial solutions. Let $u_{j,k}$ denote the ullage of bin j, in the kth dimension, after we have placed items 1 through i into bins. Let a denote the greatest possible sum of a subset of the sizes of items $i+1$ through n, in the kth dimension, that does not exceed $u_{j,k}$. If $a < u_{j,k}$, then we may consider the ullage of bin j, in the kth dimension, to be a rather than $u_{j,k}$ without changing the set of feasible completions. If the order of the items is fixed, the relevant sets of possible sums of remaining items can be computed once at the beginning of the MDD construction in $O(nc_{\max}^2)$ time, where $c_{\max}$ is the largest bin capacity in a single dimension. Using this technique of "rounding down" the ullages across all bins in all dimensions, we can sometimes identify additional equivalent partial solutions (their states may be the same after they are rounded down, even if they were not the same before). Moreover, after rounding down ullages, we may discover that the total ullage in all bins is not enough for the remaining items; then we know that the current state has no feasible completions.

If, after we have placed items 1 through i into bins, there is any bin that is so small that none of the remaining items will fit, we can declare that bin *dead* and remove it from further consideration. This is potentially stronger than rounding down, because it may be that in each dimension, considered separately, there is some remaining item that will fit into the bin; but no remaining item is small enough in every dimension to fit into the bin. Conversely, if after we have placed items 1 through i into bins, there is some bin that is large enough in every dimension that all of the remaining items will fit in it, then we know that the instance is feasible. We call such a bin *free*. Once we discover a free bin, we can immediately return a feasible solution: extend a partial solution corresponding to the current node to a complete solution by packing all remaining items into the free bin. The ideas underlying the concepts of dead and free bins are present in Behle's threshold BDD algorithm [2].

In Fig. 1c we apply the rounding-down technique to the ullage MDD. If we additionally check for dead and free bins, we will discover a free bin in the second layer (the bin with rounded-down ullage of 6).

Variable Ordering. The variable ordering used in an MDD can have a very significant effect on the size of the MDD. Behle [2] investigated the optimal variable ordering problem for threshold BDDs. In general, the problem of determining whether a given variable ordering of a BDD can be improved is NP-complete [5].

For the MBP problem, we take a simple heuristic approach. We observe that identifying dead bins and free bins is beneficial, and we would like to make such identifications as soon as possible. If we pack the largest items first, then the total size of the remaining unpacked items will decrease quickly in the beginning, which suggests that we may reach free bins early; additionally, we will tend to fill bins quickly in the beginning, which suggests that we may exhaust the bins' capacity quickly and reach dead bins early. However, these ideas are somewhat contradictory, and the latter idea is opposed by the observation that the unpacked items will be small, so they can fit into small spaces.

We therefore use an "interleaved" ordering, in which the largest item is packed first, then the smallest item, then the second largest item, then the second smallest item, and so on, packing the median-sized item last. (Our item sizes are multidimensional, so we use the total size of the item in all dimensions: $s_i = \sum_{k=1}^{d} s_{i,k}$.) This ordering seemed to work well for our experimental instances. This straightforward approach means that we can implement variable ordering by sorting the items in this manner as a preprocessing step.

5   Experimental Results

We implemented the MDD-based algorithms described above in Java, using the exploratory construction method described in Section 2, the approximation methods from Section 3, and the ullage MDD representation and the other techniques described in Section 4.

Our test instances were generated as follows. Given the parameters d (the number of dimensions), n (the number of items), m (the number of bins), and β (percentage bin slack), we first generate a list of n item sizes $(s_1, \ldots, s_n)$, each of which is a d-tuple whose coordinates are integers chosen uniformly and independently at random from $\{0, \ldots, 1000\}$. Then the sum $t = \sum_{i=1}^{n} s_i$ is computed, and the m bin capacities are all taken to be $\lceil (1 + \beta/100)t/m \rceil$; these computations are done componentwise. (If $\beta = 20$, for example, then the total bin capacity, in each dimension, will be 20% more than the total item size.) An instance is rejected and regenerated if it contains any single item that is too large to be placed into a bin, as such an instance is obviously infeasible. Our test instances have 6 dimensions, 18 items, and 6 bins; we generated 52 such instances for each integer value of β from 0 to 35. These instances are available at http://www.math.cmu.edu/~bkell/6-18-6-instances.txt or by request.

By their construction, these instances have identical bins. The ullage MDD representation can exploit this symmetry effectively to reduce the number of branches in the search tree. This is especially evident in the infeasible instances, where infeasibility must be established by some kind of exhaustive search.

The experiments were run on a 32-bit Intel Pentium 4 CPU at 3.00 GHz with 1 GiB of RAM using Windows 7 Professional. The maximum Java heap size was set to 512 MiB. We used AIMMS 3.13 with CPOptimizer 12.4 as the CP solver and CPLEX 12.4 as the MIP solver, with their default settings. The CP model has n variables $x_1, \ldots, x_n$, each with domain $\{1, \ldots, m\}$; the assignment

$x_i = j$ indicates that item i is packed into bin j. These variables are subject to d independent `cp::BinPacking` constraints. The MIP model has mn binary variables $x_{i,j}$, for $i \in \{1, \ldots, n\}$ and $j \in \{1, \ldots, m\}$; the assignment $x_{i,j} = 1$ indicates that item i is packed into bin j. The MIP model also has a nonnegative "overflow" variable ω_j for each bin, representing the maximum amount by which the bin is overfull in any dimension, and there is a nonnegative "total overflow" variable $\Omega = \sum_{j=1}^{m} \omega_j$. The MIP model appears below. It is formulated as a minimization problem only because that is the form the solver requires; the constraint $\Omega = 0$ means it is really just a feasibility problem.

$$\min \ \Omega$$

$$\text{s.t.} \ \sum_{j=1}^{m} x_{i,j} = 1;$$

$$\sum_{i=1}^{n} s_{i,k} x_{i,j} \leq c_{j,k} + \omega_j \quad \text{for all } k \in \{1, \ldots, d\}, j \in \{1, \ldots, m\};$$

$$\Omega = \sum_{j=1}^{m} \omega_j;$$

$$x_{i,j} \in \{0, 1\}, \quad \omega_j \geq 0, \quad \Omega = 0.$$

We compared the performance of CP and MIP to our MDD approaches: the exact MDD (using depth-first, heuristic-driven exploratory construction), a relaxation MDD using the relaxation merge operation, and restriction MDDs using the restriction merge operation or deletion. All instances were run to completion using each method. The maximum width for the approximation MDDs was set to 5000 nodes. With this width, the approximation MDDs returned "feasible" or "infeasible" correctly in all instances except two: the restriction merge MDD returned "infeasible" incorrectly for one instance with 25% bin slack and one instance with 26% bin slack. The combination of the relaxation merge MDD and the deletion (restriction) MDD was enough to correctly solve all 1872 instances.

Fig. 2 shows a clear feasibility phase transition centered around approximately 20% bin slack, with a corresponding hardness peak. In the infeasible region, on instances having bin slack between about 2% and 22%, the average run time of the exact MDD method is consistently less than that of MIP and significantly less than that of CP (by over three orders of magnitude at 20% bin slack). On the other hand, in the feasible region, on instances having bin slack more than about 25%, CP and MIP both tend to outperform the exact MDD method. A notable exception (visible as a spike in the hardness profile) occurs at 27% bin slack, for which one of the 52 generated instances happened to be infeasible; this single infeasible instance greatly increased the average run time of CP and MIP without noticeably affecting the performance of the exact MDD.

We investigated the instances at the hardness peak, i.e., those having 20% bin slack, in more detail. A performance profile for these instances appears in Fig. 3, including CP, MIP, the exact MDD, and the combination of the relaxation merge MDD and the deletion (restriction) MDD. The CP solver required

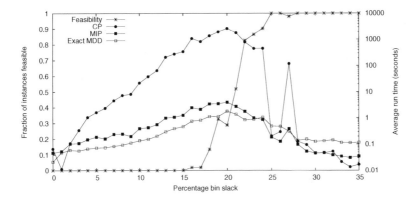

Fig. 2. Feasibility and hardness profiles for instances having 6 dimensions, 18 items, and 6 bins

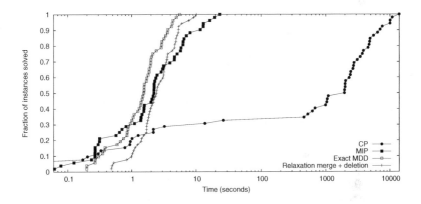

Fig. 3. Performance profile on the subset of instances having 20% bin slack

over 400 seconds for 35 instances (67%), taking almost 14000 seconds in the extreme case. The MIP solver did much better, solving every instance in less than 12 seconds. The exact MDD method, which solved each instance in less than 6 seconds, was faster than MIP in 32 instances (62%), while the relaxation MDD and the deletion MDD together (sufficient in all 52 instances to establish feasibility or infeasibility) were faster than MIP in 24 instances (46%).

When we look only at the 37 infeasible instances with 20% bin slack, as seen in Fig. 4a, the difference between CP/MIP and the MDD approaches becomes clearer. (Restriction MDDs do not give useful results for infeasible instances, so they are omitted from this plot merely for clarity. All of the approximate MDD methods we implemented ran about equally fast on all instances with 20% bin slack, so using a restriction MDD together with the relaxation approximately doubles the run time.) On the other hand, in the performance profile on the 15 feasible instances with 20% bin slack, shown in Fig. 4b (with the relaxation MDD omitted), the various methods are not as clearly separated.

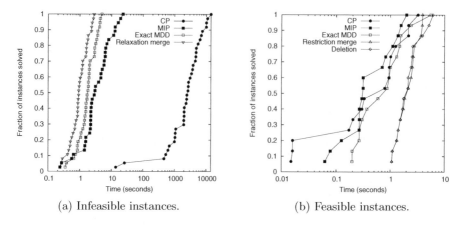

(a) Infeasible instances.   (b) Feasible instances.

Fig. 4. Performance profiles on infeasible and feasible instances having 20% bin slack

The advantage of the ullage MDD representation on infeasible instances comes from its ability to exploit the symmetry among identical bins in order to reduce the number of branches taken in an exhaustive search. However, on feasible instances, our Java code, which is not particularly optimized, does not find solutions as quickly as the commercial CP and MIP solvers do. The depth-first, heuristic-driven algorithm tends to solve feasible instances more quickly than the layer-by-layer approximation algorithms, but limited-width MDDs tend to be faster than exact MDDs on infeasible instances.

6   Conclusions

Our aim was to investigate the use of MDDs for the MBP problem. We described several variations of a generic algorithm for the construction of exact and approximate MDDs representing sets of feasible solutions to CSPs, including a heuristic-driven depth-first method to construct an exact MDD and an application of a clustering algorithm to construct approximate MDDs. We also examined several techniques to work with MBP instances effectively with MDDs, including the ullage MDD representation to handle symmetry, a rounding-down technique to more reliably detect equivalent nodes, and the identification of free and dead bins to quickly recognize feasibility and infeasibility. Experimental results show that our MDD algorithms, when combined with these representation techniques, can significantly outperform currently used CP techniques and can also consistently outperform MIP.

References

1. Andersen, H.R., Hadzic, T., Hooker, J.N., Tiedemann, P.: A constraint store based on multivalued decision diagrams. In: Bessière, C. (ed.) CP 2007. LNCS, vol. 4741, pp. 118–132. Springer, Heidelberg (2007)

2. Behle, M.: On threshold BDDs and the optimal variable ordering problem. Journal of Combinatorial Optimization 16, 107–118 (2008)
3. Bergman, D., Cire, A.A., van Hoeve, W.-J., Yunes, T.: BDD-based heuristics for binary optimization (submitted, 2013)
4. Bergman, D., van Hoeve, W.-J., Hooker, J.N.: Manipulating MDD relaxations for combinatorial optimization. In: Achterberg, T., Beck, J.C. (eds.) CPAIOR 2011. LNCS, vol. 6697, pp. 20–35. Springer, Heidelberg (2011)
5. Bollig, B., Wegener, I.: Improving the variable ordering of OBDDs is NP-complete. IEEE Transactions on Computers 45(9), 993–1002 (1996)
6. Hadzic, T., Hooker, J.N., O'Sullivan, B., Tiedemann, P.: Approximate compilation of constraints into multivalued decision diagrams. In: Stuckey, P.J. (ed.) CP 2008. LNCS, vol. 5202, pp. 448–462. Springer, Heidelberg (2008)
7. Heckbert, P.: Color image quantization for frame buffer display. In: SIGGRAPH 1982, pp. 297–307. ACM, New York (1982)
8. Hoda, S., van Hoeve, W.-J., Hooker, J.N.: A systematic approach to MDD-based constraint programming. In: Cohen, D. (ed.) CP 2010. LNCS, vol. 6308, pp. 266–280. Springer, Heidelberg (2010)
9. Knuth, D.E.: The Art of Computer Programming, vol. 4, fascicle 1: Bitwise Tricks & Techniques; Binary Decision Diagrams. Addison-Wesley (2009)
10. Lodi, A., Martello, S., Monaci, M.: Two-dimensional packing problems: A survey. European Journal of Operational Research 141(2), 241–252 (2002)
11. Martello, S., Pisinger, D., Vigo, D.: The three-dimensional bin packing problem. Operations Research 48, 256–267 (2000)
12. Schaus, P., Van Hentenryck, P., Monette, J.-N., Coffrin, C., Michel, L., Deville, Y.: Solving steel mill slab problems with constraint-based techniques: CP, LNS, and CBLS. Constraints 16(2), 125–147 (2011)
13. Wegener, I.: Branching Programs and Binary Decision Diagrams: Theory and Applications. SIAM, Philadelphia (2000)

A Synchronized Sweep Algorithm
for the *k-dimensional cumulative* Constraint

Arnaud Letort[1,*], Mats Carlsson[2], and Nicolas Beldiceanu[1]

[1] TASC team, (EMN-INRIA,LINA) Mines de Nantes, France
{arnaud.letort,nicolas.beldiceanu}@mines-nantes.fr
[2] SICS, P.O. Box 1263, SE-164 29 Kista, Sweden
matsc@sics.se

Abstract. This paper presents a sweep based algorithm for the *k-dimensional cumulative* constraint, which can operate in filtering mode as well as in greedy assignment mode. Given n tasks and k resources, this algorithm has a worst-case time complexity of $O(kn^2)$ but scales well in practice. In greedy assignment mode, it handles up to 1 million tasks with 64 resources in one single constraint in SICStus. In filtering mode, on our benchmarks, it yields a speed-up of about $k^{0.75}$ when compared to its decomposition into k independent *cumulative* constraints.

1 Introduction

In the 2011 Panel of the Future of CP [5], one of the identified challenges for CP was the need to handle large scale problems. Multi-dimensional bin-packing problems were quoted as a typical example [9], particularly relevant in the context of cloud computing. Indeed, the importance of multi-dimensional bin-packing problems was recently highlighted in [10], and was part of the topic of the 2012 Roadef Challenge [11]. Till now, the tendency is to use dedicated algorithms and metaheuristics [13] to cope with large instances. Various reasoning methods can be used for *cumulative* constraints, including Time-Table, Edge-Finding, Energetic Reasoning [2,12,6,15,1], and recently Time-Table and Edge-Finding combined [16]. A comparison between these methods can be found in [1]. These filtering algorithms focus on having the best possible deductions rather than on scalability issues. This explains why they usually focus on small size problems (i.e., typically less than 200 tasks up to 10000 tasks) but leave open the scalability issue. Following up on our preliminary work on a scalable sweep based filtering algorithm for a single *cumulative* constraint [8], this paper considers how to deal with multiple resources in an integrated way such as in [3]. In this new setting, each task uses several cumulative resources and the challenge is to come up with an approach that scales well. We should quote that the number of resources may be significant in many situations:

- For instance, in the 2012 Roadef Challenge we have up to 12 distinct resources per item to pack.

* Partially founded by the SelfXL project (contract ANR-08-SEGI-017).

C. Gomes and M. Sellmann (Eds.): CPAIOR 2013, LNCS 7874, pp. 144–159, 2013.
© Springer-Verlag Berlin Heidelberg 2013

– A new resource r' can also be introduced for modeling the fact that a given subset of tasks is subject to a *cumulative* or *disjunctive* constraint. The tasks that do not belong to the subset have their consumption of resource r' set to 0. Since we potentially can have a lot of such constraints on different subsets of tasks, this can lead to a large number of resources.

Having in the same problem multiple *cumulative* constraints that systematically share variables (i.e., the origin of a task is mentioned in several *cumulative* constraints) leads to the following source of inefficiency. In a traditional setting, each *cumulative* constraint is propagated independently on all its variables and, because of the shared variables, each *cumulative* constraint should be rerun several times to reach the fixpoint. One should quote that a *single* update of a bound of a variable by one *cumulative* constraint will trigger *all* the other *cumulative* constraints again.

The first theoretical contribution of this paper is a synchronized filtering algorithm that reaches a fixpoint for pruning the minimum (resp. maximum) of the variables *in one single step* wrt. several *cumulative* constraints and to their corresponding cumulated profiles of compulsory parts. We have an efficient, scalable k-dimensional version of the Time-Table method which achieves exactly the same pruning as k instances of the 1-dimensional version reported in [8]. The second practical contribution of this paper is the observation that the time needed to find a first solution by our new filtering algorithm for problems where the filtering is strong enough to solve the given problem without backtracking seems to depend much less on the number of resources k, at least when $k \leq 64$ and $n \leq 16000$, compared to using k independent *cumulative* constraints. In fact, we observe a speed-up by nearly $k^{0.75}$. Details are given in Sect. 3.

We now introduce the k-*dimensional cumulative* constraint and present the outline of the paper. Given k resources and n tasks, where each resource r $(0 \leq r < k)$ is described by its maximum capacity $limit_r$, and each task t $(0 \leq t < n)$ is described by its start s_t, its fixed duration d_t $(d_t \geq 0)$, its end e_t and its fixed resource consumptions $h_{t,0}, \ldots, h_{t,k-1}$ $(h_{t,i} \geq 0, i \in [0, k-1])$ on the k resources, the k-*dimensional cumulative* constraint with the two arguments

- $\langle s_0, d_0, e_0, \langle h_{0,0}, \ldots, h_{0,k-1} \rangle, \ldots, \langle s_{n-1}, d_{n-1}, e_{n-1}, \langle h_{n-1,0}, \ldots, h_{n-1,k-1} \rangle \rangle$,
- $\langle limit_0, \ldots, limit_{k-1} \rangle$

holds if and only if conditions (1) and (2) are true:

$$\forall t \in [0, n-1] : s_t + d_t = e_t \tag{1}$$

$$\forall r \in [0, k-1], \forall i \in \mathbb{Z} : \sum_{\substack{t \in [0,n-1]: \\ i \in [s_t, e_t)}} h_{t,r} \leq limit_r \tag{2}$$

Section 2 presents the new k-dimensional dynamic sweep algorithm. Section 3 evaluates the new algorithm on random multi-dimensional bin-packing and cumulative problems involving up to 64 distinct resources and one million tasks, as well on RCPSP instances taken from the PSPlib [7]. Finally Section 4 concludes.

2   The k-Dimensional Dynamic Sweep Algorithm

This section presents our contribution, a new sweep algorithm that handles several cumulative resources in one single sweep. We first introduce a running example and some general design decisions of the new sweep algorithm as well as the property the algorithm maintains, and then describe it. We present the data structures used by the new sweep algorithm, the algorithm itself and its worst-case time complexity.

Like the CP2012 sweep algorithm [8], the new algorithm only deals with domain bounds for the start and end variables, which is a good compromise to reduce the memory consumption for the representation of variables. This algorithm is also split into two distinct sweep phases. A first phase called *sweep_min* tries to filter the earliest start of each task by performing a synchronized sweep from left to right on all resources. A second phase called *sweep_max* tries to filter the latest ends by performing a synchronized sweep from right to left on all resources i.e. a sweep that for each time point considers all resources in parallel.

Before introducing a running example that will be used throughout this paper for illustrating the new filtering algorithm, let us recall the notion of *pessimistic cumulated resource consumption profile* (i.e. PCRCP). Given n tasks and k cumulative resources, the PCRCP$_r$ $(0 \leq r < k)$ is the aggregation of the compulsory parts of the n tasks on the resource r. The compulsory part of a task t $(0 \leq t < n)$ is the intersection of all its feasible instances. On the one hand, the height of the compulsory part of a task t, on a given resource r, at a given time point i is defined by $h_{t,r}$ if $i \in [\overline{s_t}, \underline{e_t})$ and 0 otherwise.[1] On the other hand, the height of the PCRCP$_r$ at a given time point i is given by:

$$\sum_{t \in [0,n), i \in [\overline{s_t}, \underline{e_t})} h_{t,r}$$

Note that the propagator needs to iterate the two phases until fixpoint. Suppose that *sweep_min* has run, and that *sweep_max* extends PCRCP$_r$ for some r. Then *sweep_min* may no longer be at fixpoint, and needs to run again, and so on. W.l.o.g., we focus from now on *sweep_min* since *sweep_max* is completely symmetric.

Example 1. Consider two resources r_0, r_1 ($k = 2$) with $limit_0 = 2$ and $limit_1 = 3$ and three tasks t_0, t_1, t_2 which have the following restrictions on their *start, duration, end* and *heights*:

- $t_0 : s_0 \in [1,1], d_0 = 1, e_0 \in [2,2], h_{0,0} = 1, h_{0,1} = 2,$
- $t_1 : s_1 \in [0,3], d_1 = 2, e_1 \in [2,5], h_{1,0} = 1, h_{1,1} = 2,$
- $t_2 : s_2 \in [0,5], d_2 = 2, e_2 \in [2,7], h_{2,0} = 2, h_{2,1} = 1.$

Since task t_1 cannot overlap t_0 without exceeding the resource limit on resource r_1, the earliest start of t_1 is adjusted to 2. Since t_1 occupies the interval $[3, 4)$ and since, on resource r_0, t_2 can neither overlap t_0, nor t_1, its earliest start is

---

[1] $\underline{v}$ and $\overline{v}$ respectively denote the minimum and maximum value of variable v.

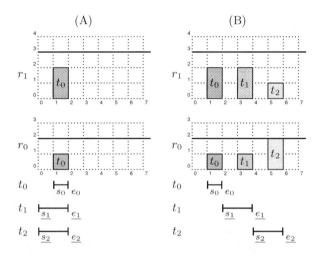

Fig. 1. This figure represents the earliest positions of the tasks and the PCRCP on resources r_0 and r_1, (A) of the initial problem described in Example 1, (B) the fixpoint.

adjusted to 4. The purpose of the k-dimensional sweep algorithm is to perform such filtering in an efficient way, i.e. in one single sweep. □

We now show how to achieve such filtering by decomposing the *2-dimensional cumulative* constraint into two *cumulative* constraints on resources r_0 and r_1.

Continuation of Example 1 (Illustrating the decomposition). The instance given in Example 1 can naturally be decomposed into two *cumulative* constraints:

- c_0 : *cumulative*$(\langle\langle s_0, d_0, e_0, h_{0,0}\rangle, \langle s_1, d_1, e_1, h_{1,0}\rangle, \langle s_2, d_2, e_2, h_{2,0}\rangle\rangle, limit_0)$.
- c_1 : *cumulative*$(\langle\langle s_0, d_0, e_0, h_{0,1}\rangle, \langle s_1, d_1, e_1, h_{1,1}\rangle, \langle s_2, d_2, e_2, h_{2,1}\rangle\rangle, limit_1)$.

- During a first sweep wrt. constraint c_0 (see Part (A) of Fig. 2), the compulsory part of task t_0 on resource r_0 (on resource r_0 and on interval $[1, 2)$ task t_0 uses one resource unit) permits to adjust the earliest start of task t_2 to 2 since the gap on top of interval $[1, 2)$ is strictly less than the resource consumption of task t_2 on resource r_0, i.e. $h_{2,0}$.
- A second sweep wrt. constraint c_1 (see Part (B) of Fig. 2) adjusts the earliest start of task t_1 to 2. This comes from the compulsory part on resource r_1 of task t_0 and from the fact that, on resource r_1 tasks t_1 and t_0 can not overlap. This last sweep causes the creation of a compulsory part for task t_1 on $[3, 4)$.
- During a third sweep wrt. constraint c_0 (see Part (C) of Fig. 2), the new compulsory part of task t_1 allows to adjust the earliest start of task t_2 since on resource r_0, the gap on top of the compulsory part of task t_1 is strictly less than the resource consumption of task t_2, i.e. $h_{2,0}$.

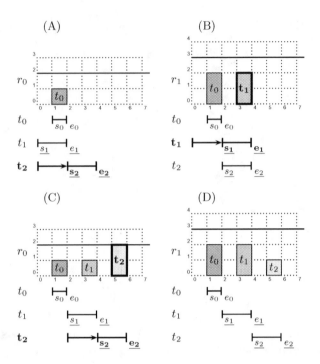

Fig. 2. Parts (A), (B), (C) and (D) respectively represent the earliest positions of the tasks and the PCRCP of the problem described in Example 1, after a first sweep on r_0, after a second sweep on r_1, after a third sweep on r_0, and a fourth sweep on r_1

- Finally, a last sweep wrt. constraint c_1 is performed to find out that the fixpoint was reached (see Part (D) of Fig. 2).

Our new *sweep_min* filtering algorithm will perform such deductions in one single step. □

We now give the property that will be achieved by our new *sweep_min* filtering algorithm.

Property 1. Given a *k-dimensional cumulative* constraint with n tasks and k resources, *sweep_min* ensures that:

$$\forall r \in [0, k-1], \forall t \in [0, n-1], \forall i \in [\underline{s_t}, \overline{e_t}) \; : \; h_{t,r} + \sum_{\substack{t' \neq t: \\ i \in [\overline{s_{t'}}, \overline{e_{t'}})}} h_{t',r} \leq limit_r \quad (3)$$

Property 1 ensures that, for any task t of the *k-dimensional cumulative* constraint, one can schedule t at its earliest start without exceeding for any resource r $(0 \leq r < k)$ its resource limit wrt. the PCRCP on resource r of the tasks of $\mathcal{T} \backslash \{t\}$.

Note that we can construct from Property 1 a relaxed solution of the k-dimensional cumulative constraint by:

① setting the resource consumption to 0 for all k resources for the tasks that do not have any compulsory part,
② setting the duration to $\underline{e_t} - \overline{s_t}$ for the tasks that do have a compulsory part, and
③ assigning each task to its earliest start.

2.1   Event Point Series

In the context of the sweep algorithms, events are specific time points where the status of the sweep-line changes. Since events are only related to the temporal aspect, they do not depend on how many resources we have, and can therefore be factored out. As a consequence, events are kept unchanged wrt. the CP2012 sweep algorithm. In order to build the PCRCP on each resource and to filter the earliest start of each task, the algorithm considers the following types of events.

- The event type $\langle SCP, t, \overline{s_t} \rangle$ for the *Start of Compulsory Part* of task t (i.e. the latest start of task t). This event is generated iff task t has a compulsory part, i.e. iff $\overline{s_t} < \underline{e_t}$.
- The event type $\langle ECPD, t, \underline{e_t} \rangle$. The date of such event corresponds to the *End of the Compulsory Part* of task t (i.e. the earliest end of task t) and may increase due to the adjustment of the earliest start of t. This event is generated iff task t has a compulsory part, i.e. iff $\overline{s_t} < \underline{e_t}$.
- The event type $\langle CCP, t, \overline{s_t} \rangle$, where CCP stands for *Conditional Compulsory Part*, is created iff task t does not have any compulsory part, i.e. iff $\overline{s_t} \geq \underline{e_t}$. At the latest, once the sweep-line reaches position $\overline{s_t}$, it adjusts the earliest start of t. Consequently the conditional event can be transformed into an SCP and an $ECPD$ events, reflecting the creation of compulsory part.
- The event type $\langle PR, t, \underline{s_t} \rangle$ where PR stands for *Pruning Event*, corresponds to the earliest start of task t. This event is generated iff task t is not yet scheduled, i.e. iff $\underline{s_t} \neq \overline{s_t}$.

Events are stored in the heap *h_events* where the top event is the event with the smallest date.

Continuation of Example 1 (Generated Events). The following events are generated and sorted according to their date: $\langle PR, 1, 0 \rangle$, $\langle PR, 2, 0 \rangle$, $\langle SCP, 0, 1 \rangle$, $\langle ECPD, 0, 2 \rangle$, $\langle CCP, 1, 3 \rangle$, $\langle CCP, 2, 5 \rangle$.   □

2.2   Sweep-Line Status

In order to build the PCRCP and to filter the earliest start of the tasks, the sweep-line jumps from event to event, maintaining the following information:

- The current sweep-line position δ, initially set to the date of the first event.
- For each resource $r \in [0, k-1]$, the amount of available resource at instant δ denoted by gap_r (i.e. the difference between the resource limit $limit_r$ and the height of the PCRCP on resource r at instant δ) and its previous value denoted by gap'_r.
- For each resource $r \in [0, k-1]$, an array denoted by a_check_r which stores tasks that have been encountered but whose earliest feasible start is yet to be determined, i.e. all tasks that potentially can overlap δ. Such tasks will be called *active tasks* in the rest of the paper. The ith entry stores a subset of the tasks t for which $h_{t,r} = i$.
- For each task t, the number of conflicts at instant δ denoted by $conflicts_t$. We say that a task t is *in conflict on the resource* r iff $t \in a\_check_r \wedge h_{t,r} > gap_r$. We know that a task t cannot overlap the current sweep-line position if the number of conflicts related to task t is greater than 0.[2]

Our sweep algorithm first creates and sorts the events wrt. their date. Then, the sweep-line moves from one event to the next event, updating the amount of available space on each resource (i.e. gap_r, $0 \le r < k$), and the list of active tasks. Once all events at δ have been handled, the sweep-line tries to filter the earliest start of the active tasks wrt. gap and the interval $[\delta, \delta_{next})$. In order to update the number of conflicts related to the tasks, for each resource r we scan the tasks that are in a_check_r and for which the height on the resource r is between gap_r and gap'_r, i.e. tasks that become in conflict on the resource r, or tasks that are no longer in conflict on the resource r.

In addition to the information maintained by the sweep-line, we define two functions $\text{pred}_r : \mathbb{N} \to \mathbb{N}$ and $\text{succ}_r : \mathbb{N} \to \mathbb{N}$:

$$\text{pred}_r(h) = \begin{cases} \max_{0 \le t < n}\{h_{t,r} \mid h_{t,r} < h\} & \text{, if such } h_{t,r} \text{ exists} \\ 0 & \text{, otherwise} \end{cases}$$

$$\text{succ}_r(h) = \begin{cases} \min_{0 \le t < n}\{h_{t,r} \mid h_{t,r} > h\} & \text{, if such } h_{t,r} \text{ exists} \\ +\infty & \text{, otherwise} \end{cases}$$

2.3   Algorithm

The *sweep_min* algorithm performs one single sweep over the event point series in order to adjust the earliest start of the tasks wrt. Property 1. It consists of a main loop that is decomposed into an event processing part and a filtering part. The processing part reads the events and updates the gaps. It also adds tasks in the list of active tasks and handles dynamic (ECPD) and conditional (CCP) events. The filtering part adjusts the earliest start of the active tasks wrt. the interval $[\delta, \delta_{next})$ and the gaps.

---

[2] There exists a specific case that we will introduce later for which this is false.

Main Algorithm. The top level algorithm (Algorithm 1) consists of:

- [CREATING EVENTS] (line 2). The events are generated wrt. the start and end variables of each task and inserted into h_events. Also, throughout our algorithm, for each task t, the relation $s_t + d_t = e_t$ is maintained while sweeping (i.e. each time the sweep algorithm adjusts the earliest start of a task, its earliest end is immediately updated accordingly).
- [INITIALIZATION] (lines 4 to 7). The events are generated and inserted into h_events. The arrays a_check_r $(0 \leq r < k)$ are initialized as empty arrays. gap_r and gap'_r are respectively initialized to the resource limit of r and to the maximum height consumed by a task on resource r.
- [MAIN LOOP] (lines 9 to 11). For each date the main loop processes all the corresponding events. It consists of the following steps:
  - The first step (line 10) handles all events at the top of h_events related to the date δ, and determines the next position of the sweep-line δ_{next}.
  - The second step (line 11) detects if a resource limit is exceeded, and processes the active tasks in order to adjust their earliest start.

The Event Processing Part. In order to update the sweep-line status, Alg. 2 (process_events) reads and processes all events related to the date δ and determines the sweep interval $[\delta, \delta_{next})$. We introduce two functions, $extract\_min(h)$ which extracts and returns all the top elements of the heap h and their common related key value (event date), and $peek\_key(h)$ which returns the key of the top elements of the heap h. Alg. 2 consists in the following parts:

- [PROCESSING START COMPULSORY PART (SCP) EVENTS] (lines 3 to 4). For each resource r updates the gap on top of the PCRCP of resource r on the current sweep interval.

```
ALGORITHM sweep_min() : bool
 1: [CREATING EVENTS]
 2: h_events ← generate SCP, ECPD, CCP and PR events wrt. task t (0 ≤ t < n)
 3: [INITIALIZATION]
 4: for r = 0 to k − 1 do
 5:     gap_r ← limit_r
 6:     gap'_r ← pred_r(gap_r + 1)    // initialized to the maximum height consumed on resource r
 7:     a_check_r ← empty array of size equal to the number of distinct heights on resource r
 8: [MAIN LOOP]
 9: while ¬empty(h_events) do                    // while the heap of events is not empty
10:     ⟨δ, δ_next⟩ ← process_events()
11:     if ¬filter(δ, δ_next) then return false
12: return true
```

Algorithm 1. Main sweep algorithm. Return *false* iff a resource overflow occurs, *true* otherwise.

- [PROCESSING DYNAMIC (ECPD) EVENTS] (lines 6 to 9). Once the sweep-line reaches the end of the compulsory part of a task t (i.e. its earliest end), we have to determine whether or not it is its final position, i.e. whether the earliest start of task t can still be updated. This requires to consider the following steps:

- If task t is in conflict on at least one resource (i.e. $conflicts_t > 0$, line 7), then t cannot be scheduled before its latest position.
- If the earliest start of task t has already been updated since the creation of the event, then this ECPD event is pushed back in the heap h_events to its correct date (line 8). Otherwise, for each resource r we update the gap on top of the PCRCP of resource r at position δ (line 9).

— [PROCESSING CONDITIONAL (CCP) EVENTS] (lines 10 to 14). When the sweep-line reaches the latest start of a task t, initially without compulsory part, we need to know whether or not a compulsory part for t was created. We have to consider the two following steps:

- If task t is in conflict on at least one resource, then t cannot be scheduled before its latest position (line 11).
- If the earliest start of task t has already been updated and its earliest end is greater than the current sweep-line position, meaning a compulsory part was created, then the event related to the end of its compulsory part is added to h_events (line 13) and the available resources are updated (line 14).

— [DETERMINE THE NEXT EVENT DATE] (lines 16). In order to process the pruning (PR) events, we need to know the next position δ_{next} of the sweep line.

— [PROCESSING EARLIEST START (PR) EVENTS] (lines 18 to 22). If the current sweep interval is too small wrt. to the duration of task t or there is at least one conflict (i.e. $d_t > \delta_{next} - \delta \vee \exists r \mid h_{t,r} > gap_r$, line 19) then for each resource r the task t is added into a_check_r and its number of conflicts is updated.

The Filtering Part. Algorithm 3 processes the active tasks (i.e. tasks in each a_check_r) in order to adjust the earliest start of the tasks. It consists of the following parts:

— [CHECK RESOURCE OVERFLOW] (line 2). If one of the gap_r is negative on the current sweep interval $[\delta, \delta_{next})$ (i.e. one of the resource limit is exceeded), Alg. 3 returns false meaning a failure.

— [TASK ENTERING INTO CONFLICT] (lines 4 to 13). Scans each resource where the current available resource is less than the previous available resource (i.e. $gap'_r > gap_r$, line 5). It has to consider each task $t \in a\_check_r$ such that $gap'_r \geq h_{t,r} > gap_r$, i.e. tasks that were not in conflict at the previous sweep-line position but that are in conflict at instant δ on resource r. If the current earliest position of a task t without compulsory part is feasible, i.e. $conflicts_t = 0 \wedge \delta \geq e_t$, task t is removed from a_check_r. For a task t with a compulsory part, we do not need to keep task t in a_check_r once the sweep-line has reached the start of the compulsory part of t.

— [TASKS NO LONGER IN CONFLICT] (lines 15 to 29). Scans each resource where the current available resource is greater than the previous available resource (i.e. $succ_r(gap'_r) < gap_r$, line 16). For each resource r, it scans each task $t \in a\_check_r$ such that $gap'_r < h_{t,r} \leq gap_r$, i.e. tasks that were in

ALGORITHM process_events() : $\langle \delta, \delta_{next} \rangle$
1: $\langle \delta, \mathcal{E} \rangle \leftarrow$ extract_min(h_events)   // extracts top elements
2: [PROCESSING START COMPULSORY PART (SCP) EVENTS]
3: **for all** tasks t that belong to an event of type SCP in $\mathcal{E}$ **do**
4:      **for** $r = 0$ **to** $k - 1$ **do** $gap_r \leftarrow gap_r - h_{t,r}$ // updates available spaces
5: [PROCESSING DYNAMIC (ECPD) EVENTS]
6: **for all** tasks t that belong to an event of type ECPD in $\mathcal{E}$ **do**
7:      **if** $conflicts_t > 0$ **then** $(\underline{s_t}, \underline{e_t}) \leftarrow (\overline{s_t}, \overline{e_t})$
8:      **if** $\underline{e_t} > \delta$ **then** add $\langle \text{ECPD}, t, \underline{e_t} \rangle$ to h_events // push back the ECPD event
9:      **else for** $r = 0$ **to** $k - 1$ **do** $gap_r \leftarrow gap_r + h_{t,r}$ // updates available spaces
        [PROCESSING CONDITIONAL (CCP) EVENTS]
10: **for all** tasks t that belong to an event of type CCP in $\mathcal{E}$ **do**
11:     **if** $conflicts_t > 0$ **then** $(\underline{s_t}, \underline{e_t}) \leftarrow (\overline{s_t}, \overline{e_t})$
12:     **if** $\underline{e_t} > \delta$ **then**
13:         add $\langle \text{ECPD}, t, \underline{e_t} \rangle$ to h_events // a compulsory part appears,
14:         **for** $r = 0$ **to** $k - 1$ **do** $gap_r \leftarrow gap_r - h_{t,r}$ // and starts at instant δ
15: [DETERMINE THE NEXT EVENT DATE]
16: $\delta_{next} \leftarrow$ peek_key(h_events)   // set to latest end of all tasks if h_events is empty
17: [PROCESSING EARLIEST START (PR) EVENTS]
18: **for all** tasks t that belong to an event of type PR in $\mathcal{E}$ **do**
19:     **if** $d_t > \delta_{next} - \delta \vee \exists r$ such that $h_{t,r} > gap_r$ **then** // unless task t holds in $[\delta, \delta_{next})$,
20:         **for** $r = 0$ **to** $k - 1$ **do**   // task t is added into a_check
21:             $a\_check_r[h_{t,r}] \leftarrow a\_check_r[h_{t,r}] \cup \{t\}$
22:             **if** $h_{t,r} > gap'_r$ **then** $conflicts_t \leftarrow conflicts_t + 1$
23: **return** $\langle \delta, \delta_{next} \rangle$

Algorithm 2. Called every time the sweep-line moves. Extract and process all events at current time point δ, returns both the current δ and the next time point δ_{next}.

conflict at the previous sweep-line position but that are no longer in conflict at instant δ on resource r.

We consider the three following cases:

- If δ has passed the latest start of t (i.e. $\delta > \overline{s_t}$, line 22) we know that t cannot be scheduled before its latest position.
- Else if the sweep interval $[\delta, \delta_{next})$ is long enough and if there is no conflict on other resources (i.e. $conflicts_t = 0 \wedge \delta_{next} - \delta \geq d_t$, line 24), the earliest start of task t is adjusted to δ. Note that this condition does not depend on the resource r.
- Otherwise (line 26), the task is left in a_check_r and its earliest start is adjusted to δ.

Continuation of Example 1 (Illustrating the k-dimensional sweep). The k-dimensional sweep first reads the events $\langle PR, 1, 0 \rangle$, $\langle PR, 2, 0 \rangle$ and set δ_{next} (i.e. the next event date) to value 1. Since tasks t_1 and t_2 have a duration longer than the sweep interval, they are added into the list of active tasks (see Alg. 2 line 21). Alg. 3 is run, there is no resource overflow and nothing can be deduced. Then the sweep-line moves to position 1 (i.e. $\delta = 1$), event $\langle SCP, 0, 1 \rangle$ is processed, gap_0 is updated

---

ALGORITHM filter(δ, δ_{next}) : *bool*

1: [CHECK RESOURCE OVERFLOW]
2: **for** $r = 0$ to $k - 1$ **do if** $gap_r < 0$ **then return false**
3: [TASKS ENTERING INTO CONFLICT]
4: **for** $r = 0$ to $k - 1$ **do**
5:    **while** $gap'_r > gap_r$ **do**   // the gap is smaller than previously
6:       $\mathcal{T} \leftarrow a\_check_r[gap'_r]$
7:       $\mathcal{U} \leftarrow \emptyset$
8:       **for all** $t \in \mathcal{T}$ **do**
9:          **if** $\neg(conflicts_t = 0 \wedge (\overline{s_t} \le \delta \vee \underline{e_t} \le \delta))$ **then**
10:            $conflicts_t \leftarrow conflicts_t + 1$
11:            $\mathcal{U} \leftarrow \mathcal{U} \cup \{t\}$
12:       $a\_check_r[gap'_r] \leftarrow \mathcal{U}$
13:       $gap'_r \leftarrow pred_r(gap'_r)$
14: [TASKS NO LONGER IN CONFLICT]
15: **for** $r = 0$ to $k - 1$ **do**
16:    **while** $succ_r(gap'_r) \le gap_r$ **do**   // the gap is greater than previously
17:       $gap'_r \leftarrow succ_r(gap'_r)$
18:       $\mathcal{T} \leftarrow a\_check_r[gap'_r]$
19:       $\mathcal{U} \leftarrow \emptyset$
20:       **for all** $t \in \mathcal{T}$ **do**
21:          $conflicts_t \leftarrow conflicts_t - 1$
22:          **if** $\delta \ge \overline{s_t}$ **then**
23:            $(\underline{s_t}, \underline{e_t}) \leftarrow (\overline{s_t}, \overline{e_t})$ // task t cannot start before its latest position
24:          **else if** $conflicts_t = 0 \wedge \delta_{next} - \delta \ge d_t$ **then**
25:            $(\underline{s_t}, \underline{e_t}) \leftarrow (\delta, \delta + d_t)$ // task t can start at instant δ
26:          **else**
27:            $(\underline{s_t}, \underline{e_t}) \leftarrow (\delta, \delta + d_t)$
28:            $\mathcal{U} \leftarrow \mathcal{U} \cup \{t\}$
29:       $a\_check_r[gap'_r] \leftarrow \mathcal{U}$
30: **return true**

Algorithm 3. Called every time the sweep-line moves from δ to δ_{next} in order to try to filter the earliest start of the tasks wrt. the available gap on each resource.

to 1 and gap_1 is updated to 1 (see Alg. 2 line 4). The next sweep-line position is set to 2 (i.e. $\delta_{next} = 2$). During the call of Alg. 3 tasks t_1 and t_2 are scanned because of the decrease of the gap on the two resources (see line 5). Since the height of task t_2 on resource r_0 is now greater than gap_0 and since the sweep-line has not passed its latest start, the condition stated at line 9 is true and $conflicts_2$ is updated to 1. On resource 1, for the same reason, the number of conflicts for task t_1 is updated to 1 (i.e. $conflicts_1 = 1$). Then the sweep-line moves to position 2 and processes event $\langle ECPD, 0, 2 \rangle$, i.e. $gap_0 = 2$ and $gap_1 = 3$. Consequently Alg. 3 notices that task t_2 is no longer in conflict on resource r_0 and adjusts its earliest start to 2 and its earliest end to 4 (see line 27). In the same way, task t_1 is no longer in conflict on resource r_1 and its earliest start is adjusted to 2 and its earliest end to 4. Then the sweep-line moves to position 3 and processes event $\langle CCP, 1, 3 \rangle$. Since task t_1 is not in conflict and since the sweep-line has not reached the earliest end of task t_1, a compulsory part appears. An event corresponding to the end of the compulsory part of task t_1 is created and gaps are decreased (see Alg. 2 lines 13 and 14) because of the start of the compulsory part of task t_1. Now, $gap_0 = gap_1 = 1$ and the next event date is equal to 4. Alg. 3 is run and task t_2 is now in conflict on the resource r_0 since $gap_0 < h_{0,0}$. Then the sweep-line moves to position 4 and processes event $\langle ECPD, 1, 4 \rangle$, i.e. $gap_0 = 2$ and $gap_1 = 3$. The next event date, 5, is given by the CCP event of task t_2. Alg. 3 is run and since gap_0 is now greater than the height of task t_2 on resource r_0, the number of conflicts associated to task t_2 is decremented. Since the sweep-line has not reached the latest start of task t_2 (i.e. $\delta < \overline{s_2}$) and since the sweep interval is not long enough for task t_2 (i.e. $\delta_{next} - \delta < d_2$), the earliest start of task t_2 is adjusted to 4 (see Alg. 3 lines 27). Then the sweep-line moves to position 5 and processes event $\langle CCP, 2, 5 \rangle$. Since task t_2 is not in conflict and since the sweep-line has not reached the earliest end of task t_2, a compulsory part appears. An event corresponding to the end of the compulsory part of task t_2 is created and gaps are decreased (see Alg. 2 lines 13 and 14) because of the start of the compulsory part of task t_1. Now, $gap_0 = 0$ and $gap_1 = 2$ and the next event date is equal to 6. Finally the sweep-line moves to position 6 and processes event $\langle ECPD, 2, 6 \rangle$, i.e. $gap_0 = 2$ and $gap_1 = 3$. There is no more event, consequently δ_{next} is set to the latest end of all tasks (see Alg. 2 line 16). We have now reached the same fixpoint as the decomposition into 2 cumulative constraint but in one single synchronized sweep.   □

2.4   Complexity

This section presents the worst-case time complexity of the k-dimensional sweep algorithm in a constraint involving n tasks and k resources. To achieve this, we first introduce the following lemma and prove it.

Lemma 1. *During a run of Alg. 1, we generate at most four events per task.*

Proof. Initially, at most three events are generated per task. In addition, at most one dynamic ECPD event can be generated per task, as shown below.

① Suppose that we have an initial ECPD event, and that line 8 of Alg. 2 is reached and its condition is true, generating a dynamic ECPD event. Then either $\underline{e_t} = \overline{e_t}$ or $conflicts_t = 0$ must hold according to line 7. Continue with "Common case".

② Suppose that we have an initial CCP event, and that line 12 of Alg. 2 is reached and its condition is true, generating a dynamic ECPD event. Then either $\underline{e_t} = \overline{e_t}$ or $conflicts_t = 0$ must hold according to line 11. Continue with "Common case".

Common case: If $\underline{e_t} = \overline{e_t}$, then $\underline{e_t}$ cannot be further updated, and there is no scope for a second dynamic ECPD event. So assume now that $\underline{e_t} < \overline{e_t}$. Then $conflicts_t = 0$ must hold. Now the only way to update $\underline{e_t}$ before handling the dynamic ECPD event is to reach line 20 of Alg. 3. But to make it there, we must first reach line 9 of Alg. 3. But the first time we reach that line, $\overline{s_t} \leq \delta$ will be true, otherwise we would not have seen a CCP/ECPD event in the first place. So the condition on line 9 will be false, t will be removed from the array slot, and line 20 will never be reached.

In either case, at most one dynamic ECPD event is generated.   □

Processing one event in Alg. 2 costs $O(k + \log n)$. Each event date costs $O(kn)$ in Alg. 3. Since there is $O(n)$ events, the worst-case time complexity of the k-dimensional sweep algorithm is $O(kn^2)$.

2.5   Greedy Assignment Mode

In the same way as for the CP2012 sweep algorithm, we implemented a greedy assignment mode related to the *sweep_min* part of the filtering algorithm. The greedy mode is an opportunistic mode which tries to build a solution in one single sweep from left to right. It has the same structure as *sweep_min*, but whenever *sweep_min* has found the earliest start of some task, the greedy mode also fixes its start time, and updates the data structures according to the new compulsory part of the task that was just fixed.

3   Evaluation

We implemented the dynamic sweep algorithm on Choco [14] and SICStus [4]. Choco benchmarks were run with an Intel i7 at 2.93 GHz processor on one single core, memory limited to 14GB under Mac OS X 64 bits. SICStus benchmarks were run on a quad core 2.8 GHz Intel Core i7-860 machine with 8MB cache per core, running Ubuntu Linux (using only one processor core).

In a first experiment, we ran random instances of bin-packing (unit duration) and cumulative (duration ≥ 1) problems, with k varying from 1 to 64 and n from 1000 to 1 million. Instances were randomly generated with a density close to 0.7. For a given number of tasks, we generated two different instances with the average number of tasks overlapping a time point equal to 5. We measured the time needed to find a first solution. As a search heuristic, the variable with the smallest minimal value was chosen, and for that variable, the domain values were tried in increasing order. All instances were solved without backtracking. The times reported are total SICStus execution time, not just the time spent in the dynamic sweep algorithm. The Choco results paint a similar picture.

In a first set of runs, *decomposed*, we modeled each instance with k *1-dimensional cumulative* constraints. In a second set of runs, *synchronized*, each instance was modeled with 1 *k-dimensional cumulative* constraint. In both sets, the new dynamic sweep algorithm was used. In a third set of runs, the greedy assignment algorithm was used.

The results are shown in Fig. 3 (first three rows). Note in particular that the greedy version could handle up to 1 million tasks in one *64-dimensional cumulative* constraint in SICStus in one hour and 40 minutes.

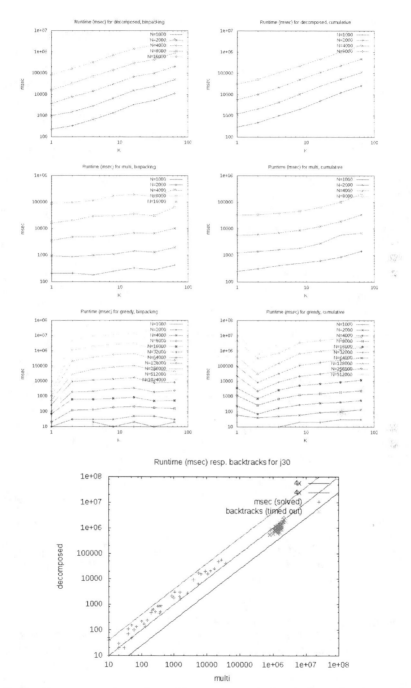

Fig. 3. Runtimes on random bin-packing and cumulative *k-dimensional cumulative* constraint (second row), greedy assignment (third row). Runtimes and backtrack counts for PSPLib instances (bottom).

A preliminary analysis of the observed runtimes as a function of n and k suggest that *decomposed* solves instances in time proportional to approx. $k \times n^{2.10}$ whereas *synchronized* solves them in time proportional to $k^{0.25} \times n^{2.25}$. In other words, we observe a speed-up by nearly $k^{0.75}$.

In a second experiment, we ran the J30 single-mode resource-constrained project scheduling benchmark suite from PSPLib [3], comparing *decomposed* with *synchronized* as above. Each instance involves 30 tasks, 4 resources and several precedence constraints. The same search heuristic was used as in the first experiment. The initial domains of the start times corresponded to the optimal makespan, which is known for all instances. The results are shown in Fig. 3 (bottom). The scatter plot compares run times for instances that were solved within a 1 minute time-out by both algorithms, as well as backtrack counts for instances that timed out in at least one of the two algorithms. Since a higher backtrack count means that propagation is faster, the plot shows that the new constraint allows up to 4 times speed-up for solved instances and up to 3 times speed-up for timed-out ones.

4   Conclusion

We have presented a synchronized sweep based algorithm for the k-*dimensional cumulative* constraint over n tasks and k resources, with a worst-case time complexity of $O(kn^2)$. The algorithm can operate in filtering mode as well as in greedy assignment mode. In the performance evaluation, we have demonstrated the importance of considering all the resources of the problem in one single sweep. On our benchmarks, the new filtering algorithm yields a speed-up of about $k^{0.75}$ over its decomposition into k independent *cumulative* constraints. The greedy mode yields another two orders of magnitude of speed-up. When considering a single resource, the new filtering algorithm is slightly faster (20%) than the dedicated CP2012 algorithm.

References

1. Baptiste, P., Le Pape, C., Nuijten, W.: Constraint-Based Scheduling: Applying Constraint Programming to Scheduling Problems. International Series in Operations Research and Management Science. Kluwer (2001)
2. Beldiceanu, N., Carlsson, M.: Sweep as a generic pruning technique applied to the non-overlapping rectangles constraint. In: Walsh, T. (ed.) CP 2001. LNCS, vol. 2239, pp. 377–391. Springer, Heidelberg (2001)
3. Beldiceanu, N., Carlsson, M., Thiel, S.: Sweep synchronisation as a global propagation mechanism. Computers and Operations Research 33(10), 2835–2851 (2006)
4. Carlsson, M., et al.: SICStus Prolog User's Manual. SICS, 4.2.1 edn. (2012), http://www.sics.se/sicstus
5. Freuder, E., Lee, J., O'Sullivan, B., Pesant, G., Rossi, F., Sellman, M., Walsh, T.: The future of CP. Personal communication (2011)

---

[3] http://129.187.106.231/psplib/, (480 instances).

6. Kameugne, R., Fotso, L.P., Scott, J., Ngo-Kateu, Y.: A quadratic edge-finding filtering algorithm for cumulative resource constraints. In: Lee, J. (ed.) CP 2011. LNCS, vol. 6876, pp. 478–492. Springer, Heidelberg (2011)
7. Kolisch, R., Sprecher, A.: PSPLIB – a project scheduling problem library. European Journal of Operational Research 96, 205–216 (1996)
8. Letort, A., Beldiceanu, N., Carlsson, M.: A scalable sweep algorithm for the *cumulative* constraint. In: Milano, M. (ed.) CP 2012. LNCS, vol. 7514, pp. 439–454. Springer, Heidelberg (2012)
9. O'Sullivan, B.: CP panel position - the future of CP. Personal communication (2011)
10. Régin, J.C., Rezgui, M.: Discussion about constraint programming bin packing models. In: AI for Data Center Management and Cloud Computing. AAAI (2011)
11. ROADEF: Challenge 2012 machine reassignment (2012), http://challenge.roadef.org/2012/en/index.php
12. Schutt, A., Feydy, T., Stuckey, P.J., Wallace, M.G.: Why `cumulative` decomposition is not as bad as it sounds. In: Gent, I.P. (ed.) CP 2009. LNCS, vol. 5732, pp. 746–761. Springer, Heidelberg (2009)
13. Shaw, P.: Using constraint programming and local search methods to solve vehicle routing problems. In: Maher, M.J., Puget, J.-F. (eds.) CP 1998. LNCS, vol. 1520, pp. 417–431. Springer, Heidelberg (1998)
14. Choco Team: Choco: an open source Java CP library. Research report 10-02-INFO, Ecole des Mines de Nantes (2010), http://choco.emn.fr/
15. Vilím, P.: Edge finding filtering algorithm for discrete cumulative resources in $\mathcal{O}(kn \log n)$. In: Gent, I.P. (ed.) CP 2009. LNCS, vol. 5732, pp. 802–816. Springer, Heidelberg (2009)
16. Vilím, P.: Timetable edge finding filtering algorithm for discrete cumulative resources. In: Achterberg, T., Beck, J.C. (eds.) CPAIOR 2011. LNCS, vol. 6697, pp. 230–245. Springer, Heidelberg (2011)

Enumerating Infeasibility:
Finding Multiple MUSes Quickly

Mark H. Liffiton and Ammar Malik

Illinois Wesleyan University, Bloomington IL 61701, USA
{mliffito,amalik}@iwu.edu
http://www.iwu.edu/~mliffito/

Abstract. Methods for analyzing infeasible constraint sets have proliferated in the past decade, commonly focused on finding maximal satisfiable subsets (MSSes) or minimal unsatisfiable subsets (MUSes). Most common are methods for producing a single such subset (one MSS or one MUS), while a few algorithms have been presented for enumerating all of the interesting subsets of a constraint set. In the case of enumerating MUSes, the existing algorithms all fall short of the best methods for producing a single MUS; that is, none come close to the ideals of 1) producing the first output as quickly as a state-of-the-art single-MUS algorithm and 2) finding each successive MUS after a similar delay. In this work, we present a novel algorithm, applicable to any type of constraint system, that enumerates MUSes in this fashion. In fact, it is structured such that one can easily "plug in" any new single-MUS algorithm as a black box to immediately match advances in that area. We perform a detailed experimental analysis of the new algorithm's performance relative to existing MUS enumeration algorithms, and we show that it avoids some severe intractability issues encountered by the others while outperforming them in the task of quickly enumerating MUSes.

1 Introduction

The most common applications of constraint systems (of any type) involve finding satisfying variable assignments for satisfiable constraint sets. As such, a huge range of algorithms exist for finding such assignments and potentially optimizing objective functions over them. Constraint sets for which no satisfying assignments exist, on the other hand, can be processed with the tools of "infeasibility analysis," a smaller but growing field of study.

Broadly, the algorithms of infeasibility analysis can be placed into two categories by the information they seek: 1) how much of an unsatisfiable constraint set can be satisfied, and 2) where in the constraint set the "problem" lies. These two categories and their solutions go by various names in the different fields where constraint systems are studied: Maximum Satisfiability (MaxSAT), Maximum Feasible Subsystem (MaxFS), and MaxCSP for the former, and Minimal[ly] Unsatisfiable Subset (MUS), Irreducible/Irredundant Infeasible/Inconsistent Subsystem (IIS), and Minimal[ly] Unsatisfiable Core (MUC) for the latter.

C. Gomes and M. Sellmann (Eds.): CPAIOR 2013, LNCS 7874, pp. 160–175, 2013.

These two types of information are diametrically opposed ("Max" and "SAT" vs "Min" and "UNSAT"), and yet a strong connection between the two has been known since at least 1987. Specifically, researchers in the field of diagnosis identified a hitting set relationship between the two [11,17]; it is based on the fact that any satisfiable subset of a constraint system cannot fully contain any unsatisfiable subset, and thus the satisfiable set must *exclude* at least one constraint from every unsatisfiable set. This relationship is occasionally exploited in infeasibility analysis by using one type of result to guide searches for the other, such as in algorithms that enumerate MUSes by way of MaxSAT solutions [1,13], solving MaxSAT with the assistance of unsatisfiable cores [6,16], and even finding [non-minimal] unsatisfiable cores to boost MaxSAT to then produce MUSes [14].

In this work, we introduce a new algorithm for infeasibility analysis inspired by this strong connection that fills a gap in the previous work and provides fertile ground for further developments in several directions. The "gap" we address is the lack of algorithms that quickly enumerate MUSes. While several approaches for enumerating MUSes exist, all suffer from severe scalability issues, and none currently match the performance of state-of-the-art algorithms for extracting a *single* MUS from an unsatisfiable constraint set. Ideally, MUS enumeration should produce the first MUS in roughly the same time T_{MUS} taken by the best algorithms for finding a single MUS, and each additional MUS should be produced as quickly as possible, with a reasonable incremental delay being roughly that same time period T_{MUS}. The new algorithm we present here, dubbed MARCO, achieves both of these goals.

In the following sections, we first define terms and describe concepts underlying this work (Section 2), followed by a discussion of past research on enumerating MUSes (Section 3). We then present the MARCO algorithm (Section 4) and an extensive empirical analysis (Section 5) before finally concluding and outlining several research paths that continue from here (Section 6).

2 Preliminaries

In this work, we discuss problems, results, and algorithms in terms of generic sets of constraints for which the constraint type and the variable domain are not specified. Generally, then, we will be discussing an ordered set of n constraints:

$$C = \{C_1, C_2, C_3, \ldots, C_n\}$$

A given constraint C_i places restrictions on assignments to a problem's variables, and C_i is *satisfied* by any assignment that meets its restrictions. If there exists some assignment to C's variables that satisfies every constraint, C is said to be *satisfiable* or *SAT*; otherwise, it is *unsatisfiable, infeasible*, or *UNSAT*. Most of the algorithms we describe in this paper, and especially our own algorithm, can be applied to any set of constraints given that there exists a solving method capable of returning SAT or UNSAT for that set of constraints; we call these *constraint-agnostic* algorithms.

An infeasible constraint set C can be analyzed in a variety of ways, often in terms of producing useful subsets of C. The most common analysis is likely the maximum satisfiability problem (MaxSAT, MaxFS, MaxCSP), which produces a satisfiable subset of C with the greatest possible cardinality. Generalizing MaxSAT by considering maximality instead of maximum cardinality yields the concept of a *Maximal Satisfiable Subset* (MSS):

$$M \subseteq C \text{ is an MSS } \iff M \text{ is SAT and } \forall c \in C \setminus M : M \cup \{c\} \text{ is UNSAT}$$

An MSS is essentially a satisfiable subset of C that cannot be expanded without becoming unsatisfiable. While any solution to the MaxSAT problem is an MSS, some MSSes may be smaller than that maximum size. The complement of an MSS is often more directly useful, and we call such a minimal set (whose removal from C makes it satisfiable or "corrects" it) a *Minimal Correction Set* (MCS):

$$M \subseteq C \text{ is an MCS } \iff C \setminus M \text{ is SAT and } \forall c \in M : (C \setminus M) \cup \{c\} \text{ is UNSAT}$$

Again, the minimality is not in terms of cardinality, but rather it requires that no proper subset of M be capable of "correcting" the infeasibility. A related concept, the focus of this work, is the *Minimal Unsatisfiable Subset* (MUS):

$$M \subseteq C \text{ is an MUS } \iff M \text{ is UNSAT and } \forall c \in M : M \setminus \{c\} \text{ is SAT}$$

MUSes are most commonly considered in terms of minimizing an unsatisfiable constraint set down to a "core" reason for its unsatisfiability. In some work, they are called "unsatisfiable cores," but that is also used to refer to any unsatisfiable subset of a constraint system, regardless of its minimality. Note that the definition of an MUS need not reference the constraint set C of which it is a subset; it is really a free-standing property of any set of constraints, as it does not depend on the existence or the structure of any other constraints. However, as it is most commonly encountered in terms of finding such a minimal subset of some larger constraint set, naming it with "subset" is traditional. In OR, the concept of the *Irreducible Inconsistent Subsystem* (IIS) [15] is equivalent to that of the MUS.

Example 1. Consider the following unsatisfiable set of Boolean clauses:

$$C = \{\ \underset{C_1}{(a)}\ ,\ \underset{C_2}{(\neg a \vee b)}\ ,\ \underset{C_3}{(\neg b)}\ ,\ \underset{C_4}{(\neg a)}\ \}$$

C has two MUSes and three MSS/MCS pairs:

MUSes	MSSes	MCSes
$\{C_1, C_2, C_3\}$	$\{C_2, C_3, C_4\}$	$\{C_1\}$
$\{C_1, C_4\}$	$\{C_1, C_3\}$	$\{C_2, C_4\}$
	$\{C_1, C_2\}$	$\{C_3, C_4\}$

Simple constraint-agnostic algorithms for finding MSSes and MUSes of a constraint set C are shown in Figure 1, and their behavior is quite similar. To find

grow(*seed*, *C*)	shrink(*seed*, *C*)
input: unsatisfiable constraint set *C*	input: unsatisfiable constraint set *C*
input: satisfiable subset *seed* ⊂ *C*	input: unsatisfiable subset *seed* ⊆ *C*
output: an MSS of *C*	output: an MUS of *C*
1. **for** *c* ∈ *C* \ *seed*:	1. **for** *c* ∈ *seed*:
2. **if** *seed* ∪ {*c*} **is satisfiable:**	2. **if** *seed* \ {*c*} **is unsatisfiable:**
3. *seed* = *seed* ∪ {*c*}	3. *seed* = *seed* \ {*c*}
4. **return** *seed*	4. **return** *seed*

Fig. 1. The basic grow and shrink methods for finding an MSS or an MUS, respectively, of a constraint set

an MSS (MUS), the grow (shrink) method starts from some satisfiable (unsatisfiable) subset *seed* ⊆ *C* and iteratively attempts to add (remove) constraints, checking each new set for satisfiability and keeping any changes that leave the set satisfiable (unsatisfiable). These algorithms are not novel (for example, shrink was described by Dravnieks in 1989 as "deletion filtering" [4]), nor are they particular efficient as shown (many improvements can be made to both), but they serve as simple illustrative examples for the purposes of this work.

Note that the input *seed* can take simple default values if no particular subset is given. The grow method can begin its construction with *seed* = ∅ (guaranteed to be satisfiable), while shrink can start with *seed* = *C* (guaranteed UNSAT). Therefore, *seed* can be considered an optional parameter for both, and each method is also a generic method for finding an MSS or MUS of a constraint set *C* without any additional information. For any given constraint type and solver, both shrink and grow can be optimized to exploit characteristics of the constraints or features of the solver; most fields have a great deal of research on efficient shrink implementations, but grow is less often studied.

3   Related Work

The existing work on algorithms for enumerating MUSes is limited, especially when compared to the amount of work on extracting single MUSes and unsatisfiable cores. Some all-MUS algorithms have been developed for specific constraint types. For example, there are many methods for computing all IISes of a linear program such as the original work by van Loon [15], later work by Gleeson and Ryan [8], etc.; however, these approaches are quite specific to linear programming, constructing a polytope and using the simplex method, and they do not generalize well. Additionally, Gasca, et al. developed methods for computing all MUSes of overconstrained numerical CSPs (NCSPs) [7]. Their approach explores all subsets of a constraint system while pruning unnecessary collections of subsets with rules based on structure specific to NCSPs. In the space of *constraint-agnostic* algorithms for enumerating MUSes, three different approaches have been presented. As with the work in this paper, all of the following algorithms are easily applied to any type of constraint system, from CSP to IP to SAT, and none rely on specific features of any constraint type or solving method.

Subset Enumeration. The technique of explicitly enumerating and checking every subset of the unsatisfiable constraint system was first explored in the field of diagnosis by Hou [10], who presented a technique for enumerating subsets in a tree structure along with pruning rules to reduce its size and avoid unnecessary work. Starting from the complete constraint set C, the algorithm searches the power set $\mathcal{P}(C)$, branching to explore all subsets. Each subset is checked for satisfiability, and any subset found to be unsatisfiable and whose children (proper subsets) are all satisfiable is an MUS. Han and Lee corrected an error in one of the pruning rules and presented additional improvements [9], and further optimizations and enhancements were made by de la Banda et al. [2].

CAMUS. A later algorithm for enumerating MUSes by Liffiton and Sakallah [12,13,14] avoids an explicit search of the power set of C by exploiting the relationship between MCSes and MUSes [11,17]. CAMUS works in two phases, first computing all MCSes of the constraint set, then finding all MUSes by computing the minimal hitting sets of those MCSes. The two-phase method can be applied with any technique for enumerating MCSes and any minimal hitting set algorithm. The authors provide an algorithm for the first phase that gives a constraint solver the ability to search for satisfiable subsets of constraints without having to feed each subset to the solver individually. With this ability, the algorithm then searches for satisfiable subsets in decreasing order of size, blocking any solutions found before continuing its search, thus guaranteeing it finds only maximal satisfiable sets whose complements are the MCSes it seeks. The second phase of CAMUS, as a purely set theoretic problem, operates independently of any constraint solver.

Due to the complexity and potential intractability of the first phase (the number of MCSes may be exponential in the size of the instance), CAMUS is unsuitable for enumerating MUSes in many applications that require multiple MUSes quickly. Variations on the core algorithm can relax its completeness and adapt it to such situations [13], but the control they provide, essentially a tradeoff between time and completeness, is crude. In any case, CAMUS is not able to be run in an *incremental* fashion, with short, consistent delays between each MUS, such that one can make a decision about the time/completeness tradeoff dynamically *while* the algorithm runs.

DAA. Closer to the goal of this work, providing a much more incremental approach than CAMUS, is the Dualize and Advance algorithm (DAA) by Bailey and Stuckey [1]. It exploits the same relationship between MCSes and MUSes, but it discovers both types of sets throughout its execution. Therefore, like our algorithm and unlike CAMUS, it is capable of producing MUSes "early" in its execution. Pseudocode for DAA is shown in Figure 2. It repeatedly computes MCSes by growing MSSes from seeds with the grow method and taking their complements. The initial seed is the empty set. It then computes the minimal

---

DAA

input: unsatisfiable constraint set C

output: MSSes and MUSes of C as they are discovered

···

1. $MCSes, MUSes, seed \leftarrow \emptyset$
2. $haveSeed \leftarrow$ True
3. **while** $haveSeed$:
4. $MSS \leftarrow \text{grow}(seed, C)$
5.   **yield** MSS
6. $MCSes \leftarrow MCSes \cup \{C \setminus MSS\}$ ◁ *the complement of an MSS is an MCS*
7. $haveSeed \leftarrow$ False
8.   **for** $candidate \in (\text{hittingSets}(MCSes) \setminus MUSes)$:
9.   **if** $candidate$ **is satisfiable**:
10. $seed \leftarrow candidate$ ◁ *if SAT, candidate is a new MSS seed*
11. $haveSeed \leftarrow$ True
12.   **break**
13.   **else**:
14.   **yield** $candidate$ ◁ *if UNSAT, candidate is an MUS*
15. $MUSes \leftarrow MUSes \cup \{candidate\}$

Fig. 2. The DAA algorithm for enumerating MSSes & MUSes of a constraint set

hitting sets of the MCSes found thus far, as CAMUS does once it has the *complete* set of MCSes. With an incomplete set of MCSes, some of the hitting sets may be unsatisfiable, and these are guaranteed to be MUSes. DAA therefore checks each for satisfiability, reporting every unsatisfiable set as an MUS, and the first set found to be satisfiable is taken as the next seed for the algorithm to repeat.

Comparisons. Bailey and Stuckey found that DAA performed much better than the subset enumeration algorithm as presented by de la Banda, et al. in their experimental evaluation [1], while somewhat limited experiments in [13] indicated that CAMUS outperformed DAA for finding all MUSes of a constraint system. However, the incremental nature of DAA is not matched by CAMUS, and so comparisons to both are warranted here. We contrast the features of DAA and CAMUS with our new algorithm following its description in Section 4, and the experimental results in Section 5 further illustrate the differences.

4   Exploring Infeasibility with the **MARCO** Algorithm

Here, we present a novel algorithm for enumerating all MUSes of an unsatisfiable constraint set C. (As with CAMUS and DAA, it also enumerates all MSSes of C, but they are not the focus of this work.) It efficiently explores the power set $\mathcal{P}(C)$ by exploiting the idea that any power set can be analyzed and manipulated as a Boolean algebra; that is, one can perform set operations within $\mathcal{P}(C)$ by manipulating Boolean functions as propositional formulas. Specifically, we note that any function $f : \mathcal{P}(C) \rightarrow \{0,1\}$ can be represented by a propositional formula over $|C|$ variables.

Our algorithm maintains a particular function $f : \mathcal{P}(C) \rightarrow \{0,1\}$ that tracks "unexplored" subsets $C' \subseteq C$ such that $f(C') = 1$ iff the satisfiability of C' is unknown and it remains to be checked. This function, stored as a propositional formula, can be viewed as a "map" of $\mathcal{P}(C)$ showing which "regions" have been explored and which have not. Named after the Venetian explorer Marco Polo, we have dubbed the algorithm MARCO (**Ma**pping **R**egions of **Co**nstraint sets) and the general technique of maintaining a power set map as a propositional logic formula POLO (**Po**wer set **Lo**gic). Overall, MARCO enumerates MUSes by repeatedly selecting an unexplored subset $C' \in \mathcal{P}(C)$ from the map, checking whether C' is satisfiable, minimizing or maximizing it into an MUS or an MSS, and marking a region of the map as explored based on that result.

MARCO

input: unsatisfiable constraint set $C = \{C_1, C_2, C_3, \ldots, C_n\}$
output: MSSes and MUSes of C as they are discovered

1. $Map \leftarrow \text{BoolFormula}(nvars = |C|)$ ◁ *Empty formula over $|C|$ Boolean variables*
2. **while** Map **is satisfiable**:
3. $m \leftarrow \text{getModel}(Map)$
4. $seed \leftarrow \{C_i \in C : m[x_i] = \text{True}\}$ ◁ *Project the assignment m onto C*
5.   **if** $seed$ **is satisfiable**:
6. $MSS \leftarrow \text{grow}(seed, C)$
7.   **yield** MSS
8. $Map \leftarrow Map \wedge \text{blockDown}(MSS)$
9.   **else**:
10. $MUS \leftarrow \text{shrink}(seed, C)$
11.   **yield** MUS
12. $Map \leftarrow Map \wedge \text{blockUp}(MUS)$

Fig. 3. The MARCO algorithm for enumerating MSSes & MUSes of a constraint set

Figure 3 contains pseudocode for the MARCO algorithm. The formula Map is created to represent the "mapping" function described above, with a variable x_i for every constraint C_i in C. Initially, the formula is a tautology, true in every model, meaning every subset of C is still unexplored. Given its semantics, any model of Map can be projected onto C (lines 3 and 4) to identify a yet-unexplored element of C's power set whose satisfiability is currently unknown. If this subset, $seed$, is satisfiable, then it must be a subset of some MSS, and it can be "grown" into an MSS. Likewise, if it is unsatisfiable, $seed$ must contain at least one MUS, and it can be "shrunk" to produce one. In either case, the result is reported (via **yield** in the pseudocode, indicating that the result is returned but the algorithm may continue).

Each result provides information about some region of $\mathcal{P}(C)$ that is either satisfiable or unsatisfiable, and so a clause is added to Map to represent that region as "explored." For an MSS M, all subsets of M are now known to be satisfiable, and so models corresponding to any subset of M are eliminated by requiring that later models of Map include at least one constraint not in M:

$$\text{blockDown}(M) \equiv \bigvee_{i\,:\,C_i \notin M} x_i$$

Similarly, all supersets of any MUS M are known to be unsatisfiable; supersets of M are blocked by requiring models to exclude at least one of its constraints:

$$\text{blockUp}(M) \equiv \bigvee_{i\,:\,C_i \in M} \neg x_i$$

Eventually, all MSSes and MUSes are enumerated, the satisfiability of every element in $\mathcal{P}(C)$ is known, and MARCO terminates when Map has no further models. We discuss implementation details after an example.

Example 2. Suppose we run MARCO on the constraint set from Example 1:

$$C = \{ \underset{C_1}{(a)} , \underset{C_2}{(\neg a \vee b)} , \underset{C_3}{(\neg b)} , \underset{C_4}{(\neg a)} \}$$

Initialization:	$Map \leftarrow$ [empty formula over $\{x_1, x_2, x_3, x_4\}$]
Iteration 1:	$Map = \top$: **SAT**
	getModel $\rightarrow [x_1, x_2, x_3, x_4]$
	seed $\leftarrow \{C_1, C_2, C_3, C_4\}$: **UNSAT**
	shrink $\rightarrow$ MUS: $\{C_1, C_2, C_3\}$
	$Map \leftarrow Map \wedge (\neg x_1 \vee \neg x_2 \vee \neg x_3)$
Iteration 2:	$Map = (\neg x_1 \vee \neg x_2 \vee \neg x_3)$: **SAT**
	getModel $\rightarrow [\neg x_1, x_2, x_3, x_4]$
	seed $\leftarrow \{C_2, C_3, C_4\}$: **SAT**
	grow $\rightarrow$ MSS: $\{C_2, C_3, C_4\}$ — equiv. MCS: $\{C_1\}$
	$Map \leftarrow Map \wedge (x_1)$
Iteration 3:	$Map = (\neg x_1 \vee \neg x_2 \vee \neg x_3) \wedge (x_1)$: **SAT**
	getModel $\rightarrow [x_1, \neg x_2, x_3, x_4]$
	seed $\leftarrow \{C_1, C_3, C_4\}$: **UNSAT**
	shrink $\rightarrow$ MUS: $\{C_1, C_4\}$
	$Map \leftarrow Map \wedge (\neg x_1 \vee \neg x_4)$

At this point, MARCO has found all MUSes. It must ensure completeness, however, and so it exhaustively explores all remaining subsets.

Iteration 4: $Map = (\neg x_1 \lor \neg x_2 \lor \neg x_3) \land (x_1)$
$\land (\neg x_1 \lor \neg x_4) : \textbf{SAT}$

getModel $\to [x_1, \neg x_2, x_3, \neg x_4]$

$seed \leftarrow \{C_1, C_3\} : \textbf{SAT}$

grow $\to$ MSS: $\{C_1, C_3\}$ — equiv. MCS: $\{C_2, C_4\}$

$Map \leftarrow Map \land (x_2 \lor x_4)$

Iteration 5: $Map = (\neg x_1 \lor \neg x_2 \lor \neg x_3) \land (x_1)$
$\land (\neg x_1 \lor \neg x_4) \land (x_2 \lor x_4) : \textbf{SAT}$

getModel $\to [x_1, x_2, \neg x_3, \neg x_4]$

$seed \leftarrow \{C_1, C_2\} : \textbf{SAT}$

grow $\to$ MSS: $\{C_1, C_2\}$ — equiv. MCS: $\{C_3, C_4\}$

$Map \leftarrow Map \land (x_3 \lor x_4)$

Iteration 6: $Map = (\neg x_1 \lor \neg x_2 \lor \neg x_3) \land (x_1)$
$\land (\neg x_1 \lor \neg x_4) \land (x_2 \lor x_4) \land (x_3 \lor x_4) : \textbf{UNSAT}$

In the final iteration, all MSSes and all MUSes have been found; therefore, every model of Map is blocked, Map is UNSAT, and the algorithm terminates.

4.1   Implementation and Efficiency

An implementation of MARCO requires solvers for both C and Map. It is constraint-agnostic, as it only needs a solver that can take a set of constraints (some subset of C) and return a SAT/UNSAT result. The solver for Map is separate, and any engine for obtaining a model of a Boolean formula can be used; an incremental interface such as provided by modern SAT solvers or Binary Decision Diagram (BDD) engines will be most efficient.

MARCO's efficiency depends primarily on the implementation of the grow and shrink subroutines, as they are the most expensive steps. Constraint-agnostic methods for both are described in Section 2, but algorithms specific to a particular constraint type will be able to leverage details of those constraints for better performance. Due to the difficulty and broad applicability of extracting MUSes (the shrink method), much research has been done on the problem, and efficient algorithms for shrink exist for many constraint types. The grow method is less studied, and far more work is done on MaxSAT, MaxFS, etc. than on the easier problem of finding an MSS. Note that the solvers for C and Map and the methods for grow and shrink are black boxes as far as MARCO is concerned; an advance in the state-of-the-art for any one of the four can be immediately "plugged in" to boost the algorithm's performance.

4.2   Impact of the Map Solver

Another important factor for performance is the behavior of the solver for Map. The particular model returned by getModel cannot affect correctness, but it can impact the work done by grow and shrink. For example, imagine a simple constraint set that is an MUS itself. If the first model found for Map corresponds to the empty set, grow will be called, and it will have to proceed through several steps to reach an MSS, which in this case must contain all but one constraint of C. If the models of Map continue to represent very small subsets of C, grow will continue to require a large number of steps in each call. On the other hand, if the first model of Map corresponds to C itself, with all constraints included, shrink will be called on an "easy" seed (an MUS that will not be shrunk farther). And now, if models of Map are generally large subsets of C, the MSSes will be found by much faster calls to grow, as it will have much less "distance" to cover to reach an MSS in each case.

Returning to the earlier stated goal of producing the first MUS as fast as a state-of-the-art single-MUS algorithm, we see that we can achieve this simply by ensuring that the first model found for Map sets all x_i variables to True, resulting in $seed = C$, the entire constraint set. This takes negligible time, and the algorithm will immediately call shrink on C to produce the first MUS. Given that shrink can be any state-of-the-art MUS extraction algorithm, MARCO can thus find the first MUS as quickly as any other algorithm.

While we cannot then guarantee that each successive model of Map will also correspond to an unsatisfiable subset of C, which would trigger further calls to shrink immediately, it *is* possible to bias a solver in that direction. If the solver for Map favors assigning variables to True, then it will be more likely to produce models corresponding to large, nearly-complete subsets of C, which are the subsets most likely to be unsatisfiable. In Example 2, the model m found in each iteration is biased in this way, and the first $seed$ is thus C itself. The next model, even if maximizing the number of variables assigned True, will not necessarily correspond to an unsatisfiable $seed$, as illustrated in Iteration 2, but it is still likely to locate other unsatisfiable subsets of C quickly.

4.3   Comparison to CAMUS and DAA

With regards to tractability, CAMUS suffers from the fact that its first phase may produce an intractably large set of MCSes, with no good way to make progress on MUSes until the MCSes are all found. DAA also faces a severe tractability issue in the intermediate collections of hitting sets it computes; these collections can be exponential in size even if the number of MUSes is not [1]. MARCO, on the other hand, faces no such issues; the only information stored outside of its black box solvers is the formula it maintains in Map, which grows linearly with the number of results found.

The intractability of the first phase of CAMUS also impacts the time until its first MUS output, which can be effectively infinite even for small problems. DAA fares better, but it still must find at least k MCSes before it might output an MUS of size k, meaning it may face a lengthy delay before outputting its first MUS. The very first step of MARCO, however, finds an MUS directly using an efficient MUS algorithm, and each subsequent MUS can be found in roughly the same amount of time. At an algorithmic level, MARCO is better suited to finding multiple MUSes quickly than either CAMUS or DAA.

5   Empirical Analysis

To evaluate the MARCO algorithm and to compare it to the previous approaches for MUS enumeration, CAMUS and DAA, we ran all three on a set of 300 benchmarks from the Boolean satisfiability domain. Compared to analyzing decision or optimization problems, addressing the indefinite nature of enumerating potentially intractable sets requires a more detailed analysis, and so we present a variety of analyses to illustrate each algorithm's strengths and weaknesses.

Each algorithm was implemented in C++ using the MiniSAT solver [5]. We used the most recent release of CAMUS for Boolean SAT, which is built on MiniSAT 1.12b, while we implemented MARCO and DAA using MiniSAT 2.2[1]. Both MARCO and DAA were written in the same framework so that each would share as much code as possible, including the implementation of the grow method. For the shrink method in MARCO, we used the MUSer2 algorithm [3], a state-of-the-art MUS extraction algorithm for Boolean SAT. The solver for Map in MARCO was biased toward models representing larger subsets of C, as described in Section 4.2. All experiments were run on 3.4GHz AMD Phenom II CPUs with a 3600 second timeout and a 1.8 GB memory limit.

We used the 300 benchmarks selected for the MUS track of the recent 2011 SAT Competition[2]. These benchmarks were drawn from a wide variety of applications and cover a range of sizes, from 26 clauses (constraints) up to 4.4 million. Of the 300 instances, our experiments found that 219 contained more than one MUS, 17 had exactly one MUS, and the remaining 64 were indeterminate (i.e., on these instances, every algorithm ran out of time or memory and output only zero or one MUS before it was terminated).

Table 1. Number of instances in which each algorithm found all, multiple, or at least one MUS

	n	CAMUS	DAA	MARCO
All instances	300			
Found all MUSes		**41**	24	25
Found ≥ 1 MUS		113	51	**244**
Instances w/ >1 MUS	219			
Found all MUSes		**26**	8	11
Found > 1 MUS		98	32	**215**
Found ≥ 1 MUS		98	35	**217**

An overview of the number of instances for which each algorithm reached certain thresholds of enumerating MUSes is shown in Table 1. The results are broken out for the complete set of 300 benchmarks and for the set of 219 benchmarks that are known to contain more than one MUS. For the goal of enumerating

---

[1] While this may disadvantage CAMUS, our experiments have shown that it does not perform substantially better when built on top of MiniSAT 2.2.

[2] http://www.satcompetition.org/2011/

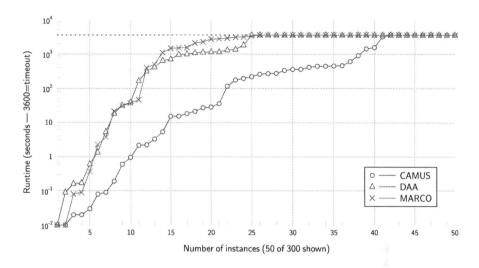

Fig. 4. Logarithmic cactus plot of each algorithm's runtime (to completion; i.e., enumerating *all* MUSes)

all MUSes, CAMUS outperforms the other two algorithms by a wide margin, completing within the time and memory limits for 41 benchmarks, compared to 24 for DAA and 25 for MARCO. Figure 4 shows a cactus plot[3] of the runtimes of the three algorithms, further supporting this point; CAMUS appears to be the best option for enumerating the *complete* set of MUSes.

It is also clear, however, that enumerating the complete set of MUSes is generally intractable (in fact, the set's cardinality may be exponential in the size of the instance), and CAMUS is outperformed by MARCO in the task emphasized by this work: enumerating *some*, but not all, MUSes. The number of instances for which MARCO can find a single MUS or multiple MUSes within the resource limits is more than twice that of CAMUS. This is consistent with the fact that the first phase of CAMUS is potentially intractable, and it often times out before reaching the second phase and producing even a single MUS. The DAA algorithm is outperformed by CAMUS in enumerating all MUSes (which agrees with earlier, more limited results [13]), and DAA produces no output at all in far more instances than either algorithm, especially MARCO. DAA most commonly exhausts its memory limit, due primarily to the number of intermediate hitting sets it generates in every iteration, and in many cases memory is exhausted before a single MUS has been found. Therefore, the remainder of the analysis will focus primarily on comparing MARCO and CAMUS.

---

[3] Cactus plots are created by sorting and plotting values in order within each series, showing distributions of values within a series, but not allowing pairwise comparisons between them. Each point (x, y) can be read as, "x instances have a value [e.g., runtime] of y or less."

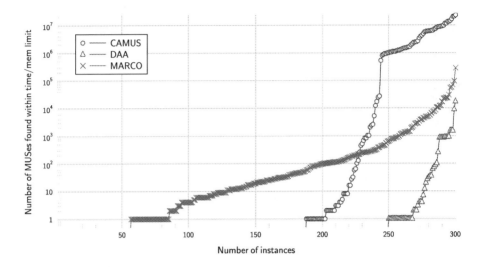

Fig. 5. Logarithmic cactus plot of the number of MUSes found by each algorithm within time / memory limits

Another view of the difference between MARCO and CAMUS is shown in a cactus plot of the numbers of MUSes found by each algorithm in Figure 5. This chart echoes some of the information in Table 1, showing the number of instances in which each algorithm is able to find one or more than one MUS, but it adds additional information about other output counts as well. For example, we can see that MARCO produces 10 or more MUSes in more than 170 instances, while CAMUS does so in only about 80 instances. However, CAMUS finds much larger sets of MUSes within the timeout in many instances, returning over 10^6 results in more than 50 instances, while MARCO only reaches above 10^5 results in one instance. This suggests that CAMUS will produce many more MUSes than MARCO, *when* it produces any, but MARCO is more robust in terms of scaling to produce some MUSes for more instances overall.

Figure 6 explores this further with pairwise comparisons of the number of MUSes found. DAA never produces more MUSes than MARCO. CAMUS, on the other hand, often produces orders of magnitude more MUSes. However, the chart also shows the large set of instances for which CAMUS outputs nothing and MARCO produces multiple MUSes; the reverse is true in only two instances.

Finally, to further contrast the performance of CAMUS and MARCO, we can look at anytime charts of their output over time, showing how many MUSes will be produced if execution is stopped at any particular time. The anytime charts in Figure 7 contain one trace for each instance that had 10 or more outputs, plotting the number of MUSes produced on the y-axis against time on the x-axis. For the sake of comparison, the data have been normalized to a scale of 0.0 to 1.0 such that 1.0 represents 100% of each algorithm's runtime on each instance (on the x-axis) or 100% of the MUSes it found in that time (y-axis). On these charts, we can see that CAMUS typically outputs the great majority of an

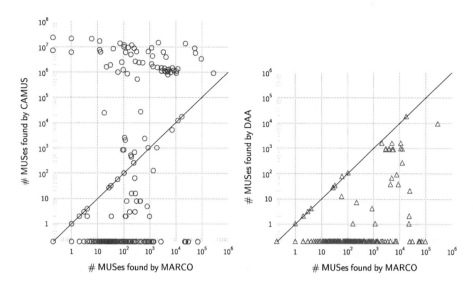

Fig. 6. Comparing MARCO to CAMUS (left) and DAA (right): number of MUSes found within time / memory limits (counts of 0 remapped to 0.2 to lie on the axis)

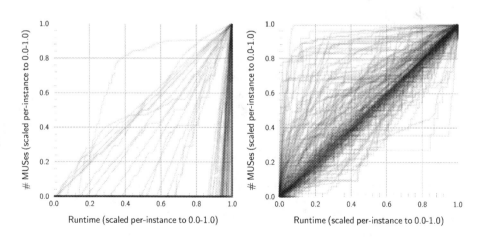

Fig. 7. Normalized anytime charts for CAMUS (left) and MARCO (right)

instance's MUSes "late," in the last 10% or less of its runtime (the dark, nearly vertical band on the right of the chart). This is consistent with its operation in two phases, where MUSes are only output in the second phase, and it shows that there is typically a long delay before any output is produced. In contrast, MARCO most often produces MUSes in a fairly steady progression (seen in the darkest band of traces along the diagonal) with a regular delay between each output, and it produces MUSes "early" (traces above the diagonal) more often than "late" (traces below). Overall, CAMUS can produce MUSes more quickly than MARCO, *if* it produces any at all (i.e., if it is given enough time), while MARCO outputs them at a much more steady pace from the beginning, making it more suitable for computing *some* MUSes quickly.

6   Conclusion

We have presented MARCO (**Ma**pping **R**egions of **Co**nstraint sets), a novel algorithm for enumerating MUSes in any type of constraint system, with the focus on producing *multiple* MUSes *quickly*, and empirical results show that MARCO outperforms other MUS enumeration algorithms at this task. It operates within the POLO framework (**Po**wer set **Lo**gic), maintaining a "map" of a constraint set's power set in a propositional logic formula, marking "explored" areas of the power set as it progresses, and using the map to find new subsets to check. Experimental results show that the CAMUS algorithm can enumerate the *complete* set of MUSes faster than MARCO, but the faster *early* results of MARCO are preferable in any application for which one wants multiple MUSes within some time limit and for which all MUSes are not needed, especially if they number in the millions or higher. MARCO can be implemented on top of any existing constraint solver, and in fact its critical shrink method can be implemented by "plugging in" any state-of-the art single-MUS extraction algorithm; therefore, it can always mirror the performance of any advances in MUS extraction.

Future research directions include exploring the effects of biasing and other heuristics in the solver for the "map" formula, as well as changes to the algorithm that focus it on enumerating MUSes alone, at the expense of missing some MSSes, or vice versa. Additionally, there are many opportunities to relax its completeness and/or optimality to be able to produce results in cases that remain too difficult for finding minimal unsatisfiable subsets as opposed to just small unsatisfiable cores. And finally, the general POLO technique provides a foundation from which new infeasibility analyses may be developed, such as new algorithms for finding a smallest MUS (SMUS) of a constraint set or for solving the MaxSAT/MaxFS/MaxCSP problem.

Acknowledgements. Many thanks to Anton Belov for providing assistance with the MUSer2 source code and to the anonymous reviewers for their helpful comments.

References

1. Bailey, J., Stuckey, P.J.: Discovery of minimal unsatisfiable subsets of constraints using hitting set dualization. In: Hermenegildo, M.V., Cabeza, D. (eds.) PADL 2005. LNCS, vol. 3350, pp. 174–186. Springer, Heidelberg (2005)
2. de la Banda, M.J.G., Stuckey, P.J., Wazny, J.: Finding all minimal unsatisfiable subsets. In: Proceedings of the 5th ACM SIGPLAN International Conference on Principles and Practice of Declaritive Programming (PPDP 2003), pp. 32–43 (2003)
3. Belov, A., Marques-Silva, J.: MUSer2: An efficient MUS extractor. Journal on Satisfiability, Boolean Modeling and Computation 8, 123–128 (2012)
4. Dravnieks, E.W.: Identifying minimal sets of inconsistent constraints in linear programs: deletion, squeeze and sensitivity filtering. Master's thesis, Carleton University (1989), https://curve.carleton.ca/theses/22864
5. Eén, N., Sörensson, N.: An extensible SAT-solver. In: Giunchiglia, E., Tacchella, A. (eds.) SAT 2003. LNCS, vol. 2919, pp. 502–518. Springer, Heidelberg (2004)
6. Fu, Z., Malik, S.: On solving the partial MAX-SAT problem. In: Biere, A., Gomes, C.P. (eds.) SAT 2006. LNCS, vol. 4121, pp. 252–265. Springer, Heidelberg (2006)
7. Gasca, R.M., Del Valle, C., Gómez-López, M.T., Ceballos, R.: NMUS: Structural analysis for improving the derivation of all MUSes in overconstrained numeric CSPs. In: Borrajo, D., Castillo, L., Corchado, J.M. (eds.) CAEPIA 2007. LNCS (LNAI), vol. 4788, pp. 160–169. Springer, Heidelberg (2007)
8. Gleeson, J., Ryan, J.: Identifying minimally infeasible subsytems. ORSA Journal on Computing 2(1), 61–67 (1990)
9. Han, B., Lee, S.J.: Deriving minimal conflict sets by CS-trees with mark set in diagnosis from first principles. IEEE Transactions on Systems, Man, and Cybernetics, Part B 29(2), 281–286 (1999)
10. Hou, A.: A theory of measurement in diagnosis from first principles. Artificial Intelligence 65(2), 281–328 (1994)
11. de Kleer, J., Williams, B.C.: Diagnosing multiple faults. Artificial Intelligence 32(1), 97–130 (1987)
12. Liffiton, M.H., Sakallah, K.A.: On finding all minimally unsatisfiable subformulas. In: Bacchus, F., Walsh, T. (eds.) SAT 2005. LNCS, vol. 3569, pp. 173–186. Springer, Heidelberg (2005)
13. Liffiton, M.H., Sakallah, K.A.: Algorithms for computing minimal unsatisfiable subsets of constraints. Journal of Automated Reasoning 40(1), 1–33 (2008)
14. Liffiton, M.H., Sakallah, K.A.: Generalizing core-guided Max-SAT. In: Kullmann, O. (ed.) SAT 2009. LNCS, vol. 5584, pp. 481–494. Springer, Heidelberg (2009)
15. van Loon, J.: Irreducibly inconsistent systems of linear inequalities. European Journal of Operational Research 8(3), 283–288 (1981)
16. Marques-Silva, J., Planes, J.: Algorithms for maximum satisfiability using unsatisfiable cores. In: Proceedings of the Conference on Design, Automation, and Test in Europe, DATE 2008 (March 2008)
17. Reiter, R.: A theory of diagnosis from first principles. Artificial Intelligence 32(1), 57–95 (1987)

Tuning Parameters of Large Neighborhood Search for the Machine Reassignment Problem[*]

Yuri Malitsky, Deepak Mehta, Barry O'Sullivan, and Helmut Simonis

Cork Constraint Computation Centre, University College Cork, Ireland
{y.malitsky,d.mehta,b.osullivan,h.simonis}@4c.ucc.ie

Abstract. Data centers are a critical and ubiquitous resource for providing infrastructure for banking, Internet and electronic commerce. One way of managing data centers efficiently is to minimize a cost function that takes into account the load of the machines, the balance among a set of available resources of the machines, and the costs of moving processes while respecting a set of constraints. This problem is called the *machine reassignment problem*. An instance of this online problem can have several tens of thousands of processes. Therefore, the challenge is to solve a very large sized instance in a very limited time. In this paper, we describe a constraint programming-based Large Neighborhood Search (LNS) approach for solving this problem. The values of the parameters of the LNS can have a significant impact on the performance of LNS when solving an instance. We, therefore, employ the Instance Specific Algorithm Configuration (ISAC) methodology, where a clustering of the instances is maintained in an offline phase and the parameters of the LNS are automatically tuned for each cluster. When a new instance arrives, the values of the parameters of the closest cluster are used for solving the instance in the online phase. Results confirm that our CP-based LNS approach, with high quality parameter settings, finds good quality solutions for very large sized instances in very limited time. Our results also significantly outperform the hand-tuned settings of the parameters selected by a human expert which were used in the runner-up entry in the 2012 EURO/ROADEF Challenge.

1 Introduction

Data centers are a critical and ubiquitous resource for providing infrastructure for banking, Internet and electronic commerce. They use enormous amounts of electricity, and this demand is expected to increase in the future. For example, a report by the *EU Stand-by Initiative* stated that in 2007 Western European data centers consumed 56 Tera-Watt Hours (TWh) of power, which is expected to almost double to 104 TWh per year by 2020.[1] A typical optimization challenge

---

[*] This work is supported by Science Foundation Ireland Grant No. 10/IN.1/I3032 and by the European Union under FET grant ICON (project number 284715).

[1] http://re.jrc.ec.europa.eu/energyefficiency/html/standby_initiative_data_centers.htm

C. Gomes and M. Sellmann (Eds.): CPAIOR 2013, LNCS 7874, pp. 176–192, 2013.

in the domain of data centres is to consolidate machine workload to ensure that machines are well utilized so that energy costs can be reduced. In general, the management of data centers provides a rich domain for constraint programming, and combinatorial optimization [13, 16–18].

Context. Given the growing level of interest from the optimization community in data center optimization and virtualization, the 2012 ROADEF Challenge was focused on machine reassignment, a common task in virtualization and service configuration on data centers.[2] Informally, the machine reassignment problem is defined by a set of machines and a set of processes. Each machine is associated with a set of available resources, e.g. CPU, RAM etc., and each process is associated with a set of required resource values and a currently assigned machine. There are several reasons for reassigning one or more processes from their current machines to different machines. For example, if the load of the machine is high, then one might want to move some of the processes from that machine to other machines. Similarly, if the machine is about to shut down for maintenance then one might want to move all processes from the machine. Also, if there is a different location where the electricity price is cheaper then one might want to reassign some processes to the machines at that location such that the total cost of electricity consumption is reduced. In general, the task is to reassign the processes to machines while respecting a set of hard constraints in order to improve the usage of the machines, as defined by a complex cost function.

Contributions of This Paper. The machine reassignment problem is one that needs to be solved in an online manner. The challenge is to solve a very large size problem instance in a very limited time. In order to do so, we formulate the problem using Constraint Programming (CP) as described in [10], and use Large Neighborhood Search (LNS) [15] to solve it. The basic idea of CP-based LNS is to repeatedly consider a subproblem, which defines a candidate neighborhood, and re-optimize it using CP. In the machine reassignment problem context, we select a subset of processes and reassign machines to them. In this paper we describe our CP-based LNS approach in detail.

There are several parameters to choose when implementing LNS, e.g., size of the neighborhood, when to change the neighborhood size, threshold in terms of time/failures for solving a subproblem etc. The values of these parameters can have a significant impact on the efficiency of LNS. We expose the parameters of our CP-based LNS approach, and study the impact of these parameters on LNS.

It is well known that manually tuning a parameterized solver can be very tedious and often consume a significant amount of time. Moreover, manual tuning rarely achieves maximal potential in terms of performance gains. Therefore, we study the application of a Gender-based Genetic Algorithm (GGA) for configuring the parameters automatically [2]. Experimental results show that the performance of the LNS solver tuned with GGA improves significantly, as

---

[2] http://challenge.roadef.org/2012/en/index.php

compared with the manually tuned LNS solver.[3] Furthermore, it is important to note that while tuning the parameters of GGA requires significant computational resources, it is still far faster than manual tuning. Additionally, GGA is an automated process that can be run in the background, thus releasing developers to focus their efforts on developing new algorithms rather than manually experimenting with parameter settings. Finally, the initial computational expenditure is further mitigated by the fact that the machine reassignment problem will be solved repeatedly in the future, so the costs tuning are amortized over time as the system is used in practice.

In the real world setting one can anticipate that the instances of the machine reassignment problem may differ from time to time. Thus, it is possible that one setting of parameters might not result in the best possible performance of the LNS solver across all possible scenarios. We, therefore, propose a methodology whereby in the offline phase a system continuously maintains a clustering of the instances and the LNS solver is tuned for each cluster of instances. In the online phase, when a new instance arrives the values of the parameters of the closest cluster are used for solving the instance. For this we study the application of Instance-Specific Algorithm Configuration (ISAC) [9]. Experimental results confirm that this further improves the performance of the LNS solver when compared with the solver tuned for all the instances with GGA. Overall the experimental results suggest that the proposed CP-based LNS approach with the aid of learning high quality parameter settings can find a good quality solution for a very large size instance in a very limited time.

The current computer industry trend is toward creating processor chips that contain multiple computation cores. If tuning the parameters of an LNS solver manually for single-core machine is tedious, then tuning for multiple parameterizations that would work harmoniously on multi-core machine is even more complex. We present an approach that can exploit multiple cores and can provide an order-of-magnitude improvement over manually configured parameters.

The paper is organized as follows. The machine reassignment problem is briefly described in Section 2 followed by the LNS used for solving this problem in Section 3. Section 4 describes how the parameters of LNS are tuned, and Section 5 presents experimental results followed by conclusions in Section 6.

2   Machine Reassignment Problem

In this section, we briefly describe the machine reassignment problem of ROADEF-EURO Challenge 2012 in collaboration with Google.[4] Let M be the set of machines and P be the set of processes. A solution of the machine reassignment problem is an assignment of each process to a machine subject to a set of constraints. Let o_p be the original machine on which process p is currently

---

[3] The manually tuned solver was runner up in the 2012 ROADEF-EURO Challenge and the difference between the first and the second was marginal.

[4] http://challenge.roadef.org/2012/files/problem_definition_v1.eps

running. The objective is to find a solution that minimizes the cost of the re-assignment. In the following we describe the constraints, various types of costs resulting from the assignment, and the objective function.

2.1   Constraints

Capacity Constraints. The usage by a machine m of resource r, denoted by u_{mr}, is equal to the sum of the amount of resource required by processes that are assigned to machine m. The usage by a machine of a resource should not exceed the capacity of the resource.

Conflict Constraints. A service is a set of processes, and a set of services partition the set of processes. The constraint is that the processes of a service should be assigned to different machines.

Spread Constraints. A location is a set of machines, and a set of locations partition the set of machines. These constraints ensure that processes of a service should be assigned to the machines such that their corresponding locations are spread over at least a given number of locations.

Dependency Constraints. A neighborhood is a set of machines and a set of neighborhoods also partition the machines. The constraint states that if service s depends on service s', then the set of the neighborhoods of the machines assigned to the processes of service s must be a subset of the set of the neighborhoods of the machines assigned to the processes of service s'.

Transient Usage Constraints. When a process is moved from one machine to another machine, some resources, e.g., hard disk space, are required in both source and target machines. These resources are called transient resources. The transient usage of a machine m for a transient-resource r is the sum of the amount of resource required by processes whose original or current machine is m. The transient usage of a machine for a resource should not exceed its capacity.

2.2   Costs

The objective is to minimize the weighted sum of load, balance, and move costs.

Load Cost. The safety capacity limit provides a soft limit, any use above that limit incurs a cost. Let sc_{mr} be the safety capacity of machine m for resource r. The load cost for a resource r is equal to $\sum_{m \in M} \max(0, u_{mr} - sc_{mr})$.

Balance Cost. To balance the availability of resources, a balance b is defined by a triple which consists of a pair of resources r_b^1 and r_b^2, and a multiplier t_b. For a given triple b, the balance cost is $\sum_{m \in M} \max(0, t_b \cdot A(m, r_b^1) - A(m, r_b^2))$ with $A(m, r) = c_{mr} - u_{mr}$.

Move Cost. A process move cost is defined as the cost of moving a process from one machine to another machine. The service move cost is defined as the maximum number of processes moved among services.

2.3   Instances

Table 1 shows the features of the machine reassignment problem and their limits on the instances of the problem that we are interested in solving. As this is an online problem, the challenge is to solve very large sized instances in a very limited time. Although the time limit for the competition was 300 seconds, we restrict the runtime to 100 seconds as we are solving more than 1000 instances with numerous parameter settings of the LNS solver.

Table 1. Features of the problems instances

Feature	Machines	Processes	Resources	Services	Locations	Neighbourhoods	Dependencies
Limit	5000	50000	20	5000	1000	1000	5000

3   Large Neighborhood Search

We formulated the machine reassignment problem using Constraint Programming (CP), which is described in [10]. We used Large Neighborhood Search for solving the instances of the problem. In this paper we omit the details of the CP model and we focus on the details of our LNS approach for solving this problem. In particular we focus on the parameters of the LNS solver that are carefully tuned for solving the problem instances efficiently.

LNS combines the power of systematic search with the scaling of local search. The overall solution method for CP-based LNS is shown in Figure 1. We maintain a current assignment, which is initialized to the initial solution given as input. At every iteration step, we select a subset of the processes to be reassigned, and accordingly update the domains of the variables of the CP model. We solve the resulting CP problem with a threshold on the number of failures, and keep the best solution found as our new current assignment.

3.1   Selecting a Subproblem

In this section we describe how a subproblem is selected. Our observation is that selecting a set of processes for reassignment from only some machines is better

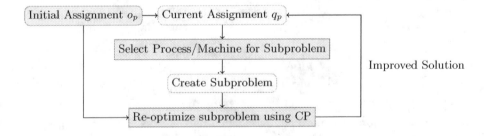

Fig. 1. Principles of the CP-based LNS approach

than selecting them from many machines. The reason is that if we select only a few processes from many machines, then we might not free enough resources from the machines for moving the processes with large resource requirements from their original machines. Therefore, our subproblem selection is a two step process. First we select a set of machines and then, from each selected machine, we select a set of processes to be reassigned.

The number of machines that are selected in a given iteration is denoted by k_m. The maximum number of processes that are selected for reassignment from each selected machine is denoted by k_p. Both k_m and k_p are non-zero positive integers. The values of k_m and k_p can change during iterations of LNS. They are decided based on the several parameters of the LNS solver.

The number of processes that can be selected from one machine is bounded by an integer parameter, which is denoted by u_p. Therefore, $k_p \leq u_p$. The total number of processes that can be selected for reassignment is bounded by an integer parameter, which is denoted by t_p. Therefore, $k_m \cdot k_p \leq t_p$.

If all the processes of a machine are selected then many of them might be reassigned to their original machines again. Therefore, we enforce that the number of processes selected from a given machine should be less than or equal to some factor of the average number of processes on a machine. More precisely, $k_p \leq r_p \cdot (|P|/|M|)$. Here r_p is a continuos parameter.

Initially k_m is set to 1. As search progresses, it is incremented when a certain number of iterations in LNS are performed consecutively without any improvement in the quality of the solution [15]. The maximum value that can be set to k_m is denoted by t_m. We re-initialize k_m to 1 when it exceeds t_m. Depending on the value of k_m the value of k_p can change.

Notice that fewer processes on a machine implies fewer combinations of the processes that can be selected from the machine for reassignment and hence fewer possible subproblems that can be created from the selected machines. Therefore, the bound on the number of consecutive non-improving iterations is obtained by multiplying the average number of processes on a machine (i.e., $|P|/|M|$) by a continuous parameter, which is denoted by r_m. The value of r_m is bounded within 0.1 and 10. Notice that u_p, t_p, r_p, t_m, and r_m are constant parameters of the LNS algorithm and different values of these parameters can have a significant impact on the efficiency of CP-based LNS approach.

3.2   Creating the Subproblem

When solving a problem using LNS, the conventional way of creating a subproblem is to reinitialize all the domains of all the variables, reassign the machines to the processes that are not chosen for reassignment, and perform constraint propagation. This way of creating a subproblem can be a bottleneck for LNS when we are interested in solving a very large sized problem in a very limited time. Furthermore, if the size of the subproblem is considerably smaller than the size of the full problem then the number of iterations that one would like to perform will increase in which case the time spent in creating the subproblems will also increase further.

For example, let us assume that the total number of processes is 50000, the number of machines is 1000, and the maximum number of processes selected for reassignment for each iteration is 100. If we want to consider each process at least once for reassignment then we need at least 500 iterations. Assuming that the time-limit is 100 seconds, an average of 0.2 seconds will be spent on each iteration. Each iteration would involve selecting 100 processes for reassignment, resetting more than 50000 variables to their original domains, freezing 49900 variables, performing constraint propagation, and finally re-optimizing the subproblem. One can envisage that in this kind of scenario the time spent on creating subproblems can be a significant part of solving the problem.

This drawback is mainly because a CP solver is typically designed for systematic search. At each node of the search tree, a CP solver uses constraints to remove inconsistent values from the domains, and it uses trailing or copying with recomputation for restoring the values. Both trailing and copying with recomputation techniques are efficient for restoring the domains when an ordering is assumed on the way decisions are undone. However, in LNS one can move from one partial assignment to another by undoing any subset of decisions in an arbitrary order. Therefore, if an existing CP solver is used for LNS then un-assigning a set of processes would result in backtracking to the node in the search tree where all the decisions made before that node are still in place, and in the worst-case this could be the root node.

We propose a novel approach for creating the subproblem. The general idea is to use constraints to determine whether a removed value can be added to the current domain when a set of assignments are undone. Instead of using trailing or copying, we maintain two sets of values for each variable: (1) the set of values that are in the current domain of the variable, and (2) the set of values that are removed from the original domain. Additionally, we maintain two sets of variables: (1) a set of variables whose domains can be replenished, and (2) and a set of variables whose domains cannot be replenished. In each iteration, the former is initialized by the variables whose assignments are undone and the latter is initialized by the variables whose assignments are frozen. A variable whose domain cannot be replenished is also called a reduced variable.

The domain of a variable is replenished by checking the consistency of each removed value with respect to the constraints that involve reduced variables. Whenever a domain is replenished the variable is tagged as a reduced variable, its neighbors that are not reduced-variables are considered for replenishing their domains, and constraint propagation is performed on the problem restricted to the set of reduced variables. A fixed point is reached when no variable is left for replenishment. This approach is called replenishing the domains via incremental recomputation. Notice that existing CP-solvers do not provide support for replenishing domains via incremental re-computation. The main advantage of this approach is that it is not dependent on the order of the assignments and therefore it can be used efficiently to create subproblems during LNS.

3.3   Reoptimizing the Subproblem

We use systematic branch and bound search with a threshold on the number of failures for solving a given subproblem. The value of the threshold is obtained by multiplying the number of processes that are selected for reassignment with the value of a continuous parameter, which is denoted by t_f. The value of t_f ranges between 0.1 and 10. At each node of the search tree constraint propagation is performed to reduce the search space. The variable ordering heuristic used for selecting a process is based on the sum of the increment in the objective cost when assigning a best machine to the process and the total weighted requirement of the process which is the sum of the weighted requirements of all resources divided by the number of machines available for the process. The value ordering heuristic for selecting a machine for a given process is based on the minimum increment in the objective cost while ties are broken randomly.

4   Tuning Parameters of LNS

While it is possible to reason about certain parameters and their effect on the overall performance individually, there are numerous possible configurations that these parameters can take. The fact that these effects might not be smoothly continuous or that there may be subtle non-linear interactions between parameters complicates the problem further. Augment this with the time incurred at trying certain parameterizations on a collection of instances, and it becomes clear why one cannot be expected to tune the parameters of the solver manually.

Table 2. Parameters of LNS for Machine Reassignment Problem

Notation	Type	Range	Description
u_p	Integer	$[1, 50]$	Upper bound on the number of processes that can be selected from one machine for reassignment
t_p	Integer	$[1, 100]$	Upper bound on the total number of processes that can be selected for reassignment
r_p	Continuous	$[0.1, 1]$	Ratio between the average number of processes on a machine
t_m	Integer	$[2, 25]$	Upper bound on the number of machines selected for subproblem selection
r_m	Continuous	$[0.1, 10]$	Ratio between the upper bound on the consecutive non-improving iterations and the average number of processes on a machine
t_f	Continuous	$[0.1, 10]$	Ratio between the threshold on the number of failures and the total number of processes selected for reassignment

In the case of our LNS solver, Table 2 lists and explains the parameters that can be controlled. Although there are only six parameters, half of them are continuous and have large domains. Therefore, it is impractical to try all possible configurations. Furthermore, the parameters are not independent of each other. To test this, we gathered a small set of 200 problem instances and evaluated 400 randomly selected parameter settings on this test set. Then, using the average performance on the data set as our evaluation metric and the parameter settings as attributes, we ran feature selection algorithms from Weka [5]. All the attributes were found as important. Adding polynomial combinations of the parameter settings, further revealed that some pairs of parameters were more important than others when predicting expected performance.

Because of the difficulty of fully extracting the interdependencies of the parameters and covering the large possible search space, a number of automated algorithm configuration and parameter tuning approaches have been proposed over the last decade. These approaches range from gradient-free numerical optimization [3], gradient-based optimization [4], iterative improvement techniques [1], and iterated local search techniques like FocusedILS [7].

One of the more successful of these approaches is the Gender-based Genetic Algorithm (GGA) [2], a highly parallelizable tool that is able to handle continuous, discrete, and categorical parameters. Being a genetic algorithm, GGA starts with a large, random, population of possible parameter configurations. This population is then randomly split into two even groups: competitive and noncompetitive. The members of the competitive set are further randomly broken up into tournaments where the parameterizations in each tournament are raced on a subset of training instances. The best performing parameter settings of each tournament get the chance to crossover with the members of the noncompetitive population and continue to subsequent generations. In the early generations, each tournament has only a small subset of the instances, but the set grows at each generation as the bad parameter settings get weeded out of consideration. In the final generation of GGA, when all training instances are used, the best parameter setting has been shown to work very well on these and similar instances.

General tuning of a solver's parameters with tools like GGA, however, ignores the common finding that there is often no single solver that performs optimally on every instance. Instead, different parameter settings tend to do well on different instances. This is the underlying reason why algorithm portfolios have been so successful in SAT [8, 20], CP [12], QBF [14], and many other domains. These portfolio algorithms try to identify the structure of an instance beforehand and predict the solver that will have the best performance on that instance.

ISAC [9] is an example of a very successful non-model based portfolio approach. Unlike similar approaches, such as Hydra [19] and ArgoSmart [11], ISAC does not use regression-based analysis. Instead, it computes a representative feature vector in order to identify clusters of similar instances. The data is therefore clustered into non-overlapping groups and a single solver is selected for each group based on some performance characteristic. Given a new instance, its features are computed and it is assigned to the nearest cluster. The instance is then evaluated with the solver assigned to that cluster.

Algorithm 1. Instance-Specific Algorithm Configuration

1: **ISAC-Learn**(A, T, F, κ) 1: **ISAC-Run**(A, x, k, P, C, d, s, t)
2: $(\bar{F}, s, t) \leftarrow$ Normalize(F) 2: $f \leftarrow$ Features(x)
3: $(k, C, S) \leftarrow$ Cluster $(T, \bar{F}, \kappa)$ 3: $\bar{f}_i \leftarrow 2(f_i/s_i) - t_i \; \forall \; i$
4: **for all** $i = 1, \ldots, k$ **do**   4: $i \leftarrow \min_i(||\bar{f} - C_i||)$
5: $\quad P_i \leftarrow GGA(A, S_i)$ 5: **return** $A(x, P_i)$
6: **return** (k, P, C, s, t)

ISAC works as follows (see Algorithm 1). In the learning phase, ISAC is provided with a parameterized solver A, a list of training instances T, their corresponding feature vectors F, and the minimum cluster size κ. First, the gathered features are normalized so that every feature ranges from $[-1, 1]$, and the scaling and translation values for each feature (s, t) is memorized. This normalization helps keep all the features at the same order of magnitude, and thereby keeps the larger values from being given more weight than the lower values. Then, the instances are clustered based on the normalized feature vectors. Clustering is advantageous for several reasons. First, training parameters on a collection of instances generally provides more robust parameters than one could obtain when tuning on individual instances. That is, tuning on a collection of instances helps prevent over-tuning and allows parameters to generalize to similar instances. Secondly, the found parameters are "pre-stabilized," meaning they are shown to work well *together*.

To avoid specifying the desired number of clusters beforehand, the g-means [6] algorithm is used. Robust parameter sets are obtained by not allowing clusters to contain fewer than a manually chosen threshold, a value which depends on the size of the data set. In our case, we restrict clusters to have at least 50 instances. Beginning with the smallest cluster, the corresponding instances are redistributed to the nearest clusters, where proximity is measured by the Euclidean distance of each instance to the cluster's center. The final result of the clustering is a number of k clusters S_i, and a list of cluster centers C_i. Then, for each cluster of instances S_i, favorable parameters P_i are computed using the instance-oblivious tuning algorithm GGA.

When running algorithm A on an input instance x, ISAC first computes the features of the input and normalize them using the previously stored scaling and translation values for each feature. Then, the instance is assigned to the nearest cluster. Finally, ISAC runs A on x using the parameters for this cluster.

In practice however, the tuning step of the ISAC methodology can be very computationally expensive, requiring on the order of a week on an 8 core machine. Fortunately it is not necessary to retune the algorithms too frequently. In many cases, even when new instances become available, the clusters are not likely to shift. We therefore propose the methodology shown in Figure 2. Here, given a set of initial instances, we perform the ISAC methodology to find a set of clusters for which the algorithm is tuned. When we observe new instances, we evaluate them according to the ISAC approach as shown in Figure 3, and

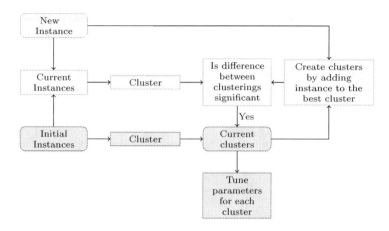

Fig. 2. Offline Phase: Tuning LNS solver

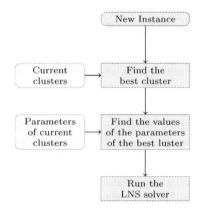

Fig. 3. Online phase: Using tuned LNS solver for solving a given instance

afterwards add the instance to the appropriate cluster. But we also try re-clustering the entire set of instances. In most cases, the two clusterings will be similar, so nothing needs to be changed. But as we gather more data, we might see that one of our clusters can be refined into two or more new clusters. When this occurs, we can then retune the LNS solver as needed.

5   Experimental Results

In this section we present results of solving the machine reassignment problem using CP-based LNS approach. We study three different ways of tuning the LNS solver. The first approach is the LNS solver, denoted by Default, which was runner up in the challenge. Here Default stands for a single set of parameters resulting from manual tuning of LNS solver on the 20 instances provided by 2012 ROADEF challenge organizers. The second approach is the LNS solver tuned using the GGA algorithm, which is denoted by GGA, and the final approach is the LNS solver tuned using ISAC, which is denoted by ISAC. The LNS solver for machine reassignment problem is implemented in C.[5]

In order to perform experiments in training and test phases we generated 1245 instances which were variations on the set B instances. Notice that the instances of set B are very large and they were used in the final phase of the ROADEF competition.[6] For each original instance we perturb the number of resources, the number of transient resources, the number of balances, weights of the load costs for resources, weights of balance costs, weights of process-move, machine-move cost and service-move costs. More precisely, for each original instance with $|R|$ number of resources, we randomly select k resources. The value of k is also chosen randomly such that $3 \leq k \leq |R|$. Out of k selected resources, a set of t resources are selected to be transient such that $0 \leq t \leq k/3$. The set of balances is also modified in a similar way. The original weight associated with each load-cost, balance-cost or any move-cost is randomly multiplied with a value selected randomly from the set $\{0.5, 1, 2\}$. Note that the uniform distribution is used to select the values randomly. All problems instances used in our experiments are available online.[7] The generated dataset was split to contain 745 training instances and 500 test instances. All the experiments were run on Linux 2.6.25 x64 on a Dual Quad Core Xeon CPU machine with overall 12 GB of RAM and processor speed of 2.66GHz.

For evaluation of a solver's performance, we used the metric utilized for the ROADEF competition:

$$Score_S(I) = 100 * (Cost(S) - Cost(B))/Cost(R).$$

Here, I is the instance, B is the best observed solution using any approach, R is the original reference solution, and S is the solution using a particular solver. The benefit of this evaluation function is that it is not influenced by some instances having a higher cost than others, and instead focuses on a normalized value that ranks all of the competing approaches. We rely on using the best observed cost because for most of the instances it is not possible to find the optimal cost. On average, the best observed cost reduces the initial cost by 65.78%, and if we use the lower-bound then it reduces the initial cost by 66.98%. This demonstrates the effectiveness of our CP-based LNS approach for finding good quality solutions in 100 seconds. The lower-bound for an instance is obtained by aggregating the resource requirements over all processes and (safety) capacities over all machines and then computing the sum of the load and balance costs.

When tuning the LNS solver using GGA/ISAC, we used the competition's evaluation metric as the optimization criterion. However, to streamline the evaluations, we approximated the best performance using the performance achieved

---

[5] http://sourceforge.net/projects/machinereassign/
[6] http://challenge.roadef.org/2012/en/index.php
[7] http://4c.ucc.ie/~ymalitsky/

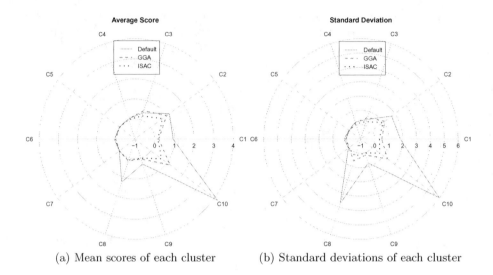

(a) Mean scores of each cluster   (b) Standard deviations of each cluster

Fig. 4. Performance of the learned parameterizations from the Default, GGA, and ISAC medodologies

by running the LNS solver with default parameters for 1 hour. While this caused some of the scores to be negative during training, this approximation still correctly differentiated the best solver while avoiding placing more weight on instances with higher costs. In order to cluster the instances for ISAC, we used the features listed in Table 1. All these features are available in the problem definition so there is no overhead in their computation. When we clustered the instances using g-means with a minimal cluster size of 50, we found 10 clusters in our training data.

The performances of the learned parameterizations from the Default, GGA and ISAC methodologies is compared in Figure 4. Specifically, we plot the average performance of each method on the instances in each of the 10 discovered clusters (Figure 4(a)). What we observe is that even though the default parameters perform very close to as well as they can for clusters 4, 5, 6 and 7, for clusters 2, 8 and 10 the performance is very poor. Tuning the solver using GGA can improve the average performance dramatically. Furthermore, we see that if we focus on each cluster separately, we can further improve performance, highlighting that different parameters should be employed for different types of instances. Interestingly, we observe that ISAC also dramatically improves on the standard deviation of the scores (Figure 4(b)), suggesting that the tuned solvers are consistently better than the default and GGA tuned parameters.

Multi-Core Results. The current trend is to create computers with an ever increasing number of cores. It is unusual to find single core machines still in use, with 2, 4 or even 8 cores becoming commonplace. It is for this reason that we also experimented scenarios where more than one core is available for running

Table 3. Mean scores of the test data using Default, GGA, and ISAC approaches evaluated for 1, 2 and 4 cores. The standard deviations are presented in parentheses.

Approach	Number of Cores		
	1	2	4
Default	0.931 (2.759)	0.843 (2.596)	0.784 (2.541)
GGA	0.357 (0.808)	0.283 (0.663)	0.214 (0.529)
ISAC	**0.259 (0.623)**	**0.151 (0.363)**	**0.095 (0.237)**

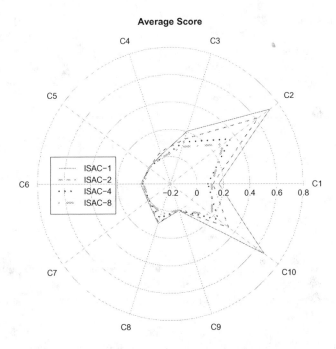

Fig. 5. Mean scores of each cluster ISAC approach for 1,2,4 and 8 cores

the experiments. For Default and GGA, we ran the same parameters multiple times using different seeds, taking the best performance of 1, 2 and 4 trials. For ISAC however, we used the training data to pick which 1, 2, or 4 parameter settings associated with which clusters should be used for running the LNS solver in parallel. ISAC had the opportunity to choose from the 10 parameterizations found for each cluster plus the parameters of Default and GGA tuned solvers.

Table 3 shows that ISAC always dominates. While adding more cores is not particularly helpful for Default and GGA, ISAC can dramatically benefit from the additional computing power. And as can be seen from the reduction of the standard deviation, the ISAC tuned solvers are consistently better. Running t-tests on all the results, the benefits of GGA over default are always statistically significant (below 0.0001), as are the gains of ISAC over GGA. The detailed mean scores per cluster for various numbers of cores for ISAC are presented in Figure 5.

Table 4. Average score on set B instances using Default, GGA, and ISAC trained parameterizations for 1, 2 and 4 cores. The standard deviations are in parentheses.

Approach	Number of Cores		
	1	2	4
Default	0.171 (0.268)	0.159 (0.261)	0.119 (0.211)
GGA	0.296 (0.416)	0.288 (0.417)	0.202 (0.285)
ISAC	**0.137 (0.224)**	**0.109 (0.184)**	**0.065 (0.120)**

Table 4 present results for 10 instances of set B which were used in the final phase of the competition. As the Default LNS is manually trained for set B instances, it is not surprising to see that the average score for set B is signficantly less than that obtained for 500 instances of test data. This demonstrates that the Default parameters have been over-tuned and because of that the performance of Default is poor on the test data. On the other hand the average score of GGA for test instances and for set B instances are very close. This demonstrates that the parameters of GGA are more stabilized, and therefore overall they work well for both test instances and set B instances. Table 3 shows that ISAC always dominates for set B instances. This confirms that for different types of instances different values of LNS parameters can help in solving problems more efficiently.

6   Conclusions

We have presented an effective constraint programming based Large Neighborhood Search (LNS) approach for the machine reassignment problem. Results show that our approach is scalable, and is suited for solving very large instances, and has good anytime behavior which is important when solutions must be reported subject to a time limit.

We have shown that by exposing parameters we are able to create an easily configurable solver. The benefits of such a development strategy are made evident through the use of automatic algorithm configuration. We show that an automated approach is able to set the parameters that out-perform a human expert. We further show that not all machine reassignment instances are the same, and that by employing the Instance-Specific Algorithm Configuration methodology we are able to improve the performance of our proposed approach. The tuning step is an initial cost that is quickly mitigated by the repeated usage of the learned parameters over a prolonged period of time.

Finally, we show that by taking advantage of the increasing number of cores available on machines, we can provide an order-of-magnitude improvement over using manually configured parameters.

References

1. Adenso-Diaz, B., Laguna, M.: Fine-tuning of algorithms using fractional experimental designs and local search. Oper. Res. 54(1), 99–114 (2006)
2. Ansótegui, C., Sellmann, M., Tierney, K.: A Gender-Based Genetic Algorithm for the Automatic Configuration of Algorithms. In: Gent, I.P. (ed.) CP 2009. LNCS, vol. 5732, pp. 142–157. Springer, Heidelberg (2009)
3. Audet, C., Orban, D.: Finding optimal algorithmic parameters using derivative-free optimization. SIAM J. on Optimization 17(3), 642–664 (2006)
4. Coy, S.P., Golden, B.L., Runger, G.C., Wasil, E.A.: Using experimental design to find effective parameter settings for heuristics. Journal of Heuristics 7, 77–97 (2001)
5. Hall, M., Frank, E., Holmes, G., Pfahringer, B., Reutemann, P., Witten, I.H.: The weka data mining software: an update. SIGKDD Explor. Newsl. 11(1), 10–18 (2009)
6. Hamerly, G., Elkan, C.: Learning the k in k-means. In: Neural Information Processing Systems, p. 2003. MIT Press (2003)
7. Hoos, H.H.: Autonomous Search (2012)
8. Kadioglu, S., Malitsky, Y., Sabharwal, A., Samulowitz, H., Sellmann, M.: Algorithm selection and scheduling. In: Lee, J. (ed.) CP 2011. LNCS, vol. 6876, pp. 454–469. Springer, Heidelberg (2011)
9. Kadioglu, S., Malitsky, Y., Sellmann, M., Tierney, K.: Isac - instance-specific algorithm configuration. In: Coelho, H., Studer, R., Wooldridge, M. (eds.) ECAI. Frontiers in Artificial Intelligence and Applications, vol. 215, pp. 751–756. IOS Press (2010)
10. Mehta, D., O'Sullivan, B., Simonis, H.: Comparing solution methods for the machine reassignment problem. In: Milano, M. (ed.) CP 2012. LNCS, vol. 7514, pp. 782–797. Springer, Heidelberg (2012)
11. Nikolić, M., Marić, F., Janičić, P.: Instance-based selection of policies for SAT solvers. In: Kullmann, O. (ed.) SAT 2009. LNCS, vol. 5584, pp. 326–340. Springer, Heidelberg (2009)
12. O'Mahony, E., Hebrard, E., Holland, A., Nugent, C., O'Sullivan, B.: Using case-based reasoning in an algorithm portfolio for constraint solving. In: Proceedings of Artificial Intelligence and Cognitive Science, AICS 2008 (2008)
13. Petrucci, V., Loques, O., Mosse, D.: A dynamic configuration model for power-efficient virtualized server clusters. In: Proceedings of the 11th Brazilian Workshop on Real-Time and Embedded Systems (2009)
14. Pulina, L., Tacchella, A.: A multi-engine solver for quantified boolean formulas. In: Bessière, C. (ed.) CP 2007. LNCS, vol. 4741, pp. 574–589. Springer, Heidelberg (2007)
15. Shaw, P.: Using constraint programming and local search methods to solve vehicle routing problems. In: Maher, M., Puget, J.-F. (eds.) CP 1998. LNCS, vol. 1520, pp. 417–431. Springer, Heidelberg (1998)
16. Srikantaiah, S., Kansal, A., Zhao, F.: Energy aware consolidation for cloud computing. In: Proceedings of HotPower (2008)
17. Steinder, M., Whalley, I., Hanson, J.E., Kephart, J.O.: Coordinated management of power usage and runtime performance. In: NOMS, pp. 387–394. IEEE (2008)

18. Verma, A., Ahuja, P., Neogi, A.: pMapper: Power and migration cost aware application placement in virtualized systems. In: Issarny, V., Schantz, R. (eds.) Middleware 2008. LNCS, vol. 5346, pp. 243–264. Springer, Heidelberg (2008)
19. Xu, L., Hoos, H.H., Leyton-Brown, K.: Hydra: Automatically configuring algorithms for portfolio-based selection. In: Fox, M., Poole, D. (eds.) AAAI. AAAI Press (2010)
20. Xu, L., Hutter, F., Hoos, H.H., Leyton-Brown, K.: Satzilla: Portfolio-based algorithm selection for SAT. J. Artif. Int. Res. 32(1), 565–606 (2008)

Improved Discrete Reformulations
for the Quadratic Assignment Problem

Axel Nyberg, Tapio Westerlund, and Andreas Lundell

Process Design and Systems Engineering
Åbo Akademi University
Biskopsgatan 8
FIN-20500 Åbo, Finland

Abstract. This paper presents an improved as well as a completely new version of a mixed integer linear programming (MILP) formulation for solving the quadratic assignment problem (QAP) to global optimum. Both formulations work especially well on instances where at least one of the matrices is sparse. Modification schemes, to decrease the number of unique elements per row in symmetric instances, are presented as well. The modifications will tighten the presented formulations and considerably shorten the computational times. We solved, for the first time ever to proven optimality, the instance esc32b from the quadratic assignment problem library, QAPLIB.

Keywords: Combinatorial optimization, Quadratic assignment problem, Mixed integer programming, Global optimization.

1 Introduction

The QAP is a well known NP-hard problem first presented in 1957 [1] as a facility location problem arising in economics. Today QAPs can be found in a vast number of different fields with applications in DNA microarray design [2], coding of signals [3] and image processing [4] to name a few. Various methods and formulations [5] have been suggested by researchers around the world, but still even relatively small problems from the quadratic assignment problem library, QAPLIB [6], remain unsolved. Recently, in [7], Nyberg and Westerlund solved four of the previously unsolved esc instances [8], to optimality. Other interesting studies have been presented recently by Fischetti et al. [9], who solved two previously unsolved instances, in addition to three of the four solved earlier, in short time exploiting the symmetries in these instances. Even the esc128 instance was solved in a few seconds which is the largest QAP from the QAPLIB solved to date. Neither of the methods, however, managed to solve the instance esc32b which seems to be surprisingly difficult compared to the rest of the esc32 instances. With the methods presented in this paper, all the instances solved in [7] were solved again in a much shorter time. For the first time the only remaining unsolved esc instance, esc32b, was solved as well.

C. Gomes and M. Sellmann (Eds.): CPAIOR 2013, LNCS 7874, pp. 193–203, 2013.
© Springer-Verlag Berlin Heidelberg 2013

In [7], we showed that when considering two permutation vectors, $\mathbf{p}$ and $\tilde{\mathbf{p}}$ where $p_i = k$ if facility i is at location k and $\tilde{p}_i = k$ if facility k is at location i, the quadratic assignment problem, $\min \sum_{i=1}^{n} \sum_{j=1}^{n} a_{p_i j} b_{i \tilde{p}_j}$, can be written in the following useful non-linear form:

$$\min \sum_{i=1}^{n} \sum_{j=1}^{n} a'_{ij} b'_{ij} \tag{1}$$

subject to

$$a'_{ij} = \sum_{k=1}^{n} a_{kj} x_{ik} \qquad \forall i, j, \tag{2}$$

$$b'_{ij} = \sum_{k=1}^{n} b_{ik} x_{kj} \qquad \forall i, j. \tag{3}$$

Here, $x_{ij} \in \{0, 1\}$ are the elements of a permutation matrix $\mathbf{X}$ where $\sum_{i=1}^{n} x_{ij} = 1 \forall j$ and $\sum_{j=1}^{n} x_{ij} = 1 \forall i$. The constants, a_{ij} and b_{ij}, are the given elements in the flow and distance matrices, $\mathbf{A}$ and $\mathbf{B}$, respectively. This formulation turned out to be very useful when creating the discrete linear reformulation (DLR), presented in the paper [7]. In the current paper we will present some improvements to the DLR as well as two alternative MILP formulations. All the presented formulations are suited for instances with a few unique elements per row in one of the matrices. Therefore, a modification scheme to decrease the number of unique elements, for instances with symmetric matrices, is also presented in Section 4.

2   Improvements to the Discrete Linear Reformulation

We found that the DLR method presented in [7] can be slightly improved. The complete original linear reformulation of Eqs. (1) to (3) was given by:

$$\min \sum_{i=1}^{n} \sum_{j=1}^{n} \sum_{m=1}^{M_i} B_i^m z_{ij}^m \tag{4}$$

subject to

$$\sum_{i=1}^{n} x_{ij} = 1 \qquad \forall j, \tag{5}$$

$$\sum_{j=1}^{n} x_{ij} = 1 \qquad \forall i, \tag{6}$$

$$z_{ij}^m \leq \overline{A}_j \sum_{k \in \mathbf{K}_i^m} x_{kj} \qquad m = 1, ..., M_i \quad \forall i, j, \tag{7}$$

$$\sum_{m=1}^{M_i} z_{ij}^m = \sum_{k=1}^{n} a_{kj} x_{ik} \qquad \forall i, j, \tag{8}$$

$$x_{ij} \in \{0,1\} \quad z_{ij}^m \in [0, \overline{A_j}] \quad \forall i, j \wedge m = 1, ..., M_i, \tag{9}$$

where

$$\overline{A_j} = \max_i a_{ij} \quad \forall j, \tag{10}$$

$$\mathbf{K}_i^m = \{j | b_{ij} = B_i^m\} \quad \forall i, j \wedge m = 1, ..., M_i, \tag{11}$$

$$b_{ij} \in \left\{ B_i^1, B_i^2, ..., B_i^{M_i} \right\} \forall i, j, \tag{12}$$

where M_i is the number of unique elements in row i, B_i^m the value of these elements and K_i^m the index sets corresponding to the elements. It should, however, be observed that every bilinear term $a'_{ij} b'_{ij}$ in Eq. (1), indirectly includes the variable x_{ij} in both the a'_{ij} (when $k = j$ in Eq. (2)) and the b'_{ij} (when $k = i$ in Eq. (3)) variables. Since $x_{ij}^2 = x_{ij}$ for binary variables, the term $a_{jj} b_{ii} x_{ij}^2$ can be replaced with $a_{jj} b_{ii} x_{ij}$. This is a well-known observation and can be handled such that the diagonal in both matrices are set to zero and a linear cost matrix is added instead [10]. Using matrix notation, the objective function Eq. (1), when including Eqs. (2) and (3), can be written as $\mathbf{XA} \bullet \mathbf{BX}$, where $\bullet$ is the scalar product of the matrices. If the diagonal elements of $\mathbf{A}$ and $\mathbf{B}$ are considered separately and the off-diagonal elements are considered in the matrices $\mathbf{A}^0$ and $\mathbf{B}^0$ (with zero diagonal elements), then the objective function can be written as

$$\mathbf{XA}^0 \bullet \mathbf{B}^0 \mathbf{X} + \sum_{i=1}^n \sum_{j=1}^n a_{jj} b_{ii} x_{ij}. \tag{13}$$

Then Eq. (4) in the DLR will take the form

$$\min \sum_{i=1}^n \sum_{j=1}^n \sum_{m=1}^{M_i} B_i^m z_{ij}^m + \sum_{i=1}^n \sum_{j=1}^n a_{jj} b_{ii} x_{ij}, \tag{14}$$

and the variables corresponding to the diagonal elements are now excluded from Eq. (11),

$$\mathbf{K}_i^m = \{j | b_{ij} = B_i^m\} \quad \forall i, j : i \neq j; m = 1, ..., M_i. \tag{15}$$

The same variables are therefore also left out from the first term $(m = 1)$ in Eq. (7), which will lower the upper bound for the variable z_{ij}^1. The effect of this change can be seen on instances having multiple elements equal to zero per row.

The esc instances, which have many zeros in both matrices, benefit from this change greatly. The root node relaxation values for the esc instances are shown in Table 1. Also, we found that Eq. (7) can be left out for the last $(z_{ij}^{M_i})$ variables because of linear dependency. This will not affect the solution times much but the final model will have n^2 fewer constraints.

Another improvement for this modified DLR formulation is to bound the z_{ij}^m variables, from below, as follows

$$z_{ij}^m \geq \underline{A_j} \sum_{k \in \mathbf{K}_i^m} x_{kj} \quad \forall i, j, m, \tag{16}$$

Table 1. Root node relaxation values for the esc instances with the improved and normal DLR formulation

Instance	DLR	Improved DLR
esc32a	3	11
esc32b	0	48
esc32c	254	309
esc32d	44	70
esc32h	119	156
esc64a	29	38
esc128	1	2

where $\underline{A}_j$ is the lowest value in column j of the matrix $\mathbf{A}$ with the diagonal element excluded. The idea with the lower bound in Eq. (16) is to tighten the relaxation in the branch and bound tree, where the binary variables take fractional values.

3   New Discrete Formulations

By using the formulation in Eq. (1), we propose two alternative formulations to the DLR method in Section 2 by reformulating the bilinear terms in different manners. The objective function in Eq. (17), as well as Eqs. (18) and (19), are the same for all these formulations. The w_{ij} variables correspond to each reformulated bilinear term $a'_{ij}b'_{ij}$ in Eq. (1), and therefore we get the following common form:

$$\min \sum_{i=1}^{n} \sum_{j=1}^{n} w_{ij} \tag{17}$$

subject to

$$\sum_{i=1}^{n} x_{ij} = 1, \qquad j = 1, ..., n, \tag{18}$$

$$\sum_{j=1}^{n} x_{ij} = 1, \qquad i = 1, ..., n. \tag{19}$$

In the formulations below, the variables a'_{ij} and b'_{ij} are used for clarity but can also be implicitly included in the model by replacing them with their corresponding expressions in Eqs. (2) and (3) respectively. Since Eq. (17) is the objective function to be minimized, the variables w_{ij} can also be replaced with the right hand side of their relaxed expressions in Eqs. (20) and (22).

In a branch and bound method the integer variables are relaxed and therefore take continuous values throughout the search tree. The tightness of the different formulations is dependent on the values of the elements in the matrices. That said, all of the presented formulations are exact in the discrete values of the variables and can therefore be used to solve QAPs to global optimum but with completely different search trees and time requirements.

In the DLR in [7] b'_{ij} was discretized so that each value corresponded to a unique element in that row. In the below two formulations the discretization is made on the deviation between the values of the elements. In these formulations, which are similar to the nf7r formulation for continuous bilinear terms in pooling problems presented in [11], the z^m_{ij} variables are dependent on every x_{kj} variable that makes $b'_{ij} \geq B^m_i$. In contrast to the DLR, multiple z^m_{ij} variables from the same bilinear term can be active and in fact, for each term $a'_{ij}b'_{ij}$ where $b'_{ij} = B_i^{M_i}$, all z^m_{ij} variables will be active.

Formulation ver2

$$w_{ij} \geq B_i^1 a'_{ij} + \sum_{m=2}^{M_i}(B_i^m - B_i^{(m-1)})z^m_{ij}, \quad \forall i,j, \qquad (20)$$

$$z^m_{ij} \geq a'_{ij} - \overline{A}_j + \overline{A}_j \sum_{\substack{k \in \mathbf{K}_i^{m'} \\ m' \geq m}} x_{kj} \quad \forall i,j \wedge m = 1, ..., M_i. \qquad (21)$$

Formulation ver2b. Here we take advantage of $\underline{A}_j$, which is the lower bound for a'_{ij}. For a shorter expression, we define $B_i^0 = 0$.

$$w_{ij} \geq \overline{A}_j b'_{ij} + \sum_{m=1}^{M_i}((B_i^m - B_i^{(m-1)})z^m_{ij} - (\overline{A}_j - \underline{A}_j) \sum_{\substack{k \in \mathbf{K}_i^{m'} \\ m' \geq m}} x_{kj}), \quad \forall i,j, \qquad (22)$$

$$z^m_{ij} \geq a'_{ij} - \overline{A}_j + (\overline{A}_j - \underline{A}_j) \sum_{\substack{k \in \mathbf{K}_i^{m'} \\ m' \geq m}} x_{kj} \quad \forall i,j \wedge m = 1, ..., M_i. \qquad (23)$$

The first linear parts of Eqs. (20) and (22) can be included in the sums following them instead. However, leaving them the way they are, the formulations are at least as tight, but with fewer variables and constraints. This is particularly useful on instances with only a few different values in one of the matrices. For example the border length minimization problems [2] contain only ones and zeros in the flow matrices, and therefore no z^m_{ij} variables will be needed when applying this formulation on them.

We can further take advantage of the variable x_{ij}, corresponding to the diagonal element, by adding it to every sum of variables in Eqs. (21) and (23).

3.1   Numerical Examples

We illustrate each formulation by linearizing one bilinear term, $a'_{23}b'_{23}$ where $i = 2$ and $j = 3$, for the QAP shown in Table 2. From Eqs. (2) and (3) we get the following:

Table 2. Matrix **A** to the left and **B** to the right for a small example

$$
\begin{array}{ccccc}
0 & 4 & 2 & 2 & 3 \\
4 & 0 & 5 & 1 & 2 \\
2 & 5 & 0 & 5 & 8 \\
2 & 1 & 5 & 0 & 1 \\
3 & 2 & 8 & 1 & 0
\end{array}
\qquad
\begin{array}{ccccc}
0 & 0 & 5 & 2 & 6 \\
0 & 0 & 3 & 3 & 7 \\
5 & 3 & 0 & 2 & 2 \\
2 & 3 & 2 & 0 & 4 \\
6 & 7 & 2 & 4 & 0
\end{array}
$$

$$a'_{23} = 2x_{21} + 5x_{22} + (0x_{23}) + 5x_{24} + 8x_{25},$$

$$b'_{23} = 0x_{13} + (0x_{23}) + 3x_{33} + 3x_{43} + 7x_{53}.$$

Example 1. The improved DLR for the above terms will now take the following form:

$$w_{23} \geq 3z_{23}^2 + 7z_{23}^3,$$

$$z_{23}^1 \leq 8x_{13},$$

$$z_{23}^2 \leq 8(x_{33} + x_{43}),$$

$$z_{23}^1 + z_{23}^2 + z_{23}^3 = a'_{23}.$$

Since we are minimizing over w_{23}, the z_{23}^3 variable is defined in the last constraint. The constant terms in the expression for w_{23} correspond directly to the terms in b_{23}. From these examples it can easily be seen that the fewer unique values per row in **B**, the fewer z_{ij}^m variables will be added in the complete formulation.

Example 2. Ver2

$$w_{23} \geq (0a'_{23}) + 3z_{23}^1 + 4z_{23}^2,$$

$$z_{23}^1 \geq a'_{23} - 8 + 8(x_{33} + x_{43} + x_{53}),$$

$$z_{23}^2 \geq a'_{23} - 8 + 8x_{53}.$$

Example 3. Ver2b

$$w_{23} \geq 8b'_{23} + 3(z_{23}^1 - 6(x_{33} + x_{43} + x_{53})) + 4(z_{23}^2 - 6x_{53}),$$

$$z_{23}^1 \geq a'_{23} - 8 + 6(x_{33} + x_{43} + x_{53}),$$

$$z_{23}^2 \geq a'_{23} - 8 + 6x_{53}.$$

The constant 6 results from subtracting the lower bound from the upper bound of a'_{23}. If there is another constant equal to zero than the diagonal in a'_{23} the two latter constraints in both `ver2` and `ver2b` will be equal. If on the other hand, the lower and upper bound of a'_{23} are the same, it is easy to see that all z_{23}^m variables in `ver2b` will be equal to zero and the only remaining constraint in the objective function is the value of a'_{23} times b'_{23}. As can be seen from all the examples above, the number of unique values as well as the difference of the largest and smallest value per row or column have a direct impact on the tightness and compactness of these formulations. Thus, in the following section we will show how to take advantage of these properties by minimizing the number of unique elements in the matrices.

4   Modification of the Matrices

In this section we propose a modification for QAPs where at least one of the matrices is symmetric. As has been shown, e.g. in [7], instances with symmetric matrices can be rewritten in such a way that one of the matrices is an upper or lower triangular matrix. This can be done on one of the matrices as long as the other matrix is symmetric: When $\mathbf{B}$ is symmetric (i.e. $\mathbf{B} = \mathbf{B}^T$)

$$\mathbf{XA} \bullet \mathbf{BX} = \mathbf{A} \bullet \mathbf{X}^T \mathbf{BX} = \mathbf{A}^T \bullet \mathbf{X}^T \mathbf{B}^T \mathbf{X} = \mathbf{A}^T \bullet \mathbf{X}^T \mathbf{BX}. \qquad (24)$$

We now express $\mathbf{A} = \mathbf{A}_1 + \mathbf{A}_2$, apply Eq. (24) to $\mathbf{A}_2$ and obtain:

$$(\mathbf{A}_1 + \mathbf{A}_2) \bullet \mathbf{X}^T \mathbf{BX} = (\mathbf{A}_1 + \mathbf{A}_2^T) \bullet \mathbf{X}^T \mathbf{BX} = \tilde{\mathbf{A}} \bullet \mathbf{X}^T \mathbf{BX}. \qquad (25)$$

From Eq. (25) we observe that each element in matrix $\mathbf{A}$ can be changed to any value, without changing the objective value, as long as $\tilde{a}_{ij} + \tilde{a}_{ji} = a_{ij} + a_{ji}$. This is useful for the MILP formulations presented in this paper since each term z_{ij}^m in the formulations is dependent on the number of unique values in the discretized matrix. Therefore, minimizing the unique elements per row in one of the matrices minimizes the number of variables and constraints as well. The tightness of the formulations presented in this paper are dependent on the difference between the biggest and smallest elements per row. One might argue that rewriting a symmetric QAP as an asymmetric one is not very beneficial. However, breaking the symmetric structure of one of the matrices does not break the symmetries in the QAP itself. Therefore, for example orbital branching [12], which is implemented in the commercial solver Gurobi, is still applicable to these problems.

To minimize the difference between the largest and smallest element in a row we solve the following LP problem and change the matrices accordingly before applying any of the formulations to the QAP:

$$\min \sum_i (\overline{y}_i - \underline{y}_i)$$

subject to

$$\begin{aligned}
\overline{y}_i &\geq y_{ij} \quad \forall i, j, \\
\underline{y}_i &\leq y_{ij} \quad \forall i, j, \\
y_{ij} + y_{ji} &= a_{ij} + a_{ji} \quad \forall i, j,
\end{aligned} \qquad (26)$$

where y_{ij} are variables that will correspond to the new elements. The variables $\overline{y}_i$ and $\underline{y}_i$ represent the new largest and smallest values in row i. Using Eq. (26) we minimize the difference between the largest and smallest element per row (with the diagonal elements excluded). This formulation does not minimize the number of unique elements in the model. In order to get fewer elements and therefore also fewer variables and constraints in the final DLR model, we use a different LP formulation.

Table 3. The distance matrix of test instance had8 unchanged to the left and after modification with less unique values per row to the right

```
0 1 2 2 3 4 4 5        0 2 2 2 6 6 6 6
1 0 1 1 2 3 3 4        0 0 0 0 4 4 4 4
2 1 0 2 1 2 2 3        2 2 0 2 2 2 2 2
2 1 2 0 1 2 2 3        2 2 2 0 2 2 2 2
3 2 1 1 0 1 1 2        0 0 0 0 0 0 0 0
4 3 2 2 1 0 2 3        2 2 2 2 2 0 2 2
4 3 2 2 1 2 0 1        2 2 2 2 2 2 0 2
5 4 3 3 2 3 1 0        4 4 4 4 4 4 0 0
```

$$\min \sum_{i,j,k} \Delta_{ijk}$$

subject to

$$\Delta_{ijk} \geq y_{ij} - y_{ik} \quad \forall i, j, k,$$
$$\Delta_{ijk} \geq y_{ik} - y_{ij} \quad \forall i, j, k, \qquad (27)$$
$$y_{ij} + y_{ji} = a_{ij} + a_{ji} \quad \forall i, j,$$
$$i, j, k = 1, \ldots, n.$$

Here the Δ_{ijk} variables represent the difference between two elements in row i. The formulation in Eq. (27) has more variables and takes slightly longer to solve than Eq. (26). However, solving this problem for the esc32 instances, for example, still take less than half a second. One of the matrices in esc32b contains only the elements $0, 1, 2$ and this matrix can therefore be written as a matrix with every element equal to either 0 or 2. In Table 3 the modification of an instance had8[1] is shown. The other matrix remains unchanged and is not shown here. In this particular example the resulting matrix is the same using both formulations. Since the diagonal can be disregarded it can be seen that the row with the most unique elements has only two different values, while five of the rows now have the same constant value for each element. By modifying the matrices in this manner one can alter the lower bound for the problem to be as high as possible for a certain method. In Table 4 the lower bound values in the root nodes for some instances are shown. Considering that solving these modification LP problems take under a second, the improvement in the lower bounds of the root nodes are huge. Since these modifications do not have to be optimal, an heuristic method could be used as well which might give even faster solution times. This would probably be useful if these modifications are done more times than once per problem.

Remark. While experimenting with the modifications we noticed that the method can also be used to tighten the well known Gilmore-Lawler Bound [13]. Table 5 shows improvement in these bounds on some instances in the QAPLIB when Eq. (27) is applied. In this case we minimize the elements in each column instead.

---

[1] The example instance had8 is constructed from the instance had12 by taking the first 8 rows and columns

Table 4. Lower bounds in the root node before and after the modification on some instances from the QAPLIB

Instance	Opt	impDLR	Eq. (26)	Eq. (27)
nug12	578	409.3	477.3	456.5
esc16a	68	26	34	36
esc16b	292	196	220	230
esc32a	130	11	18	25
esc32b	168	48	60	48
esc32c	642	309	314	328
esc32d	200	70	70	96
esc32h	438	156	212	222
esc64a	116	38	39	39

Table 5. Gilmore-Lawler bound for some instances with and without matrix modification

Instance	Opt	GLB	Eq. (27)	Instance	Opt	GLB	Eq. (27)
nug12	578	493	492	esc32c	642	350	381
esc16a	68	38	42	esc32d	200	106	112
esc16b	292	220	240	esc32h	438	257	264
esc16c	160	83	84	ste36a	9526	7124	7586
esc32a	130	35	40	ste36b	15852	8653	9638
esc32b	168	96	96	ste36c	8239110	6393629	6691054

5 Results

All numerical results and solution times reported in this paper were obtained using a single 2.8 GigaHertz 4-core Intel i7 processor, running Gurobi 4.5.1 with default parameters. Table 6 shows solution times for the different formulations on some of the easier small problems from the QAPLIB. The chr, scr and esc instances, are all sparse while both the nug and had instances are dense. It can clearly be seen from the solution times in Table 6 that these formulations are suitable for sparse instances. We tested the new formulations in this paper on the instance esc32b which was the most difficult instance for both the old DLR method as well as for the methods in [9]. Even after weeks of computing the esc32b seemed unsolvable with the old DLR method. When solving esc32b with the improved DLR the solution process was aborted after five weeks with the new best lower bound 148. The optimal value 168 was finally proven in 58 hours using the formulation `ver2` as well as the partitioning scheme presented in Section 4. The best known value reported in the QAPLIB, once again, turned out to be optimal. In Table 7 the solution times for some esc instances are shown. As can be noted, compared to the formulation presented in [7], the new formulations are much more effective. The instance esc32d was solved in only 10 minutes, approximately 60 times faster than the time reported in [7] for the original DLR model. On the instance esc32a the speed up factor was about 13.

Table 6. Solution times for the different formulations on some small easy instances from the QAPLIB

Instance	impDLR (s)	ver2 (s)	ver2b (s)
nug12	4.0	2.5	**1.3**
esc16a	0.9	**0.6**	1.0
had12	7.2	4.4	**3.2**
chr12a	**0.3**	**0.3**	**0.3**
scr12	**0.4**	**0.4**	0.5

Table 7. Solution times for some of the esc32 instances from the QAPLIB

Instance	Old DLR (s)	New best (s)	Model	Symmetry
esc32a	1 618 580	117 850	ver2	Eq. (27)
esc32b	***	210 045	ver2	Eq. (26)
esc32c	24 365	7 801	ver2	Eq. (27)
esc32d	36 256	610	impDLR	Eq. (27)
esc64a	16 370	2 899	impDLR	Eq. (27)

6   Conclusions

The formulations presented in this paper have shown to be effective on some large scale, sparse QAPs. Future work will address the lower bounding of the z_{ij}^m variables in Eq. (22) in a manner that will tighten the relaxation. Also, the modifications for the matrices presented in Section 4 could be very useful for other methods for solving QAPs as well.

Acknowledgments. Financial support from the Academy of Finland project 127992 as well as from the Foundation of Åbo Akademi University, as part of the grant for the Center of Excellence in Optimization and Systems Engineering, are gratefully acknowledged.

References

1. Koopmans, T., Beckmann, M.: Assignment problems and location of economic activities. Econometrica 25, 53–76 (1957)
2. de Carvalho Jr., S., Rahmann, S.: Microarray layout as a quadratic assignment problem. In: Proceedings of the German Conference on Bioinformatics (GCB). Lecture Notes in Informatic, vol. P-83, pp. 11–20 (2006)
3. Ben-David, G., Malah, D.: Bounds on the performance of vector-quantizers under channel errors. IEEE Transactions on Information Theory 51(6), 2227–2235 (2005)
4. Taillard, E.: Comparison of iterative searches for the quadratic assignment problem. Location Science 3(2), 87–105 (1995)
5. Loiola, E., Abreu, N., Boaventura-Netto, P., Hahn, P., Querido, T.: A survey for the quadratic assignment problem. European Journal of Operational Research 176(2), 657–690 (2007)

6. Burkard, R., Çela, D.E., Karisch, S., Rendl, F.: QAPLIB - a quadratic assignment problem library. Journal of Global Optimization 10, 391–403 (1997)
7. Nyberg, A., Westerlund, T.: A new exact discrete linear reformulation of the quadratic assignment problem. European Journal of Operational Research 220(2), 314–319 (2012)
8. Eschermann, B., Wunderlich, H.J.: Optimized synthesis of self-testable finite state machines. In: 20th International Symposium on Fault-Tolerant Computing, FTCS-20. Digest of Papers, pp. 390–397 (June 1990)
9. Fischetti, M., Monaci, M., Salvagnin, D.: Three ideas for the quadratic assignment problem. Operations Research 60, 954–964 (2012)
10. Çela, E.: The Quadratic Assignment Problem Theory and Algorithms. Kluwer Academic Publishers (1998)
11. Gounaris, C.E., Misener, R., Floudas, C.A.: Computational comparison of piecewise linear relaxations for pooling problems. Industrial & Engineering Chemistry Research 48(12), 5742–5766 (2009)
12. Ostrowski, J., Linderoth, J., Rossi, F., Smriglio, S.: Orbital branching. Mathematical Programming 126, 147–178 (2011), doi:10.1007/s10107-009-0273-x
13. Gilmore, P.C.: Optimal and suboptimal algorithms for the quadratic assignment problem. Journal of the Society for Industrial and Applied Mathematics 10, 305–313 (1962)

Orbital Shrinking: A New Tool for Hybrid MIP/CP Methods

Domenico Salvagnin

DEI, University of Padova
salvagni@dei.unipd.it

Abstract. Orbital shrinking is a newly developed technique in the MIP community to deal with symmetry issues, which is based on aggregation rather than on symmetry breaking. In a recent work, a hybrid MIP/CP scheme based on orbital shrinking was developed for the multi-activity shift scheduling problem, showing significant improvements over previous pure MIP approaches. In the present paper we show that the scheme above can be extended to a general framework for solving arbitrary symmetric MIP instances. This framework naturally provides a new way for devising hybrid MIP/CP decompositions. Finally, we specialize the above framework to the multiple knapsack problem. Computational results show that the resulting method can be orders of magnitude faster than pure MIP approaches on hard symmetric instances.

1 Introduction

We consider a integer linear optimization problem P of the form

$$\min cx \tag{1}$$
$$Ax \geq b \tag{2}$$
$$x \in \mathbb{Z}_+^n \tag{3}$$

For ease of explanation, we assume that the feasible set $f(P)$ of P is bounded and non-empty. Let Π^n be the set of all permutations π of the ground set $I^n = \{1, \ldots, n\}$. The symmetry group G of P is the set of all permutations π_i such that if x is a feasible solution of P then $\pi_i(x)$ is again a feasible solution of P of the same cost. Clearly, G is a permutation group of I^n, i.e., a subgroup of Π^n. In addition, G naturally induces a partition $\Omega = \{V_1, \ldots, V_K\}$ of the set of variables of P, called *orbital partition*. Intuitively, two variables x_i and x_j are in the same orbit V_k if and only if there exists $\pi \in G$ such that $\pi(i) = j$. Integer programs with large symmetry groups occur naturally when formulating many combinatorial optimization problems, such as graph coloring, scheduling, packing and covering design.

Symmetry has long been recognized as a curse for the traditional enumeration approaches used in both the MIP and CP communities—we refer to [1,2] for recent surveys on the subject. The reason is that many subproblems in the

C. Gomes and M. Sellmann (Eds.): CPAIOR 2013, LNCS 7874, pp. 204–215, 2013.
© Springer-Verlag Berlin Heidelberg 2013

enumeration tree are isomorphic, with a clear waste of computational resources. Various techniques for dealing with symmetric problems have been studied by different research communities and the usual approach to deal with symmetry is to try to eliminate it by introducing artificial symmetry-breaking constraints and/or by using ad-hoc search strategies.

In [3], a new technique called *orbital shrinking* for dealing with symmetric problems was presented, which is based on aggregation rather than on symmetry breaking. Let $G = \{\pi^1, \ldots, \pi^M\}$ be the symmetry group of P. Given an arbitrary feasible point $x \in f(P)$, we can construct the average point $\bar{x}$

$$\bar{x} = \frac{1}{M} \sum_{\pi^i \in G} \pi^i(x)$$

Trivially, $c\bar{x} = cx$. It is also easy to prove that $\bar{x}$ must have $\bar{x}_j$ constant within each orbit V_k, and that it can be efficiently computed by taking averages within each orbit, i.e.

$$\bar{x}_j = \frac{1}{|V_k|} \sum_{i \in V_k} x_i \quad \text{where } j \in V_k$$

If P were a convex optimization problem (e.g. a linear program), then $\bar{x}$ would be feasible for P, as it is a convex combination of feasible points of P. Thus, if we wanted to optimize over P, the only unknowns would be the averages within each orbit or, equivalently, their sums $y_k = \sum_{j \in V_k} x_j$, and we could derive an *equivalent* shrunken reformulation Q by

i) introducing sum variables y_k
ii) replacing x_j, $j \in V_k$ with $y_k/|V_k|$ in each constraint and in the objective function.

However, Q is not in general an equivalent reformulation of P when P is an arbitrary integer program, since the average point $\bar{x}$ may not satisfy the integrality requirements. However, it is still possible to prove (see [3] for details) that, if we impose the integrality requirements on the aggregated variables y_k, Q is an equivalent reformulation of the problem obtained from P by relaxing the integrality constraints on x with the surrogate integrality constraints on the sums over the orbits, and thus Q itself is a relaxation of P. Note that the LP relaxation of Q is equivalent to the LP relaxation of P and thus Q cannot be weaker than the standard LP relaxation of P, and can be quite stronger.

Example 1. Let us consider the very tiny Steiner Triple System (STS) instance of size 7

$$\min x_1 + x_2 + x_3 + x_4 + x_5 + x_6 + x_7$$
$$x_1 + x_2 + x_4 \geq 1$$
$$x_2 + x_3 + x_5 \geq 1$$
$$x_3 + x_4 + x_6 \geq 1$$
$$x_4 + x_5 + x_7 \geq 1$$
$$x_5 + x_6 + x_1 \geq 1$$
$$x_6 + x_7 + x_2 \geq 1$$
$$x_7 + x_1 + x_3 \geq 1$$
$$x \in \{0,1\}^7$$

It is easy to see that all variables belong to the same orbit, and thus, after introducing the sum variable $y = \sum_{j=1}^{7} x_j$, the orbital shrinking model reads

$$\min y$$
$$3/7y \geq 1$$
$$y \leq 7$$
$$y \in \mathbb{Z}_+$$

Its linear programming relaxation has value $7/3$, which is of course the same value as the LP relaxation of the original model. However, imposing the integrality requirement on y, we can increase the value of the relaxation to 3. In this case, this is also the value of an optimal integral solution of the original model, so orbital shrinking closes 100% of the integrality gap. Unfortunately, this is not always the case, even for instances from the same class. Indeed, if we move to the STS instance of size 9, orbital shrinking is not able to improve over the LP bound of 3, while the optimum is 5.   □

The STS example above shows that the orbital shrinking relaxation Q might be a oversimplified approximation of P (in the worst case, reducing to a trivial integer program with only one variable), thus providing no useful information for solving P. This is however not always the case. In some particular cases, Q is indeed an exact reformulation of P, even if P is an integer program. For example, consider a knapsack problem with identical items: an orbital shrinking relaxation would replace the binary variables x_j associated with the identical items with a general integer variable y_k that counts how many items of type k need to be taken in the solution, and this is clearly equivalent to the original formulation (but symmetry free). In other cases, Q, although not a reformulation, still retains enough structure from P such that that solving Q provides useful insights for solving P. Of course the question is how to exploit this information to obtain a sound and complete method for solving P.

A partial answer to this question was given in [4], where a hybrid MIP/CP scheme based on orbital shrinking was developed for the multi-activity shift scheduling problem, which is the problem of covering the demands of a finite set of activities over a time horizon by assigning them to a set of employees. In real-world applications, the set of feasible shifts (i.e., the set of feasible sequences of activity assignments to a single employee) is defined by many regulation constraints. In this case, many constraints of the problem (in particular those formulated with the aid of formal languages, such regular expressions or context-free grammars) are preserved by the shrinking process, while some others are not. In particular, cardinality constraints (e.g., the number of allowed working hours for a single employee in a single day) are replaced by surrogate versions in Q. Still, Q provides a very strong dual bound on P and its aggregated solutions can often be turned into feasible solutions for P. In order to get a complete method, the following strategy was proposed in [4]: solve the orbital shrinking model Q with a black box MIP solver and, whenever an (aggregated) integer feasible solution y^* is found, check with a CP solver if it can be turned into a feasible solution x^* for P. The scheme is akin to a logic-based Benders decomposition [5], although the decomposition is not based on a traditional variable splitting, but on aggregation.

In the present paper we show that the scheme above can be extended to a general framework for solving arbitrary symmetric MIP instances. This framework, described in Section 2, naturally provides a new way for devising hybrid MIP/CP decompositions. Then, in Section 3, we specialize the above framework to the multiple knapsack problem. Computational results in Section 4 show that the resulting method can be orders of magnitude faster than pure MIP approaches on hard symmetric instances. Conclusions are finally drawn in Section 5.

2   A General Orbital Shrinking Based Decomposition Method

Let P be an integer linear program as in the previous section and let G be the symmetry group of P. Note that if G is unknown to the modeler then the whole scheme can be applied starting from a subgroup G' of G, such as, for example, the symmetry group G_{LP} of the formulation, which is defined as

$$G_{LP} = \{\pi \in \Pi^n | \pi(c) = c \wedge \exists \sigma \in \Pi^m \ s.t. \ \sigma(b) = b, A(\pi, \sigma) = A\}$$

where $A(\pi, \sigma)$ is the matrix obtained by A permuting the columns with π and the rows with σ. Intuitively, a variable permutation π defines a symmetry if there exists a constraint permutation σ such that the two together leave the formulation unchanged. Note that if the permutation group G_{LP} is used, also constraints of P are partitioned into *constraint orbits*: in this case, two constraints are in the same orbit if and only if there exists a σ (as defined above) mapping one to the other. G_{LP} can be computed with any graph isomorphism package such as Nauty [6] or Saucy [7], which perform satisfactorily in practice.

Using G (or G_{LP}) we can compute the orbital partition Ω of P and construct the shrunken model Q

$$\min dy \tag{4}$$
$$By \geq r \tag{5}$$
$$y \in \mathbb{Z}_+^K \tag{6}$$

where $y_k = \sum_{j \in V_k} x_j$ and constraints (5) are obtained from the constraints (2) by replacing each occurrence of variable x_j with $y_k/|V_k|$, where k is the index of the orbit to which x_j belongs. It is easy to show that all constraints in the same orbit will be mapped to the same constraint in Q, so in practice Q has one variable for each variable orbit and one constraint for each constraint orbit in P. In the small STS example of the previous section, all constraints are in the same constraint orbit, and indeed they are all are mapped to $3/7y \geq 1$ in the orbital shrinking reformulation (the other constraint, $y \leq 7$, is derived from the upper bounds of the binary variables). Note that model Q acts like a master problem in a traditional Benders decomposition scheme.

For each integer feasible solution y^* of Q, we can then define the following (slave) feasibility check problem $R(y^*)$

$$Ax \geq b \tag{7}$$
$$\sum_{j \in V_k} x_j = y_k^* \quad \forall k \in K \tag{8}$$
$$x \in \mathbb{Z}_+^n \tag{9}$$

If $R(y^*)$ is feasible, then the aggregated solution y^* can be *disaggregated* into a feasible solution x^* of P, with the same cost. Otherwise, y^* must be rejected, in either of the following two ways:

1. Generate a *nogood* cut that forbids the assignment y^* to the y variables. As in logic-based Benders decomposition, an ad-hoc study of the problem is needed to derive stronger nogood cuts.
2. Branching. Note that in the (likely) event that the solution y^* is the integral LP relaxation of a node, then branching on non-fractional y variables is needed, and y^* will still be a feasible solution in one of the two child nodes. However, the method would still converge, because the number of variables is finite and the tree has a finite depth. Note that in this case the method may repeatedly check for feasibility the same aggregated solution: in practice, this can be easily avoided by keeping a list (cache) of recently checked aggregated solutions with the corresponding feasibility status.

It is important to note that, by construction, problem $R(y^*)$ has the same symmetry group of P, so symmetry may still be an issue while solving $R(y^*)$. This issue is usually solvable because (i) the linking constraints (8) may make the model much easier to solve and (ii) the (easier) structure of the problem may allow for more effective symmetry breaking techniques. Note also that $R(y^*)$ is a pure feasibility problem, so a CP solver may be a better choice than a MIP solver.

3   Application to the Multiple Knapsack Problem

In the present section, we specialize the general framework of the previous section to the multiple knapsack problem (MKP) [8,9]. This a natural generalization of the traditional knapsack problem [10], where multiple knapsack are available. Given a set of n items with weights w_j and profits p_j, and m knapsacks with capacity C_i, MKP reads

$$\max \sum_{i=1}^{m} \sum_{j=1}^{n} p_j x_{ij} \tag{10}$$

$$\sum_{j=1}^{n} w_j x_{ij} \leq C_i \quad \forall i = 1, \ldots, m \tag{11}$$

$$\sum_{i=1}^{m} x_{ij} \leq 1 \quad \forall j = 1, \ldots, n \tag{12}$$

$$x \in \{0, 1\}^{m \times n} \tag{13}$$

where binary variable x_{ij} is set to 1 if and only if item j is loaded into knapsack i. Since we are interested in symmetric instances, we will assume that all m knapsacks are identical and have the same capacity C, and that also some items are identical.

When applied to problem MKP, the orbital shrinking reformulation Q reads

$$\max \sum_{k=1}^{K} p_k y_k \tag{14}$$

$$\sum_{k=1}^{K} w_k y_k \leq mC \tag{15}$$

$$0 \leq y_k \leq |V_k| \quad \forall k = 1, \ldots, K \tag{16}$$

$$y \in \mathbb{Z}_+^K \tag{17}$$

Intuitively, in Q we have a general integer variable y_k for each set of identical items and a single knapsack with capacity mC. Given a solution y^*, the corresponding $R(y^*)$ is thus a one dimensional bin packing instance, whose task is to check whether the selected items can indeed be packed into m bins of capacity C.

To solve the bin-packing problem above, we propose two different approaches. The first approach is to deploy a standard compact CP model based on the global **binpacking** constraint [11] and exploiting the CDBF [12] branching scheme for search and symmetry breaking. Given an aggregated solution y^*, we construct a vector s with the sizes of the items picked by y^*, and sort it in non-decreasing order. Then we introduce a vector of variables b, one for each item: the value of b_j is the index of the bin where item j is placed. Finally, we introduce a variable l_i for each bin, whose value is the load of bin i. The domain of variables l_i is $[0..C]$. With this choice of variables, the model reads:

$$\text{binpacking}(b, l, s) \tag{18}$$

$$b_{j-1} \le b_j \quad \text{if } s_{j-1} = s_j \tag{19}$$

where (19) are symmetry breaking constraints.

The second approach is to consider an extended model, akin to the well known Gilmore and Gomory column generation approach for the cutting stock problem [13]. Given the objects in y^*, we generate all feasible packings p of a single bin of capacity C. Let P denote the set of all feasible packings and, given packing p, let a_{pk} denote the number of items of type k picked. The corresponding model is

$$\sum_{p \in P} a_{pk} x_p = y_k^* \tag{20}$$

$$\sum_{p \in P} x_p = m \tag{21}$$

$$x_p \in \mathbb{Z}_+ \tag{22}$$

where integer variables x_p count how many bins are filled according to packing p. In the following, we will denote this model with BPcg. Model BPcg is completely symmetry free, but it needs an exponential number of columns in the worst case.

4   Computational Experiments

We implemented our codes in C++, using IBM ILOG Cplex 12.4 [14] as black box MIP solver and Gecode 3.7.3 [15] as CP solver. All tests have been performed on a PC with an Intel Core i5 CPU running at 2.66GHz, with 8GB of RAM (only one CPU was used by each process). Each method was given a time limit of 1 hour per instance.

In order to generate hard MKP instances, we followed the systematic study in [16]. According to [16], difficult instances can be obtained introducing some correlation between profits and weights. Among the hardest instances presented in [16] are the so-called *almost strongly correlated* instances, in which weights w_j are distributed—say uniformly—in the range $[1, R]$ and the profits p_j are distributed in $[w_j + R/10 - R/500, w_j + R/10 + R/500]$. These instances correspond to real-life situations where the profit is proportional to the weight plus some fixed charge value and some noise. Given this procedure, a possibility for generating hard-enough instances is to construct instances where the coefficients are of moderate size, but where all currently used upper bounds have a bad performance. Among these difficult classes, we consider the *spanner instances*: these instances are constructed such that all items are multiples of a quite small set of items—the so-called spanner set. The spanner instances span(v, l) are characterized by the following three parameters: v is the size of the spanner set, l is the multiplier limit, and we may have any distribution of the items in the spanner set. More formally, the instances are generated as follows: a set of v items is generated with weights in the interval $[1, R]$, with $R = 1000$, and profits according to the distribution. The items (p_k, w_k) in the spanner set are normalized by dividing the profits and weights by $l + 1$, with $l = 10$. The n items are then constructed by repeatedly choosing an item (p_k, w_k) from the spanner set, and a multiplier a randomly generated in the interval $[1, l]$. The constructed item has profit and weight (ap_k, aw_k). Capacities are computed as $C = \frac{\sum_{i=1}^{n} w_i}{8}$.

In order to have a reasonable test set, we considered instances with a number of items n in $\{30, 40, 50\}$ and number of knapsacks m in $\{3, 4, 5, 6\}$. For each pair of (n, m) values, we generated 10 random instances following the procedure described above, for a total of 120 instances. All the instances are available from the author upon request. For each set of instances, we report aggregate results comparing the shifted geometric means of the number of branch-and-cut nodes and the computation times of the different methods. Note that we did not use specialized solvers, such as ad-hoc codes for knapsack or bin packing problems, because the overall scheme is very general and using the same (standard) optimization packages in all the methods allows for a clearer comparison of the different approaches.

As a first step, we compared 2 different pure MIP formulations. One is the natural formulation $(10)-(13)$, denoted as `cpxorig`. The other is obtained by aggregating the binary variables corresponding to identical items. The model, denoted as `cpx`, reads

$$\max \sum_{i=1}^{m} \sum_{k=1}^{K} p_j z_{ik} \tag{23}$$

$$\sum_{k=1}^{K} w_j z_{ik} \leq C \quad \forall i = 1, \ldots, m \tag{24}$$

$$\sum_{i=1}^{m} z_{ik} \leq U_k \quad \forall k = 1, \ldots, K \tag{25}$$

$$z \in \mathbb{Z}_+^{m \times K} \tag{26}$$

where U_k is the number of items of type k. Note that `cpx` would be obtained automatically from formulation `cpxorig` by applying the orbital shrinking procedure if the capacities of the knapsacks were different. While one could argue that `cpxorig` is a modeling mistake, the current state-of-the-art in preprocessing is not able to derive `cpx` automatically, while orbital shrinking would. A comparison of the two formulations is shown in Table 1. As expected, `cpx` clearly outperforms `cpxorig`, solving 82 instances (out of 120) instead of 65. However, `cpx` performance is rapidly dropping as the number of items and knapsacks increases.

Then, we compared three variants of the hybrid MIP/CP procedure described in Section 3, that differs on the models used for the feasibility check. The first variant, denoted by `BPstd`, is based on the compact model $(18)-(19)$. The second and the third variants are both based on the extended model $(20)-(22)$, but differs on the solver used: a CP solver for `BPcgCP` and a MIP solver for `BPcgMIP`. All variants use model $(14)-(17)$ as a master problem, which is fed to Cplex and solved with dual reductions disabled, to ensure the correctness of the method.

Table 1. Comparison between `cpxorig` and `cpx`

n	m	# solved		time (s)		nodes	
		cpxorig	cpx	cpxorig	cpx	cpxorig	cpx
30	3	10	10	1.16	0.26	3,857	1,280
30	4	9	10	12.28	3.42	65,374	16,961
30	5	6	8	291.75	79.82	2,765,978	1,045,128
30	6	7	7	108.83	48.05	248,222	164,825
40	3	9	10	19.48	2.72	103,372	9,117
40	4	8	8	351.07	35.56	3,476,180	421,551
40	5	2	3	2,905.70	1,460.95	25,349,383	23,897,899
40	6	3	5	308.29	234.19	626,717	805,007
50	3	6	9	70.73	12.44	259,099	32,310
50	4	2	7	1,574.34	254.58	8,181,128	4,434,707
50	5	0	2	3,600.00	700.69	26,017,660	4,200,977
50	6	3	3	308.29	307.98	586,400	1,025,907

Table 2. Comparison between hybrid methods

n	m	# solved			time (s)			nodes		
		BPstd	BPcgCP	BPcgMIP	BPstd	BPcgCP	BPcgMIP	BPstd	BPcgCP	BPcgMIP
30	3	10	10	10	0.07	0.05	0.05	245	270	270
30	4	10	10	10	0.18	0.12	0.08	157	160	160
30	5	10	10	10	1.28	0.26	0.14	90	88	88
30	6	10	10	10	1.24	0.25	0.13	42	40	40
40	3	10	10	10	0.64	0.42	0.17	502	540	540
40	4	10	10	10	0.54	0.20	0.17	225	224	224
40	5	9	10	10	8.63	1.20	0.62	202	225	225
40	6	8	10	10	17.96	1.65	0.46	48	60	60
50	3	10	10	10	1.59	0.93	0.44	837	914	914
50	4	10	10	10	4.06	1.11	0.60	337	335	335
50	5	6	8	10	137.52	23.97	3.58	172	245	335
50	6	7	7	10	17.15	12.73	2.85	17	16	140

Cplex callbacks are used to implement the decomposition. A comparison of the three methods is given in Table 2. Note that the number of nodes reported for hybrid methods refers to the master only—the nodes processed to solve the feasibility checks are not added to the count, since they are not easily comparable, in particular when a CP solver is used. Of course the computation times refer to the whole solving process (slaves included). According to the table, even the simplest model `BPstd` clearly outperforms `cpx`, solving 110 instances (28 more) and with speedups up to two orders of magnitude. However, as the number of knapsacks increases, symmetry can still be an issue for this compact model, even though symmetry breaking is enforced by constraints (19) and by CDBF. Replacing the compact model with the extended model, while keeping the same solver, shows some definite improvement, increasing the number of solved instances from 110 to 115 and further reducing the running times. Note that for the instances in our testbed, the number of feasible packings was always manageable (at most a few thousands) and could always be generated by Gecode in a fraction of a second. Still, on some instances, the CP solver was not very effective in solving the feasibility model. The issue is well known in the column generation community: branching on variables x_p yields highly unbalanced trees, because fixing a variable x_p to a positive integer value triggers a lot of propagations, while fixing it to zero has hardly any effect. In our particular case, replacing the CP solver with a MIP solver did the trick. Indeed, just solving the LP relaxation was sufficient in most cases to detect infeasibility. Note that if infeasibility is detected by the LP relaxation of model (20)−(22), then standard LP duality can be used to derive a (Benders) nogood cut violated by the current aggregated solution y^*, without any ad-hoc study. In our implementation, however, we did not take advantage of this possibility, and just stuck to the simpler strategy of branching on integer variables. BPcgMIP is able to solve all 120 instances, in less than four seconds (on average) in the worst case. The reduction in the number of nodes is particularly significant: while cpx requires millions of nodes for some classes, BPcgMIP is always solving the instances in fewer than 1,000 nodes.

Finally, Table 3 shows the average gap closed by the orbital shrinking relaxation with respect to the initial integrality gap, and the corresponding running times (obtained by solving the orbital shrinking relaxation with a black box MIP solver, without the machinery developed in this section). According to the table, orbital shrinking yields a much tighter relaxation than standard linear programming, while still being very cheap to compute.

Table 3. Average gap closed by orbital shrinking and corresponding time

n	m	gap closed	time (s)
30	3	45.3%	0.007
30	4	46.6%	0.004
30	5	42.8%	0.004
30	6	54.4%	0.002
40	3	48.4%	0.013
40	4	67.2%	0.007
40	5	55.3%	0.005
40	6	58.6%	0.003
50	3	52.7%	0.031
50	4	64.5%	0.030
50	5	61.1%	0.006
50	6	76.7%	0.003

5   Conclusions

In this paper we presented a general framework for deploying hybrid MIP/CP decomposition methods for symmetric optimization problems. This framework is similar in spirit to logic-based Benders decomposition schemes, but it is based on aggregation rather than on the usual variable splitting argument, thus being applicable to a completely different class of problems. The overall scheme can be obtained as a generalization of a recent approach developed for the multi-activity shift scheduling problem, where it showed significant improvements over previous pure MIP approaches. In order to further test its effectiveness, we specialized the general scheme to the multiple knapsack problem, giving a clear example on how to apply the method in practice and some recommendations on how to solve the possible pitfalls of the approach. Computational results confirmed that the resulting method can be orders of magnitude faster than standard pure MIP approaches on hard symmetric instances.

References

1. Margot, F.: Symmetry in integer linear programming. In: Jünger, M., Liebling, T., Naddef, D., Nemhauser, G., Pulleyblank, W., Reinelt, G., Rinaldi, G., Wolsey, L. (eds.) 50 Years of Integer Programming 1958-2008, pp. 647–686. Springer, Heidelberg (2010)
2. Gent, I.P., Petrie, K.E., Puget, J.F.: Symmetry in constraint programming. In: Rossi, F., van Beek, P., Walsh, T. (eds.) Handbook of Constraint Programming, pp. 329–376. Elsevier (2006)
3. Fischetti, M., Liberti, L.: Orbital shrinking. In: Mahjoub, A.R., Markakis, V., Milis, I., Paschos, V.T. (eds.) ISCO 2012. LNCS, vol. 7422, pp. 48–58. Springer, Heidelberg (2012)
4. Salvagnin, D., Walsh, T.: A hybrid MIP/CP approach for multi-activity shift scheduling. In: Milano, M. (ed.) CP 2012. LNCS, vol. 7514, pp. 633–646. Springer, Heidelberg (2012)
5. Hooker, J.N., Ottosson, G.: Logic-based Benders decomposition. Mathematical Programming 96(1), 33–60 (2003)
6. McKay, B.D.: Practical graph isomorphism (1981)
7. Katebi, H., Sakallah, K.A., Markov, I.L.: Symmetry and satisfiability: An update. In: Strichman, O., Szeider, S. (eds.) SAT 2010. LNCS, vol. 6175, pp. 113–127. Springer, Heidelberg (2010)
8. Scholl, A., Klein, R., Jürgens, C.: Bison: A fast hybrid procedure for exactly solving the one-dimensional bin packing problem. Computers & OR 24(7), 627–645 (1997)
9. Pisinger, D.: An exact algorithm for large multiple knapsack problems. European Journal of Operational Research 114(3), 528–541 (1999)
10. Martello, S., Toth, P.: Knapsack Problems: Algorithms and Computer Implementations. Wiley (1990)
11. Shaw, P.: A constraint for bin packing. In: Wallace, M. (ed.) CP 2004. LNCS, vol. 3258, pp. 648–662. Springer, Heidelberg (2004)

12. Gent, I.P., Walsh, T.: From approximate to optimal solutions: Constructing pruning and propagation rules. In: IJCAI, pp. 1396–1401. Morgan Kaufmann (1997)
13. Gilmore, P.C., Gomory, R.E.: A linear programming approach to the cutting-stock problem. Operations Research 9, 849–859 (1961)
14. IBM ILOG: CPLEX 12.4 User's Manual (2011)
15. Gecode Team: Gecode: Generic constraint development environment (2012), http://www.gecode.org
16. Pisinger, D.: Where are the hard knapsack problems? Computers & Operations Research 32, 2271–2284 (2005)

On Solving Mixed-Integer Constraint Satisfaction Problems with Unbounded Variables

Hermann Schichl*, Arnold Neumaier,
Mihály Csaba Markót, and Ferenc Domes

Faculty of Mathematics, University of Vienna, Austria

Abstract. Many mixed-integer constraint satisfaction problems and global optimization problems contain some variables with unbounded domains. Their solution by branch and bound methods to global optimality poses special challenges as the search region is infinitely extended. Many usually strong bounding methods lose their efficiency or fail altogether when infinite domains are involved. Most implemented branch and bound solvers add artificial bounds to make the problem bounded, or require the user to add these. However, if these bounds are too small, they may exclude a solution, while when they are too large, the search in the resulting huge but bounded region may be very inefficient. Moreover, if the global solver must provide a rigorous guarantee (as for the use in computer-assisted proofs), such artificial bounds are not permitted without justification by proof.

We developed methods based on compactification and projective geometry as well as asymptotic analysis to cope with the unboundedness in a rigorous manner. Based on projective geometry we implemented two different versions of the basic idea, namely (i) projective constraint propagation, and (ii) projective transformation of the variables, in the rigorous global solvers COCONUT and GloptLab. Numerical tests demonstrate the capability of the new technique, combined with standard pruning methods, to rigorously solve unbounded global problems. In addition, we present a generalization of projective transformation based on asymptotic analysis.

Compactification and projective transformation, as well as asymptotic analysis, are fruitless in discrete situations but they can very well be applied to compute bounded relaxations, and we will present methods for doing that in an efficient manner.

Keywords: Mixed-integer CSPs, constraint propagation, relaxation methods, unbounded variables, interval analysis, directed acyclic graphs.

1 Introduction

1.1 Mixed Integer Constraint Satisfaction Problems

Many real-world problems lead to mixed-integer and numerical constraint satisfaction problems (MICSPs). Every MICSP is a triplet $(\mathcal{V}, \mathcal{C}, \mathcal{D})$ consisting of

---

* This work was supported by the Austrian Science Fund (FWF) grant P22239.

C. Gomes and M. Sellmann (Eds.): CPAIOR 2013, LNCS 7874, pp. 216–233, 2013.

a finite set $\mathcal{V}$ of variables taking their values in domains $\mathcal{D}$ over the reals (possibly restricted to the subset of integers) subject to a finite set $\mathcal{C}$ of *numerical* or purely *combinatorial* constraints. A tuple of values assigned to the variables such that all the constraints are satisfied is called a solution. The set of all the solutions is called the solution set. When dealing with a MICSP, depending on the application, it might suffice to find *one* solution, but in some cases it might be necessary to identify the whole solution set.

In practical problems, numerical constraints are often expressed as equations and inequalities in *factorable* form, that is, they are described by functions that are recursively composed of elementary functions such as arithmetic operators $(+, -, *, /)$, and univariate (sometimes bivariate) basic functions like log, exp, sin, cos,... In other words, such an MICSP can be expressed as

$$F(x) \in \mathbf{b}, \ x \in \mathbf{x}, \ x_I \in \mathbb{Z}^{|I|}, \tag{1}$$

where $F : \mathbb{R}^n \to \mathbb{R}^m$ is a factorable function, x is a vector of n real variables, $\mathbf{x}$ and $\mathbf{b}$ are interval vectors of sizes n and m respectively, and I is the set of integer variables.

Many solution techniques have been proposed in *Constraint Programming* and *Mathematical Programming* to solve MICSPs. A difficulty when dealing with continuous variables is roundoff errors. For achieving full rigor, almost all solution techniques for MICSPs use *interval arithmetic* (see [12, 16–18]) or some of its variants (affine arithmetic [30], Taylor arithmetic [4, 5, 19], etc.). During the last two decades, a lot of work has been put into the development of *inclusion tests* and *contractors* based on interval arithmetic. In addition, numerous *relaxation techniques* (many of them based on interval arithmetic combined with algorithmic differentiation methods [11, 25]) have been devised (see [13, 20]).

The function of an inclusion test is to check whether the domain of a variable is included in the projection of the solution set. A contractor, also called a *narrowing operator* [2, 10] or *contracting operator* [1, 29, 32], is a method that computes a (hopefully proper) subset of the variable domains such that all solutions are retained. Various basic inclusion tests and contractors have been described in [13] and [20].

In particular, a contraction operator approach called *interval constraint propagation* was developed [2, 3, 31], which associates *constraint propagation/local consistency* techniques, as defined in artificial intelligence, with interval analytic methods. Advanced contractors, such as the *forward-backward contractor* [2, 13], result from the interval constraint propagation (CP) approach. It is a way to propagate domain reductions forwards and backwards through the computational trees of the constraints. Based on the fundamental framework for interval analysis on directed acyclic graphs (DAGs) [27], a high performance constraint propagator FBPD for continuous CSPs has been developed in [34].

In practical constraint solvers inclusion tests and contractors are interleaved with some form of *exhaustive search* to compute a representation of the solution set. Search by *bisection* or more advanced branching is the most commonly used technique. In the context of MICSPs this leads to the branch and bound class of algorithms, which generate a search graph consisting of subproblems that are subsequently solved or further subdivided.

A *relaxation* is a (usually much easier solvable) replacement MICSP whose solution set provably contains all solutions of problem (1). There are several classes of relaxations, although linear and convex ones are mostly used, see [20, 21]. Relaxations usually are an efficient tool for fathoming nodes of the search graph during the search procedure.

1.2   Unbounded Variables

An especially difficult class of MICSPs are those which contain variables whose domain set is unbounded. Their solution by branch and bound methods poses special challenges as the search region is infinitely extended. On the one hand, the unboundedness cannot be removed by splitting. On the other hand most inclusion and contraction operators become inefficient or dysfunctional when applied to unbounded domains.

Most branch and bound solvers add artificial bounds to make the problem bounded, or require the user to add these by forbidding unbounded problems altogether. However, if these artificial bounds are too small, they may exclude a solution, even render the problem infeasible, while when they are too large, the search in the resulting huge but bounded region may be very inefficient. Moreover, if the global solver must provide a rigorous guarantee (as for the use in computer-assisted proofs), such artificial bounds are not permitted without justification.

The contribution of this paper is twofold. Firstly, we developed methods based on compactification and projective geometry to cope with the unboundedness in a rigorous manner. We implemented two different versions of the basic idea, namely

1. projective transformation of the variables, and
2. projective constraint propagation.

They are implemented in the rigorous global solvers GloptLab [6–8] and COCONUT [23, 24], respectively. Numerical tests demonstrate the capability of the new technique, combined with standard pruning methods, to rigorously solve unbounded global problems.

Secondly, these projective transformations are most efficient for those MICSPs, whose unbounded variables are continuous and all constraints involving them are rational. Although the method is still applicable when transcendental functions are involved, the effectiveness is significantly reduced. Therefore, we developed an extension using asymptotic analysis that is more efficient in the presence of transcendental functions. This is based on ideas from the unpublished thesis [9]. Since for discrete variables the transformation method is not applicable, we shortly describe asymptotic relaxations for improved node fathoming in the unbounded case.

In Section 2 we will explain explicit projective transformation and projective constraint propagation. Section 3 generalizes that to asymptotic transformations.

Some information on projective and asymptotic relaxations are given in Section 4, and numerical results are provided in Section 5.

Throughout this paper we will need some notation: a real interval $\mathbf{a} \in \overline{\mathbb{IR}}$ is defined as $[\underline{a}, \overline{a}] = \{a \in \mathbb{R} \mid \underline{a} \leq a \leq \overline{a}\}$, with $\underline{a} \in \mathbb{R} \cup \{-\infty\}$ and $\overline{a} \in \mathbb{R} \cup \{\infty\}$. In case both bounds $\underline{a}$ and $\overline{a}$ of $\mathbf{a}$ are finite, we call $\mathbf{a}$ a finite or bounded interval, otherwise $\mathbf{a}$ is an infinite or unbounded interval. We will also need the set $\overline{\mathbb{UR}}$ of all finite disjoint unions of intervals. Real arithmetic and elementary functions can be extended to intervals, see [18], and to interval unions. An n-dimensional real box (union box) $\mathbf{x} \in \overline{\mathbb{IR}}^n (\overline{\mathbb{UR}}^n)$ is a vector of n real intervals (interval unions). If all components of $\mathbf{x}$ are finite, then $\mathbf{x}$ is a finite or bounded box or interval union, otherwise $\mathbf{x}$ is infinite or unbounded.

2 Projective Transformation

Throughout this section we will consider factorable MICSPs of the form (1). We will assume that the variables x_J have unbounded domains and that the other variables x_K have bounded domains. Furthermore, we will for the moment require that all integer variables are bounded, i.e., that $I \subset K$.

Since all functions involved in (1) are factorable, the problem can be represented as a reduced computational directed acyclic graph $\Gamma = (V(\Gamma), E(\Gamma))$, see [27]. All nodes $\nu \in V(\Gamma)$ represent intermediate expressions y_ν of some constraints. The local sources of Γ correspond to constants and variables, i.e., $x_k = y_{\nu_k}$ for all k and some $\nu_k \in V(\Gamma)$, and the local sinks correspond to the constraints.

The basic idea of the projective transformation is the natural embedding of $\mathbb{R}^{|J|} \times \mathbf{x}_K$, which contains the feasible set, into the compact manifold with boundary $P\mathbb{R}^{|J|} \times \mathbf{x}_K$, where $P\mathbb{R}^{|J|}$ is the projective space over $\mathbb{R}^{|J|}$. For the transformation we represent each intermediate node y_ν for $\nu \in V(\Gamma)$ in the form

$$y_\nu = \widehat{y}_\nu / t^{m_\nu}, \tag{2}$$

where m_ν is a rational number and t is a scaling factor to be chosen. The new variable t and the exponents m_ν are defined such that $t \in [0,1]$ and the $\widehat{y}_\nu$ are well-bounded. (Actually, that can only be guaranteed in the case of a rational MICSP. In the presence of transcendental functions some intermediate $\widehat{y}_\nu$ may still be unbounded. This is the motivation for the generalization in Section 3.) Note that while these transformations are singular, the transformed problem has no singularities, and the solution set is preserved with full mathematical rigor.

The transformation is achieved by a recursive construction, implemented in a forward walk through Γ. For the original variables x_k and all interval constants, we define

$$m_{\nu_k} = \begin{cases} 0 & \text{if } k \in K, \\ 1 & \text{if } k \in J. \end{cases} \tag{3}$$

For practical reasons we put in the implementation also those indices j into J, for which the bounds $\mathbf{x}_j$ are huge, e.g., bigger than 10^7 but this limit is problem and scaling dependent. Those bounds, in general, are artificial in the first place and in a branch and bound context pose similar problems as unbounded variables.

For constructing the $\widehat{y}_\nu$ we choose a real number $0 \le s \le 1$ and set

$$t := \left(1 - s + \sum_{k \in K} d_k x_k^2\right)^{-1/2} \tag{4}$$

with scaling factors $d_k > 0$. This leads to the constraint

$$(1 - s)t^2 + \sum d_k \widehat{x}_k^2 = 1, \tag{5}$$

from which we deduce the bounds

$$t \in \mathbf{t} := [0, \max(0, 1 - s)^{-1/2}],$$
$$|\widehat{x}_k| \le d_k^{-1/2} \quad \text{for } k \in K. \tag{6}$$

To guarantee that t is real, we need to choose s such that

$$\sum_{k \in K} d_k x_k^2 \ge s$$

is a valid constraint. For example,

$$s := \inf \sum_{k \in K} d_k \mathbf{x}_k^2$$

qualifies (if necessary, rescale the d_k to have $s \le 1$), but better bounds might be available. A possible choice is $s = 0$, however in general this is suboptimal. Then

$$t \in [0, 1].$$

Since $\widehat{y}_{\nu_k} = x_k$ for the well-bounded variables, we have expressed all variables in terms of bounded ones.

The exponent m_ν for an intermediate variable y_ν depends on the operation that creates it. If $y = \sum \alpha_\nu y_\nu$ then $y = \widehat{y}/t^m$ with

$$m := \max m_\nu, \quad \widehat{y} := \sum \alpha_\nu t^{m-m_\nu} \widehat{y}_\nu, \tag{7}$$

and we get the finite enclosure

$$\widehat{y} \in \widehat{\mathbf{y}} := \sum \alpha_\nu \mathbf{t}^{m-m_\nu} \widehat{\mathbf{y}}_\nu. \tag{8}$$

If $y = \prod y_\nu^{\alpha_\nu}$ with rational α_ν then $y = \widehat{y}/t^m$ with

$$m := \sum \alpha_\nu m_\nu, \quad \widehat{y} := \prod \widehat{y}_\nu^{\alpha_\nu}, \tag{9}$$

and we get the finite enclosure

$$\widehat{y} \in \widehat{\mathbf{y}} := \prod \widehat{\mathbf{y}}_\nu^{\alpha_\nu}. \tag{10}$$

This accounts for all elementary operations and powers with fixed exponent.

For other elementary functions, one can derive similar formulas, though their derivation and implementation is more complex. For example, if $y = \log y_\nu$ then $y = \widehat{y}/t^m$ with

$$m := 1, \quad \widehat{y} := t(\log \widehat{y}_\nu - m_\nu \log t),$$

and we get a finite enclosure derivable by monotony considerations.

For some transcendental functions $y = \varphi(y_\nu)$ even a projective transformation cannot in general guarantee boundedness of $\widehat{y}$ (one example is $\varphi = \exp$). In that case, we define $\widehat{y} := \varphi(t^{-m_\nu}\widehat{y}_\nu)$ and get a possibly unbounded enclosure of $\varphi(\mathbf{t}^{-m_\nu}\mathbf{y}_\nu)$ for $\widehat{y}$.

There are two possibilities to utilize projective transformations. The problem can be explicitly transformed (see Section 2.1), or the projective transformation can be used implicitly during constraint propagation (see Section 2.2) and relaxation calculation (see Section 4).

2.1 Explicit Projective Transformation

We have implemented the explicit transformation method in the software package GLOPTLAB [6–8], a constraint satisfaction package for enclosing all solutions of systems of quadratic equations and inequalities.

The special quadratic structure allows one to implement projective transformations explicitly by rewriting the original equations after a projective transformation (2) on the variables x_k using (4). Then all linear inequality constraints

$$Ax \geq b$$

are transformed into the homogeneous linear constraints

$$A\widehat{x} - bt \geq 0.$$

Bound constraints are treated as linear constraints, too. Nonlinear quadratic constraints

$$x^T G x + c^T x \geq \gamma$$

are transformed into the homogeneous quadratic constraints

$$\widehat{x}^T G \widehat{x} + tc^T \widehat{x} - \gamma t^2 \geq 0.$$

In all cases, equations and inequalities with the opposite sign are handled analogously. The additional constraints (5) and (6) are also quadratic and linear, respectively, and thus the transformed problem is again quadratic but bounded. Hence, it can be solved with traditional methods. After solving the transformed problem, one can recover the original solution from

$$x_i = \widehat{x}_i/t, \quad \text{if } t \neq 0.$$

Solutions of the transformed problem with $t = 0$ correspond to limiting solutions at infinity of the original problem. They can be discarded in general.

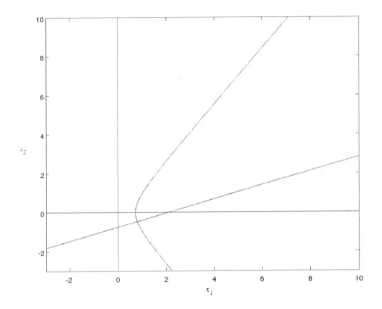

Fig. 1. Example 1

Alternatively, one can solve a bigger constraint satisfaction problem containing both the original and the transformed variables and constraints. In that case, however, the transformation equations themselves have to be added as additional quadratic constraints

$$x_i t - \widehat{x}_i = 0.$$

This allows one to exploit the features of both the original and the transformed problem at the same time, at the cost of doubling the problem size.

Example 1. The constraint satisfaction problem

$$0.36x_1 - x_2 = 0.75,$$
$$2x_1^2 - x_2^2 = 1,$$
$$x_1 \geq 0, \quad x_2 \geq 0$$

is infeasible but the equations have a solution at

$$x_1 \approx 0.6491, \quad x_2 \approx -0.5163,$$

slightly outside the defining box, see Fig. 1.

The problem is difficult to solve with standard CP and branch and bound, since no box $\mathbf{x}$ of the form $\mathbf{x}_1 = [a, \infty]$, $\mathbf{x}_2 = [b, \infty]$ can be reduced by CP.

The projective transformation leads to the problem

$$0.36\widehat{x}_1 - \widehat{x}_2 - 0.75t = 0,$$
$$2\widehat{x}_1^2 - \widehat{x}_2^2 - t^2 = 0,$$
$$\widehat{x}_1^2 + \widehat{x}_2^2 + t^2 = 1,$$
$$\widehat{x}_1, \widehat{x}_2, t \in [0, 1]$$

This problem is easily found to be infeasible by CP.

2.2 Projective Constraint Propagation

For more general problems explicit transformation becomes more cumbersome. For the implementation in the COCONUT Environment [23, 24], a software platform for global optimization, we have therefore chosen a different approach. Frequently, in the non-quadratic case a transformed MICSP is not easier to solve than the original problem. Therefore, we utilize the projective transformation together with a special split into a bounded subproblem and its complement.

For that observe that the transformation (2) on the variables x_k has the following property. If $\|\widehat{x}\|_p = \alpha$ and $t \in [0, 1]$ then $\|x\|_p \geq \alpha$, for every $p \in [1, \infty]$. Adding the constraint $\|x\|_p \leq \alpha$ to problem (1) makes it bounded, so it can be solved by standard methods.

The complement of that ball, described by the complementary constraint $\|x\|_p \geq \alpha$, must be handled as well and can be projectively transformed using (2) and the constraint $\|\widehat{x}\|_p = \alpha$. The choice of the constant α is application specific.

However, this transformation is never performed explicitly. Rather, many of the bounding tools, and foremost CP, implicitly make use of the transformation by calculating in $\mathbb{IPR}$, the set of so-called *projective intervals*. Those are pairs $(\widehat{\mathbf{x}}, r; \mathbf{t})$ of intervals and rational numbers together with a common interval $\mathbf{t}$ representing the range of the scaling parameter. The operations of projective intervals are defined according to (7–10) with extensions for transcendental elementary functions.

Not performing the transformation explicitly has an additional advantage which is connected to the following important observation: There exist CSPs for which CP proceeds with range reduction on the original problem but where it has no reducing effect on the projectively transformed problem.

Example 2. Consider the constraints (with $\alpha = 1$)

$$y = x^2 - 3 \tag{11}$$
$$x^2 + y^2 \geq 1. \tag{12}$$

From (11) we get $y \in [-\infty, \infty] \cap ([-\infty, \infty]^2 - 3) = [-3, \infty]$, which reduces the range of y.

The projectively transformed problem associated to (11–12) is as follows:

$$\widehat{y}t = \widehat{x}^2 - 3t \tag{13}$$
$$\widehat{x}^2 + \widehat{y}^2 = 1 \tag{14}$$

$$\hat{x} \in [-1,1], \ \hat{y} \in [-1,0], \ t \in (0,1]. \tag{15}$$

From (13) we get

$$y \in [-1,0] \cap ([-1,1]^2 - 3 \cdot (0,1]) = [-1,0].$$

Also from (13),

$$x^2 = \hat{y}t + 3t \in [-1,0] + (0,3] = (-1,3],$$

so the current range $[-1,1]$ of $\hat{x}$ cannot be reduced; (14) yields no improvement as well, thus, CP cannot reduce any of the initial variable ranges of the transformed problem.

The effect that no range reduction is possible on the original problem whereas CP works on the transformed problem was already demonstrated in Example 1.

Consequently, we need to utilize CP simultaneously on the original and on the transformed problem. For that we developed the algorithm *Projective Forward and Backward Propagation on DAGs* (PFBPD), which is an advanced version of FBPD from [33, 34]. It is based on propagating enclosures of the form $(\mathbf{x}, (\hat{\mathbf{x}}, r; \mathbf{t})) \in \overline{\mathbb{UR}}^n \times \mathbb{IPR}^n$, of pairs of interval unions (not projectively transformed) and projective intervals in parallel, in order to get the advantages of both approaches. These pairs are interwoven since after each forward or backward propagation step an internal intersection between the two enclosures is performed for additional reduction.

For this algorithm let $D(\mathbf{G})$ be a DAG with the ground $\mathbf{G}$, $\mathcal{C}$ the set of active constraints, and $\mathcal{D}$ the variable domains. Furthermore, for every node ν of the DAG we introduce the set $\mathcal{N}(\nu) \subseteq \overline{\mathbb{UR}} \times \mathbb{IPR}$, or $\mathcal{N}(\nu) \subseteq \mathbb{UZ} \times \mathbb{IPR}$ for the integer variables, containing the current enclosure of the range of ν. Note that we keep the integer information only in the untransformed problem since the transformation destroys the integrality information.

Algorithm PFBPD($\mathbf{in} : D(\mathbf{G}), \mathcal{C}, \alpha; \mathbf{in/out} : \mathcal{D}$)

```
00:     L_f := ∅; L_b := ∅; V_oc := (0, . . . , 0); V_ch := (0, . . . , 0); t_o := [0, 1];
01:     Set the node ranges N_{ν_k} of every variable x_k to (D_k, (x̂_k, m_k; t_o));
02:     V_lvl := (0, . . . , 0);
03:     for each node C representing an active constraint in C do
04:             NodeOccurrences(C, V_oc);
05:             NodeLevel(C, V_lvl); /* this can be made optional */
06:     end-for
07:     Add a virtual node V with maximal node level (for constraint ||x̂||²₂ = α²).
08:     C := C ∪ {V};
09:     for each node C representing an active constraint in C \ {V} do
10:             FindVirtualEdges(C, V, V);
11:             ForwardEvaluation(C, V_ch, L_b);
12:     end-for
13:     while L_b ≠ ∅ ∨ L_f ≠ ∅ do
14:             N := getNextNode(L_b, L_f);
```

```
15:        if N was taken from 𝓛ᵦ then
16:            for each child C of N do
17:                BP(N, C);
18:                if 𝒩(C) = ∅ then return infeasible;
19:                if 𝒩(C) changed enough for forward evaluation then
20:                    for each P ∈ parents(C) \ {N, G} do
21:                        if V_oc[P] > 0 then put P into 𝓛_f;
22:                end-if
23:                if 𝒩(C) changed enough for backward propagation then
24:                    Put C into 𝓛ᵦ;
25:            end-for
26:        else /* N was taken from 𝓛_f */
27:            FE(N, [f]); /* f is the operator at N */
28:            if 𝒩(N) = ∅ then return infeasible;
29:            if 𝒩(N) changed enough for forward evaluation then
30:                for each P ∈ parents(N) \ {G} do
31:                    if V_oc[P] > 0 then put P into 𝓛_f;
32:            end-if
33:            if 𝒩(N) changed enough for backward propagation then
34:                Put N into 𝓛ᵦ;
35:        end-if
36:        if t ≠ t_o then
37:            if t changed enough for forward propagation then
38:                Put all nodes C ∈ 𝒱 into 𝓛_f;
39:            end-if
40:            t_o := t;
41:        end-if
42:    end-while
43:    Update 𝒟 with the ranges of the nodes representing the variables;
end
```

procedure ForwardEvaluation(**in** : N; **in/out** : $V_{ch}, \mathcal{L}_b$)
```
01:    if N is a leaf or V_ch[N] = 1 then return;
02:    for each child C of node N do ForwardEvaluation(C, V_ch, 𝓛ᵦ);
03:    if N = G then return;
04:    FE(N, [f]); /* f is the operator at N */
05:    V_ch[N] := 1; /* the range of this node is cached */
06:    if 𝒩(N) = ∅ then return infeasible;
07:    if 𝒩(N) changed enough for backward propagation then put C into 𝓛ᵦ;
end
```

procedure FindVirtualEdges(**in** : N, V; **in/out** : $\mathcal{V}$)
```
01:    if N is a leaf and m_N ≠ 0 then put N into the set of children of V;
02:    if t is explicitly needed in the calculation of y_N then add N to 𝒱;
03:    for each child C of node N do FindVirtualEdges(C, V, 𝒱);
end
```

procedure NodeLevel(**in** : **N**; **out** : V_{lvl})
01:        **for each** child **C** of node **N do**
02:            $V_{lvl}[\mathbf{C}] := \max\{V_{lvl}[\mathbf{C}], V_{lvl}[\mathbf{N}] + 1\}$;
03:            NodeLevel($\mathbf{C}, V_{lvl}$);
04:        **end-for**
end

Apart from the virtual nodes and constraints the layout of the PFBPD algorithm is analogous to the FBPD algorithm from [33, 34]. The main difference lies in the forward and backward propagation operators FE($\mathbf{N}, [f]$) and BP($\mathbf{N}, \mathbf{C}$). The aim of *forward evaluation* FE is the reduction of $\mathcal{N}(\mathbf{N})$ of the node **N** based on the known $\mathcal{N}(\mathbf{C})$ for all children **C** of **N**. It is performed by first calculating the $\overline{\mathbb{UR}}$ and the $\mathbb{IPR}$ parts of $\mathcal{N}(\mathbf{N}) = (\mathbf{x}, (\widehat{\mathbf{x}}, m; \mathbf{t}))$ separately by interval extension functions of f. Immediately thereafter the internal intersection $\mathcal{N}'(\mathbf{N}) = (\mathbf{x}', (\widehat{\mathbf{x}}', m; \mathbf{t}'))$ is computed as follows:

$$
\begin{aligned}
\mathbf{x}' &= \mathbf{x} \cap \mathbf{t}^{-m}\widehat{\mathbf{x}} \\
\widehat{\mathbf{x}}' &= \widehat{\mathbf{x}} \cap \mathbf{t}^{m}\mathbf{x} \\
\mathbf{t}' &= \mathbf{t} \cap (\mathbf{x}/\widehat{\mathbf{x}})^{-1/m} \cap (\widehat{\mathbf{x}}/\mathbf{x})^{1/m}.
\end{aligned}
\tag{16}
$$

The backward propagation BP is concerned with reducing the sets $\mathcal{N}(\mathbf{C}_i)$ of all children $\mathbf{C}_i$ of **N** using $\mathcal{N}(\mathbf{N})$ and all $\mathcal{N}(\mathbf{C}_j)$ for all other children $\mathbf{C}_j$ with $j \neq i$. Again, first the $\overline{\mathbb{UR}}$ and $\mathbb{IPR}$ parts are calculated separately by inclusion extensions of the partial inverse functions, followed by an internal intersection operation (16).

Like FBPD the PFBPD algorithm is contractive and complete in the following sense.

Proposition 1. *We define a function* $P : (\overline{\mathbb{UR}} \times \mathbb{IPR})^n \times 2^{\mathbb{R}^n} \to (\overline{\mathbb{UR}} \times \mathbb{IPR})^n$ *to represent the* PFBPD *algorithm. This function takes as input the variable domains* **B** *(in form of a combined interval-union enclosure and an enclosure of the projective transformation) and the exact solution set* S *of the input problem. The function* P *returns an enclosure, denoted by* $P(\mathbf{B}, S)$, *that represents the variable domains of the output of the* PFBPD *algorithm, again for the original and the projectively transformed problem. If the input problem is factorable, then the* PFBPD *algorithm stops after a finite number of iterations and the following properties hold:*

(*i*) $P(\mathbf{B}, S) \subseteq \mathbf{B}$ *(Contractiveness)*
(*ii*) $P(\mathbf{B}, S) \supseteq \mathbf{B} \cap S$ *(Completeness)*

The proof is completely analogous to [34, Proposition 2].

3   Asymptotic Transformation

As mentioned in Section 2 for MICSPs involving transcendental functions, the projective transformation does not necessarily lead to bounded internal variables $\widehat{y}_\nu$ for all nodes $\nu \in V(\Gamma)$. Therefore, we have developed a more general transformation, based on asymptotic analysis.

Let $\Psi \subseteq C(\mathbb{R}^2, \mathbb{R})$ be a subset of functions $\psi(x, t; \alpha)$ depending on the real parameter vector $\alpha \in \mathbb{R}^n$.

For the asymptotic transformation we enclose each intermediate node y_ν for $\nu \in V(\Gamma)$ in the form

$$\psi(\widehat{y}_\nu, t; \underline{\alpha}_\nu) \le y_\nu \le \psi(\widehat{y}_\nu, t; \overline{\alpha}_\nu). \tag{17}$$

The new variable t and the parameters $\underline{\alpha}_\nu$ and $\overline{\alpha}_\nu$ are defined such that $t \in [0, 1]$ and the $\widehat{y}_\nu$ are well-bounded. Clearly, the projective transformation is a special case of that scheme, by choosing $\psi(x, t, \alpha) := x/t^\alpha$ with $\underline{\alpha} = \overline{\alpha} = \alpha \in \mathbb{Q}$.

The transformation is like in the projective case achieved by a recursive construction, implemented in a forward walk through Γ. For the original variables x_k and all interval constants, we define a map $f : \overline{\mathbb{IR}} \to \overline{\mathbb{IR}} \times \mathbb{R}^n \times \mathbb{R}^n$ with the property that for all $t \in [0, 1]$ there exists a $\widehat{x}_k \in f_1(\mathbf{x}_k)$ with

$$\psi(\widehat{x}_k, t; f_2(\mathbf{x}_k)) \le x_k \le \psi(\widehat{x}_k, t; f_3(\mathbf{x}_k)), \tag{18}$$

for all $x_k \in \mathbf{x}_k$.

The parameters $\underline{\alpha}_\nu$ and $\overline{\alpha}_\nu$ for an intermediate variable y_ν depend on the operation that creates it. If $y = g(y_{\nu_1}, \dots, y_{\nu_\ell})$ then we must have $\widehat{\mathbf{y}}$ such that

$$\psi(\widehat{y}, t; \underline{\alpha}) \le g(\psi(\widehat{y}_{\nu_1}, t; \mathbf{a}_{\nu_1}), \dots, \psi(\widehat{y}_{\nu_\ell}, t; \mathbf{a}_{\nu_\ell})) \le \psi(\widehat{y}, t; \overline{\alpha}), \tag{19}$$

for $\mathbf{a}_{\nu_i} = [\underline{\alpha}_{\nu_i}, \overline{\alpha}_{\nu_i}]$, all $\widehat{y}_\nu \in \widehat{\mathbf{y}}_\nu$, and some $\widehat{y} \in \widehat{\mathbf{y}}$, and the inequalities should be as tight as possible, $\widehat{\mathbf{y}}$ should be bounded, and there should be a simple way to calculate it for all elementary operations g.

If we have a constraint $y \in \mathbf{y}$ we transform it to the two constraints

$$\psi(\widehat{y}, t; \underline{\alpha}) \le \overline{\mathbf{y}}$$
$$\psi(\widehat{y}, t; \overline{\alpha}) \ge \underline{\mathbf{y}},$$

ensuring that the transformed problem is a relaxation of the original problem.

Example 3. A very useful set of functions is $\Psi := \{\psi(x, t; \alpha) := xt^{-\alpha_1} e^{\alpha_2 t^{-\alpha_3}} \mid \alpha \in \mathbb{R}^3\}$. The corresponding transformation is then

$$y_\nu = \widehat{y}_\nu t^{-\alpha_{\nu,1}} e^{\alpha_{\nu,2} t^{-\alpha_3}},$$

where $\overline{\alpha} = \underline{\alpha} = \alpha$. For constructing the $\widehat{y}_\nu$ we choose again a real number $0 \le s \le 1$ and set

$$t := \left(1 - s + \sum_{k \in K} d_k x_k^2\right)^{-1/2} \tag{20}$$

with scaling factors $d_k > 0$, like in the projective case. This again leads to the constraint (5).

The parameters α_ν for an intermediate variable y_ν depend on the operation that creates it. If, e.g., $y = \sum \beta_\nu y_\nu$ then $y = \widehat{y} t^{-\alpha_1} e^{\alpha_2 t^{-\alpha_3}}$ with

$$\alpha_1 := \max \alpha_{\nu,1}, \quad \alpha_3 := \max \alpha_{\nu,3}, \quad \alpha_2 := \max\{\alpha_{\nu,2} \mid \alpha_{\nu,3} = \alpha_3\},$$
$$\widehat{y} := \sum \beta_\nu t^{\alpha_1 - \alpha_{\nu,1}} e^{-\alpha_2 t^{-\alpha_3}(1 - \frac{\alpha_{\nu,2}}{\alpha_2} t^{\alpha_3 - \alpha_{\nu,3}})} \widehat{y}_\nu, \tag{21}$$

and we get the finite enclosure

$$\widehat{y} \in \widehat{\mathbf{y}} := \sum \beta_\nu \mathbf{t}^{\alpha_1 - \alpha_{\nu,1}} e^{-\alpha_2 t^{-\alpha_3}} (1 - \tfrac{\alpha_{\nu,2}}{\alpha_2} \mathbf{t}^{\alpha_3 - \alpha_{\nu,3}}) \widehat{\mathbf{y}}_\nu. \tag{22}$$

Note that the lower bound of the exponential term is 0 by construction.

If $y = \prod y_\nu^{\beta_\nu}$ with real β_ν then $y = \widehat{y} t^{-\alpha_1} e^{\alpha_2 t^{-\alpha_3}}$ with

$$
\begin{aligned}
\alpha_1 &:= \sum \beta_\nu \alpha_{\nu,1}, \quad \alpha_3 := \max \alpha_{\nu,3}, \\
\alpha_2 &:= \max \beta_\nu \alpha_{\nu,2}, \quad \widehat{y} := e^{\sum \frac{\beta_\nu \alpha_{\nu,2}}{\alpha_2} t^{\alpha_3 - \alpha_{\nu,3}}} \prod \widehat{y}_\nu,
\end{aligned} \tag{23}
$$

and we get the finite enclosure

$$\widehat{y} \in \widehat{\mathbf{y}} := e^{\sum \frac{\beta_\nu \alpha_{\nu,2}}{\alpha_2} \mathbf{t}^{\alpha_3 - \alpha_{\nu,3}}} \prod \widehat{\mathbf{y}}_\nu. \tag{24}$$

This accounts for all elementary operations and powers with fixed exponent.

For other elementary functions, one can again derive similar formulas, which are rather complex. E.g., if $y = \log y_\nu$ then $y = \widehat{y} t^{-\alpha_1} e^{\alpha_2 t^{-\alpha_3}}$ with

$$
\begin{aligned}
\alpha_1 &:= \alpha_{\nu,3} + \delta, \quad \alpha_2 := 0 \\
\alpha_3 &:= 0, \qquad \widehat{y} := t^{\alpha_1 + \delta}(\log \widehat{y}_\nu - \alpha_{\nu,1} \log t + \alpha_{\nu,2}),
\end{aligned}
$$

and $\delta = 0$ for $\alpha_1 > 0$, and $\delta = \varepsilon - \alpha_1$ for some small $\varepsilon > 0$, if $\alpha_1 \leq 0$, providing a finite enclosure for $\widehat{y}$.

For $y = \exp(\beta y_\nu^\gamma)$ we find $y = \widehat{y} t^{-\alpha_1} e^{\alpha_2 t^{-\alpha_3}}$ with

$$\alpha_1 = 0, \quad \alpha_2 = \sup(\widehat{\mathbf{y}}_\nu^\gamma), \quad \alpha_3 = \alpha_{\nu,1} \gamma$$

$$\widehat{y} = e^{t^{-\alpha_1}(\widehat{y}^\gamma e^{\gamma \alpha_{\nu,2} t^{-\alpha_{\nu,3}}} - \alpha_2)},$$

giving the enclosure

$$\widehat{y} \in \widehat{\mathbf{y}} := e^{\mathbf{t}^{-\alpha_1}(\widehat{\mathbf{y}}^\gamma e^{\gamma \alpha_{\nu,2} \mathbf{t}^{-\alpha_{\nu,3}}} - \alpha_2)},$$

which is finite for $\alpha_2 \leq 0$ or $\alpha_3 \leq 0$.

This asymptotic transformation, therefore, can also cope with exponentials, as long as they are not nested.

An analysis of the DAG Γ can provide information about which asymptotic transformation is most useful for transforming the MICSP to a bounded form. Of course, a generalization of $\mathbb{IPR}$ to a more general set of *asymptotic intervals* implementing the above operations provides an algorithm analogous to PFBPD.

4   Projective and Asymptotic Relaxations

A very useful tool for solving MICSPs are relaxations of all kind. There are many different classes of relaxations utilized—linear, mixed-integer linear, convex quadratic, semidefinite, general convex to name only the most important

Table 1. Comparison of PFBPD, FBPD, and HC4, easy problems

#var	#problems	PFBPD		PFBPD $I\overline{PR}$ only		PFBPD $\overline{UR}$ only		FBPD		HC4	
		#solved	Σmsec	#solved	Σmsec	#solved	Σmsec	#solved	Σmsec	#solved	Σmsec
1	4	4	1.30	4	1.70	2	0.70	1	0.20	0	20.20
2	112	112	40.90	102	26.00	22	21.20	5	8.00	7	545.60
3	104	57	75.80	28	38.50	20	33.10	7	14.90	8	1354.10
4	70	41	87.60	24	41.70	22	33.20	4	12.10	6	1212.60
5	68	42	89.30	17	43.80	14	34.20	7	16.20	8	1294.20
6	48	23	90.20	13	38.30	11	28.80	6	10.10	6	1404.40
7	18	6	34.90	2	12.90	1	12.00	1	3.90	1	444.50
8	43	20	178.70	5	43.60	5	38.90	2	22.70	3	1697.20
9	24	16	52.20	10	25.60	10	18.90	0	6.80	1	949.60
10	36	24	66.00	10	28.60	9	20.80	7	9.30	7	980.60
11–15	32	24	54.20	6	26.90	7	21.50	5	7.40	5	667.50
16–20	17	13	38.90	3	37.20	1	16.20	1	10.10	0	737.30
21–30	25	20	116.50	3	49.60	3	40.70	2	17.00	2	1421.70
31–46	15	12	80.10	2	174.00	3	32.90	3	40.50	3	2186.10

ones, see [20]. Most of these relaxations come in two flavors: They can be of reformulation type, like reformulation linearization [14, 15, 22], and be much higher dimensional than the original problem. They can also be dimension preserving, like the ones in [21, 27]. However, usually the computation of the relaxations requires that all variables are bounded.

This problem can be overcome by computing a relaxation of a suitably transformed problem, like the projectively transformed problem of Section 2 or the asymptotically transformed problem of Section 3. Even for mixed-integer problems the relaxations have the additional advantage that they are continuous problems. Hence, the transformations can be readily applied.

Reformulation type relaxations can be computed directly from the structure of the operators separately for each node $\nu \in V(\Gamma)$. Dimension preserving relaxations are usually computed by algorithmic differentiation techniques. Those can be generalized to projective or asymptotic intervals by careful examination of the differentiation rules and the properties of first and second order slopes [26].

5   Numerical Results

We tested all global optimization and constraint satisfaction problems of the COCONUT test set [28] of dimensions up to 50. They are of general structure

$$\min f(x)$$
$$\text{s.t. } F(x) \in \mathbf{F},\ x \in \mathbf{x}.$$

Of those 865 test problems 15 failed for various reasons (e.g. missing operators, local optimization failed, . . .). Of the remaining 850 problems 663 contained at least one unbounded variable. We used local optimization to find at least one local minimum $\tilde{x}$ with objective function value $\tilde{f}$. Then we added the constraint $f(x) \leq \tilde{f}$ on the objective function for converting the global optimization problem to a CSP. Then we tried to exclude the region where all unbounded variables are outside the box $[-1000, 1000]$.

Table 2. Comparison of PFBPD, FBPD, and HC4, complex problems

Name/Lib.	#var	PFBPD		PFBPD IPR only		PFBPD UR only		FBPD		HC4	
		Res.	msec	Res.	msec	Res.	msec	Res.	msec	Res.	msec
esfl/3	2	I	4909.60	I	1702.70		1610.00	I	181.20	I	11.40
pt/2	2	I	27.60	I	12.60		11.20		3.10		818.10
sipow1/2	2	I	514.20	I	264.10		201.30		58.90		314635.00
sipow1m/2	2	I	517.60	I	259.70		206.70		56.80		300657.00
sipow2/2	2	I	253.60	I	121.10		94.90		23.60		73982.50
sipow2m/2	2	I	256.50	I	127.00		99.80		22.60		71689.80
gulf/2	3		55.10		21.90		18.20		2.70		222.20
oet1/2	3	I	83.20		40.20		31.00	.	92.80		3403.70
oet2/2	3	I	80.30		34.50		28.60		28.70		3312.80
tfi2/2	3		849.70		402.90		358.80		84.20		288961.00
fourbar/3	4		20.20		6.90		7.00		1.90		101.00
oet3/2	4	I	126.60		49.40		45.30		19.60		3848.10
sipow3/2	4	I	1037.30		391.30		317.90		263.50		318190.00
sipow4/2	4		1036.70		459.80		439.00		112.10		315615.00
cpdm5/3	5		17.10		6.60		5.40		1.60		90.90
expfitb/2	5		29.10		11.30		9.10		2.10		333.30
expfitc/2	5		400.20		158.50		115.70		19.40		1959.40
rbpl/3	6	I	22.50		7.80		7.00		4.30		90.90
oet7/2	7		173.30		83.90		67.90		17.10		6352.90
arglinb/2	10		14.60		1.60		1.40		1.50		90.90
fir_convex/3	11		181.10		60.10		52.90		17.60		1343.30
osborneb/2	11		13.40		7.70		5.80		1.30		464.60
watson/3	12	I	35.20		4.00		3.60		7.50		272.70
ex2_1_10/1	20	I	23.40		8.20		6.70		0.70		70.70
ex2_1_7/1	20	I	18.30		6.60		5.40		1.60		60.60
ksip/2	20		2174.20		695.40		732.40		233.30		11392.80
antenna2/3	24		3552.70		1135.30		1040.90		365.60		14725.80
himmelbk/2	24	I	47.00		16.00		16.30		5.00		141.40
3pk/2	30	I	17.20		3.90		3.60		1.90		80.80
loadbal/2	31	I	16.70		7.50		6.10		2.80		101.00
lowpass/3	31		2760.00		884.80		843.40		285.40		7908.30
watson/2	31	I	183.70		6.60		6.40		48.30		717.10
hs088/2	32	I	1358.10		408.30		443.80		253.60		323.20
hs089/2	33	I	1379.60		453.90		474.20		262.30		363.60
hs090/2	34	I	1373.10		427.40		420.60		243.10		484.80
hs091/2	35	I	1390.70		496.00		458.40		237.80		484.80
hs092/2	36	I	1275.40		436.80		437.00		238.80		676.70
chemeq/3	38	I	15.70		6.10		5.20		2.50		151.50
polygon2/3	38	I	41.40		13.20		12.00		3.90		414.10
srcpm/1	38	I	13.20		4.60		3.80		0.60	I	0.70
gridnetg/2	44	I	49.60		15.20		13.60		4.60		141.40
chnrosnb/2	50	I	53.10		17.80		16.10		2.70		80.80
errinros/2	50	I	53.90		17.70		16.20		4.10		70.70
hilbertb/2	50		9529.80		2986.50		3034.10		898.80		818.10
qp1/1	50	I	9407.00		3139.90		2906.00		976.70		575.70
qp2/1	50	I	9589.70		3115.90		2910.20		993.30		606.00
tointqor/2	50	I	39.70		12.80		11.50		2.00		70.70

These 663 constructed CSPs constitute our test set. As can be deduced from Table 3, for 446 of the problems PFBPD was able to prove infeasibility of the problem, effectively reducing the problem to the standard search box $[-1000, 1000]^n$ of global optimization algorithms like BARON. Using only projective intervals solved just 235 of the 675 problems, while pure interval union arithmetic proved infeasibility of only 130 problems. Of those problems 122 are solved by both methods, so they can be considered easy. It is thus indeed important to combine interval unions and projective intervals performing internal intersection after each operation, as described in Algorithm PFBPD. All calculations are performed in a completely rigorous way with full rounding error control.

Table 3. Result summary for PFBPD

case	1	2	3	4	5
$\mathbb{UR}$ only	+	+	−	−	−
$\mathbb{IPR}$ only	+	−	+	−	−
PFBPD	+	+	+	+	−
#problems	122	8	113	203	217

Overall performance of PFBPD is very strong; it is comparable to FBPD being just a factor of 5-10 on average slower than the interval version and about half as fast as the interval union version of FBPD. It is still orders of magnitude faster than HC4 [2] and many other numerical CP algorithms, as they were tested in [34]. However, there are exceptions like srcpm where HC4 performs faster and can still prove infeasibility. Detailed results can be found in Table 1, summarizing all easy problems with solution times up to 12 ms, and Table 2, containing the remaining problems. A result of I in Table 2 means that infeasibility was proved by the corresponding propagator for the respective problem. The running times were measured on an Intel Core i7 Q 720 running at 1.60GHz running Linux 3.6.11.

6   Conclusion

We provided several methods for solving MICSPs for which some variables have unbounded domains. In a large numerical test we showed effectiveness of this new approach.

References

1. Benhamou, F., Goualard, F.: Universally Quantified Interval Constraints. In: Dechter, R. (ed.) CP 2000. LNCS, vol. 1894, pp. 67–82. Springer, Heidelberg (2000)
2. Benhamou, F., Goualard, F., Granvilliers, L., Puget, J.F.: Revising Hull and Box Consistency. In: Proceedings of the International Conference on Logic Programming (ICLP 1999), Las Cruces, USA, pp. 230–244 (1999)
3. Benhamou, F., Older, W.J.: Applying Interval Arithmetic to Real, Integer and Boolean Constraints. Journal of Logic Programming, 32–81 (1997)
4. Berz, M., Makino, K.: Verified integration of odes and flows using differential algebraic methods on high-order taylor models. Reliable Computing 4, 361–369 (1998)
5. Berz, M.: COSY INFINITY version 8 reference manual. Technical report, National Superconducting Cyclotron Lab., Michigan State University, East Lansing, Mich., MSUCL–1008 (1997)
6. Domes, F.: Gloptlab-a configurable framework for solving continuous, algebraic CSPs. In: IntCP, Int. WS on Interval Analysis, Constraint Propagation, Applications, at CP Conference, pp. 1–16 (2009)
7. Domes, F.: Gloptlab: a configurable framework for the rigorous global solution of quadratic constraint satisfaction problems. Optimization Methods & Software 24(4-5), 727–747 (2009)

8. Domes, F., Neumaier, A.: Verified global optimization with gloptlab. PAMM 7(1), 1020101–1020102 (2008)

9. Eiermann, M.C.: Adaptive Berechnung von Integraltransformationen mit Fehlerschranken. PhD thesis, Institut für Angewandte Mathematik der Albert–Ludwigs–Universität Freiburg im Breisgau (October 1989)

10. Granvilliers, L., Goualard, F., Benhamou, F.: Box Consistency through Weak Box Consistency. In: Proceedings of the 11th IEEE International Conference on Tools with Artificial Intelligence (ICTAI 1999), pp. 373–380 (November 1999)

11. Griewank, A., Corliss, G.F.: Automatic Differentiation of Algorithms. SIAM Publications, Philadelphia (1991)

12. Hansen, E.: Global Optimization using Interval Analysis. Marcel Dekker, New York (1992)

13. Jaulin, L., Kieffer, M., Didrit, O., Walter, E.: Applied Interval Analysis, 1st edn. Springer (2001)

14. Kearfott, R.: Decomposition of arithmetic expressions to improve the behavior of interval iteration for nonlinear systems. Computing 47(2), 169–191 (1991)

15. McCormick, G.: Computability of global solutions to factorable nonconvex programs: Part iconvex underestimating problems. Mathematical Programming 10(1), 147–175 (1976)

16. Moore, R.E.: Interval Arithmetic and Automatic Error Analysis in Digital Computing. PhD thesis, Appl. Math. Statist. Lab. Rep. 25. Stanford University (1962)

17. Moore, R.E.: Interval Analysis. Prentice-Hall, Englewood Cliffs (1966)

18. Neumaier, A.: Interval Methods for Systems of Equations. Cambridge University Press, Cambridge (1990)

19. Neumaier, A.: Taylor forms - use and limits. Reliable Computing 9, 43–79 (2002)

20. Neumaier, A.: Complete search in continuous global optimization and constraint satisfaction. Acta Numerica 13(1), 271–369 (2004)

21. Ninin, J., Messine, F., Hansen, P.: A reliable affine relaxation method for global optimization (2010); Optimzation Online

22. Ryoo, H., Sahinidis, N.: A branch-and-reduce approach to global optimization. Journal of Global Optimization 8(2), 107–138 (1996)

23. Schichl, H.: Global optimization in the COCONUT project. In: Alt, R., Frommer, A., Kearfott, R.B., Luther, W. (eds.) Numerical Software with Result Verification. LNCS, vol. 2991, pp. 243–249. Springer, Heidelberg (2004)

24. Schichl, H., Markót, M.C., et al.: The COCONUT Environment. Software, http://www.mat.univie.ac.at/coconut-environment

25. Schichl, H., Markót, M.C.: Algorithmic differentiation techniques for global optimization in the coconut environment. Optimization Methods and Software 27(2), 359–372 (2012)

26. Schichl, H., Neumaier, A.: Exclusion regions for systems of equations. SIAM Journal on Numerical Analysis 42(1), 383–408 (2004)

27. Schichl, H., Neumaier, A.: Interval analysis on directed acyclic graphs for global optimization. Journal of Global Optimization 33(4), 541–562 (2005)

28. Shcherbina, O., Neumaier, A., Sam-Haroud, D., Vu, X.H., Nguyen, T.V.: Benchmarking global optimization and constraint satisfaction codes. In: Bliek, C., Jermann, C., Neumaier, A. (eds.) COCOS 2002. LNCS, vol. 2861, pp. 211–222. Springer, Heidelberg (2003)

29. Silaghi, M.-C., Sam-Haroud, D., Faltings, B.V.: Search Techniques for Non-linear Constraint Satisfaction Problems with Inequalities. In: Stroulia, E., Matwin, S. (eds.) AI 2001. LNCS (LNAI), vol. 2056, pp. 183–193. Springer, Heidelberg (2001)

30. Stolfi, J., Andrade, M., Comba, J., Van Iwaarden, R.: Affine arithmetic: a correlation-sensitive variant of interval arithmetic, Web document (1994)
31. Van Hentenryck, P.: Numerica: A Modeling Language for Global Optimization. In: Proceedings of IJCAI 1997 (1997)
32. Vu, X.H., Sam-Haroud, D., Silaghi, M.C.: Numerical Constraint Satisfaction Problems with Non-isolated Solutions. In: Bliek, C., Jermann, C., Neumaier, A. (eds.) COCOS 2002. LNCS, vol. 2861, pp. 194–210. Springer, Heidelberg (2003)
33. Vu, X., Schichl, H., Sam-Haroud, D.: Using directed acyclic graphs to coordinate propagation and search for numerical constraint satisfaction problems. In: 16th IEEE International Conference on Tools with Artificial Intelligence, ICTAI 2004, pp. 72–81. IEEE (2004)
34. Vu, X., Schichl, H., Sam-Haroud, D.: Interval propagation and search on directed acyclic graphs for numerical constraint solving. Journal of Global Optimization 45(4), 499–531 (2009)

Explaining Time-Table-Edge-Finding Propagation for the Cumulative Resource Constraint

Andreas Schutt, Thibaut Feydy, and Peter J. Stuckey

Optimisation Research Group, National ICT Australia, and Department of
Computing and Information Systems, The University of Melbourne,
Victoria 3010, Australia
{andreas.schutt,thibaut.feydy,peter.stuckey}@nicta.com.au

Abstract. Cumulative resource constraints can model scarce resources
in scheduling problems or a dimension in packing and cutting problems.
In order to efficiently solve such problems with a constraint programming solver, it is important to have strong and fast propagators for cumulative resource constraints. Time-table-edge-finding propagators are
a recent development in cumulative propagators, that combine the current resource profile (time-table) during the edge-finding propagation.
The current state of the art for solving scheduling and cutting problems involving cumulative constraints are lazy clause generation solvers,
i.e., constraint programming solvers incorporating nogood learning, have
proved to be excellent at solving scheduling and cutting problems. For
such solvers, concise and accurate explanations of the reasons for propagation are essential for strong nogood learning. In this paper, we develop a time-table-edge-finding propagator for `cumulative` that explains
its propagations. We give results using this propagator in a lazy clause
generation system on resource-constrained project scheduling problems
from various standard benchmark suites. On the standard benchmark
suite PSPLib, we are able to improve the lower bound of about 60% of
the remaining open instances, and close 6 open instances.

1 Introduction

A cumulative resource constraint models the relationship between a scarce
resource and activities requiring some part of the resource capacity for their
execution. Resources can be workers, processors, water, electricity, or, even, a
dimension in a packing and cutting problem. Due to its relevance in many industrial scheduling and placement problems, it is important to have strong and
fast propagation techniques in constraint programming (CP) solvers that detect inconsistencies early and remove many invalid values from the domains of
the variables involved. Moreover, when using CP solvers that incorporate "fine-
grained" nogood learning it is also important that each inconsistency and each
value removal from a domain is explained in such a way that the full strength
of nogood learning is exploited.

C. Gomes and M. Sellmann (Eds.): CPAIOR 2013, LNCS 7874, pp. 234–250, 2013.

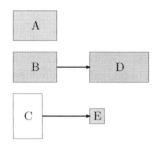

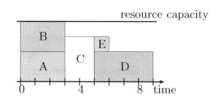

Fig. 1. Five activities with precedence relations

Fig. 2. A possible schedule of the activities

In this paper, we consider *renewable* resources, *i.e.*, resources with a constant resource capacity over time, and *non-preemptive* activities, *i.e.*, whose execution cannot be interrupted, with fixed processing times and resource usages. In this work, we develop explanations for the time-table-edge-finding (TTEF) propagator [34] for use in lazy clause generation (LCG) solvers [22,10].

Example 1. Consider a simple cumulative resource scheduling problem. There are 5 activities A, B, C, D, and E to be executed before time period 10. The activities have processing times 3, 3, 2, 4, and 1, respectively, with each activity requiring 2, 2, 3, 2, and 1 units of resource, respectively. There is a resource capacity of 4. Assume further that there are precedence constraints: activity B must finish before activity D begins, written $B \ll D$, and similarly $C \ll E$. Figure 1 shows the five activities and precedence relations, while Fig. 2 shows a possible schedule, where the start times are: 0, 0, 3, 5, and 5 respectively.

In CP solvers, a cumulative resource constraint can be modelled by a decomposition or, more successfully, by the global constraint `cumulative` [2]. Since the introduction of this global constraint, a great deal of research has investigated stronger and faster propagation techniques. These include time-table [2], (extended) edge-finding [21,33], not-first/not-last [21,29], and energetic-reasoning propagation [4,6]. Time-table propagation is usually superior for *highly disjunctive* problems, *i.e.*, in which only some activities can run concurrently, while (extended) edge-finding, not-first/not-last, and energetic reasoning are more appropriate for *highly cumulative* problems, *i.e.*, in which many activities can run concurrently [4]. The reader is referred to [6] for a detailed comparison of these techniques.

Vilím [34] recently developed TTEF propagation which combines the time-table and (extended) edge-finding propagation in order to perform stronger propagation while having a low runtime overhead. Vilím [34] shows that on a range of highly disjunctive open resource-constrained project scheduling problems from the well-established benchmark library PSPLib,[1] TTEF propagation can generate lower bounds on the project deadline (*makespan*) that are superior to those

---

[1] See http://129.187.106.231/psplib/

found by previous methods. He uses a CP solver without nogood learning. This result, and the success of LCG on such problems, motivated us to study whether an explaining version of this propagation yields an improvement in performance for LCG solvers.

In general, nogood learning is a resolution step that infers redundant constraints, called *nogoods*, given an inconsistent solution state. These nogoods are permanently or temporarily added to the initial constraint system in order to reduce the search space and/or to guide the search. Moreover, they can be used to short circuit propagation. How this resolution step is performed is dependent on the underlying system.

LCG solvers employ a "fine-grained" nogood learning system that mimics the learning of modern Boolean satisfiability (SAT) solvers (see e.g. [20]). In order to create a strong nogood, it is necessary that each inconsistency and value removal is explained concisely and in the most general way possible. For LCG solvers, we have previously developed explanations for time-table and (extended) edge-finding propagation [26]. Moreover, for time-table propagation we have also considered the case when processing times, resource usages, and resource capacity are variable [24]. Explanations for the time-table propagator were successfully applied on resource-constraint project scheduling problems [26,27] and carpet cutting [28] where in both cases the state-of-the-art of exact solution methods were substantially improved.

Explanations for the propagation of the `cumulative` constraint have also been proposed for the PaLM [14,13] and SCIP [1,7,12] frameworks. In the PaLM framework, explanations are only considered for time-table propagation, while the SCIP framework additionally provides explanations for energetic reasoning propagation and a restricted version of edge-finding propagation. Neither framework consider bounds widening in order to generalise these explanations as we do in this paper. Other related works include [32], which presents explanations for different propagation techniques for problems only involving disjunctive resources, *i.e.*, cumulative resources with unary resource capacity, and generalised nogoods [15]. A detailed comparison of explanations for the propagation of `cumulative` resource constraints in LCG solvers can be found in [24].

The contributions of this paper are:

- We define a new simpler TTEF propagator for `cumulative`.
- We define how to explain the propagation of this propagator.
- We compare the performance of the TTEF propagator with explanation, against time-table propagation with explanation.
- We improve the lower bounds of a large proportion of the open instances in the well studied PSPlib, and close 6 instances from PSPlib.
- We improve the lower bounds and close many more instances on less studied highly cumulative benchmarks.

2   Cumulative Resource Scheduling

In cumulative resource scheduling, a set of (non-preemptive) activities V and one cumulative resource with a (constant) resource capacity R is given where an *activity* i is specified by its *start time* S_i, its *processing time* p_i, its *resource usage* r_i, and its *energy* $e_i := p_i \cdot r_i$. In this paper we assume each S_i is an integer variable and all others are assumed to be integer constants. Further, we define est_i (ect_i) and lst_i (lct_i) as the *earliest* and *latest* start (completion) time of i.

In this setting. the cumulative resource scheduling problem is defined as a constraint satisfaction problem that is characterised by the set of activities $\mathcal{V}$ and a cumulative resource with resource capacity R. The goal is to find a solution that assigns values from the domain to the start time variables S_i ($i \in \mathcal{V}$), so that the following conditions are satisfied.

$$est_i \leq S_i \leq lst_i, \qquad\qquad \forall i \in \mathcal{V}$$

$$\sum_{i \in \mathcal{V}: \tau \in [S_i, S_i + p_i)} r_i \leq R \qquad\qquad \forall \tau$$

where τ ranges over the time periods considered. Note that this problem is NP-hard [5].

We shall tackle problems including cumulative resource scheduling using CP with nogood learning. In a CP solver, each variable $S_i, i \in \mathcal{V}$ has an initial domain of possible values $D^0(S_i)$ which is initially $[est_i, lst_i]$. The solver maintains a current domain D for all variables. CP search interleaves propagation with search. The constraints are represented by propagators that, given the current domain D, creates a new smaller domain D' by eliminating infeasible values. The current *lower* and *upper bound* of the domain $D(S_i)$ are denoted by $lb(S_i)$ and $ub(S_i)$, respectively. For more details on CP see e.g. [23].

For a learning solver we also represent the domain of each variable S_i using Boolean variables $[\![S_i \leq v]\!], est_i \leq v < lst_i$. These are used to track the reasons for propagation and generate nogoods. For more details see [22]. We use the notation $[\![v \leq S_i]\!], est_i < v \leq lst_i$ as shorthand for $\neg[\![S_i \leq v - 1]\!]$, and treat $[\![v \leq S_i]\!], v \leq est_i$ and $[\![S_i \leq v]\!], v \geq lst_i$ as synonyms for *true*. Propagators in a learning solver must explain each reduction in the domain by building a clausal explanation using these Boolean variables.

3   TTEF Propagation

TTEF propagation was developed by Vilim [34]. The idea of TTEF propagation is to splits the treatment of activities into a fixed and free part. The former results from the activities' compulsory part whereas the latter is the remainder. The fixed part of an activity i is characterised by the length of its *compulsory part* $p_i^{TT} := \max(0, ect_i - lst_i)$ and its fixed energy $e_i^{TT} := r_i \cdot p_i^{TT}$. The free part has a processing time $p_i^{EF} := p_i - p_i^{TT}$, a latest start time $lst_i^{EF} := lst_i + p_i^{TT}$, and a free energy of $e_i^{EF} := e_i - e_i^{TT}$. An illustration of this is shown in Figure 3.

TTEF propagation reasons about the energy available from the resource and energy required for the execution of activities in specific time windows. Let $\mathcal{V}^{EF}$ be the set of activities with a non-empty free part $\{i \in \mathcal{V} \mid p_i^{EF} > 0\}$. The start and end times of these windows are determined by the earliest start and

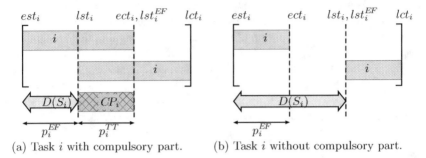

(a) Task i with compulsory part.   (b) Task i without compulsory part.

Fig. 3. A diagram illustrating an activity i when started at est_i or lst_i, and its possible range of start times, as well as the compulsory part CP_i (the hatched box), and the fixed and free parts of the processing time

the latest completion times of two activities $\{a, b\} \subseteq \mathcal{V}^{EF}$. These time windows $[begin, end)$ are characterised by the so-called *task intervals* $\mathcal{V}^{EF}(a, b) := \{i \in \mathcal{V}^{EF} \mid est_a \leq est_i \wedge lct_i \leq lct_b\}$ where $a, b \in \mathcal{V}^{EF}$, $begin := est_a$, and $end := lct_b$.

It is not only the free energy of activities in the task interval $\mathcal{V}^{EF}(a, b)$ that is considered, but also the energy resulting from the compulsory parts in the time window $[est_a, lct_b)$. This energy is defined by $ttEn(a, b) := ttAfter[est_a] - ttAfter[lct_b]$ where $ttAfter[\tau] := \sum_{t \geq \tau} \sum_{i \in \mathcal{V}: lst_i \leq t < ect_i} r_i$ is the total energy of all compulsory parts occurring at time τ and after.

Furthermore, we also consider activities $i \in \mathcal{V}^{EF} \setminus \mathcal{V}^{EF}(a, b)$ in which a portion of their free part must be run within the time window as described in [34]. Suppose activity i starts after est_a, i.e., $est_a \leq est_i$. Then activity i's free part consumes at least $r_i \cdot (lct_b - lst_i^{EF})$ energy units in $[est_a, lct_b)$ assuming $lst_i^{EF} < lct_b$. We define the energy contributed by such activities by $rsEn(a, b) := \sum_{i \in \mathcal{V}^{EF} \setminus \mathcal{V}^{EF}(a,b): est_a \leq est_i} r_i \cdot max(0, lct_b - lst_i^{EF})$. Note that this is a special case of energetic reasoning that is cheaper to compute.

In summary, TtEf propagation considers three ways in which an activity i can contribute to energy consumption within a time window determined by a task interval $\mathcal{V}^{EF}(a, b)$. First, the free parts that must fully be executed in the time window; second, some free parts that must partially be run in the time window, and third, the compulsory parts that must lie within the time window; Thus, the considered length of an activity i is

$$
p_i(a, b) := \begin{cases} p_i & i \in \mathcal{V}^{EF}(a, b) \\ max(0, lct_b - lst_i) & i \notin \mathcal{V}^{EF}(a, b) \wedge est_a \leq est_i \\ max(0, min(lct_b, ect_i) - max(est_a, lst_i)) & others \end{cases}
$$

The considered energy consumption is $e_i(a, b) := r_i \cdot p_i(a, b)$ in the time window. An illustration of the three cases is shown in Fig. 4(a).

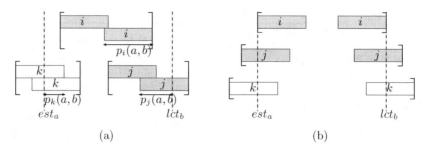

Fig. 4. (a) Diagram explaining the cases of energy contribution: task i is completely included in the task interval and its entire length is considered; task j starts after est_a, and the length from its latest start time to lct_b is considered; and task k has the intersection of its compulsory part with $[est_a, lct_b)$ considered. (b) Diagram illustrating how the bounds can be weakened in explanation, to still ensure that at least $p_l(a, b)$ for $l \in \{i, j, k\}$ is used with $[est_a, lct_b)$.

3.1   Explanation for the TTEF Consistency Check

The consistency check is one part of TTEF propagation that checks whether there is a resource overload in any task interval.

Proposition 1 (Consistency Check). *The cumulative resource scheduling problem is inconsistent if*

$$R \cdot (lct_b - est_a) - energy(a, b) < 0$$

where $energy(a, b) := \sum_{i \in \mathcal{V}^{EF}(a,b)} e_i^{EF} + ttEn(a, b) + rsEn(a, b)$.

This check can be done in $\mathcal{O}(l^2 + n)$ runtime, where $l = |\mathcal{V}^{EF}|$, if the resource profile is given.

The algorithm we use for the consistency check is shown in Alg. 1. It is different from that proposed by Vilim [34]. The main difference is that by iterating over the end times in decreasing order we can calculate the minimal available free energy $minAvail$ from the previous iteration. If the reduction in this free energy for the next iteration cannot make it negative we know that none of the task intervals in this iteration can lead to resource overload, and we can skip the entire set of task intervals. This optimization is highly useful high up in the search tree when there is little chance of resource overload. In preliminary experiments on instances with 30 activities, the number of checked task intervals could be reduced about 60% on average.

The algorithm iterates on each end time in decreasing order. For each end time the algorithm first checks if no propagation is possible with this end time (lines 5-6), and if so skips to the next. Otherwise it examines each possible start time, updating the free energy used E for the new task interval (lines 13-14), and calculating the energy available $avail$ in the task interval (line 15). If this is negative it explains the overload in the interval and returns *false*. If not

---

Algorithm 1. TTEF consistency check

Input: X an array of activities sorted in non-decreasing order of the earliest
        start time.
Input: Y an array of activities sorted in non-decreasing order of the latest
        completion time.

1 $end := \infty$; $minAvail := \infty$;
2 **for** $y := n$ **down to** 1 **do**
3       $b := Y[y]$;
4   **if** $lct_b = end$ **then** continue;
5   **if** $end \neq \infty$ **and** $minAvail \neq \infty$ **and**
        $minAvail \geq R \cdot (end - lct_b) - ttAfter[lct_b] + ttAfter[end]$ **then**
6   continue;
7       $end := lct_b$;
8       $E := 0$; $minAvail := \infty$;
9   **for** $x := n$ **down to** 1 **do**
10          $a := X[x]$;
11   **if** $end \leq est_a$ **then** continue;
12          $begin := est_a$;
13   **if** $lct_a \leq end$ **then** $E := E + e_a^{EF}$;
14   **else if** $lst_a^{EF} < end$ **then** $E := E + r_a \cdot (end - lst_a^{EF})$;
15          $avail := R \cdot (end - begin) - E - ttEn(a,b)$;
16   **if** $avail < 0$ **then**
17   explainOverload($begin, end$);
18   **return** *false*;
19   **if** $avail < minAvail$ **then** $minAvail := avail$;
20 **return** *true*;

---

it updates the minimum available energy and examines the next task interval
(line 19).

A naïve explanation for a resource overload in the time window $[est_a, lct_b)$
only considers the current bounds on activities' start times S_i.

$$\bigwedge_{i \in \mathcal{V}:p_i(a,b)>0} [\![est_i \leq S_i]\!] \wedge [\![S_i \leq lst_i]\!] \rightarrow \bot$$

However, we can easily generalise this explanation by only ensuring that at least
$p_i(a,b)$ time units are executed in the time window. This results in the following
explanation.

$$\bigwedge_{i \in \mathcal{V}:p_i(a,b)>0} [\![est_a + p_i(a,b) - p_i \leq S_i]\!] \wedge [\![S_i \leq lct_b - p_i(a,b)]\!] \rightarrow \bot$$

Figure 4(b) shows how the explanations are weakened for the tasks shown in
Figure 4(a). Note that this explanation expresses a resource overload with respect
to energetic reasoning propagation which is more general than TTEF.

Let $\Delta := energy(a,b) - R \cdot (lct_b - est_a) - 1$. If $\Delta > 0$ then the resource
overload has extra energy. We can use this extra energy to further generalise the

explanation, by reducing the energy required to appear in the time window by up to Δ. For example, if $r_i \leq \Delta$ then the lower and upper bound on S_i can simultaneously be decreased and increased by a total amount in $\{1, 2, ..., \min(\lfloor \Delta/r_i \rfloor, p_i(a, b))\}$ units without resolving the overload. If $r_i \cdot p_i(a, b) \leq \Delta$ then we can remove activity i completely from the explanation. In a greedy manner, we try to maximally widen the bounds of activities i where $p_i(a, b) > 0$, first considering activities with non-empty free parts. If Δ_i denotes the time units of the widening then it holds that $p_i(a, b) \geq \Delta_i \geq 0$ and $\sum_{i \in \mathcal{V}: p_i(a,b) > 0} \Delta_i \cdot r_i \leq \Delta$ and we create the following explanation.

$$\bigwedge_{i \in \mathcal{V}: p_i(a,b) - \Delta_i > 0} [\![est_a + p_i(a, b) - p_i - \Delta_i \leq S_i]\!] \wedge [\![S_i \leq lct_b - p_i(a, b) + \Delta_i]\!] \rightarrow \perp$$

The last generalisation mechanism can be performed in different ways, e.g. we could widen the bounds of activities that were involved in many recent conflicts. By default we generalise the tasks in order. We experimented with different policies, but found any reasonable generalization policy was equally effective.

3.2 Explanation for the TTEF Start Times Propagation

Propagation on the lower and upper bounds of the start time variables S_i are symmetric; consequently we only present the case for the lower bounds' propagation. To prune the lower bound of an activity u, TTEF bounds propagation tentatively starts the activity u at its earliest start time est_u and then checks whether that causes a resource overload in any time window $[est_a, lct_b)$ ($\{a, b\} \subseteq \mathcal{V}^{EF}$). Thus, bounds propagation and its explanation are very similar to that of the consistency check.

The work of Vilim [34] considers four positions of u relative to the time window: *right* ($est_a \leq est_u < lct_b < ect_u$), *inside* ($est_a \leq est_u < ect_u \leq lct_b$), *through* ($est_u < est_a \wedge lct_b < ect_u$), and *left* ($est_u < est_a < ect_u \leq lct_b$). The first two of these positions correspond to edge-finding propagation and the last two to extended edge-finding propagation. In this work we fully consider the *right* and *inside* positions, i.e., $est_a \leq est_u$ (note that a could be u), and only opportunistically consider the *through* and *left* positions.

The calculation of a *right* or *inside* bounds update of u with respect to the time interval $[est_a, lct_b)$ are identical. Then, the bounds update rule is

$$R \cdot (lct_b - est_a) - energy(a, b, u) < 0 \rightarrow est_a + \left\lceil \frac{rest(a, b, u)}{r_u} \right\rceil \leq S_u \quad (1)$$

where $energy(a, b, u) := energy(a, b) - e_u(a, b) + r_u \cdot (\min(lct_b, ect_u) - est_u)$ and

$$rest(a, b, u) := energy(a, b, u) - (R - r_u) \cdot (lct_b - est_a)$$
$$- r_u \cdot (\min(lct_b, ect_u) - est_u) \ .$$

The first two terms in the sum of $energy(a, b, u)$ give the energy consumption in the time window $[est_a, lct_b)$ of all considered activities except u, whereas the last term is the required energy of u in $[est_a, lct_b)$ if it is scheduled at est_u. The propagation, including explanation generation, can be performed in $\mathcal{O}(l^2 + k \cdot n)$ runtime, where $l = |\mathcal{V}^{EF}|$ and k is the number of bounds' updates, if the resource profile is given. Moreover, TTEF propagation does not necessarily consider each $u \in \mathcal{V}^{EF}$, but those only that maximise $\min(e_u^{EF}, r_u \cdot (lct_b - est_a)) - r_u \cdot \max(0, lct_b - lst_u^{EF})$ and satisfy $est_a \leq est_u$.

The pseudo-code for lower bounds propagation is shown in Algorithm 2. Similarly to the consistency check the task intervals are explored in an order using the latest completion time and all decreasing start times, before considering the next completion time.

If the global variable *opportunistic* is set to *true* then the algorithm first (lines 11-17) opportunistically searches for and records an upper bound change of the first task a if possible by using the calculated minimum available energy *minAvail* in the task interval $[minBegin, lct_b)$ where $compIn(minBegin, end, a)$ (line 12) is the energy of the compulsory part of a in that task interval. This upper bound change is an extended edge finding propagation. It then updates the free energy used E for the new task interval, and updates the task u which requires maximum energy $enReqU$ in the new task interval (lines 18-23). It then calculates the energy available *avail* in the task interval $[est_a, lct_b)$ (line 24), updates the interval with minimum available energy (lines 25-26) needed for the extended edge finding propagation, and records a lower bound change of the task u requiring most energy if this is possible (lines 27-33). Only after all task intervals are visited are the bounds actually changed by *updateBound*.

The procedure $explainUpdate(begin, end, v, oldbnd, newbnd)$ explains the bound change of v to $newbnd$ where $oldbnd$ is the old bound (the difference allows calculating which bound is being updated). A naïve explanation for a lower bound update from est_u to $newLB := \lceil rest(a, b, u)/r_u \rceil$ with respect to the time window $[est_a, lct_b)$ additionally includes the previous and new lower bound on the left and right hand side of the implication, respectively, in comparison to the naïve explanation for a resource overload.

$$[\![est_u \leq S_u]\!] \wedge \bigwedge_{i \in \mathcal{V} \setminus \{u\} : p_i(a,b) > 0} [\![est_i \leq S_i]\!] \wedge [\![S_i \leq lst_i]\!] \rightarrow [\![newLB \leq S_u]\!]$$

As we discussed in the case of resource overload, we perform a similar generalisation for the activities in $\mathcal{V} \setminus \{u\}$, and for u we decrease the lower bound on the left hand side as much as possible so that the same propagation holds when u is executed at that decreased lower bound.

$$[\![est_a + lct_b - newLB + 1 - p_u \leq S_u]\!] \wedge$$
$$\bigwedge_{i \in \mathcal{V} \setminus \{u\} : p_i(a,b) > 0} [\![est_a + p_i(a,b) - p_i \leq S_i]\!] \wedge [\![S_i \leq lct_b - p_i(a,b)]\!]$$
$$\rightarrow [\![newLB \leq S_u]\!] \quad (2)$$

Again this more general explanation expresses the energetic reasoning propagation and the bounds of activities in $\{i \in \mathcal{V} \setminus \{u\} \mid p_i(a, b) > 0\}$ can further be

---

Algorithm 2. TTEF lower bounds propagator on the start times

Input: X an array of activities sorted in non-decreasing order of the earliest start time.

Input: Y an array of activities sorted in non-decreasing order of the latest completion time.

1 **for** $i \in \mathcal{V}^{EF}$ **do** $est'_i := est_i$; $lst'_i := lst_i$;
2 $end := \infty$; $k := 0$;
3 **for** $y := n$ **down to** 1 **do**
4     $b := Y[y]$;
5   **if** $lct_b = end$ **then** continue;
6     $end := lct_b$; $E := 0$; $minAvail := \infty$; $minBegin := \infty$ $u := -\infty$; $enReqU := 0$;
7   **for** $x := n$ **down to** 1 **do**
8        $a := X[x]$;
9   **if** $end \leq est_a$ **then** continue;
10        $begin := est_a$;
11   **if** *opportunistic* **and** $minAvail \neq \infty$ **and** $minAvail < r_a \cdot (\min(end, lct_a) - \max(minBegin, lst_a^{EF}))$ **then**
12           $rest := minAvail + compIn(minBegin, end, a)$;
13           $ubA := minBegin + \lfloor rest/r_a \rfloor - p_a$;
14   **if** $lst'_a > ubA$ **then**
15              $expl := explainUpdate(minBegin, end, a, lst'_a, ubA)$;
16              $Update[++k] := (a, \mathtt{ub}, ubA, expl)$;
17              $lst'_a := ubA$;
18   **if** $lct_a \leq end$ **then** $E := E + e_a^{EF}$;
19   **else**
20           $enIn := r_a \cdot \max(0, end - lst_a^{EF})$;
21           $E := E + enIn$;
22           $enReqA := \min(e_a^{EF}, r_a \cdot (end - est_a)) - enIn$;
23   **if** $enReqA > enReqU$ **then** $u := a$; $enReqU := enReqA$;
24        $avail := R \cdot (end - begin) - E - ttEn(a, b)$;
25   **if** *opportunistic* **and** $avail < minAvail$ **then**
26           $minAvail := avail$; $minBegin := begin$;
27   **if** $enReqU > 0$ **and** $avail - enReqU < 0$ **then**
28           $rest := E - avail - r_a \cdot \max(0, end - lst_a)$;
29           $lbU := begin + \lceil rest/r_u \rceil$;
30   **if** $est'_u < lbU$ **then**
31              $expl := explainUpdate(begin, end, u, est'_u, lbU)$;
32              $Update[++k] := (u, \mathtt{lb}, lbU, expl)$;
33              $est'_u := lbU$;
34 **for** $z := 1$ **to** k **do** $updateBound(Update[z])$;

---

generalised in the same way as for a resource overload. But here the available energy units Δ for widening the bounds is $rest(a, b, u) - r_u \cdot (newLB - 1) + 1$. Hence, $0 \leq \Delta < r_u$ indicate that the explanation only can further be generalised a little bit. We perform this generalisation as for the overload case.

Table 1. Specifications of the benchmark suites

suite	sub-suites	#inst	#act	p_i	#res	notes
AT [3]	ST27/ST51/ST103	48 each	25/49/101	1–12	6 each	
PSPLib [16]	J30 [17]/J60/J90	480 each	30/60/90	1–10	4 each	
	J120	600	30	1–10	4	
BL [4]	BL20/BL25	20 each	20/25	1–6	3 each	
PACK [8]		55	15–33	1–19	2–5	
KSD15_D [18]		480	15	1–250	4	based on J30
PACK_D [18]		55	15–33	1–1138	2–5	based on PACK

4   Experiments on Resource-Constrained Project Scheduling Problems

We carried out extensive experiments on RCPSP instances comparing our solution approach using both time-table and/or TTEF propagation. We compare the obtained results on the lower bounds of the makespan with the best known so far. Detailed results are available at http://www.cs.mu.oz.au/~pjs/rcpsp.

We used six benchmark suites for which an overview is given in Table 1 where #inst, #act, p_i, and #res are the number of instances, number of activities, range of processing times, and number of resources, respectively. The first two suites are highly disjunctive, while the remainder are highly cumulative.

The experiments were run on a X86-64 architecture running GNU/Linux and a Intel(R) Core(TM) i7 CPU processor at 2.8GHz. The code was written in Mercury [30] using the G12 Constraint Programming Platform [31].

We model an instance as in [26] using global cumulative constraints cumulative and difference logic constraints ($S_i + p_i \leq S_j$), resp. In addition, between two activities i, j in disjunction, *i.e.*, two activities which cannot concurrently run without overloading some resource, the two half-reified constraints [9] $b \rightarrow S_i + p_i \leq S_j$ and $\neg b \rightarrow S_j + p_j \leq S_i$ are posted where b is a Boolean variable.

We run cumulative constraint propagation using different phases:

(a) time-table consistency check in $\mathcal{O}(n + p \log p)$ runtime,
(b) TTEF consistency check in $\mathcal{O}(l^2 + n)$ runtime as defined in Section 3.1,
(c) time-table bounds' propagation in $\mathcal{O}(l \cdot p + k \cdot \min(R, n))$ runtime, and
(d) TTEF bounds' propagation in $\mathcal{O}(l^2 + k \cdot n)$ runtime as defined in Section 3.2

where k, l, n, p are the numbers of bounds' updates, unfixed activities, all activities, and height changes in the resource profile, respectively. Note that in our setup phase (d) TTEF bounds' propagation does not take into account the bounds' changes of the phase (c) time-table bounds' propagation. For the experiments, we consider four settings of the cumulative propagator: tt executes phases (a) and (c), ttef(c) (a–c), ttef (a–d) with *opportunistic* set to *false*, and ttef+ (a–d) with *opportunistic* set to *true*. Each phase is run once for each execution of the propagator. The propagator is itself run multiple times in the usual propagation fixpoint calculation. Note that phases (c) and (d) are not run if either phase (a) or (b) detects inconsistency.

4.1   Upper Bound Computation

For solving RCPSP we use the same branch-and-bound algorithm as we used in [26], but here we limit ourselves to the search heuristic HOTRESTART which was the most robust one in our previous studies [25,26]. It executes an adapted search of [4] using serial scheduling generation for the first 500 choice points and, then, continues with an activity based search (a variant of VSIDS [20]) on the Boolean variables representing a lower part $x \leq v$ and upper part $v < x$ of the variable x's domain where x is either a start time or the makespan variable and v a value of x's initial domain. Moreover, it is interleaved with a geometric restart policy [35] on the number of node failures for which the restart base and factor are 250 failures and 2.0, respectively. The search was halted after 10 minutes.

The results are given in Tables 2 and 3. For each benchmark suite, the number of solved instances (#svd) is given. The column cmpr(a) shows the results on the instances solved by all methods, where a is the number of such instances. The left entry in that column is the average runtime on these instances in seconds, and the right entry is the average number of failures during search. The entries in column all(a) have the same meaning, but here all instances are considered where a is the total number of instances. For unsolved instances, the number of failures after 10 minutes is used.

Table 2 shows the results on the highly disjunctive RCPSPs. As expected, the stronger propagation (ttef(c), ttef) reduces the search space overall in comparison to tt, but the average runtime is higher by a factor of about 5%–70% for ttef(c) and 50%–100% for ttef. Interestingly, ttef(c) and ttef solved respectively 1 and 2 more instances on J60 and closed the instance j120_1_1 on J120 which has an optimal makespan 105. This makespan corresponds to the best known upper bound. However, the stronger propagation does not generally pay off for a CP solver with nogood learning on highly disjunctive RCPSPs. The opportunistic extended edge finding ttef+ does not pay off on the highly disjunctive problems.

Table 3 presents the results on highly cumulative RCPSPs which clearly shows the benefit of TTEF propagation, especially on BL for which ttef(c) and ttef reduce the search space and the average runtime by a factor of 8, and PACK for which they solved 23 instances more than tt. On PACK_D, ttef(c) is about 50% faster on average than tt while ttef is slightly slower on average than tt. The opportunistic extended edge finding is beneficial on the the hardest highly cumulative problems PACK and PACK_D. No conclusion can be drawn on KSD15_D because the instances are too easy for LCG solvers.

4.2   Lower Bound Computation

The lower bound computation tries to solve RCPSPs in a destructive way by converging to the optimal *makespan* from below, *i.e.*, it repeatedly proves that there exists no solution for current makespan considered and continues with an

Table 2. UB results on highly disjunctive RCPSPs

	J30				J60			
	#svd	cmpr(480)		all(480)		#svd	cmpr(429)		all(480)	
tt	480	0.12	1074	0.12	1074	430	1.82	5798	64.25	93164
ttef(c)	480	0.20	1103	0.20	1103	431	2.00	4860	64.39	80845
ttef	480	0.23	991	0.23	991	432	3.04	5191	64.87	62534
ttef+	480	0.28	1045	0.28	1045	430	3.58	5172	66.63	58577

	J90				J120			
	#svd	cmpr(398)		all(480)		#svd	cmpr(278)		all(600)	
tt	400	4.01	7540	104.09	132234	283	8.92	13636	322.35	398941
ttef(c)	400	4.90	7263	105.69	104297	282	11.13	14387	324.73	297562
ttef	400	6.57	7277	106.66	72402	283	13.30	11881	324.66	186597
ttef+	398	6.05	6165	107.52	70436	282	12.53	11016	325.41	168803

	AT				
	#svd	cmpr(129)		all(144)	
tt	132	8.90	19997	66.22	87226
ttef(c)	130	9.36	16466	69.41	72056
ttef	129	13.55	17239	74.60	63554
ttef+	129	15.82	18060	76.68	61665

Table 3. UB results on highly cumulative RCPSPs

	BL					PACK				
	#svd	cmpr(40)		all(40)		#svd	cmpr(16)		all(55)	
tt	40	0.16	2568	0.16	2568	16	77.65	245441	447.69	699615
ttef(c)	40	0.02	370	0.02	370	39	37.22	122038	186.79	292101
ttef	40	0.02	269	0.02	269	39	44.44	105751	188.23	257747
ttef+	40	0.06	484	0.06	484	39	36.42	95704	185.69	262506

	KSD15_D					PACK_D				
	#svd	cmpr(480)		all(480)		#svd	cmpr(37)		all(55)	
tt	480	0.01	26	0.01	26	37	32.72	42503	218.26	184293
ttef(c)	480	0.01	26	0.01	26	37	23.96	32916	212.37	170301
ttef	480	0.01	26	0.01	26	37	36.93	37004	221.11	157015
ttef+	480	0.13	26	0.13	26	37	23.13	28489	212.14	152950

incremented *makespan* by 1. If a solution is found then it is the optimal one. For these experiments we use the search heuristic HOTSTART as we did in [25,26]. This heuristic is HOTRESTART (as described earlier) but no restart. We used the same parameters as for HOTRESTART. For the starting *makespan*, we choose the best known lower bounds on J60, J90, and J120 recorded in the PSPLib at http://129.187.106.231/psplib/ and [34] at http://vilim.eu/petr/cpaior2011-results.txt. On the other suites, the search starts from *makespan* 1. Due to the tighter *makespan*, it is expected that the TTEF propagation will perform better than for upper bound computation on the highly disjunctive instances. The search was cut off at 10 minutes as in [25,26].

Table 4. LB results on AT, PACK, and PACK_D

	AT (12)	PACK (16)	PACK_D (18)
ttef(c)	5/4/3 +52	0/4/12 +100	0/7/11 +632
ttef	7/2/3 +44	1/4/11 +101	2/6/10 +618
ttef+	7/2/3 +45	0/2/14 +105	3/5/10 +638

Table 5. LB results on J60, J90, and J120

		J60			J90					J120									
		+1	+2	+3	+1	+2	+3	+4	+5	+1	+2	+3	+4	+5	+6	+7	+8	+9	+10
1 min	ttef(c)	4	1	-	12	1	-	-	-	27	8	4	-	-	-	2	-	-	-
	ttef	7	5	-	25	14	3	1	-	90	20	10	5	2	-	-	2	-	-
	ttef+	10	5	-	29	12	3	1	-	83	20	7	8	2	-	-	1	1	-
10 mins	ttef(c)	21	2	-	25	7	-	-	-	68	16	4	4	2	-	-	1	1	-
	ttef	13	6	3	35	17	6	3	1	116	39	9	9	4	1	-	-	1	1
	ttef+	19	7	3	33	17	6	2	1	111	35	9	9	5	1	-	-	1	1

Table 4 shows the results on AT, PACK, and PACK_D restricted to the instances that none of the methods could solve using the upper bound computation, The number of instances for each class is shown in parentheses in the header. An entry $a/b/c + d$ for method x means that x achieved respectively a-times, b-times and c-times a worse, the same and a better lower bound than tt, while the $+d$ is the sum of lower bounds' differences of method x to tt. On PACK and PACK_D, ttef(c) and ttef clearly perform better than tt. On the highly disjunctive instances in AT, ttef(c) and tt are almost balanced whereas tt could generate better lower bounds on more instances than ttef. The lower bounds' differences on AT are dominated by the instance st103_4 for which ttef(c) and ttef retrieved a lower bound improvement of 54 and 53 time periods with respect to tt. Opportunistic extended edge finding ttef+ is beneficial on the highly disjunctive benchmarks, but can only better tt on PACK.

The more interesting results are presented in Tab. 5 because the best lower bounds are known for all the remaining open instances (48, 77, 307 in J60, J90, J120).[2] An entry in a column $+d$ shows the number of instances for that the corresponding method could improve the lower bound by d time periods. On these instances, we run at first the experiments with a runtime limit of one minute as it was done in the experiments for TTEF propagation in [34] but he used a CP solver without nogood learning. tt could not improve any lower bound because its corresponding results are already recorded in the PSPLib. ttef(c), ttef, and ttef+ improved the lower bounds of 59, 183 and 173 instances, respectively, which is about 13.7%, 42.3% and 40.0% of the open instances. Although, the experiments in [34] were run on a slower machine[3] the results

---

[2] Note that the PSPLib still lists the instances j60_25_5, j90_26_5, j120_8_3, j120_48_5, and j120_35_5 as open, but we closed the first four ones in [26] and [19] closed the last one.

[3] Intel(R) Core(TM)2 Duo CPU T9400 on 2.53GHz.

confirm the importance of nogood learning. For the experiments with 10 minutes runtime, ttef(c), ttef and ttef+ could improve the lower bounds of more instances, namely 151, 264 and 258 instances, respectively, which is about 35.0%, 61.1% and 59.7%. Again for the highly disjunctive instances the opportunistic extended edge finding does not pay off, although interestingly it gives the best results on J60. Moreover, 3, 1, and 1 of the remaining open instances on J60, J90, and J120, respectively, could be solved optimally.

5   Conclusion and Outlook

We present explanations for the recently developed TTEF propagation of the global cumulative constraint for lazy clause generation solvers. These explanations express an energetic reasoning propagation which is a stronger propagation than the TTEF one.

Our implementation of this propagator was compared to time-table propagation in lazy clause generation solvers on six benchmark suites. The preliminary results confirms the importance of energy-based reasoning on highly disjunctive RCPSPs for CP solvers with nogood learning.

Moreover, our approach with TTEF propagation was able to close six open instances. It also improves the best known lower bounds for 264 of the remaining 432 remaining open instances on RCPSPs from the PSPLib.

In the future, we want to integrate the extended edge-finding propagation into TTEF propagation as it was originally proposed in [34], to perform experiments on cutting and packing problems, and to study different variations of explanations for TTEF propagation. Furthermore, we want to look at a more efficient implementation of the TTEF propagation as well as an implementation of energetic reasoning.

Acknowledgements. NICTA is funded by the Australian Government as represented by the Department of Broadband, Communications and the Digital Economy and the Australian Research Council through the ICT Centre of Excellence program. This work was partially supported by Asian Office of Aerospace Research and Development grant 10-4123.

References

1. Achterberg, T.: SCIP: solving constraint integer programs. Mathematical Programming Computation 1, 1–41 (2009)
2. Aggoun, A., Beldiceanu, N.: Extending CHIP in order to solve complex scheduling and placement problems. Mathematical and Computer Modelling 17(7), 57–73 (1993)
3. Alvarez-Valdés, R., Tamarit, J.M.: Heuristic algorithms for resource-constrained project scheduling: A review and an empirical analysis. In: Advances in Project Scheduling, pp. 113–134. Elsevier (1989)

4. Baptiste, P., Le Pape, C.: Constraint propagation and decomposition techniques for highly disjunctive and highly cumulative project scheduling problems. Constraints 5(1-2), 119–139 (2000)

5. Baptiste, P., Le Pape, C., Nuijten, W.: Satisfiability tests and time-bound adjustments for cumulative scheduling problems. Annals of Operations Research 92, 305–333 (1999)

6. Baptiste, P., Le Pape, C., Nuijten, W.: Constraint-Based Scheduling. Kluwer Academic Publishers, Norwell (2001)

7. Berthold, T., Heinz, S., Lübbecke, M., Möhring, R., Schulz, J.: A constraint integer programming approach for resource-constrained project scheduling. In: Lodi, A., Milano, M., Toth, P. (eds.) CPAIOR 2010. LNCS, vol. 6140, pp. 313–317. Springer, Heidelberg (2010)

8. Carlier, J., Néron, E.: On linear lower bounds for the resource constrained project scheduling problem. European Journal of Operational Research 149(2), 314–324 (2003)

9. Feydy, T., Somogyi, Z., Stuckey, P.J.: Half reification and flattening. In: Lee, J. (ed.) CP 2011. LNCS, vol. 6876, pp. 286–301. Springer, Heidelberg (2011)

10. Feydy, T., Stuckey, P.J.: Lazy clause generation reengineered. In: Gent (ed.) [11], pp. 352–366

11. Gent, I.P. (ed.): CP 2009. LNCS, vol. 5732. Springer, Heidelberg (2009)

12. Heinz, S., Schulz, J.: Explanations for the cumulative constraint: An experimental study. In: Pardalos, P.M., Rebennack, S. (eds.) SEA 2011. LNCS, vol. 6630, pp. 400–409. Springer, Heidelberg (2011)

13. Jussien, N.: The versatility of using explanations within constraint programming. Research Report 03-04-INFO, École des Mines de Nantes, Nantes, France (2003)

14. Jussien, N., Barichard, V.: The PaLM system: explanation-based constraint programming. In: Proceedings of TRICS: Techniques foR Implementing Constraint Programming Systems, a Post-conference Workshop of CP 2000, Singapore, pp. 118–133 (2000)

15. Katsirelos, G., Bacchus, F.: Generalized nogoods in CSPs. In: Veloso, M.M., Kambhampati, S. (eds.) Proceedings on Artificial Intelligence – AAAI 2005, pp. 390–396. AAAI Press/The MIT Press (2005)

16. Kolisch, R., Sprecher, A.: PSPLIB – A project scheduling problem library. European Journal of Operational Research 96(1), 205–216 (1997)

17. Kolisch, R., Sprecher, A., Drexl, A.: Characterization and generation of a general class of resource-constrained project scheduling problems. Management Science 41(10), 1693–1703 (1995)

18. Koné, O., Artigues, C., Lopez, P., Mongeau, M.: Event-based MILP models for resource-constrained project scheduling problems. Computers & Operations Research 38(1), 3–13 (2011)

19. Liess, O., Michelon, P.: A constraint programming approach for the resource-constrained project scheduling problem. Annals of Operations Research 157(1), 25–36 (2008)

20. Moskewicz, M.W., Madigan, C.F., Zhao, Y., Zhang, L., Malik, S.: Chaff: Engineering an efficient SAT solver. In: Proceedings of Design Automation Conference – DAC 2001, pp. 530–535. ACM, New York (2001)

21. Nuijten, W.P.M.: Time and Resource Constrained Scheduling. Ph.D. thesis. Eindhoven University of Technology (1994)

22. Ohrimenko, O., Stuckey, P.J., Codish, M.: Propagation via lazy clause generation. Constraints 14(3), 357–391 (2009)

23. Schulte, C., Stuckey, P.J.: Efficient constraint propagation engines. ACM Transactions on Programming Languages and Systems 31(1), Article No. 2 (2008)
24. Schutt, A.: Improving Scheduling by Learning. Ph.D. thesis, The University of Melbourne (2011), http://repository.unimelb.edu.au/10187/11060
25. Schutt, A., Feydy, T., Stuckey, P.J., Wallace, M.G.: Why cumulative decomposition is not as bad as it sounds. In: Gent (ed.) [11], pp. 746–761
26. Schutt, A., Feydy, T., Stuckey, P.J., Wallace, M.G.: Explaining the cumulative propagator. Constraints 16(3), 250–282 (2011)
27. Schutt, A., Feydy, T., Stuckey, P.J., Wallace, M.G.: Solving RCPSP/max by lazy clause generation. Journal of Scheduling, 1–17 (2012), online first
28. Schutt, A., Stuckey, P., Verden, A.: Optimal carpet cutting. In: Lee, J. (ed.) CP 2011. LNCS, vol. 6876, pp. 69–84. Springer, Heidelberg (2011)
29. Schutt, A., Wolf, A.: A new $\mathcal{O}(n^2 \log n)$ not-first/not-last pruning algorithm for cumulative resource constraints. In: Cohen, D. (ed.) CP 2010. LNCS, vol. 6308, pp. 445–459. Springer, Heidelberg (2010)
30. Somogyi, Z., Henderson, F., Conway, T.: The execution algorithm of Mercury, an efficient purely declarative logic programming language. The Journal of Logic Programming 29(1-3), 17–64 (1996)
31. Stuckey, P.J., de la Banda, M.G., Maher, M.J., Marriott, K., Slaney, J.K., Somogyi, Z., Wallace, M., Walsh, T.: The G12 project: Mapping solver independent models to efficient solutions. In: Gabbrielli, M., Gupta, G. (eds.) ICLP 2005. LNCS, vol. 3668, pp. 9–13. Springer, Heidelberg (2005)
32. Vilím, P.: Computing explanations for the unary resource constraint. In: Barták, R., Milano, M. (eds.) CPAIOR 2005. LNCS, vol. 3524, pp. 396–409. Springer, Heidelberg (2005)
33. Vilím, P.: Edge finding filtering algorithm for discrete cumulative resources in $\mathcal{O}(kn \log n)$. In: Gent (ed.) [11], pp. 802–816
34. Vilím, P.: Timetable edge finding filtering algorithm for discrete cumulative resources. In: Achterberg, T., Beck, J.C. (eds.) CPAIOR 2011. LNCS, vol. 6697, pp. 230–245. Springer, Heidelberg (2011)
35. Walsh, T.: Search in a small world. In: Proceedings of Artificial intelligence – IJCAI 1999, pp. 1172–1177. Morgan Kaufmann (1999)

A Lagrangian Relaxation for Golomb Rulers

Marla R. Slusky[1] and Willem-Jan van Hoeve[2]

[1] Department of Mathematical Sciences, Carnegie Mellon University
mslusky@andrew.cmu.edu
[2] Tepper School of Business, Carnegie Mellon University
vanhoeve@andrew.cmu.edu

Abstract. The Golomb Ruler Problem asks to position n integer marks on a ruler such that all pairwise distances between the marks are distinct and the ruler has minimum total length. It is a very challenging combinatorial problem, and provably optimal rulers are only known for n up to 26. Lower bounds can be obtained using Linear Programming formulations, but these are computationally expensive for large n. In this paper, we propose a new method for finding lower bounds based on a Lagrangian relaxation. We present a combinatorial algorithm that finds good bounds quickly without the use of a Linear Programming solver. This allows us to embed our algorithm into a constraint programming search procedure. We compare our relaxation with other lower bounds from the literature, both formally and experimentally. We also show that our relaxation can reduce the constraint programming search tree considerably.

1 Introduction

For some positive integer n, let $x_1, \ldots, x_n$ represent the integer positions of n marks on a ruler. We can assume that $x_i < x_j$ for all $1 \leq i < j \leq n$ and that $x_1 = 0$. A *Golomb ruler* has pairwise distinct distances between the marks, i.e., $x_j - x_i$ for all $1 \leq i < j \leq n$ are distinct. Given n, the Golomb ruler problem asks to find a Golomb ruler with minimum length x_n .

Practical applications of the Golomb ruler problem include radio communications, X-ray crystallography, coding theory, and radio astronomy [1, 2, 3, 4]. The problem continues to be very difficult to solve in practice, although it is still unknown whether it is NP-hard. Optimal Golomb rulers are only known up to $n = 26$. The optimality of rulers of 24, 25 and 26 marks was proven by a massively parallel search coordinated by distributed.net/ogr. The 27-mark search started in March 2009, and as of October 2012, only 65% of this project is complete.[1]

The Golomb ruler problem is a popular benchmark for discrete optimization, and for constraint programming methods in particular (it is problem prob006 in CSPLib). Several exact methods based on constraint programming have been proposed in the literature, (e.g., [5, 6]). Other solution methods include algebraic methods (affine and projective plane constructions, [7, 8]), evolutionary

---

[1] See http://stats.distributed.net/projects.php?project_id=27

C. Gomes and M. Sellmann (Eds.): CPAIOR 2013, LNCS 7874, pp. 251–267, 2013.
© Springer-Verlag Berlin Heidelberg 2013

algorithms [9], and hybrid methods combining constraint programming and local search [10, 11].

A crucial component of exact solution methods is producing lower bounds, which appears to be more challenging than providing upper bounds (feasible rulers). Lower bounds can help to dramatically prune an exhaustive search, but only if they can be found quickly enough. Lower bounds based on linear programming formulations were proposed in [12, 13, 14]. These three formulations were proved equivalent in [15]. Another bound was discussed in [6] and applied within a constraint programming approach for solving the problem. This bound is weaker than the LP bound, but it can be computed more quickly.

In this paper, we propose a new method for producing lower bounds, based on a Lagrangian relaxation of the problem. We show that our relaxation generalizes the bounds proposed in [6], and can produce a bound that is equivalent to the LP bound. Furthermore, we present an algorithm that allows solving the relaxation in $O(n^2 \log n)$ time for fixed Lagrangian multipliers. This allows us to efficiently approximate the LP bound using a subgradient optimization method, and apply our bound within a constraint programming search procedure. We experimentally demonstrate that in practice our method can produce bounds almost as strong as the LP bound much faster than existing methods. Moreover, we demonstrate that it can decrease the search tree size up to 91%, which translates into a solving time reduction of up to 78%.

We note that Lagrangian relaxations have been applied before in the context of CP, see for example [16, 17, 18, 19, 20, 21, 22, 23]. Our results further strengthen the idea that Lagrangian relaxations are a particularly useful method from operations research for enhancing the inference process of constraint programming. In particular, Lagrangian relaxations can help improve the representation of integrating arithmetic constraints into the *alldifferent* constraint, which is a challenging issue in constraint programming [24].

The rest of the paper is organized as follows. In Section 2 we present formal models of the Golomb ruler problem. In Section 3 we present the Lagrangian formulation, our efficient algorithm to solve the relaxation, and the subgradient optimization method. Section 4 discusses exact methods to solve the Lagrangian relaxation and relates our formulation to the formulations in [13], [15], and [6]. Section 5 contains the computational results comparing our new formulation to the formulations in [13] and [14], the current state of the art. In Section 6 we present our search algorithm and demonstrate the benefit provided by the Lagrangian relaxation bound.

2   Exact Models for the Golomb Ruler Problem

We first present a formal model of the Golomb ruler problem. In the following, we will assume that the marks take their position from a range $\{1, \ldots, L\}$ for some appropriate upper bound L.

Rather than taking the marks $x_1, \ldots, x_n$ to be our variables, we will take the $\binom{n}{2}$-many *segment lengths* $d_{ij} := x_j - x_i$ to be our variables. Then the Golomb ruler problem can be expressed as the following constraint programming (CP) model:

$$\min \quad \sum_{k=1}^{n-1} d_{k,k+1}$$

$$\text{s.t.} \quad \text{alldifferent}(d_{12}, d_{13}, \ldots, d_{n-1,n}) \tag{1}$$

$$d_{ij} = \sum_{k=i}^{j-1} d_{k,k+1} \qquad \text{for all } 2 \leq i+1 < j \leq n.$$

We can alternatively express this CP model as an integer programming (IP) model, by representing the *alldifferent* constraint explicitly as a bipartite matching problem. That is, we introduce a vertex set corresponding to the pairs of marks $\{(i,j) \mid 1 \leq i < j \leq n\}$, a vertex set corresponding to the possible lengths $\{1, 2, \ldots, L\}$, and we define the complete bipartite graph between these two vertex sets. Clearly, a maximum matching in this graph corresponds to a solution to *alldifferent* [25]. For our IP model, we introduce a binary 'edge' variable such that $e_{ijv} = 1$ when the pair (i,j) induces a distance $v \in \{1, \ldots, L\}$ and $e_{ijv} = 0$ otherwise. The model thus becomes:

$$\min \quad \sum_{k=1}^{n-1} d_{k,k+1}$$

$$\text{s.t.} \quad \sum_{v=1}^{L} e_{ijv} = 1 \qquad \text{for all } 1 \leq i < j \leq n,$$

$$\sum_{i<j} e_{ijv} \leq 1 \qquad \text{for all } v = 1, \ldots, L,$$

$$\sum_{v=1}^{L} v \cdot e_{ijv} = d_{ij} \qquad \text{for all } 1 \leq i < j \leq n, \tag{2}$$

$$\sum_{k=i}^{j-1} d_{k,k+1} = d_{ij} \qquad \text{for all } 2 \leq i+1 < j \leq n,$$

$$e_{ijv} \in \{0,1\} \qquad \text{for all } 1 \leq i < j \leq n, \, v = 1, \ldots, L.$$

In this model, the first two constraints represent the bipartite matching. The third constraint establishes the relationship between the variables e_{ijv} and d_{ij}. The fourth is the requirement that each larger segment is made up of the smaller segments it contains. We note that model (2) corresponds to the formulation suggested in [12]. We will refer to it as the *matching* formulation and to its objective value as z_{matching}. We will derive our Lagrangian relaxation from this model.

3 Lagrangian Relaxation

In this section, we first present the Lagrangian formulation, which provides a relaxation for any fixed set of Lagrangian multipliers. We then show that each such relaxation can be solved efficiently. In order to find the best relaxation (corresponding to the LP lower bound), we lastly present a subgradient optimization method that approximates the optimal Lagrangian multipliers.

3.1 Formulation

We create a Lagrangian relaxation from model (2) as follows. For every pair of non-consecutive marks, that is, for all i, j such that $2 \leq i + 1 < j \leq n$, we choose a coefficient $\lambda_{ij} \in \mathbb{R}$ and consider the LP resulting from moving the last constraint of the matching formulation to the objective function:

$$
\min \quad \sum_{k=1}^{n-1} d_{k,k+1} + \sum_{i+1<j} \lambda_{ij} \left(d_{ij} - \sum_{k=i}^{j-1} d_{k,k+1} \right)
$$

$$
\text{s.t.} \quad \sum_{v=1}^{L} e_{ijv} = 1 \qquad\qquad \text{for all } 1 \leq i < j \leq n,
$$

$$
\sum_{i<j} e_{ijv} \leq 1 \qquad\qquad \text{for all } v = 1, \ldots, L, \tag{3}
$$

$$
d_{ij} = \sum_{v=1}^{L} v \cdot e_{ijv} \qquad\qquad \text{for all } 1 \leq i < j \leq n,
$$

$$
e_{ijv} \geq 0 \qquad\qquad \text{for all } 1 \leq i < j \leq n, \, v = 1, \ldots, L.
$$

In this formulation we do not enforce $\sum_{k=i}^{j-1} d_{k,k+1} = d_{ij}$, but we do incur a penalty, weighted by λ_{ij}, if we do not satisfy that constraint. Note that the optimal solution for the matching formulation is still feasible in this relaxation, and gives the same objective value. Therefore, the optimal value here is *at most* z_{matching}.

We can simplify our model further by rearranging the objective function to become

$$
\sum_{i+1<j} \lambda_{ij} d_{ij} + \sum_{k=1}^{n-1} d_{k,k+1} \left(1 - \sum_{\substack{i \leq k < j \\ i+1 \neq j}} \lambda_{ij} \right). \tag{4}
$$

Also, recall that we did not choose $\lambda_{k,k+1}$ for any k earlier, so let us take

$$
\lambda_{k,k+1} := 1 - \sum_{\substack{i \leq k < j \\ i+1 \neq j}} \lambda_{ij} . \tag{5}
$$

Then for any fixed (λ_{ij}) satisfying equation (5), we have the simpler LP:

$$
\begin{aligned}
\min \quad & \sum_{i<j} \lambda_{ij} d_{ij} \\
\text{s.t.} \quad & \sum_{v=1}^{L} e_{ijv} = 1 && \text{for all } 1 \leq i < j \leq n, \\
& \sum_{i<j} e_{ijv} \leq 1 && \text{for all } v = 1, \ldots, L, \\
& d_{ij} = \sum_{v=1}^{L} v \cdot e_{ijv} && \text{for all } 1 \leq i < j \leq n,
\end{aligned}
\tag{6}
$$

This is the LP we will refer to as the Lagrangian relaxation, and we will refer to its objective value as z_{LR}. Note that the d_{ij} variables are simply an intermediate calculation. If we replace the d_{ij} in the objective function with $\sum_{v=1}^{L} v \cdot e_{ijv}$, then we can eliminate the third constraint, and so this LP represents a matching problem. This ensures that this LP has an integer solution.

Proposition 1. *For any fixed (λ_{ij}) we have $z_{\mathrm{LR}} \leq z_{\mathrm{matching}}$, and there exists (λ_{ij}) for which $z_{\mathrm{LR}} = z_{\mathrm{matching}}$.*

Proof. The proposition follows from choosing (λ_{ij}) to be the dual variables of the last equation in (2). (see, e.g., [26]). □

3.2 A Combinatorial Algorithm for Solving the Relaxation

Proposition 2. *For any fixed (λ_{ij}), the Lagrangian relaxation can be solved in $O(n^2 \log n)$ time.*

Proof. What the Lagrangian relaxation LP actually represents is a matching problem where we are matching each λ_{ij} with a number in $\{1, \ldots, L\}$, and we are trying to minimize the sum of the product of the pairs. It is clear that if $\lambda_{ij} \geq 0$ for all $i < j$, then to minimize the objective value we must match the largest λ_{ij} with 1, the next largest λ_{ij} with 2, etc. (If we have some $\lambda_{ij} < 0$ then we will just take d_{ij} as large as possible making our objective value $-\infty$.) Thus our method for solving the Lagrangian relaxation will be as follows.

1   Sort (λ_{ij}) into decreasing order.

2   Let d_{ij} be the location of λ_{ij} in the sorted list.

Since our algorithm for solving the Lagrangian relaxation reduces to sorting $\binom{n}{2}$ elements, we can solve it in $O(n^2 \log n)$ time. □

3.3 Subgradient Optimization Method

In order to find (close to) optimal values for (λ_{ij}), we designed an iterative local search scheme similar to subgradient optimization methods as applied in, e.g., [27, 28]. To approximate good values for (λ_{ij}), recall that λ_{ij} is a penalty for not satisfying the constraint $\sum_{k=i}^{j-1} d_{k,k+1} = d_{ij}$. Therefore, if we solve the Lagrangian relaxation and do not satisfy the constraint for pair (i, j), we should increase the penalty λ_{ij}. Our algorithm is as follows:

1   Choose initial *stepsize*

2   Choose initial values for λ_{ij} with $i + 1 < j$ (for example, all 0)

3   Set $\lambda_{k,k+1} := 1 - \sum_{i<k<j} \lambda_{ij}$ for all $k \in \{1, \ldots, n-1\}$

4   Repeat until some stopping criterion {

5   Solve the Lagrangian relaxation

6   For each $i < j$ do

7             $\lambda_{ij} := \lambda_{ij} + \left(d_{ij} - \sum_{k=i}^{j-1} d_{k,k+1} \right) \dfrac{stepsize}{n^2}$

8   Adjust *stepsize* if necessary

9   }

The performance of this algorithm highly depends on the choice and adjustment of the stepsize parameter. In our implementation, we start with a stepsize of 1 (in line 1). When an iteration results in negative values for some λ_{ij}, we divide the stepsize in half to refine the search. Otherwise, after each 5 iterations of decreasing values for z_{LR}, we multiply the stepsize by 0.999 (line 8).

Unfortunately, this algorithm does not have a natural stopping condition based on optimality of the solution. In fact, even if we use the optimal (λ_{ij}) as initial data, one iteration will return different values. Nevertheless, this algorithm produces very good approximations of z_{matching} very quickly, as we will see in Section 5.

4   Relationship with Other Formulations

In this section we investigate the relationship of our Lagrangian relaxation with other, existing, formulations for obtaining lower bounds. Throughout this section we will use λ to mean $(\lambda_{ij}) \in \mathbb{R}^{\binom{n}{2}}$; $S_{\binom{n}{2}}$ to be the set of all permutations of the numbers $\{1, 2, \ldots, \binom{n}{2}\}$ indexed by pairs (i, j) with $i < j$; $\sigma = (\sigma_{ij}) \in S_{\binom{n}{2}}$; and $\lambda \cdot \sigma = \sum_{1 \leq i < j \leq n} \lambda_{ij} \sigma_{ij}$.

4.1   Permutation Formulation

Our goal in the last section was

$$\min_{\sigma} \lambda \cdot \sigma$$

for a fixed λ, because this gives us a lower bound for the length of a Golomb ruler. However, our overall goal is to strengthen this bound, that is

$$\max_{\lambda} \min_{\sigma} \lambda \cdot \sigma$$

or, expressed as an LP,

$$
\begin{aligned}
\max \quad & z \\
\text{s.t.} \quad & \sum_{i \leq k < j} \lambda_{ij} = 1 \qquad \text{for all } k = 1, \ldots, n-1, \\
& z \leq \sum_{i<j} \lambda_{ij} \cdot \sigma_{ij} \qquad \text{for all } \sigma \in S_{\binom{n}{2}}.
\end{aligned}
\tag{7}
$$

We will refer to this model as the *permutation* formulation. This formulation was also given in [13] and [15].

The correspondence between model (7) and our Lagrangian relaxation is that by solving model (7) we obtain optimal values for λ with respect to the Lagrangian relaxation, and both models will provide the same objective value. Unfortunately, solving the permutation model directly is non-trivial; we have about $\binom{n}{2}!$ constraints. However, we can apply Proposition 2 to solve it more quickly. We will iterate solving model (7) for a subset of constraints $C \subset S_{\binom{n}{2}}$:

$$
\begin{aligned}
\max \quad & z \\
\text{s.t.} \quad & \sum_{i \leq k < j} \lambda_{ij} = 1 \qquad \text{for all } k = 1, \ldots, m-1, \\
& z \leq \sum_{i<j} \lambda_{ij} \cdot \sigma_{ij} \qquad \text{for all } \sigma \in C.
\end{aligned}
\tag{8}
$$

Our algorithm is as follows:

1   Choose any initial C
2   Solve (8) and let z be the objective value
3   Sort λ into decreasing order
4   For $i < j$ let $\sigma_{ij} =$ (the position of λ_{ij} in sorted order)
5   If $(z = \lambda \cdot \sigma)$
6   Then terminate
7   Else {
8       $C := C \cup \{\sigma\}$
9   Goto 2
10 }

The sorting algorithm and the restricted permutation model provide lower and upper bounds, respectively; optimality is proved when these bounds meet (line 5). This can serve as a systematic alternative approach to our local search.

4.2   Equation Sums Bound

We next study the relationship of the Lagrangian relaxation with the lower bounds proposed in [6]. For this, we consider the constraint $\sum_{i \leq k < j} \lambda_{ij} = 1$ in models (7) and (8). We assign a coefficient λ_{ij} to each segment of the ruler, but why should we have them summing to 1 in this way? Before we answer that question, we recall the lower bounds given in [6] by illustration with an example.

Let $n = 5$, for which the length of the ruler is given by d_{15}. If we want to bound d_{15}, we can first divide this segment into sub-segments in different ways:

$$d_{15} = d_{12} + d_{23} + d_{34} + d_{45}$$
$$d_{15} = d_{13} + d_{35}$$
$$d_{15} = d_{12} + d_{24} + d_{45}$$

Multiplying each equation by $\frac{1}{3}$ and adding them together gives

$$d_{15} = \frac{2}{3}(d_{12} + d_{45}) + \frac{1}{3}(d_{23} + d_{34} + d_{13} + d_{24} + d_{35}) \ .$$

Since all these numbers will be distinct naturals, we get

$$d_{15} \geq \frac{2}{3}(1 + 2) + \frac{1}{3}(3 + 4 + 5 + 6 + 7)$$
$$d_{15} \geq 10.333$$

There are, of course, many ways we can write out d_{1n} as a sum of smaller segments, and [6] proposes some heuristics. We will refer to bounds of this form as *equation sums* bounds. Another option we have is to weight the equations differently. For example, we could have given the first two equations weights of 0.4 and the last equation a weight of 0.2 instead of giving them all a weight of $\frac{1}{3}$. This would result in the equation

$$d_{15} = 0.6(d_{12} + d_{45}) + 0.4(d_{23} + d_{34} + d_{13} + d_{35}) + 0.2(d_{24})$$

and the corresponding bound

$$d_{15} \geq 0.6(1 + 2) + 0.4(3 + 4 + 5 + 6) + 0.2(7)$$
$$d_{15} \geq 10.4$$

We will refer to bounds of this form as *generalized equation sums* bounds.

Proposition 3. *The generalized equation sums bounds are equivalent to z_{LR} for an appropriate choice of λ.*

Proof. The weights of the equations in the generalized equation sums bound must always be distributed so that they sum to 1. This way d_{1n} always gets a coefficient of 1, and we always end up with a bound of the form $d_{1n} \geq \sum \mu_{ij} d_{ij}$ for some coefficients μ_{ij}. Note that in each equation for each $k = 1, \ldots, n-1$, there is some term that encapsulates the segment $(k, k+1)$. That is, there is some d_{ij} such that $i \leq k < j$. Since the weights on each equation sum to one, the coefficients that encapsulate the pair $(k, k+1)$ should sum to 1. That is: $\sum_{i \leq k < j} \mu_{ij} = 1$. Then to find the minimum value of $\sum_{i<j} d_{ij}\mu_{ij}$ we simply sort (μ) into decreasing order and assign each d_{ij} the corresponding value. This is precisely what we did in Proposition 2.   □

This shows that although the bound from [6] is weaker than the LP bound, it can be generalized to be as strong as the LP bound, and it gives a nice intuition for our constraint on λ.

5   Computational Results for Approximating the LP Bound

The purpose of our experimental results is twofold. First, we would like to gain insight in the performance of our approximate local search scheme relative to the systematic iterative scheme based on the permutation formulation for solving the Lagrangian relaxation. Second, we wish to evaluate our Lagrangian relaxation with the state of the art for solving the LP relaxation.

5.1   Subset Formulation

The current fastest method for solving the LP relaxation for the Golomb ruler problem was proposed by [15]. It is based on the following formulation of the lower bound, proposed in [14]. Let $\mathcal{S} = \{(i, j) : i < j\}$.

$$
\begin{aligned}
\min \quad & d_{1n} \\
\text{s.t.} \quad & \sum_{k=i}^{j-1} d_{k,k+1} = d_{ij} && \text{for all } 1 \leq i < j \leq n, \\
& \sum_{(i,j) \in R} d_{ij} \geq \frac{1}{2}|R| \cdot (|R| + 1) && \text{for all } R \in \mathcal{P}(\mathcal{S}).
\end{aligned}
\tag{9}
$$

We will call this the *subset* formulation. Again, this LP is too big to solve as stated since it has $O(2^{\binom{n}{2}})$ constraints. However, [15] proposes an iterative solving method in which we only include the second constraint above for some subset $\mathcal{T} \subset \mathcal{P}(\mathcal{S})$.

1   Let $\mathcal{T} = \{\{i\} : 1 \leq i \leq n\} \cup \{\{1, \ldots, n\}\}$

2   Solve (9)

3   Sort (d_{ij})

4   For $1 \leq k \leq \binom{n}{2}\{$

5   Let $T = \{(i, j) : d_{ij}$ is in within the first k positions$\}$

6   If $(\sum_{(i,j) \in T} d_{ij} < \binom{k}{2})$ then

7           $\mathcal{T} := \mathcal{T} \cup \{T\}$

8   $\}$

This approach is the currently best known algorithm for finding the LP bound, and we will compare our proposed algorithm to this.

5.2   Implementation and Results

We implemented the Lagrangian relaxation and the subgradient method in C++, following the description in Section 3.3. It was run using C++ on an Intel core i3 processor (2.13 GHz). The times reported are the number of seconds elapsed between when the program started running and when that lower bound was found.

We implemented both the subset formulation and the permutation formulation in AIMMS. The AIMMS implementations were run on the same Intel core i3 processor. The times reported are the sums of the solve times for each call to CPLEX, i.e., we eliminate the overhead that AIMMS may add).

We ran the cases $n = 30$, 40, 50, and 60 to completion, and $n = 130$ for 600 seconds. In each case we can see from the figures that although the Lagrangian relaxation does not achieve the LP bound, it gets close to it before the subset formulation does. We also show, for reference, the constant functions $y = \text{UB}$, where UB is the best known upper bound (length of the shortest known ruler[2]) and $y = \text{LB}$ where LB is the value of the LP bound (the final value of z in all our formulations).

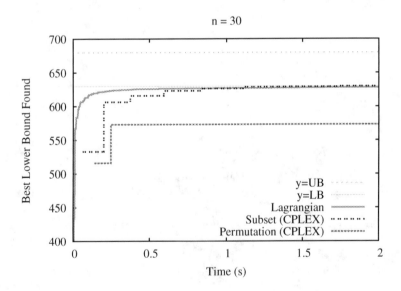

Fig. 1. Speed comparison between the permutation, subset, and Lagrangian formulations. How quickly can each find the lower bound when $n = 30$?

---

[2] See http://www.research.ibm.com/people/s/shearer/grtab.html for the list of shortest known rulers.

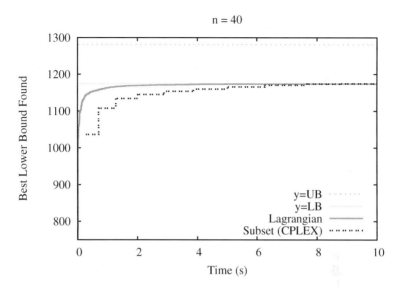

Fig. 2. Speed comparison between the subset and Lagrangian formulations. How quickly can each find the lower bound when $n = 40$?

6   CP Implementation

We implemented a CP search program that, given n and L, finds all n-mark Golomb rulers of length L. It is implemented as a set constraint problem [29] concerning two set variables: X, the set of marks in the ruler, and D, the set of distances measured by the ruler. We apply the standard subset + cardinality domain ordering, whereby we maintain a lower bound of mandatory elements (denoted by X^- and D^-) and an upper bound of possible elements (denoted by X^+ and D^+). Our constraints are as follows.

$$
\begin{aligned}
&X = \{x_1, x_2, \ldots, x_n\} \in [\{0, L\}, \{0, \ldots, L\}] \\
&|X| = n \\
&D \in [\{L\}, \{1, \ldots, L\}] \\
&|D| = \binom{n}{2} \\
&d \in D \iff \exists x_i, x_j \in X \text{ s.t. } x_j - x_i = d \\
&x_2 - x_1 < x_n - x_{n-1}
\end{aligned}
\tag{10}
$$

Our branching procedure is described in Figure 6. Line 18 ensures that between any ruler and its mirror image only one is found by this program, reflecting the last constraint in model (10).

The search strategy considers each distance $d \in \{1, \ldots, L\}$ in decreasing order and decides where and if d will be measured in the ruler.

Proposition 4. *If we have already decided if and where to measure the lengths* $\{d + 1, \ldots, L\}$*, and we have not decided if and where to place* d*, then the only place* d *can be measured is from 0 to* d *or from* $L - d$ *to* L*.*

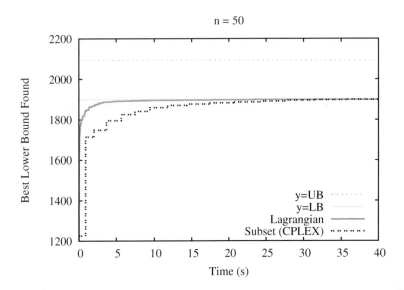

Fig. 3. Speed comparison between the subset and Lagrangian formulations. How quickly can each find the lower bound when $n = 50$?

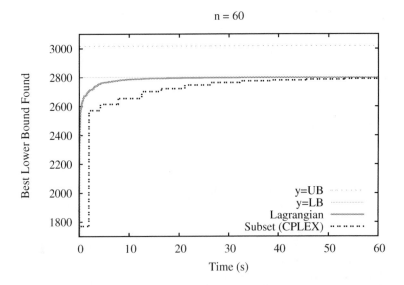

Fig. 4. Speed comparison between the subset and Lagrangian formulations. How quickly can each find the lower bound when $n = 60$?

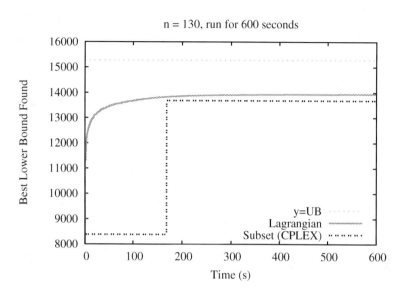

Fig. 5. Speed comparison between the subset and Lagrangian formulations. How quickly can each find the lower bound when $n = 130$?

```
1     define branch(X⁻, X⁺, D⁻, D⁺){
2         if (|X⁻| = n)
3             return X⁻
4         if (|D⁺| < (ⁿ₂))
5             return 0
6
7         D⁻ := {xⱼ − xᵢ : xᵢ, xⱼ ∈ X⁻}
8         for (x ∈ X⁻ and d ∈ D⁻){
9             if(x − d ∈ X⁺ \ X⁻)
10                X⁺ := X⁺ \ {x − d}
11            if(x + d ∈ X⁺ \ X⁻)
12                X⁺ := X⁺ \ {x + d}
13        }
14        for (x, y ∈ X⁻ with x ≡ y mod 2)
15            X⁺ := X⁺ \ {x+y/2}
16
17        d⁺ := max(D⁺ \ D⁻)
18        if (|X⁻| > 2)
19            if (d⁺ ∈ X⁺ \ X⁻)
20                branch(X⁻ ∪ {d⁺}, X⁺, D⁻, D⁺)
21                X⁺ := X⁺ \ {d⁺}
22            if (L − d⁺ ∈ X⁺ \ X⁻)
23                branch(X⁻ ∪ {L − d⁺}, X⁺, D⁻, D⁺)
24                X⁺ := X⁺ \ {L − d⁺}
25        branch(X⁻, X⁺, D⁻, D⁺ \ {d⁺})
26  }
```

Fig. 6. Our branching algorithm for finding Golomb Rulers

Table 1. The performance of the CP search (baseline), the Lagrangian relaxation applied at a third of the search nodes (LR1), and the Lagrangian relaxation applied at each search node (LR2). C or F denotes whether we are reporting the time/nodes to program Completion or the time/nodes to Find a ruler. We report the total number of search nodes and the solving time in seconds.

			baseline		LR1		LR2	
n	L	C/F	nodes	time	nodes	time	nodes	time
10	54	C	60,554	0.51	10,377	0.15	4,984	0.11
10	55	F	4,492	0.04	4,179	0.07	3,512	0.07
11	71	C	2,993,876	27.09	2,402,590	28.45	2,055,429	37.29
11	72	F	5,581	0.05	5,412	0.08	5,343	0.11
12	84	C	10,298,716	103.62	4,143,356	57.40	2,773,734	59.04
12	85	F	7,103,301	70.84	5,618,338	76.41	4,698,798	96.17
13	105	C	445,341,835	4782	323,717,500	5533	273,340,407	6618
13	106	F	205,714,305	2187	191,016,739	3309	177,429,879	4278

Proof. Without loss of generality, suppose there is $x, x + d \in X^+$ with $0 < x < L - d$. Then since we have decided if and where to place the distance $x + d$, and we know 0 will be a mark in our ruler, we already know whether we are including the mark $x + d$. Similarly, since $d < L - x$, we already know if and where we are including the distance $L - x$ and since we are including the mark L, we also know whether we are including the mark x. If we had decided to include both x and $x + d$, then we would not need to decide on the distance d. Thus if d is the largest distance we have not decided whether or not to include, we only need to consider three possibilities: the mark d is in the ruler, the mark $L - d$ is in the ruler, or the distance d is not measured by the ruler.   □

We ran three programs to test our algorithm, and the results are in Table 1. The baseline program just calls the procedure above. The other two programs start by running 2000 iterations of our subgradient optimization procedure, thus fixing our values for λ, and then call a modified version of the branch procedure, which, at line 7, uses proposition 2 to check if we have violated the LP bound. LR1 performs this check when $|X^-| \equiv 1 \bmod 3$, and LR2 performs this check at every node.

Our algorithm always reduces the size of the search tree, sometimes by as much as 91% as in Table 2. The Lagrangian relaxation does not appear to speed up the algorithm when we are searching for a ruler, but it can speed up the algorithm when we are trying to prove a ruler does not exist. Interestingly, it appears the strength of this method is correlated with the strength of the LP bound.

Table 2. Percent improvement of the Lagrangian relaxation applied at a third of the search nodes (LR1) and at each search node (LR2) over the CP search (baseline). C or F denotes whether we are reporting the time/nodes to program Completion or the time/nodes to Find a ruler. We also provide, for reference, the strength of the LP bound as $\frac{LB}{UB} = \frac{\text{LP bound}}{\text{Optimal Ruler Length}}$.

				LR1		LR2	
n	L	C/F	$\frac{LB}{UB}$	nodes	time	nodes	time
10	54	C	98%	82%	70%	91%	78%
10	55	F	98%	6%	-75%	21%	-75%
11	71	C	93%	19%	-5%	31%	-37%
11	72	F	93%	3%	-60%	4%	-120%
12	84	C	96%	59%	44%	73%	43%
12	85	F	96%	20%	-7%	33%	-35%
13	105	C	92%	27%	-15%	38%	-38%
13	106	F	92%	7%	-51%	13%	-95%

7 Conclusion

We have presented a new way to approximate the LP bound for Golomb Rulers. We have demonstrated its relationship to existing methods, and shown that we can compute the LP bound much faster using combinatorial methods.

We then used this fast computation in a search procedure, demonstrating that we can use this bound to reduce the size of the search tree and, in cases where the LP bound is strong enough, reduce the search time as well.

References

[1] Bloom, G.S., Golomb, S.W.: Applications of numbered undirected graphs. Proceedings of the IEEE 65(4), 562–570 (1977)

[2] Moffet, A.T.: Minimum-redundancy linear arrays. IEEE Transactions on Anntennas and Propagation AP-16(2), 172–175 (1968)

[3] Gagliardi, R., Robbins, J., Taylor, H.: Acquisition sequences in PPM communications. IEEE Transactions on Information Theory IT-33(5), 738–744 (1987)

[4] Robinson, J.P., Bernstein, A.J.: A class of binary recurrent codes with limited error propagation. IEEE Transactions on Information Theory IT-13(1), 106–113 (1967)

[5] Smith, B., Stergiou, K., Walsh, T.: Modelling the Golomb ruler problem. In: IJCAI Workshop on Non-binary Constraints (1999)

[6] Galinier, P., Jaumard, B., Morales, R., Pesant, G.: A constraint-based approach to the Golomb ruler problem. In: Third International Workshop on the Integration of AI and OR Techniques in Constraint Programming for Combinatorial Optimization Problems (CPAIOR) (2001), A more recent version (June 11, 2007) can be downloaded from
http://www.crt.umontreal.ca/~quosseca/pdf/41-golomb.pdf

[7] Singer, J.: A theorem in finite projective geometry and some applications to number theory. Transactions of the American Mathematical Society 43(3), 377–385 (1938)

[8] Drakakis, K., Gow, R., O'Carroll, L.: On some properties of costas arrays generated via finite fields. In: 2006 40th Annual Conference on Information Sciences and Systems, pp. 801–805. IEEE (2006)

[9] Soliday, S.W., Homaifar, A., Lebby, G.L.: Genetic algorithm approach to the search for Golomb rulers. In: 6th International Conference on Genetic Algorithms (ICGA 1995), pp. 528–535. Morgan Kaufmann (1995)

[10] Prestwich, S.: Trading completeness for scalability: Hybrid search for cliques and rulers. In: Third International Workshop on the Integration of AI and OR Techniques in Constraint Programming for Combinatorial Optimization Problems, CPAIOR (2001)

[11] Dotú, I., Van Hentenryck, P.: A simple hybrid evolutionary algorithm for finding Golomb rulers. In: The IEEE Congress on Evolutionary Computation, pp. 2018–2023. IEEE (2005)

[12] Lorentzen, R., Nilsen, R.: Application of linear programming to the optimal difference triangle set problem. IEEE Trans. Inf. Theor. 37(5), 1486–1488 (2006)

[13] Hansen, P., Jaumard, B., Meyer, C.: On lower bounds for numbered complete graphs. Discrete Applied Mathematics 94(13), 205–225 (1999)

[14] Shearer, J.B.: Improved LP lower bounds for difference triangle sets. Journal of Combinatorics 6 (1999)

[15] Meyer, C., Jaumard, B.: Equivalence of some LP-based lower bounds for the Golomb ruler problem. Discrete Appl. Math. 154(1), 120–144 (2006)

[16] Sellmann, M., Fahle, T.: Constraint programming based Lagrangian relaxation for the automatic recording problem. Annals of Operations Research 118(1-4), 17–33 (2003)

[17] Cronholm, W., Ajili, F.: Strong cost-based filtering for Lagrange decomposition applied to network design. In: Wallace, M. (ed.) CP 2004. LNCS, vol. 3258, pp. 726–730. Springer, Heidelberg (2004)

[18] Sellmann, M.: Theoretical foundations of CP-based Lagrangian relaxation. In: Wallace, M. (ed.) CP 2004. LNCS, vol. 3258, pp. 634–647. Springer, Heidelberg (2004)

[19] Gellermann, T., Sellmann, M., Wright, R.: Shorter path constraints for the resource constrained shortest path problem. In: Barták, R., Milano, M. (eds.) CPAIOR 2005. LNCS, vol. 3524, pp. 201–216. Springer, Heidelberg (2005)

[20] Khemmoudj, M.O.I., Bennaceur, H., Nagih, A.: Combining arc-consistency and dual Lagrangean relaxation for filtering CSPs. In: Barták, R., Milano, M. (eds.) CPAIOR 2005. LNCS, vol. 3524, pp. 258–272. Springer, Heidelberg (2005)

[21] Menana, J., Demassey, S.: Sequencing and counting with the multicost-regular constraint. In: van Hoeve, W.-J., Hooker, J.N. (eds.) CPAIOR 2009. LNCS, vol. 5547, pp. 178–192. Springer, Heidelberg (2009)

[22] Cambazard, H., O'Mahony, E., O'Sullivan, B.: Hybrid methods for the multileaf collimator sequencing problem. In: Lodi, A., Milano, M., Toth, P. (eds.) CPAIOR 2010. LNCS, vol. 6140, pp. 56–70. Springer, Heidelberg (2010)

[23] Benchimol, P., van Hoeve, W.J., Régin, J.C., Rousseau, L.M., Rueher, M.: Improved filtering for weighted circuit constraints. Constraints 17(3), 205–233 (2012)

[24] Régin, J.-C.: Solving problems with CP: Four common pitfalls to avoid. In: Lee, J. (ed.) CP 2011. LNCS, vol. 6876, pp. 3–11. Springer, Heidelberg (2011)

[25] Régin, J.C.: A filtering algorithm for constraints of difference in CSPs. In: Proceedings of AAAI, pp. 362–367. AAAI Press (1994)
[26] Nemhauser, G.L., Wolsey, L.A.: Integer and Combinatorial Optimization. Wiley (1988)
[27] Held, M., Karp, R.M.: The travelling salesman problem and minimum spanning trees. Operations Research 18, 1138–1162 (1970)
[28] Held, M., Wolfe, P., Crowder, H.: Validation of subgradient optimization. Mathematical Programming 6, 62–88 (1974)
[29] Gervet, C.: Constraints over structured domains. In: Rossi, F., van Beek, P., Walsh, T. (eds.) Handbook of Constraint Programming. Elsevier Science Inc. (2006)

MiniZinc with Functions

Peter J. Stuckey[1] and Guido Tack[2]

[1] National ICT Australia (NICTA) and University of Melbourne, Victoria, Australia
pstuckey@unimelb.edu.au
[2] National ICT Australia (NICTA) and Faculty of IT, Monash University, Australia
guido.tack@monash.edu

Abstract. Functional relations are ubiquitous in combinatorial problems – the Global Constraint Catalog lists 120 functional constraints. This paper argues that the ability to express functional constraints with functional syntax leads to more elegant and readable models, and that it enables a better translation of the models to different underlying solving technologies such as CP, MIP, or SAT.

Yet, most modelling languages only support built-in functions, such as arithmetic, Boolean, or array access operations. Custom, user-defined functions are either not catered for at all, or they have an ad-hoc implementation without a useful semantics in Boolean contexts and not exploiting potential optimisations.

This paper develops a translation from MiniZinc with user-defined functions to FlatZinc. The translation respects the relational semantics of MiniZinc, correctly dealing with partial functions in arbitrary Boolean contexts. At the same time, it takes advantage of the full potential of common subexpression elimination.

1 Introduction

Functions are ubiquitous in models of combinatorial problems. They appear in arithmetic expressions, array accesses, expressions over sets of variables, or more complicated relations such as sorting or channeling. While it is always possible to express a functional dependency as a relational constraint, standard functional notation makes models more elegant, concise, and self-documenting. Furthermore, functional constraints are the main source of *common subexpressions*, which can be detected and eliminated from the model to improve solving performance.

Yet, most constraint modelling languages only provide a restricted set of built-in functions, e.g. for arithmetic expressions. It is either impossible for users to define their own functions, or one must resort to functions as present in the host language that the modelling language is embedded in. The latter, however, means that the semantics of functions is dictated by the host language, and in the case of partial functions, this almost certainly clashes with the logical semantics one would expect from the constraint model. Using host language functions also prevents the detection of common subexpressions and the corresponding optimisations.

Solver independent modelling languages need powerful abstraction facilities in order to encode how models are mapped to the form required by an underlying solver. The MiniZinc [11] compiler translates a model together with the instance parameters

C. Gomes and M. Sellmann (Eds.): CPAIOR 2013, LNCS 7874, pp. 268–283, 2013.
© Springer-Verlag Berlin Heidelberg 2013

to FlatZinc, a lower-level language understood by many solvers. To specialise the generated FlatZinc for a particular solver, the compiler uses a library of solver-specific constraint decompositions. The addition of functions, with common subexpression elimination, makes the encoding of solver-specific translations much more powerful.

This paper introduces an extension to the MiniZinc modelling language that adds support for user-defined functions, respecting the relational semantics of the language, and taking full advantage of potential optimisations. The contributions of this paper are:

- An extension to MiniZinc to support user-defined functions and constraints in `let` expressions.
- The first algorithmic description of the translation from MiniZinc (with functions) to FlatZinc. Keeping FlatZinc unchanged means that any solver that can interpret FlatZinc can take advantage of the improved modelling features.
- The introduction of totality annotations and a schema for defining partial functions that makes user-defined functions more flexible and efficient.

Partial Functions and the Relational Semantics. Partial functions are undefined on some inputs. The *relational semantics* [5] of MiniZinc regards a partial function as special notation for a relation, which when applied to a value outside its domain makes its surrounding *Boolean context* false.

Let us have a look at the partial function `div`, which expresses integer division and requires the divisor to be non-zero. Consider the following simple constraint:

```
constraint y != 0 -> (x div y) + z = 0;
```

The relational semantics demands that $x = 0, y = 0, z = 0$ is a solution of the problem, because the partiality of `div` is confined to its Boolean context, the right-hand side of the implication. This means that we cannot simply decompose constraints by introducing auxiliary variables for intermediate results:

```
var int: tmp;
constraint tmp = x div y;
constraint y != 0 -> tmp + z = 0;
```

This formulation lifts the partiality of `div` to the top level, so that $y = 0$ is ruled out by propagation of the `div` constraint, and $x = 0, y = 0, z = 0$ is no longer a solution of the problem. This simple decomposition gives the *strict semantics* [5], as implemented e.g. in SICStus Prolog and OPL. But this is not usually what modellers require – indeed the example was written specifically to guard against the case that $y = 0$.

Clearly, user-defined functions should respect the same relational semantics as the built-ins. Any function that constrains its arguments must be considered partial.

Common Subexpression Elimination. A widely used optimisation for programming languages in general and constraint modelling in particular is common subexpression elimination (CSE). A compiler detects when the same expression (modulo some equivalences) appears several times in a model, and automatically keeps only one of these expressions, replacing all others by a reference to it. For example, the division in the following code occurs twice:

```
constraint (x div y) + a = 0;
constraint b * (x div y) = c;
```

So an equivalent model would hoist the division to the top level:

```
var int: tmp = x div y;
constraint tmp + a = 0;
constraint b * tmp = c;
```

User-defined functions increase the potential for automatic CSE because they introduce **syntactic equalities**. One could introduce and detect the same equalities without functional notation, by e.g. annotating "functional predicates". However, this would not simplify the translation presented in section 5 that performs CSE while maintaining the relational semantics, but it would make the syntax much less convenient.

CSE is particularly effective if the function definition is complex, it introduces additional variables, or its constraints are expensive to propagate. CSE also makes the abstraction facilities in a language more useful, as common subexpressions will be detected across different functions.

2   Adding Functions and Local Constraints to MiniZinc

We extend MiniZinc by adding functions, using the Zinc [10] syntax for functions. The other important extension to MiniZinc, not currently supported by Zinc, is to allow constraints to occur inside `let` constructs.

The most basic form of user-defined functions would be a simple macro mechanism that can be used to define abbreviations for functional compositions. For instance, we could define a function for the Manhattan distance between two points as

```
function var int: manhattan(var int:x1, var int:y1,
                            var int:x2, var int:y2) =
   abs(x2 - x1) + abs(y2 - y1);
```

Such a macro mechanism would be useful by itself and straightforward to implement; both CSE and the relational semantics would be ensured by the translation of primitive functions. However, as soon as we permit `let` expressions that introduce new variables and constraints in the function body, we have to define how these are translated in reified contexts. This is the main contribution of this paper.

Suppose we wished to add a `sqr` function to MiniZinc that squares its argument.[1] While we could do this without using local variables

```
function var int: sqr(var int:x) = x * x;
```

we can make the model propagate stronger using the following definition

```
function var int: sqr(var int: x)   =
   let { var int: y = x * x; constraint y >= 0 } in y;
```

which explicitly adds that the result is non-negative.

---

[1] This does not preclude solvers from using a more efficient built-in version of `sqr`, as they can simply override the standard definition in their solver-specific MiniZinc library.

Note that we assume that functions defined in MiniZinc are indeed pure (always give the same answer for each input). It is certainly possible to write impure functions:

```
function var int: f(var int:x) = let { var 0..1: b } in b*x;
```

We consider this a modelling error – non-functional relations like this must be expressed using predicates in MiniZinc. As purity analysis is difficult, our implementation does not enforce purity but results in undefined behaviour of impure functions. We plan to investigate a simple but incomplete purity analysis in future work.

3   Using Functions

In this section we illustrate the modelling possibilities that arise from the introduction of user-defined functions.

Big Data. The bigger the data sets involved, the more crucial it is to get common subexpression elimination. The motivating example that made us add functions to MiniZinc arose from modelling data mining problems [7].

An item set mining problem consists of a large data base of *transactions* TDB, which is a list of sets of *items* (e.g., items that were bought together in one transaction from a supermarket). Each transaction has an integer identifier.

An important concept in item set mining is the *cover* of an item set, the set of labels of transactions in which the item set occurs. It can be defined naturally as a function:

```
function var set of int: cover(var set of int: Items,
     array[int] of set of int: TDB) =
   let {   var set of index_set(TDB): Trans;
           constraint forall (t in index_set(TDB)) (
                      t in Trans <-> Items subset TDB[t])
   } in Trans;
```

Constraints involving the cover could restrict its size to at least *k* (*frequent* item set mining), or require item sets to be *closed* (i.e., maximal).

With TDB being a huge data base, it would have a catastrophic impact on translation and solver performance if we did not get CSE for different calls to cover. Without support for functions, we can only lift out the definition of cover, performing CSE by hand. The result is a much less readable model, and indeed a loss in compositionality, as predicates that use cover can no longer be defined in a self-contained way.

Functional Global Constraints. The global constraint catalog [3] lists 120 constraints (almost a third) as functional in nature, such as cycle, change, common, distance, global_cardinality, graph_crossing, indexed_sum, path.

We can define functional versions that map to the global constraints, which yields natural models and gives the MiniZinc compiler important hints for performing common subexpression elimination.

In a model for the Warehouse Location Problem (CSPLib 034) we need to constrain the number of stores supplied by each warehouse. We can use the global constraint count(x, i, c) which constrains c to be the number of times that the value i appears in the array x. Assuming an array `supplier` mapping each store to its supplying warehouse, the function `inUse`(i) returns the number of stores supplied by warehouse i.

```
function var int: inUse(int: i) =
   let { var int: use;
            constraint count(supplier,i,use) } in use;
constraint forall (i in index_set(supplier))
      ( inUse(i) <= capacity[i] );
```

Using MiniZinc functions we do not have to introduce auxiliary variables or perform CSE by hand in our model.

In recent work on reifying global constraints [2], functional global constraints play a key role, as they permit decomposition of globals into a functionally defined part that need not be reified, and a part that is easy to reify. User-defined functions make these techniques immediately accessible in MiniZinc.

Functional Tables. A convenient way to model an ad-hoc partial function is using a `table` constraint, but doing so hides the fact that the constraint is functional.

The following model fragment defines a partial function `partner` using a list of pairs `pairs`. This is used in a complex constraint (with common subexpressions).

```
array[1..10] of var 1..10: p;
array[1..10] of var bool: notused;
array[1..2][1..6] of int: pairs =
       [ 1, 4 | 2, 1 | 5, 1 | 7, 10 | 8, 3 | 10, 2  ];
function var int: partner(var int:x) =
   let { var int: r;
            constraint table([x,r], pairs) } in r;
constraint forall(i in 1..10)( notused[i] \/
   partner(p[i]) > i \/ partner(partner(p[i])) > i );
```

The function definition allows the MiniZinc compiler to detect the common subexpressions, and the user can simply use the partial function as they intended. Note that `partner` creates a new variable and a `table` constraint in a reified context. Without functions, CSE as well as guarding for partiality has to be performed manually:

```
constraint forall(i in 1..10)(
   let { var int: p_pi;
            var int: p_p_pi;
            var bool: b1 = table([p[i],p_pi],pairs);
            var bool: b2 = table([p_pi,p_p_pi], pairs);
   } in notused[i] \/ (b1 /\ p_pi > i) \/
         (b1 /\ b2 /\ p_p_pi > i));
```

This manual approach is not *compositional*. MiniZinc encourages the use of predicate definitions, which is incompatible with manual, whole-program CSE.

Solver Specific Translation / Channelling. Solver independent models in MiniZinc are translated to solver-specific versions using custom predicate definitions. In many cases these definitions need to introduce new variables that are functionally defined in terms of the original variables. We can use functions to make this straightforward and automatically achieve CSE if the translation is required more than once.

Consider the mapping of not equals for an integer programming solver. We can define a function int2array01 such that int2array01(x)[i]=1 iff x=i. Using this function, the not equals constraint is straightforward to define:[2]

```
predicate int_neq(var int: x, var int:y) = let {
    array[int] of var 0..1: bx = int2array01(x);
    array[int] of var 0..1: by = int2array01(y);
} in forall(i in max(lb(x),lb(y))..min(ub(x),ub(y)))
            (bx[i] + by[i] <= 1);
```

Crucially, CSE guarantees that for any integer variable, at most one array of 01 variables is created, so that we get an efficient translation of constraints that share variables such as int_neq(x,y) /\ int_neq(x,z). Without functions, an efficient abstraction like int_neq would not be possible, and the user would be required to carefully avoid introducing common subexpressions while modelling.

The above encoding generalises easily to the *alldifferent* constraint on an array of variables. We can use similar mechanisms for translating integer and set models for SAT solvers, and for channelling constraints between two different viewpoints of a model, such as a set-based model and a Boolean model where for each set variable x we introduce an array of Boolean variables b such that $i \in x \Leftrightarrow b[i]$.

Simplifying MiniZinc Translation. Once we have generic methods for handling functions, many of the built-in functions in MiniZinc can be simply treated as library functions. We can then use the generic translation to FlatZinc. This simplifies the translation and makes it more transparent and adaptable.

Consider the built-in abs function. We can remove it from MiniZinc and add the following library definition.

```
function var int: abs(var int: x) =
    let { var int: y; constraint int_abs(x,y) } in y;
```

The handling of this function will automatically create the FlatZinc constraint int_abs for any absolute value expressions.[3] Built-in functions that can be relegated in this way include: the trigonometric functions, their inverses and hyperbolic versions, exponentiation and logarithms, amongst others.

4 Partial and Total Functions and Negation

Negative Boolean contexts require particular care in constraint programming systems with local variables. Consider

---

[2] The lb and ub functions return a guaranteed lower and upper bound of a variable (usually the bound declared in the model).

[3] To be used in all contexts it needs to be annotated as total as explained in the next section.

```
function var int: evendiv2(var int: x) =
  let { var int: y; constraint x = 2*y } in y;
constraint not evendiv2(x)=1;
```

A logical interpretation would be $\neg\exists y.x = 2 \times y \wedge y = 1$ or $\forall y.x \neq 2 \times y \vee y \neq 1$. However, we cannot easily express universal quantification, and a naive translation of the above (simply ignoring the universal quantifier) would produce the following:

```
var int: y;
constraint not (x = 2*y /\ y = 1);
```

This does not have the desired semantics, as it permits for example the solution $x = 2, y = 3$, although $x = 2$ clearly should not be a solution. MiniZinc (and Zinc) therefore consider models illegal that contain `let` expressions in a negative context which introduce non-functionally defined variables (like y in the above example).

The upshot of this is that user-defined functions become almost useless in negative contexts, since interesting functions can usually only be defined by introducing new variables (as in most of our examples). The following technique avoids this problem.

Totality Annotations. Functions with local variables can be used safely in negative contexts if they are *total*, that is they are defined on all their inputs. Consider the following example:

```
function var int: g(var int: x) :: total =
  let { var int: y;
          constraint (x > 0) -> y = x;
          constraint (x <= 0) -> y = 10-x
  } in y;
var -10..10: u;
constraint not g(u) = 5;
```

Even though syntactically y is not functionally defined by the function, it is semantically. It is safe to translate the last line as

```
var int: y = g(u);      % evaluate g in root context
constraint not y = 5;   % only negate the equality
```

To allow the user to declare total functions, we introduce the annotation `total` which can be added to function definitions as seen above. The annotation `total` promises that the function is total for all uses, and that it does not constrain its arguments. The function will be translated *in the root context*, which means that free variable definitions are allowed. If a user annotates a function as total that is in fact partial, this effectively results in the function being translated according to the strict semantics. We do not, at this point, attempt to detect incorrect totality annotations automatically.

Recipe for Partial Functions. Totality annotations enable us to define some partial functions that introduce variables, by following a simple *recipe*. Assume that we want to implement a partial function $f(x)$, and that we can express its domain of definition by a constraint $c(x)$ that does not introduce any free variables. Then

1. create a *total extension* f' such that $f'(x) = f(x)$ if $c(x)$, and $f'(x) = 0$ otherwise. We annotate f' as `::total`, so it can introduce arbitrary free variables.
2. create a *partial guard function* $g(x) = $ `let { ` $c(x)$ ` } in ` $f'(x)$ `;`

Consider for example `evendiv2` as defined above. It can be rewritten as

```
function var int: evendiv2(var int: x) =
  let { constraint x mod 2 = 0 } in safe_ed2(x);
function var int: safe_ed2(var int: x) :: total =
  let { var int: y;
        constraint x mod 2 = 0 -> x = 2*y;
        constraint not (x mod 2 = 0) -> y = 0} in y;
```

Now `evendiv2` is a partial function, but does not introduce variables which are not functionally defined, and `safe_ed2` is a total function, which is the same as the original `evendiv2` when x mod $2 = 0$. The constraint `not evendiv2(x)=1` now translates as

```
var int: y;
constraint safe_ed2(x)=y;
constraint not (x mod 2 = 0 /\ y = 1);
```

Now $x = 2$ enforces $y = 1$, so the negation fails and only correct solutions are returned.

We use the same mechanism for the translation of built-in partial functions `div` and array access, assuming built-in total functions *safediv* and *safeelement*.

If we can express the implicit constraint $c(x)$ of a partial function on its input arguments without introducing new variables, we can freely use the partial function in any context. Of course, for some functions, $c(x)$ may be difficult to express, so this mechanism is no general solution to the problem. This is not surprising, as otherwise we could translate models with arbitrary quantifiers efficiently to MiniZinc.

5   Translation to FlatZinc

Constraint problems formulated in MiniZinc are solved by translating them to a simpler, solver-specific subset of MiniZinc, called FlatZinc. This section shows how to translate our extended MiniZinc to FlatZinc.

The complexities in the translation arise from the need to simultaneously (a) unroll array comprehensions (and other loops), (b) replace predicate and function applications by their body, and (c) flatten expressions.

Once we take into account CSE, we cannot perform these separately. In order to have names for common subexpressions we need to flatten expressions. And in order to take advantage of functions for CSE we cannot replace predicate and function applications without flattening to generate these names. And without replacing predicate and function application by their body we are unable to see all the loops to unroll.

The translation algorithm presented below generates a flat model equivalent to the original model as a global set of constraints S. We ignore the collection of variable declarations, which is also clearly important, but quite straightforward. The translation uses full reification to create the model. It can be extended to use half reification [4], but we omit this for space reasons. Common subexpression elimination is implemented using a technique similar to *hash-consing* in Lisp [1]. For simplicity we only show syntactic CSE, which eliminates expressions that are identical after parameter replacement. The extension to semantic CSE, using commutativity and other equivalences, is well understood (see [12] for a detailed discussion) but makes the pseudo-code much longer.

MiniZinc Syntax. Below is a grammar for a subset of MiniZinc as it currently stands, with enough complexity to illustrate all the main challenges in extending it to include functions. The cons nonterminal defines constraints (Boolean terms), term defines integer terms, barray defines Boolean arrays, and iarray defines integer arrays:

$$
\begin{aligned}
\text{cons} &\rightarrow \texttt{true} \mid \texttt{false} \mid \text{bvar} \mid \text{term relop term} \\
&\rightarrow \texttt{not cons} \mid \text{cons} \;\texttt{/\char92}\; \text{cons} \mid \text{cons}\;\texttt{\char92/}\;\text{cons} \mid \text{cons -> cons} \mid \text{cons <-> cons} \\
&\rightarrow \texttt{forall barray} \mid \texttt{exists barray} \mid \text{barray[term]} \\
&\rightarrow \texttt{pred(term, ..., term)} \mid \texttt{if cons then cons else cons endif} \\
&\rightarrow \texttt{let \{ decls \} in cons} \\
\text{term} &\rightarrow \texttt{int} \mid \text{ivar} \mid \text{term arithop term} \mid \text{iarray[term]} \mid \texttt{sum iarray} \\
&\rightarrow \texttt{if cons then term else term endif} \\
&\rightarrow \texttt{let \{ decls \} in term} \\
\text{barray} &\rightarrow \texttt{[cons, ..., cons]} \mid \texttt{[cons | ivar in term .. term]} \\
\text{iarray} &\rightarrow \texttt{[term, ..., term]} \mid \texttt{[term | ivar in term .. term]}
\end{aligned}
$$

The grammar uses the symbols bvar for Boolean variables, relop for relational operators { ==, <=, <, !=, >=, > }, pred for names of predicates, int for integer constants, ivar for integer variables, and arithop for arithmetic operators { +, -, *, div }.

In the `let` constructs we make use of the nonterminal decls for declarations. We define this below using idecl for integer variable declarations, bdecl for Boolean variable declarations. We also define args as a list of integer variable declarations, an item item as either a predicate declaration or a constraint, items as a list of items, and a model model as some declarations followed by items. Note that ε represents the empty string.

$$
\begin{aligned}
\text{idecl} &\rightarrow \texttt{int: ivar} \mid \texttt{var int: ivar} \mid \texttt{var term .. term : ivar} \\
\text{bdecl} &\rightarrow \texttt{bool: bvar} \mid \texttt{var bool: bvar} \\
\text{decls} &\rightarrow \varepsilon \mid \texttt{idecl; decls} \mid \texttt{idecl = term; decls} \mid \texttt{bdecl; decls} \mid \texttt{bdecl = cons; decls} \\
\text{args} &\rightarrow \texttt{var int: ivar} \mid \texttt{var int: ivar , args} \mid \texttt{int: ivar} \mid \texttt{int: ivar , args} \\
\text{item} &\rightarrow \texttt{predicate pred(args) = cons;} \mid \texttt{constraint cons;} \\
\text{items} &\rightarrow \varepsilon \mid \texttt{item items} \\
\text{model} &\rightarrow \texttt{decls items}
\end{aligned}
$$

To simplify presentation, we assume all predicate arguments are integers, but the translation can be extended to arbitrary arguments in a straightforward way.

To introduce functions in MiniZinc, the grammar above is modified as follows. We add a new item type for functions (again for simplicity all arguments are assumed to be integers). We change the form of the `let` construct for both constraints and terms to allow optional constraints (ocons).

```
item   → function var int  : func ( args ) = term ;
cons   → let  { decls ocons } in cons
term   → let  { decls ocons } in term
ocons  → ε |  ; constraint cons ocons
```

Further Notation. Given an expression e that contains the subexpression x, we denote by $e[\![x/y]\!]$ the expression that results from replacing all occurrences of subexpression x by expression y. We will also use the notation for multiple simultaneous replacements $[\![\bar{x}/\bar{y}]\!]$ where each $x_i \in \bar{x}$ is replaced by the corresponding $y_i \in \bar{y}$.

Given a cons term defining the constraints of the model we can split its cons subterms as occurring in different kinds of places: root contexts, positive contexts, negative contexts, and mixed contexts. A Boolean subterm t of constraint c is in a *root context* iff there is no solution of $c[\![t/false]\!]$, that is c with subterm t replaced by *false*.[4] Similarly, a subterm t of constraint c is in a *positive context* iff for any solution θ of c then θ is also a solution of $c[\![t/true]\!]$; and a *negative context* iff for any solution θ of c then θ is also a solution of $c[\![t/false]\!]$. The remaining Boolean subterms of c are in *mixed* contexts. While determining contexts according to these definitions is hard, there are simple syntactic rules which can determine the correct context for most terms, and the rest can be treated as mixed. Consider the constraint expression

```
constraint x > 0 /\ (i <= 4 -> x + bool2int(b) = 5);
```

then $x > 0$ is in the root context, $i \le 4$ is in a negative context, $x + \text{bool2int}(b) = 5$ is in a positive context, and b is in a mixed context.

Flattening Constraints. Flattening a constraint c in context *ctxt*, flatc$(c, ctxt)$, returns a Boolean literal b representing the constraint and as a side effect adds a set of constraints (the flattening) S to the store such that $S \models b \Leftrightarrow c$.

It uses the context (*ctxt*) to decide how to translate, where possible contexts are: *root*, at the top level conjunction; *pos*, positive context, *neg*, negative context, and *mix*, mixed context. We use the context operations $+$ and $-$ defined as:

$$+root = pos \quad +neg = neg, \quad -root = neg \quad -neg = pos,$$
$$+pos = pos \quad +mix = mix \quad -pos = neg \quad -mix = mix$$

Note that flattening in the root context always returns a Boolean b made equivalent to *true* by the constraints in S. For succinctness we use the notation **new** b (**new** v) to introduce a fresh Boolean (resp. integer) variable and return the name of the variable.

The Boolean result of flattening c is stored in a hash table, and reused if an identical constraint expression is ever flattened again. If the context for an expression e is root, the result of a successful hash should be *true*. If we have a common subexpression e in another context, then since it is true at the root it is true there. If we first meet the expression e in a non-root context and later meet expression e at the root, we simply need to set the Boolean created for the first met version to *true*. This is the role of the addition to S on the first line of flatc. Note that in this simplified presentation, if an expression introduces a fresh variable and it appears first in a negative context and only

---

[4] For the definitions of context we assume that the subterm t is uniquely defined by its position in c, so the replacement is of exactly one subterm.

later in the root context, the translation aborts. This can be fixed in a preprocessing step that sorts expressions according to their context.

The flattening proceeds by evaluating fixed Boolean expressions and returning the value. We assume fixed checks if an expression is fixed (determined during MiniZinc's type analysis), and eval evaluates a fixed expression. For simplicity of presentation, we assume that fixed expressions are never undefined (such as 3 div 0).

For non-fixed expressions we treat each form in turn. Boolean literals and variables are simply returned. Basic relational operations flatten their terms using the function flatt, which for a term t returns a tuple $\langle v, b \rangle$ of an integer variable/value v and a Boolean literal b such that $b \Leftrightarrow (v = t)$ (described in detail below). The relational operations then return a reified form of the relation. The logical operators recursively flatten their arguments, passing in the correct context. The logical array operators evaluate their array argument, then create an equivalent term using foldl and either /\ or \/ which is then flattened. A Boolean array lookup flattens its arguments, and creates a *safeelement* constraint (which does not constrain the index variable) and Boolean b'' to capture whether the array lookup was safe. Built-in predicates abort if not in the root context. They flatten their arguments and add an appropriate built-in constraint. User defined predicate applications flatten their arguments and then flatten a renamed copy of the body. *if-then-else* evaluates the condition (which must be fixed) and flattens the *then* or *else* branch appropriately. The handling of let is the most complicated. The expression is renamed with new copies of the let variables. We extract the constraints from the let expression using function flatlet which returns the extracted constraint and a rewritten term (not used in this case, but used in flatt). The constraints returned by function flatlet are then flattened. Finally if we are in the root context, we ensure that the Boolean b returned must be *true* by adding b to S.

flatc(c,*ctxt*)

    h := hash[c]; **if** ($h \neq \bot$) $S \cup$:= $\{(ctxt = root) \Rightarrow h\}$; **return** h

    if (fixed(c)) b := eval(c)

    else

        switch c

        case b' (bvar): b := b';

        case t_1 r t_2 (relop): $\langle v_1, b_1 \rangle$:= flatt(t_1, *ctxt*); $\langle v_2, b_2 \rangle$:= flatt(t_2, *ctxt*);

            $S \cup$:= {**new** $b \Leftrightarrow (b_1 \wedge b_2 \wedge$ **new** b'), $b' \Leftrightarrow v_1$ r v_2}

        case not c_1: b := $\neg$flatc(c_1, $-ctxt$)

        case c_1 /\ c_2: $S \cup$:= {**new** $b \Leftrightarrow$ (flatc(c_1, *ctxt*) $\wedge$ flatc(c_2, *ctxt*))}

        case c_1 \/ c_2: $S \cup$:= {**new** $b \Leftrightarrow$ (flatc(c_1, $+ctxt$) $\vee$ flatc(c_2, $+ctxt$))}

        case c_1 -> c_2: $S \cup$:= {**new** $b \Leftrightarrow$ (flatc(c_1, $-ctxt$) $\Rightarrow$ flatc(c_2, $+ctxt$))}

        case c_1 <-> c_2: $S \cup$:= {**new** $b \Leftrightarrow$ (flatc(c_1, *mix*) $\Leftrightarrow$ flatc(c_2, *mix*))}

        case forall ba: b := flatc(foldl(evala(ba), true, /\), *ctxt*)

        case exists ba: b := flatc(foldl(evala(ba), false, \/), *ctxt*)

        case $[c_1,...,c_n]$ [t]: **foreach**($j \in 1..n$) b_j := flatc(c_j, $+ctxt$); $\langle v, b_{n+1} \rangle$:= flatt(t, *ctxt*);

            S:={**new** $b \Leftrightarrow (b_{n+1} \wedge$ **new** $b' \wedge$ **new** b''), *safeelement*($v, [b_1,...,b_n], b'$), $b'' \Leftrightarrow v \in \{1,..,n\}$}

        case p ($t_1,...,t_n$) (pred) built-in predicate:

            if ($ctxt \neq root$) **abort**

            foreach($j \in 1..n$) $\langle v_j, \_ \rangle$:= flatt(t_j, *ctxt*);

$b := \textit{true}; \; S \cup := \{p(v_1, \ldots, v_n)\}$

case $p \; (t_1, \ldots, t_n)$ (pred): user-defined predicate

   let $p(x_1, \ldots, x_n) = c_0$ be defn of p

   foreach$(j \in 1..n) \; \langle v_j, b_j \rangle := \mathsf{flatt}(t_j, ctxt);$

   new $b' := \mathsf{flatc}(c_0[\![x_1/v_1, \ldots, x_n/v_n]\!], ctxt)$

   $S \cup := \{\textbf{new } b \Leftrightarrow b' \wedge \bigwedge_{j=1}^{n} b_j\}$

case if c_0 then c_1 else c_2 endif: **if** $(\mathsf{eval}(c_0)) \; b := \mathsf{flatc}(c_1, ctxt)$ **else** $b := \mathsf{flatc}(c_2, ctxt)$

case let $\{ \; d \; \}$ in c_1:

   let $\bar{v}'$ be a renaming of variables $\bar{v}$ defined in d

   $\langle c', \_ \rangle := \mathsf{flatlet}(d[\![\bar{v}/\bar{v}']\!], c_1[\![\bar{v}/\bar{v}']\!], 0, ctxt)$

   $b := \mathsf{flatc}(c', ctxt);$

if $(ctxt = root) \; S \cup := \{b\};$

$\mathsf{hash}[c] := b$

return b

The function evala replaces an array comprehension by the resulting array. Note that terms lt and ut must be fixed in a correct MiniZinc model.

$\mathsf{evala}(t)$

   switch t

   case $[\; t_1, \ldots, t_n \;]$: **return** $[\; t_1, \ldots, t_n \;]$

   case $[\; e \mid v \text{ in } lt \; .. \; ut \;]$: **let** $l = \mathsf{eval}(lt), u = \mathsf{eval}(ut)$

     return $[\; e[\![v/l]\!], e[\![v/l+1]\!], \ldots e[\![v/u]\!] \;]$

Handling `let` Expressions. Much of the handling of a `let` is implemented by flat-let$(d, c, t, ctxt)$ which takes the declarations d inside the `let`, the constraint c or term t of the scope of the `let` expression, as well as the context type. First flatlet replaces parameters defined in d by their fixed values. Then it collects in c all the constraints that need to stay within the Boolean context of the `let`: the constraints arising from the variable and constraint items, as well as the Boolean variable definitions. Integer variables that are defined have their right hand side flattened, and a constraint equating them to the right hand side t' added to the global set of constraints S. If variables are not defined and the context is negative or mixed the translation aborts.

$\mathsf{flatlet}(d, c, t, ctxt)$

   foreach $(item \in d)$

     switch $item$

     case int : $v = t'$: $d := d[\![v/\mathsf{eval}(t')]\!]; \; c := c[\![v/\mathsf{eval}(t')]\!]; \; t := t[\![v/\mathsf{eval}(t')]\!]$

     case bool : $b = c'$: $d := d[\![b/\mathsf{eval}(c')]\!]; \; c := c[\![b/\mathsf{eval}(c')]\!]; \; t := t[\![b/\mathsf{eval}(c')]\!]$

   foreach $(item \in d)$

     switch $item$

     case var int : v: **if** $(ctxt \in \{neg, mix\})$ **abort**

     case var int : $v = t'$: $\langle v', b' \rangle := \mathsf{flatt}(t', ctxt); \; S \cup := \{v = v'\}; \; c := c \; / \backslash \; b'$

     case var $l \; .. \; u$: v: **if** $(ctxt \in \{neg, mix\})$ **abort else** $c := c \; / \backslash \; l \mathrel{<=} v \; / \backslash \; v \mathrel{<=} u$

     case var $l \; .. \; u$: $v = t'$: $\langle v', b' \rangle := \mathsf{flatt}(t', ctxt); \; S \cup := \{v = v'\};$

                       $c := c \; / \backslash \; b' \; / \backslash \; l \mathrel{<=} v \; / \backslash \; v \mathrel{<=} u;$

     case var bool : b: **if** $(ctxt \in \{neg, mix\})$ **abort**

     case var bool : $b = c'$: $c := c \; / \backslash \; (b \mathrel{<\text{-}>} c')$

case constraint c' : $c := c \wedge c'$
return $\langle c,t \rangle$

Flattening Integer Expressions. flatt$(t,ctxt)$ flattens an integer term t in context $ctxt$. It returns a tuple $\langle v,b \rangle$ of an integer variable/value v and a Boolean literal b, and as a side effect adds constraints to S so that $S \models b \Leftrightarrow (v = t)$. Note that again flattening in the root context always returns a Boolean b made equivalent to *true* by the constraints in S.

Flattening first checks if the same expression has been flattened previously and if so returns the stored result. flatt evaluates fixed integer expressions and returns the result. For non-fixed integer expressions t each form is treated in turn. Simple integer expressions are simply returned. Operators have their arguments flattened and the new value calculated on the results. Safe division (which does not constrain its second argument to be non-zero) is used for division with the constraint being captured by the new Boolean b'. Array lookup flattens all the integer expressions involved and creates a *safeelement* constraint as in the Boolean case. sum expressions evaluate the array argument, and then replace the sum by repeated addition using foldl and flatten that. *if-then-else* simply evaluates the *if* condition (which must be fixed) and flattens the *then* or *else* branch appropriately. Functions are simply handled by flattening each of the arguments, the function body is then renamed to use the variables representing the arguments, and the body is then flattened. Importantly, if the function is declared total it is flattened *in the root context*. Let constructs are handled analogously to flatc. We rename the scoped term t_1 to t' and collect the constraints in the definitions in c'. The result is the flattening of t', with b capturing whether anything inside the let leads to failure.

flatt$(t,ctxt)$
$\quad \langle h,b' \rangle := $ hash$[t]$; **if** $(h \neq \bot)$ $S \cup := \{(ctxt = root) \Rightarrow b'\}$; **return** $\langle h,b' \rangle$
$\quad$ **if** $($fixed$(t))$ $v := $ eval(t); $b := true$
$\quad$ **else switch** t
$\quad\quad$ **case** v' (ivar): $v := v'$; $b := true$
$\quad\quad$ **case** t_1 a t_2 (arithop): $\langle v_1,b_1 \rangle := $ flatt$(t_1,ctxt)$; $\langle v_2,b_2 \rangle := $ flatt$(t_2,ctxt)$;
$\quad\quad\quad$ **if** $(a$ is not div$)$ $S \cup := \{$**new** $b \Leftrightarrow (b_1 \wedge b_2), a(v_1,v_2,$**new** $v)\}$
$\quad\quad\quad$ **else** $S \cup := \{$**new** $b \Leftrightarrow (b_1 \wedge b_2 \wedge $**new** $b'), safediv(v_1,v_2,$**new** $v), b' \Leftrightarrow v_2 \neq 0\}$
$\quad\quad$ **case** $[t_1,\ldots,t_n]$ [t_0]: **foreach**$(j \in 0..n)$ $\langle v_j,b_j \rangle := $ flatt$(t_j,ctxt)$;
$\quad\quad\quad$ $S := \{$**new** $b \Leftrightarrow ($**new** $b' \wedge \bigwedge_{j=0}^n b_j), safeelement(v_0,[v_1,\ldots,v_n],$**new** $v), b' \Leftrightarrow v_0 \in \{1,\ldots,n\}\}$
$\quad\quad$ **case** sum ia: $\langle v,b \rangle := $ flatt$($ foldl$($evala(ia), 0, $+)$, $ctxt)$
$\quad\quad$ **case** if c_0 then t_1 else t_2 endif: **if** $(eval(c_0))$ $\langle v,b \rangle := $ flatt$(t_1,ctxt)$ **else** $\langle v,b \rangle := $ flatt$(t_2,ctxt)$
$\quad\quad$ **case** f $(t_1,\ldots,t_n)$ (func): function
$\quad\quad\quad$ **foreach**$(j \in 1..n)$ $\langle v_j,b_j \rangle := $ flatt$(t_j,ctxt)$;
$\quad\quad\quad$ **let** $f(x_1,\ldots,x_n) = t_0$ be defn of f
$\quad\quad\quad$ **if** $(f$ is declared total$)$ $ctxt' = root$ **else** $ctxt' = ctxt$
$\quad\quad\quad$ $\langle v,b' \rangle := $ flatt$(t_0[\![x_1/v_1,\ldots,x_n/v_n]\!],ctxt')$
$\quad\quad\quad$ $S \cup := \{$**new** $b \Leftrightarrow b' \wedge \bigwedge_{j=1}^n b_j\}$
$\quad\quad$ **case** let $\{$ d $\}$ in t_1:
$\quad\quad\quad$ **let** $\bar{v}'$ be a renaming of variables $\bar{v}$ defined in d
$\quad\quad\quad$ $\langle c',t' \rangle := $ flatlet$(d[\![\bar{v}/\bar{v}']\!],$true$,t_1[\![\bar{v}/\bar{v}']\!],ctxt)$
$\quad\quad\quad$ $\langle v,b_1 \rangle := $ flatt$(t',ctxt)$; $b_2 := $ flatc$(c',ctxt)$; $S \cup := \{$**new** $b \Leftrightarrow (b_1 \wedge b_2)\}$

```
if (ctxt = root) S ∪:= {b};
hash[t] := ⟨v,b⟩
return ⟨v,b⟩
```

Consider the translation of the code

```
function var int: h(var int: a) =
    let { var int:d = 12 div a; constraint d < 3 } in d;
constraint not (h(c) = 1);
```

Flattening of not (h(c) = 1) requires flattening the term h(c) in a *neg* context. This requires flattening let { var int:d = 12 div c; constraint d < 3 } in d; in a *neg* context. This requires flattening 12 div c in a *neg* context which creates the constraints safediv(12, c, v), $b_1 \Leftrightarrow c \neq 0$ and returns $\langle v, b_1 \rangle$. Flattening the let adds $d = v$ and collects $b_1 \wedge d < 3$ in the constraints c which are flattened to give $b_3 \Leftrightarrow (d < 3), b_4 \Leftrightarrow (b_3 \wedge b_1)$ and returns tuple $\langle d, b_4 \rangle$. The treatment of the equality adds $b_5 \Leftrightarrow d = 1, b_6 \Leftrightarrow (b_5 \wedge b_4)$ and returns b_6. The negation returns $\neg b_6$ and asserts $\neg b_6$.

Implementation. We have implemented the above rules in a compiler that translates MiniZinc with functions into FlatZinc. This prototype can handle the complete grammar as presented here, extended with support for arbitrary function arguments (arrays, Booleans, sets). Compared to the existing mzn2fzn translator, most built-in functions (such as abs, bool2int, card) are now realised as user-defined functions in the MiniZinc standard library rather than hard coding them into the translation.

The following table shows some experimental results obtained with the prototype. We compiled five different instances of a standard model for a 16x16 Sudoku problem. Each instance has 48 *alldifferent* constraints, and we used the linearisation presented in section 3 to show the effect of common subexpression elimination. Crucially, the *alldifferent* constraints in Sudoku puzzles *overlap*, many pairs of constraints share either one or four variables. The standard decomposition without functions cannot take advantage of this sharing across constraints, as the results below will show.

Benchmark	Translation (s)		# Cons		Solving (s)		
	fn	*nofn*	*fn*	*nofn*	*fn*	*nofn*	*speedup*
Sudoku 1 (16x16)	0.2	0.3	1280	2304	0.24	0.78	3.33
Sudoku 2 (16x16)	0.2	0.3	1280	2304	0.29	3.45	11.96
Sudoku 3 (16x16)	0.2	0.3	1280	2304	0.32	15.32	47.78
Sudoku 4 (16x16)	0.2	0.3	1280	2304	0.24	0.43	1.80
Sudoku 5 (16x16)	0.2	0.3	1280	2304	0.34	6.07	17.72

The columns labelled *fn* present the results for the new prototype translator using a decomposition of *alldifferent* based on functions. The *nofn* columns use the existing mzn2fzn tool with the standard MiniZinc linearisation library (available with the command line option -G linear).

Translation time does not suffer from the additional CSE, as the results in column *Translation* show. In fact, the new translator is slightly faster. CSE clearly reduces the number of constraints by almost half (column *# Cons*). The solving time (column *Solving*) is the average of 20 runs of the Gurobi MIP solver [8] on a 3.4 GHz Intel Core i7, running on a single core under Windows. The additional, redundant constraints in *nofn* cause a noticeable overhead.

We will extend our prototype into a full replacement for the current mzn2fzn, working towards version 2.0 of the MiniZinc language and toolchain. The prototype and the benchmark problems are available from the authors upon request.

6   Related Work and Conclusion

The only modelling language that supports functions other than MiniZinc that we are aware of is Zinc [10]. Zinc functions are restricted in the same way, not allowing new variables in negative or mixed contexts. Zinc currently does not support constraints in let expressions and total annotations. We intend to extend Zinc with these features, although this will be challenging as Zinc models are compiled without instance data, making the treatment of partiality much more complicated and precluding most CSE.

Constraint-based local search languages such as Comet [15] support user-defined *objective functions*. These cannot be used as arguments in other constraints, and therefore these systems do not deal with partiality, Boolean contexts, or CSE.

Other modelling languages such as OPL [14] and Essence [6] do not include user-defined functions or local variables (let expressions), and hence the issues that we consider here do not arise. Essence supports constrained *function variables*, which are used to model problems whose *solution* is a function. Our user-defined functions, in contrast, serve a different purpose, they express the *constraints* of a problem. The approach to flattening and CSE for Essence is described in [12], but Essence includes neither let expressions, predicates or functions which are the main complicating features herein. This earlier work [12] showed the importance of CSE for modelling languages, and this is only magnified by the introduction of user-defined functions.

Modelling languages incorporated in a procedural OO language, such as IBM ILOG Concert [9] expressions, Gecode's [13] MiniModel expressions or Comet's [15] modelling constructs, do allow the use of functions in the *host language*. The problem is that such functions do not extend the modelling language, and if treated in this manner define the strict semantics.

In conclusion the addition of user-defined functions, local constraints in let constructs, and totality annotations gives a powerful modelling capability to MiniZinc. Using our schema for separating a partial function into a total extension with local variable introduction and a partial function with no local variables, user-defined functions are usable in all parts of a model. We believe functional modelling will become more and more commonplace, particularly given the prevalence of functional constraints in the global constraint catalog, given the importance of abstraction for defining solver-specific MiniZinc libraries, and given that functions can be implemented efficiently as shown in this paper.

Acknowledgements. We thank our reviewers for their constructive comments. NICTA is funded by the Australian Government as represented by the Department of Broadband, Communications and the Digital Economy and the Australian Research Council.

References

1. Allen, J.: Anatomy of LISP. McGraw-Hill, Inc., New York (1978)
2. Beldiceanu, N., Carlsson, M., Flener, P., Pearson, J.: On the reification of global constraints. Constraints 18, 1–6 (2013)
3. Beldiceanu, N., Carlsson, M., Rampon, J.X.: Global constraint catalog, working version as of October 4, 2012, http://www.emn.fr/z-info/sdemasse/gccat/
4. Feydy, T., Somogyi, Z., Stuckey, P.J.: Half reification and flattening. In: Lee, J. (ed.) CP 2011. LNCS, vol. 6876, pp. 286–301. Springer, Heidelberg (2011)
5. Frisch, A., Stuckey, P.: The proper treatment of undefinedness in constraint languages. In: Gent, I.P. (ed.) CP 2009. LNCS, vol. 5732, pp. 367–382. Springer, Heidelberg (2009)
6. Frisch, A.M., Harvey, W., Jefferson, C., Hernández, B.M., Miguel, I.: Essence: A constraint language for specifying combinatorial problems. Constraints 13(3), 268–306 (2008)
7. Guns, T.: Declarative Pattern Mining using Constraint Programming. Ph.D. thesis, Department of Computer Science, K.U.Leuven (2012)
8. Gurobi Optimization, Inc.: Gurobi Optimizer Reference Manual (2012), http://www.gurobi.com
9. IBM: ILOG Concert, part of IBM ILOG CPLEX Optimization Studio (2012), http://www-01.ibm.com/software/integration/optimization/cplex-optimizer/interfaces/
10. Marriott, K., Nethercote, N., Rafeh, R., Stuckey, P., Garcia de la Banda, M., Wallace, M.: The design of the Zinc modelling language. Constraints 13(3), 229–267 (2008)
11. Nethercote, N., Stuckey, P.J., Becket, R., Brand, S., Duck, G.J., Tack, G.: MiniZinc: Towards a standard CP modelling language. In: Bessière, C. (ed.) CP 2007. LNCS, vol. 4741, pp. 529–543. Springer, Heidelberg (2007)
12. Rendl, A.: Effective Compilation of Constraint Models. Ph.D. thesis, School of Computer Science, University of St. Andrews (2010)
13. Schulte, C., et al.: Gecode, the generic constraint development environment (2009), http://www.gecode.org/
14. Van Hentenryck, P.: The OPL Optimization Programming Language. MIT Press (1999)
15. Van Hentenryck, P., Michel, L.: Constraint-Based Local Search. MIT Press (2005)

Solving Wind Farm Layout Optimization with Mixed Integer Programming and Constraint Programming

Peter Y. Zhang, David A. Romero, J. Christopher Beck, and Cristina H. Amon

Department of Mechanical and Industrial Engineering
University of Toronto
peteryun.zhang@mail.utoronto.ca, {d.romero,cristina.amon}@utoronto.ca,
jcb@mie.utoronto.ca

Abstract. The wind farm layout optimization problem is concerned with the optimal location of turbines within a fixed geographical area to maximize energy capture under stochastic wind conditions. Previously it has been modelled as a maximum diversity (or p-dispersion-sum) problem, but such a formulation cannot capture the nonlinearity of aerodynamic interactions among multiple wind turbines. We present the first constraint programming (CP) and mixed integer linear programming (MIP) models that incorporate such nonlinearity. Our empirical results indicate that the relative performance between these two models reverses when the wind scenario changes from a simple to a more complex one. We also propose an improvement to the previous maximum diversity model and demonstrate that the improved model solves more problem instances.

1 Introduction

Wind farm layout optimization problems deal with the optimal placement of turbines in a given wind farm field. Currently this problem appears only in the engineering research literature [1–6], where much effort has been spent on developing metaheuristics [1–3, 7] for variations of this problem. Some existing heuristic methods [1, 2] and mixed integer models [4, 5] have explored this problem with discretization: land is decomposed into a set of small cells, where each accommodates one turbine. Compared with a continuous approach [6], the discrete approach is less sensitive to discontinuity in the wind farm land. Such discontinuity is common in practice due to existing infrastructure and geographic constraints [8]. In the current work, without loss of generality, our problem instances are square wind farms with equal-size square cells.

An interesting feature of this problem that sets it apart from standard location problems is the aerodynamic interaction among multiple turbines. In a basic scenario where only two turbines are present, the turbine downstream is said to be in the wake region of the upstream turbine, and it experiences a loss in energy production due to the reduction in wind speed and increase in turbulence intensity [9]. In practice, a turbine that is downstream of multiple turbines

C. Gomes and M. Sellmann (Eds.): CPAIOR 2013, LNCS 7874, pp. 284–299, 2013.

is affected by all upstream turbines simultaneously, and the overall effect is a nonlinear function of individual wakes. There are different analytical equations to describe the superposition of multiple wakes, some being closer to the physical reality than the others [10]. It is difficult to incorporate the more accurate wake equations into a mathematical programming model due to their nonlinearity: currently only heuristics [1–3, 6] include the most accurate wake models. Our goals are to computationally improve existing mixed integer programming (MIP) models and incorporate more accurate wake models into constraint programming (CP) and MIP models.

The contributions of this paper are: the proposal of two novel mathematical programming models (CP and MIP) that can describe the physics of the problem more accurately than the previous MIP models; the extension of a previous MIP model so that the solution quality and time are improved; the comparison of four models on twelve problem instances, with varying wind scenario complexity, turbine numbers, and wind farm grid resolution; and the elicitation of insights from the experiments to suggest future research directions.

2 Problem Definition and the Physics of Wake Modelling

2.1 Description of the Problem

Wind farm site selection, or wind farm siting, is based on, among other factors, meteorological conditions, topological features of the site, and accessibility for construction and grid transmission [8]. After siting, wind farm developers optimize the layout of the turbines according to prescribed objectives and constraints in a process called *micro-siting*. In a typical case, design engineers try to maximize the expected profit and minimize hazardous side-effects during wind farm construction and operation [8]. This is a challenging task because there are many objectives and constraints, and every site is different. To limit our scope, we consider the maximization of energy capture of a wind farm as our only objective, as it is closely related to the long term profit of the wind farms and it is well accepted in the wind farm optimization literature [1–3]. We further assume that the wind farm land is flat, and all turbines are of the same type.

We use the same problem setup that Mosetti et al. [1] proposed in their seminal paper. The objective is to maximize the wind farm's overall power generation capability. There are three types of constraints:

1. Proximity: turbines must be placed five diameters apart to avoid structural damage induced by strong aerodynamic interactions;
2. Boundary: Turbines must be placed within the wind farm boundaries;
3. Turbine number: The number of turbines is fixed.

The reason that the total number of turbines is fixed – instead of bounded by a maximum number of turbines – is due to practical considerations. During wind farm development, the total number of turbines is determined prior to the design process, by government regulations and the local electricity grid interconnection capacity among other factors. However, to explore the design space more fully,

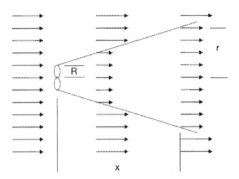

Fig. 1. Jensen [9] wake model

a given model can always be solved multiple times with different numbers of turbines.

As mentioned in the previous section, we use a discrete representation of wind farm: land is decomposed into a set of cells, where each cell can only accommodate one turbine. This approach is common in the literature [1, 2, 4].

2.2   The Physics: Wake and Energy Models

While some constraints of this problem are similar to vertex packing [11], undesirable facility location [12], and circle packing problems [13], the objective function is unique to wind farm layout optimization. In particular, the energy capture at each turbine is proportional to the cube of wind speed at that location. In turn the wind speed at a turbine is a nonlinear function of the distances to its upstream turbines. Note that "upstream" is relative to the wind direction, which varies over time.

Although wind changes speed and direction frequently, we assume that the turbine can re-orient its rotor towards the upcoming wind direction. We further assume that there is no power loss during the transient states. Overall, the yearly wind frequency data at each direction fits well into a Weibull distribution [14]. In the literature, it is a common practice to discretize the yearly wind frequency data into multiple directions and multiple speeds [1–3], so that the total energy production is the weighted sum of energy produced at each wind state (speed and direction). Then the expected power is only different from the expected energy by a scalar (the number of seconds per year). Therefore we only deal with the expected power in this work to simplify calculations.

Single Wake. The downstream region of a wind turbine, with increased level of turbulence and decreased energy, is called the wake region (Fig. 1). Equation (1), first proposed by Jensen [9], describes the propagation of a single wake.

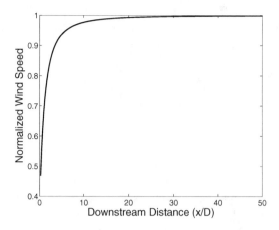

Fig. 2. Wind speed recovery after a turbine. D is the turbine diameter.

Parallel arrows in Fig. 1 represent wind direction and speed. The region with lower wind speed (shorter arrows) is the wake region. The two ellipses represent a turbine. The key idea of Jensen's wake model is momentum conservation within the wake region. In addition, wind speed is assumed to be uniform and non-turbulent across the circular wake cross-section.

Here R is the wake radius immediately after rotor; r is the downstream wake radius; r_0 is the rotor radius; u_∞ is the free stream wind speed; u_r is the wind speed immediately behind the rotor; u is the speed of wind at downstream distance x; α is the wake decay constant; z_0 is the roughness of terrain; z is the turbine height; and a is the axial induction factor (the percentage reduction in wind speed between the free stream and the turbine rotor) [15]:

$$\pi R^2 u_r + \pi \left(r^2 - R^2 \right) u_\infty = \pi r^2 u \tag{1}$$

where $u_r = (1 - a)u_\infty$, $r = R + \alpha x$, $R = r_0 \sqrt{\frac{1-a}{1-2a}}$ and $\alpha = 0.5/\ln(\frac{z}{z_0})$.

Figure 2 describes the wind speed recovery after an upstream turbine, based on the previous equations.

Multiple Wakes: Sum-of-Squares. Following Renkema [10], we write the effective wind speed in the wakes of multiple turbines as:

$$u_{id} = u_{id,\infty} \left[1 - \sqrt{\sum_{j \in \mathcal{U}_{id}} \left(1 - \frac{u_{ijd}}{u_{id,\infty}} \right)^2} \right] \tag{2}$$

u_{id} and $u_{id,\infty}$ are the wind speeds at turbine i at wind state (speed and direction) d with and without wake interactions respectively, where $d \in \mathcal{D}$, the set of all possible wind states; $\mathcal{U}_{id}$ is the set of upstream turbines for turbine i at wind state d; u_{ijd} is the wind speed at turbine i due to a single wake from upstream turbine at j, which can be obtained by (1). Currently, this is the most accurate analytical expression accounting for multiple wakes [10].

Based on this model, the expected power production of the wind farm can be calculated as:

$$\text{expected power} = \sum_{i=1}^{m} \sum_{d \in \mathcal{D}} \frac{1}{3} u_{id}^3 p_d \qquad (3)$$

where m is the total number of turbines and p_d is the probability of wind state d occurring, subject to $\sum_{d \in \mathcal{D}} p_d = 1$.

Note that we will be using average power (watts) instead of total energy (kilo-watt hours) in the objective function because they are equivalent for our purpose, and the former is easier to represent.

Multiple Wakes: Linear Superposition. Another way to account for multiple wakes in the energy production calculation is to use a direct linear superposition of power deficits. This is known to be less accurate than (2). However, it is more easily representable in the mathematical programming models [4, 5], because we can pre-calculate the pairwise interactions between two locations, then "activate" the interactions with binary variables indicating the existence of turbines at those locations, and sum up the interactions linearly:

$$\text{expected power} = \sum_{i=1}^{n} \sum_{d \in \mathcal{D}} \left(\frac{1}{3} u_{id,\infty}^3 - \sum_{j \in \mathcal{U}_{id}} \frac{1}{3} \left(u_{id,\infty}^3 - u_{ijd}^3 \right) \right) p_d \qquad (4)$$

and again u_{ijd} can be obtained from (1).

With the physics introduced, we want to make a note on the representation of these equations in our optimization models. Although the power calculation equations (2) and (4) appear to be nonlinear in wind speeds, we can actually remove some of the nonlinearity due to the choice of discrete optimization models. As illustrated by Donovan [4], the linear superposition model (4) is completely linear, because all the wind speed terms can be calculated prior the optimization, since the candidate turbine locations (i, j) and wind states $\mathcal{D}$ are known in the discrete representation.

However, linearizing the wake model given by (2) is a non-trivial task, even though the wind speed terms can be pre-calculated. The next section will first introduce a CP model that directly represents the nonlinearity in its objective function, and then describe our novel approach that can incorporate the physics of (2) into a mixed integer linear program.

$$\text{maximize} \qquad \sum_{i=1}^{n} \sum_{d \in \mathcal{D}} \frac{1}{3} x_i \left(u_{id,\infty} \left[1 - \sqrt{\sum_{j \in \mathcal{U}_{id}} x_j \left(1 - \frac{u_{ijd}}{u_{id,\infty}} \right)^2} \right] \right)^3 p_d$$

$$\text{subject to} \qquad \sum_{i=1}^{n} x_i = k$$

$$x_i + x_j \le 1 \quad \forall (i,j) \in \mathcal{E}$$

$$x_i \in \{0,1\} \quad \forall i = 1, \dots, n.$$

Fig. 3. SOM1: a constraint programming model

3 Optimization Models

3.1 Sum-of-Squares Optimization Models (SOM)

The following three optimization models are based on the more accurate way of accounting for multiple wakes (2).

CP and MIP Models. Figure 3 presents the SOM1 CP model. The binary decision variable x_i represents whether there is a turbine at location i; n is the total number of grid points; k is the total number of turbines; and $\mathcal{E} = \{(i,j) \mid$ grid i and grid j cannot both have turbines due to proximity constraint$\}$. This set is determined by the proximity constraint and the grid resolution. We choose equality for the constraint $\sum_{i=1}^{n} x_i = k$ for practical reasons – the total number of turbines is usually determined prior to the optimization based on project financing and government regulations. For the problem instances used in this work, the optimal energy production is an increasing function of k [16].

In Figure 4, we present SOM2, a MIP sum-of-squares model where the non-linearity is dealt with via a potentially exponential number of constraints. The auxiliary variable z_i represents the average power production at each location i. The key of this model are the constraints indicated by $(*)$. M is a sufficiently large constant. In general, $w_{i,\mathcal{S}_i}$ is the maximum amount of power convertible when all cells with indices in $\mathcal{S}_i$ have turbines and all cells with indices in $\mathcal{I} \setminus \mathcal{S}_i$ do not; $\mathcal{I}$ is the set of all turbine location indices $\{1, \dots, n\}$; and $\mathcal{S}_i$ is a set of turbine locations not including i. $w_{i,\mathcal{S}_i}$ is calculated according to (2) and (3).

A Decomposition Model. It is not hard to see that the number of constraints $(*)$ is exponential in n due to the requirement $(\forall \mathcal{S}_i \subset \mathcal{I} \setminus i)$. Therefore, rather than experimenting with SOM2, we propose a third model, SOM3, which can be understood as a decomposition of SOM2.

In Fig. 5, a MIP master problem is formulated that includes all constraints of SOM2 except for those indicated with $(*)$. After solving the master problem, a sub-problem calculates the actual power according to (2) and (3) as follows:

$$\text{maximize} \qquad \sum_{i=1}^{n} z_i$$

$$\text{subject to} \qquad \sum_{i=1}^{n} x_i = k$$

$$x_i + x_j \leq 1 \quad \forall (i,j) \in \mathcal{E}$$

$$z_i \leq \sum_{i=1}^{n} \sum_{d \in \mathcal{D}} \frac{1}{3} u_{id,\infty}^3 p_d x_i \quad \forall i = 1, \ldots, n.$$

$$z_i \leq M \left(|\mathcal{S}_i| - \sum_{j \in \mathcal{S}_i} x_j \right) + w_{i,\mathcal{S}_i} \quad \forall \mathcal{S}_i \subset \mathcal{I} \setminus i \quad (*)$$

$$x_i \in \{0,1\} \quad \forall i = 1, \ldots, n.$$

Fig. 4. SOM2: a mixed integer programming model

1. Evaluate the turbine layout power considering full wake effects based on $\mathbf{x}^t$ (the solution from the master problem at iteration t) by substituting it into

$$\sum_{i=1}^{n} \sum_{d \in \mathcal{D}} \frac{1}{3} x_i^t \left(u_{id,\infty} \left[1 - \sqrt{\sum_{j \in \mathcal{U}_{id}} x_j^t \left(1 - \frac{u_{ijd}}{u_{id,\infty}} \right)^2} \right] \right)^3 p_d$$

2. If it is evaluated to the same as the objective value from the master problem or the maximum solution time is reached, terminate; otherwise:
3. Generate cuts in the form of $z_i \leq g_i(\mathbf{x}^t)$, where $g_i(\mathbf{x}^t)$ is defined by (5) and (6); return to the master problem.

The master problem is then re-solved with the new cuts. In the first iteration, the master problem assumes that there is no wake interaction at all. In each subsequent iteration, the cuts refine the modeling of turbine interactions. The master problem does not represent the interaction of a specific group of turbines unless the related cuts are added. Therefore, the master problem always over-estimates the true objective value.

Instead of solving the master problem to optimality, we run it with a time limit of T seconds. In our experiment, we choose $T_0 = 30$ seconds for the first iteration. T is increased by 5 seconds each time the current best master solution is the same as the previous iteration. In other words, if the master problem produces the same solution as the previous iteration and it does not converge to the subproblem value, the algorithm will keep running with no new cuts generated, therefore getting stuck in a loop. This will happen if the master problem is unable to make any new progress in a new iteration (compared with the previous iteration) within the prescribed time limit.

Cuts. We propose two types of cut: a *no-good cut* and a *3-cut*. The former is presented in Equation (5). M is a large constant; x_j^t is the jth component of

$$\text{maximize} \qquad \sum_{i=1}^{n} z_i$$

$$\text{subject to} \qquad \sum_{i=1}^{n} x_i = k$$

$$x_i + x_j \leq 1 \quad \forall (i,j) \in \mathcal{E}$$

$$z_i \leq \sum_{i=1}^{n} \sum_{d \in \mathcal{D}} \frac{1}{3} u_{id,\infty}^3 p_d x_i \quad \forall i = 1, \ldots, n.$$

$$\text{(cuts)}$$

$$x_i \in \{0, 1\} \quad \forall i = 1, \ldots, n.$$

Fig. 5. SOM3: A mixed integer programming model of the master problem

$\mathbf{x}^t$; $w_{i,\mathcal{A}}$ is the maximum amount of power convertible when all cells with indices in $\mathcal{A}$ have turbines; $\mathcal{S}_i = \{j \mid x_j^t = 1\}$; w_i, $w_{i,j}$ and $w_{i,jk}$ are short forms for $w_{i,\emptyset}$, $w_{i,\{j\}}$ and $w_{i,\{jk\}}$, following the definition of $w_{i,\mathcal{A}}$.

$$g_i(\mathbf{x}^t) = M \left(|\mathcal{S}_i| - \sum_{j \in \mathcal{S}_i} x_j \right) + w_{i,\mathcal{S}_i} \tag{5}$$

In practice, the no-good cuts alone are inefficient in large problem instances, because an exponential number of them are required to correctly shape the feasible region and the information of each cut is minimal when there are many wind states and location cells. Therefore, we propose another type of cut to increase the speed of refinement of the representation of turbine interactions. Equation (6) presents the 3-cuts.

$$g_i(\mathbf{x}^t) = w_i + (w_{i,j} - w_i)x_j + (w_{i,jk} - w_{i,j})x_k \quad \forall j, k \in \mathcal{S}_i \ . \tag{6}$$

For each downstream turbine i, there are $\binom{|\mathcal{S}_i|}{2}$ cuts generated. The power of 3-cuts lie in their accurate description of the interaction between a group of three turbines (thus the name 3-cut). In practice, the closest few upstream turbines have the most significant influence on a downstream turbine (see Fig. 2).

The following proposition states that the three-turbine interaction accurately describes the feasible region:

Proposition 1. *The cut* $z_i \leq w_i + (w_{i,j} - w_i)x_j + (w_{i,jk} - w_{i,j})x_k$ *is tight (cutting off all infeasible values for* z_i *assuming no turbines are "on" except for* j, k*) at* $(x_i, x_j, x_k) = (1, 0, 0), (1, 1, 0),$ *and* $(1, 1, 1).$

Proof. When $(x_i, x_j, x_k) = (1, 0, 0), (1, 1, 0),$ and $(1, 1, 1),$ the cut reduces to $z_i \leq w_i$, $z_i \leq w_{i,j}$, and $z_i \leq w_{i,jk}$ respectively. These values are tight by definition

$$\text{maximize} \qquad \sum_{i=1}^{n} \sum_{d \in \mathcal{D}} \left(\frac{1}{3} u_{id,\infty}^3 x_i - \sum_{j=1}^{n} \frac{1}{3} \left(u_{id,\infty}^3 - u_{ijd}^3 \right) y_{ij} \right) p_d$$

$$\text{subject to} \qquad \sum_{i=1}^{n} x_i = k$$

$$x_i + x_j \le 1 \quad \forall (i,j) \in \mathcal{E}$$

$$x_i + x_j - 1 \le y_{ij} \quad \forall i, j = 1, \dots, n.$$

$$y_{ij} \ge 0 \quad \forall i, j = 1, \dots, n.$$

$$x_i \in \{0, 1\} \quad \forall i = 1, \dots, n.$$

Fig. 6. LSOM1 [4]

of w_i, $w_{i,j}$, and $w_{i,jk}$. When $(x_i, x_j, x_k) = (1, 0, 1)$, the cut reduces to $z_i \le w_i + w_{i,jk} - w_{i,j}$. Since $z_i \le w_{i,k}$ by the definition of $w_{i,k}$, and the combination of power deficits is sub-linear (2 and 3), $w_i + w_{i,jk} - w_{i,j} \ge w_{i,k}$. Therefore $z_i \le w_i + w_{i,jk} - w_{i,j}$ is not tight and does not cut off any feasible region.   □

Overall, the no-good cuts ensure that the problem can eventually reach the true optimality while 3-cuts increase the communication between subproblem and master to speed up convergence.

3.2   Linear Superposition Optimization Models (LSOM)

Previous MIP models use a simpler (and less accurate) [10] calculation of energy: the power deficits from individual wakes are combined linearly to account for the total power loss. The following two MIP models are based on such linear superposition technique. The first model (LSOM1) was originally proposed by Donovan [4], while the second one (LSOM2) is our extension of LSOM1.

Figure 6 presents the LSOM1 model where $\frac{1}{3} \left(u_{id,\infty}^3 - u_{ijd}^3 \right)$ is the power reduction at wind state d at turbine i due to the presence of upstream turbine j. These values can be calculated prior to running the optimization (4). Variable y_{ij} indicates whether there are turbines at both positions i and j, and so y_{ij} is 1 if both x_i and x_j are 1, and 0 otherwise. $\mathcal{E}$ is the set of cell pairs (i, j) that are too close to both host turbines. Due to the use of the simpler linear superposition model of upstream turbines, the model over-estimates the energy deficit [10].

Other location problems in the literature such as the maximum diversity problem (MDP) [17] and the p-dispersion-sum (pDS) problem [18] are similar to Donovan's model. However, we have not seen any application of the state-of-art MDP/pDS solution algorithms to this wind farm layout optimization model.

In LSOM1, the y_{ij} variables are always equal to the product of x_i and x_j, indicating if there are turbines at both places. Since $i, j \in \mathcal{I}$, there are in total $|\mathcal{I}|^2$ y_{ij} variables. As our experiments below demonstrate, a high resolution of the

$$\text{maximize} \quad \sum_{i=1}^{n} z_i$$

$$\text{subject to} \quad \sum_{i=1}^{n} x_i = k$$

$$x_i + x_j \leq 1 \quad \forall (i,j) \in \mathcal{E}$$

$$z_i \leq \sum_{i=1}^{n} \sum_{d \in \mathcal{D}} \frac{1}{3} u_{id,\infty}^3 p_d x_i \quad \forall i = 1, \ldots, n. \quad (\dagger)$$

$$z_i \leq \sum_{i=1}^{n} \sum_{d \in \mathcal{D}} \left(\frac{1}{3} u_{id,\infty}^3 - \sum_{j=1}^{n} \frac{1}{3} \left(u_{id,\infty}^3 - u_{ijd}^3 \right) x_j \right) p_d \quad \forall i = 1, \ldots, n. \quad (\dagger\dagger)$$

$$x_i \in \{0,1\} \quad \forall i = 1, \ldots, n.$$

Fig. 7. LSOM2

wind farm grid with a complicated wind regime results in too many y_{ij} variables for reasonable performance. To address this weakness we propose LSOM2.

Figure 7 presents the LSOM2 model. It does not have y_{ij} variables. Instead, we use z_i to represent the power production at location i. If there is no turbine at location i, then the right hand side of constraint ($\dagger$) is zero, and in most cases it is tighter than ($\dagger\dagger$). So, if no turbine appears at i, then there will be no power production from location i. If there is a turbine at i, then in general constraint ($\dagger\dagger$) is tighter due to the extra negative terms (deduction of power due to upstream turbines). The value that z_i is, therefore, calculated by the total available power subtracting the linear combination of power losses due to wakes.

However, LSOM2 is not equivalent to LSOM1. When $x_i = 0$ and constraint ($\dagger$) is $z_i \leq 0$, constraint ($\dagger\dagger$) may become $z_i \leq -c$, where $-c$ is a negative value. This is because the linear superposition model over-estimates power losses, making it possible for the right hand side of ($\dagger\dagger$) to be negative. Fortunately, this case does not arise during our experiments due to the proximity constraint such that turbines cannot be less than 5 rotor-diameters apart. This value is an industry standard for wind farm design.

4   Experiment Setup

All models were implemented with Microsoft Visual C++ Express 2010 and IBM ILOG CPLEX 12.3. Twelve benchmark instances [1–3, 6, 16] (referred to as WRq-n-k) were used to test the performance of models. WR1 refers to the wind regime of 1 directional wind (from west to east) and WR36 refers to the wind regime with wind coming from 36 directions at different speeds (Fig. 8).

Experiments were run on a Dell Vostro 460 with Core i5-2500 CPU (3.30GHz) and 64-bit Windows 7 OS. Since CPLEX solvers are deterministic by default,

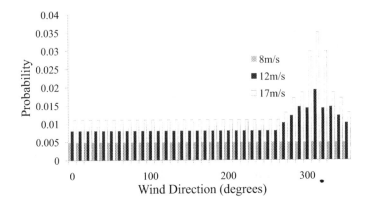

Fig. 8. Cumulative wind probability distribution for problem instance WR36

only one run of each instance was performed. Common parameters are: $z = 60$m, $z_0 = 0.3$m $R = 20$m, and wind farm is 2km by 2km.

Since the SOM models evaluate the power production by sum-of-squares (2) and the LSOM evaluation is based on linear superposition (4), the four models are not directly comparable. We therefore compare the solution quality in Table 1 by a posteriori re-evaluating the LSOM solutions based on the (more accurate) sum-of-squares method. The power production values in brackets indicate the objective function value of the LSOM solutions.

5   Results

Table 1 summarizes the performance of the four models by comparing the expected power and solution times. The MIP optimality gaps are included where applicable. Columns n and k represent the total numbers of cells and turbines. For LSOM1 and LSOM2, there are two power values: the power calculated with the sum-of-squares wake model and, in parentheses, the objective function value based on linear superposition model. Overall, LSOM2 outperforms the other models in most cases in terms of solution quality. SOM3 and LSOM2 can solve problem instances with high grid resolution and high wind data resolution (WR36-400), while the other two models cannot even initialize these instances within an hour due to the size of the model.

SOM1 vs. SOM3. Table 1 shows that SOM3 solves more instances than SOM1. In the higher resolution case (WR36-400), the SOM1 objective function expression must account for many wind directions and turbine pairs leading to memory saturation during the model creation phase and the inability to *start* search within an hour. SOM3 also performs better in WR1 cases. We believe that the simpler turbine interactions for WR1 instances are accurately captured by no-good and 3-cuts.

For a more detailed examination of these results, Figs. 9-11 present the evolution of solution quality over time for SOM1 and SOM3 for selected problem instances. In the single-direction scenario (WR1), SOM3 consistently outperforms SOM1. For the WR36 instances where SOM1 was able to run, SOM3 performs much worse than SOM1 (Fig. 11). To understand these results, recall that SOM3 is a decomposed model where the improvement from iteration to iteration is based on cuts representing information of turbine interactions. In the WR1 cases, every cell has about 10 (20) upstream cells, because the wind farm resolution is 10 by 10 (20 by 20). During the optimization, the better layouts often have turbines spaced out in the wind direction, thus the 3-cut, although only describing the interactions between a few turbines, already contains enough information for the master problem to make good decisions. However, in the WR36 cases, every turbine has $k - 1$ upstream turbines and the 3-cut only expresses the impact of the most significant two upstream turbines. Therefore, SOM3's search for better objective value in WR36 instances is often stalled due to the lack of effective cuts.

Table 1. Comparison of solutions based on sum-of-squares power calculation (○: experiments took more than one hour to setup; boldface: better objective value). Numbers in brackets are the original obj. values (based on linear superposition method) of LSOM solutions.

Wind Regime	n	k	SOM1		SOM3		LSOM1			LSOM2		
			Power (W)	Sol. Time (s)	Power (W)	Sol. Time (s)	Power (W)	Sol. Time (s)	Opt. Gap (%)	Power (W)	Sol. Time (s)	Opt. Gap (%)
WR1	100	20	10253.4	3600	**10256.0**	3600	**10256.0** (10256.0)	3.5	0	**10256.0** (10256.0)	8.0	0
	100	30	14732.5	3600	**14800.9**	3600	14795.1 (14702.2)	2.5	0	14798.6 (14702.2)	3600	1.2
	100	40	18459.5	3600	**18674.5**	3600	**18674.5** (18186.3)	1.0	0	**18674.5** (18186.3)	98.8	0
	400	20	**10368.0**	3600	**10368.0**	3600	**10368.0** (10368.0)	2.5	0	**10368.0** (10368.0)	0.4	0
	400	30	15359.6	3600	15410.9	3600	15407.6 (15409.3)	3600	0.9	**15414.2** (15414.2)	3600	0.9
	400	40	20270.0	3600	20334.7	3600	**20353.6** (20118.4)	3600	3.1	20341.2 (20172.4)	3600	2.8
WR36	100	20	16675.1	3600	16631.0	3600	16705.2 (16356.6)	3600	3.0	**16706.7** (16365.2)	3600	3.2
	100	30	24574.0	3600	24391.0	3600	24597.3 (23442.3)	3600	7.3	**24625.1** (23520.5)	3600	7.4
	100	40	32204.6	3600	31616.0	3600	32197.3 (29651.9)	3600	9.9	**32310.7** (29904.0)	3600	11.9
	400	20	○	○	16551.0	3600	○	○	○	**16762.3** (16470.1)	3600	3.1
	400	30	○	○	24426.0	3600	○	○	○	**24870.4** (24027.1)	3600	5.9
	400	40	○	○	31977.0	3600	○	○	○	**32715.4** (30844.9)	3600	10.0

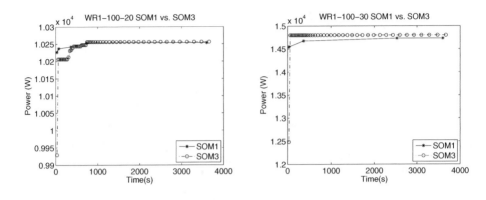

Fig. 9. Single wind direction, 100 cells

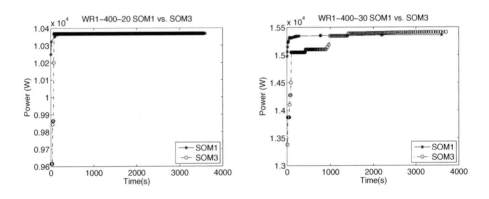

Fig. 10. Single wind direction, 400 cells

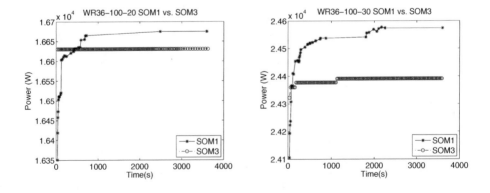

Fig. 11. Thirty-six wind directions, 100 cells

SOM3 also improves more slowly in the WR1-400 instances (Fig. 10) than in the WR1-100 instances (Fig. 9). There are more combinations of three turbines in the former instances and therefore more 3-cuts must be generated for the master problem to improve its objective value. Eventually SOM3 catches up with SOM1 because the 3-cuts describe the interactions of all turbines reasonably accurately.

LSOM1 vs. LSOM2. Table 1 shows that the power production calculated by linear superposition method (in brackets) is always lower than the sum-of-squares calculation. For some values of n and k, the problem cannot be solved to optimality within an hour, while some other instances are solved in less than 10 seconds. This observation confirms with other work on MDP and p-dispersion-sum problems [19]: increase in n with fixed k will often lead to longer solution time, and increase in k while n is fixed often leads to longer solution time too (except for when k is close to 0 or n). Current state-of-art MDP algorithms are benchmarked on problem instances with similar dimensions (in n and k) as our instances [19].

Table 1 clearly shows that LSOM2 outperforms LSOM1 in solution quality in all but one instance (WR1-400-40). In the WR1-400-40 case, LSOM1 outperforms LSOM2 in terms of the revised power calculation, however, LSOM2 is "misled" by the objective function (in brackets). Thus if we compare the true optimization objective for LSOM1 and LSOM2, LSOM2 strictly dominates LSOM1 in solution quality. In terms of computation time, LSOM2 performs similarly to LSOM1 except the WR1-100-30 and WR1-100-40 instances. A closer look at the CPLEX solution log reveals that LSOM2 arrived at the actual optimal solution (proven by LSOM1) within a few seconds for both cases, but was unable to quickly tighten the dual bound.

As with the SOM1 model, the bigger cases (WR36-400) cannot be solved by LSOM1 within 1 hour because the CPLEX solver either took a few hours to initialize or could not start at all.

6   Discussion

LSOM1 represents the state-of-art solution model for this wind farm layout optimization problem. Our LSOM2, an extension of LSOM1, outperforms it and the other two models we proposed on most instances. Since the SOMs represent the first time that the most accurate analytical wake equations [10] are modelled with constraint programming and mixed integer programming, there is much to learn about the performance and potential opportunities for the SOMs. We describe several promising research directions.

For SOM1, nonlinearity appears only in the objective function, thus we could apply nonlinear solvers that are based on linear solvers (e.g., SCIP [20]). The SOM3 cuts capture information in two ways: no-good cuts capture interaction between all k turbines in very specific layouts, while the 3-cuts capture information from a wider range of layouts, but limited to the interaction among three turbines. We can potentially apply the same idea of 3-cut and generate constraints that inform the master problem more effectively, without having to generate too many of them (e.g., 4,5-cut).

A straightforward hybridization would be the sequential application of SOM1 and SOM3, where we start the problem with low resolution (coarse grid) and progressively increase it. In this case, we could solve SOM1 in the initial stages, utilizing the constraint propagation of CP solvers for the proximity constraints, and then solve the problem with SOM3 in the later stages while fine-tuning the turbine positions, utilizing the fact that this fine-tuning focuses on clusters of closely located turbines and that such information can be effectively captured by 3-cut (or 4,5-cut).

Finally, it is interesting to observe that although the LSOMs employ less accurate power models (i.e., their optimal solutions are not the same as the optimal solutions of SOM), LSOMs can still produce good solutions even when benchmarked by the more accurate power calculations. We plan to explore this in more detail once the SOMs are improved.

7   Conclusion

We have presented the wind farm design layout problem and proposed two models to incorporate the nonlinearity in the problem. The first model (SOM1) is a direct formulation of the problem in constraint programming. While having promising performance under complex wind scenarios, the major drawback of this approach is the "curse of dimensionality" – the growth of the numbers of variables and terms quickly exceeds reasonable computational capacity. A second decomposed MIP model (SOM3) performs well in the simple wind regimes, because no-good and 3-cuts can accurately describe the turbine interactions. However with more complicated wind regimes, SOM3 is unable to improve its early feasible solutions due to the weakness of the current cuts. We also presented a novel extension of an existing LSOM model. The LSOM models are based on a less accurate model of power productions, thus having different objective functions than the SOMs. However, the models can be solved more quickly and achieve high quality solutions when a posteriori evaluated with the more accurate sum-of-squares power calculation.

In summary, we have presented two new models for the wind farm layout optimization problem. These CP and MIP models are the first mathematical programming models that capture the wind turbine interactions by modelling the sum-of-squares equations – most accurate analytical multi-wake modelling in the literature [10]. We also presented an extension (LSOM2) to a previous MIP model (LSOM1) and demonstrated improved solution quality and time. Based on the experimental study, we think that the most promising directions for future work include the strengthening of cuts for SOM3 and hybridization of SOM1 and SOM3.

References

1. Mosetti, G., Poloni, C., Diviacco, B.: Optimization of wind turbine positioning in large windfarms by means of a genetic algorithm. Journal of Wind Engineering and Industrial Aerodynamics 51(1), 105–116 (1994)
2. Grady, S.: Placement of wind turbines using genetic algorithms. Renewable Energy 30(2), 259–270 (2005)
3. Chowdhury, S., Messac, A., Zhang, J., Castillo, L., Lebron, J.: Optimizing the Unrestricted Placement of Turbines of Differing Rotor Diameters in a Wind Farm for Maximum Power Generation. In: Proceedings of the ASME 2010 International Design Engineering Technical Conference & Computers and Information in Engineering Conference, IDETC/CIE 2010, Montreal, Quebec, Canada, pp. 1–16 (2010)
4. Donovan, S.: Wind Farm Optimization. In: 40th Annual Conference, Operational Research Society of New Zealand, Wellington, New Zealand (2005)
5. Fagerfjäll, P.: Optimizing wind farm layout: more bang for the buck using mixed integer linear programming. Master's thesis. Chalmers University of Technology and Gothenburg University (2010)
6. Kwong, W.Y., Zhang, P.Y., Romero, D., Moran, J., Morgenroth, M., Amon, C.: Multi-objective Optimization of Wind Farm Layouts under Energy Generation and Noise Propagation. In: Proceedings of the ASME 2012 International Design Engineering Technical Conferences & Computers and Information in Engineering Conference, IDETC/CIE 2012, Chicago (2012)
7. Dilkina, B., Kalagnanam, J., Novakovskaia, E.: Method for designing the layout of turbines in a windfarm (2011)
8. Manwell, J.F., McGowan, J.G., Rogers, A.L.: Chapter 9 - Wind Turbine Siting, System Design, and Integration. In: Wind Energy Explained (2010)
9. Jensen, N.: A note on wind generator interaction. Technical report. Risoe National Laboratory (1983)
10. Renkema, D.J.: Validation of wind turbine wake models. Master of science thesis, Delft University of Technology (2007)
11. Atamtürk, A., Nemhauser, L.G., Savelsbergh, M.W.P.: The Mixed Vertex Packing Problem
12. Owen, S.H.: Scenario Planning Approaches to Facility Location: Models and Solution Methods. PhD thesis, Northwestern University (1998)
13. Stephenson, K.: Introduction to Circle Packing: The Theory of Discrete Analytic Functions. Cambridge University Press, Cambridge (2005)
14. Manwell, J.F., McGowan, J.G., Rogers, A.L.: Chapter 2 - Wind Characteristics and Resources. In: Wind Energy Explained (2009)
15. Manwell, J.F., McGowan, J.G., Rogers, A.L.: Chapter 3 - Aerodynamics of Wind Turbines. In: Wind Energy Explained (2009)
16. Du Pont, B.L., Cagan, J.: An Extended Pattern Search Approach to Wind Farm Layout Optimization. ASME Conference Proceedings, vol. 2010(44090), pp. 677–686 (2010)
17. Kuo, C.C., Glover, F., Dhir, K.S.: Analyzing and Modeling the Maximum Diversity Problem by Zero-One Programming. Decision Sciences 24(6), 1171–1185 (1993)
18. Pisinger, D.: Exact solution of p-dispersion problems, pp. 1–16 (December 1999)
19. Martí, R., Gallego, M., Duarte, A.: A branch and bound algorithm for the maximum diversity problem. European Journal of Operational Research 200(1), 36–44 (2010)
20. Achterberg, T.: SCIP: solving constraint integer programs. Mathematical Programming Computation 1(1), 1–41 (2009)

The Rooted Maximum Node-Weight Connected Subgraph Problem

Eduardo Álvarez-Miranda[1], Ivana Ljubić[2], and Petra Mutzel[3]

[1] Dipartimento di Elettronica, Informatica e Sistemistica, University of Bologna, Italy
e.alvarez@unibo.it
[2] Department of Statistics and Operations Research, University of Vienna, Vienna,
Austria
ivana.ljubic@univie.ac.at
[3] Faculty of Computer Science, TU Dortmund, Otto-Hahn-Str. 14, 44227 Dortmund,
Germany
petra.mutzel@tu-dortmund.de

Abstract. Given a connected node-weighted (di)graph, with a root node r, and a (possibly empty) set of nodes R, the *Rooted Maximum Node-Weight Connected Subgraph Problem* (RMWCS) is the problem of finding a connected subgraph rooted at r that connects all nodes in R with maximum total weight. In this paper we consider the RMWCS as well as its budget-constrained version, in which also non-negative costs of the nodes are given, and the solution is not allowed to exceed a given budget. The considered problems belong to the class of network design problems and have applications in various different areas such as wildlife preservation planning, forestry, system biology and computer vision.

We present three new integer linear programming formulations for the problem and its variant which are based on node variables only. These new models rely on a different representation of connectivity than the one previously presented in the RMWCS literature that rely on a transformation into the Steiner Arborescence problem. We theoretically compare the strength of the proposed and the existing formulations, and show that one of our models preserves the tight LP bounds of the previously proposed cut set model of Dilkina and Gomes. Moreover, we study the rooted connected subgraph polytope in the natural space of node variables. We conduct a computational study and (empirically) compare the theoretically strongest one of our formulations with the one previously proposed using ad-hoc branch-and-cut implementations.

1 Introduction

In this work we study a variant of the *connected subgraph problem* in which we are given a graph with a pre-specified root node (and possibly an additional set of terminals). Nodes of the graph are associated with (not necessarily positive) weights. The goal is to find a connected subgraph containing the root and the terminals that maximizes the sum of node-weights. In addition, a budget constraint may be imposed as well: in this case, each node is additionally associated

C. Gomes and M. Sellmann (Eds.): CPAIOR 2013, LNCS 7874, pp. 300–315, 2013.
© Springer-Verlag Berlin Heidelberg 2013

with a non-negative cost, and the cost of connecting the nodes is not allowed to exceed the given budget. Both problem variants are NP-hard, unless all node weights are non-negative and no budget is imposed, in which case the problem is trivial. The problem is called the *Rooted Maximum Node-Weight Connected Subgraph Problem* (RMWCS), or the RMWCS with *Budget Constraint* (B-RMWCS), respectively.

The problem has been introduced by Lee and Dooly [12] in the context of the design of fiber-optic communication networks over time, where the authors refer to the problem as the *constrained maximum weight connected graph problem*. The authors impose K-cardinality constraints, i.e., they search for a connected subgraph containing K nodes (including a predetermined root) that maximizes the collected node-weights. Obviously, K-cardinality constraints are a special form of the budget constraints in which every node is associated a cost equal to one, and the budget is equal to K.

A budgeted version arises in the wildlife conservation planning, where the task is to select land parcels for conservation to ensure species viability, also called *corridor design* (see, e.g. [4, 5]). Here, the nodes correspond to land parcels, their weights are associated with the habitat suitability, and node costs are associated with land value. The task is to design wildlife corridors that maximize the suitability with a given limited budget. Also in forest planning, the connected subgraph arises as subproblem, e.g., for designing a contiguous site for a natural reserve or for preserving large contiguous patches of mature forest [2]. Moss and Rabani [15] have proposed an $O(\log n)$ approximation algorithm for the B-RMWCS with non-negative node-weights, where n is the number of nodes in the graph. For more details on the problems related on the RMWCS, see e.g., the literature review given in [5].

In this paper we will address the RMWCS in digraphs as well. This is motivated by some applications in systems biology where regulatory networks are represented using (not necessarily bidirected) digraphs and with node weights that can also be negative. The goal is to find a rooted subgraph in which there is a directed path from the root to any other node that maximizes the sum of node weights. In systems biology, the roots are frequently referred to as "seed genes" as they are assumed to be involved in a particular disease. In Backes et al. [1], for example, the authors search for the connected subgraph in a digraph without a prespecified root node (i.e., determination of the seed gene, also called the key player, is part of the optimization process). To solve the problem of Backes et al. [1] one can, for example, iterate over a set of potential key players, solve the corresponding RMWCS and choose the best solution.

Our Contribution. Previously studied mixed integer programming (MIP) formulations for the (B-)RMWCS use arc and possibly flow variables to model the problem (see Dilkina and Gomes [5]). In this paper we propose three new MIP models for the (B-)RMWCS derived in the natural space of node variables. We first provide a theoretical comparison of the quality of lower bounds of these models. We also show that one of our models which is based on the concept of *node separators*, preserves the tight LP bounds of the previously proposed *cut set* model of Dilkina and Gomes [5]. In the second part of the paper we study the rooted connected subgraph polytope (in the natural space of node variables) and show under which conditions the node separator inequalities are facet-defining. In an extensive computational study, we compare the node-separator and the cut-set model on a set of benchmark instances for the wildlife corridor design problem used in [5] and on a set of network design instances.

Outline of the Paper. Three new MIP models for the (B-)RMWCS are proposed in Section 2. A comparison of the MIP models and results regarding the facets of the rooted connected subgraph polytope are given in Section 3 and computational results are presented in Section 4.

2   MIP Formulations for the RMWCS

In this section we present three new MIP models for the RMWCS and its budget-constrained variant. Before that, we first review the model recently proposed by Dilkina and Gomes [5] which is based on the reformulation of the problem into the (budget-constrained) Steiner arborescence problem. The latter model is derived on the space of arc variables, while the remaining ones are defined in the natural space of node variables.

Since every RMWCS on undirected graphs can be considered as the same problem on digraphs (by replacing every edge with two oppositely directed arcs), in the remainder of this paper we will present the more general results for digraphs. The corresponding results for undirected graphs can be easily derived from them.

Definitions and Notation. Formally, we define the RMWCS as follows: Given a digraph $G = (V \cup \{r\}, A)$, with a root r, a set of terminals $R \subset V$, and node weights $p : V \to \mathbb{R}$, the RMWCS is the problem of finding a connected subgraph $T = (V_T, A_T)$, that spans the nodes from $\{r\} \cup R$ and such that every node $j \in V_T$ can be reached from r by a directed path in T, and that maximizes the sum of node weights $p(T) = \sum_{v \in V_T} p_v$. Additionally, in the B-RMWCS, node costs $c : V \to \mathbb{R}^+$ and a budget limit $B > 0$ are given. The goal is to find a connected subgraph T that maximizes $p(T)$ and such that its cost does not exceed the given budget, i.e., $c(T) = \sum_{v \in V_T} c_v \le B$.

A set of vertices $S \subset V$ ($S \ne \emptyset$) and its complement $\bar{S} = V \setminus R$, induce two directed cuts: $(S, \bar{S}) = \delta^+(S) = \{(i,j) \in A \mid i \in S, j \in \bar{S}\}$ and $(\bar{S}, S) = \delta^-(S) = \{(i,j) \in A \mid i \in \bar{S}, j \in S\}$. For a set $C \subset V$, let $D^-(C)$ denote the set of nodes outside of C that have ingoing arcs into C, i.e., $D^-(C) = \{i \in V \setminus C \mid \exists (i,v) \in A, v \in C\}$.

A digraph G is called strongly connected (or simply, *strong*) if for any two distinct nodes k and ℓ from V, there exists a (k, ℓ) path in G. A node i is a *cut point* in a strong digraph G if there exists a pair of distinct nodes k and ℓ from V such that there is no (k, ℓ) path in $G - i$. A node i is a *cut point with respect to r* if there exists a node $k \ne i, r$ such that there is no (r, k) path in $G - i$. For two distinct nodes k and ℓ from V, a subset of nodes $N \subseteq V \setminus \{k, \ell\}$ is called (k, ℓ) *(node) separator* if there exists a (k, ℓ) path in G and after eliminating N from V there is no (k, ℓ) path in G. A (k, ℓ) separator N is *minimal* if $N \setminus \{i\}$ is not a (k, ℓ) separator, for any $i \in N$. Let $\mathcal{N}(k, \ell)$ denote the family of all (k, ℓ) separators. Obviously, if $\exists (k, \ell) \in A$ or if ℓ is not reachable from k, we have $\mathcal{N}(k, \ell) = \emptyset$.

For variables $\mathbf{a}$ defined on a finite set F, we denote by $a(F')$ the sum $\sum_{i \in F'} a_i$ for any subset $F' \subseteq F$. Throughout the paper, let the graph $G = (V \cup \{r\}, A)$, $n = |V|$, and $m = |A|$.

2.1   Directed Steiner Tree Model of Dilkina and Gomes [5]

Dilkina and Gomes [5] propose to solve the B-RMWCS as a budget-constrained directed Steiner tree problem rooted at r. Their models are based on the observation that it is sufficient to search for a subtree (subarborescence) since no costs are associated to arcs in G, hence every solution containing cycles can be reduced without changing the weight. It is sufficient to use arc variables to model the problem since in a directed tree, the in-degree of every node is equal to one, so that the objective function can be expressed as $\max \sum_{i \in V} p_i z(\delta^-(i))$, where z are binary variables associated with the arcs of A that encode the subarborescence. Dilkina and Gomes [5] proposed three MIP models for the B-RMWCS. Two of them are flow based formulations (a single-commodity flow and a multi-commodity flow based one). The authors showed that the flow-based formulations are computationally outperformed by the cut-set model which is presented below.

We further use a set of auxiliary binary variables y for the vertex set V, where y_i will be equal to one if node i is part of the subtree, and zero, otherwise. In other words, we basically perform the substitution $y_i = z(\delta^-(i))$. The set of feasible B-RMWCS solutions can be described using inequalities (1)-(4). Constraints (1) and (2) ensure that the solution is a Steiner arborescence rooted at r, equations (3) make sure that all terminals are connected and (4) is the budget constraint:

$$z(\delta^-(i)) = y_i \qquad \forall i \in V \setminus \{r\} \qquad (1)$$

$$z(\delta^-(S)) \geq y_k \qquad \forall k \in S, \ \forall S \subseteq V \setminus \{r\}, \ S \neq \emptyset \qquad (2)$$

$$y_i = 1 \qquad \forall i \in R \qquad (3)$$

$$\mathbf{c}^T y \leq B \qquad (4)$$

Constraints (2), also known as *cut* or *connectivity inequalities* ensure that there is a directed path from the root r to each node k such that $y_k = 1$. *In-degree* constraints (1) guarantee that the in-degree of each vertex of the arborescence is equal to one. Thus, the rooted Steiner arborescence model for the B-RMWCS (denoted by (SA_r)) is given as

$$(SA)_r \quad \max \left\{ p^T y \mid (\mathbf{y}, \mathbf{z}) \text{ satisfies } (1)\text{-}(4), \ (\mathbf{y}, \mathbf{z}) \in \{0, 1\}^{n+m} \right\}.$$

We notice that in Ljubić et al. [14] these sets of constraints and the transformation into the directed Steiner tree were used for solving the Prize-Collecting Steiner Tree problem (PCStT). A connection between the PCStT and the unrooted MWCS has been observed by Dittrich et al. [7]: the authors showed that the unrooted MWCS can be transformed into the PCStT and used the branch-and-cut approach from [14] to solve the MWCS on a large protein-protein interaction network. Consequently, the same relation holds for the rooted MWCS as well.

The previous model uses node and arc variables ($\mathbf{y}$ and $\mathbf{z}$) given that it relies on a transformation into the Steiner arborescence problem. However it seems more natural to find a formulation based only in the space of $\mathbf{y}$ variables since no arc costs are involved in the objective function. In the next section we will discuss several models that enable elimination of arc variables in the MIP models.

2.2   Node-Based Formulations for the RMWCS

We now propose three MIP models that are derived in the natural space of y variables defined as above. We search for an arborescence rooted at r, but this time, we avoid explicit use of arc variables.

Model Based on Subtour Elimination Constraints. This model is an adaptation of the model by Backes et al. [1] that was recently proposed for the unrooted MWCS on directed graphs. The following inequalities will be called the *in-degree constraints*:

$$y(D^-(i)) \geq y_i, \quad \forall i \in V \setminus (\{r\} \cup D^+(r)) \tag{5}$$

They ensure that, whenever a node i is taken into a solution, at least one of its incoming neighbors has to be in the solution as well (notice that we do not need to impose this constraint for the outgoing neighbors of the root node). Constraints (5) however do not guarantee that the obtained solution is connected to the root. Let $\mathcal{C}$ denote the family of all directed cycles in G that do not contain the root node and are not "neighbors" of the root, i.e.:

$$\mathcal{C} = \{C \mid C \text{ is a cycle in } G, \text{ s.t. } r \notin C, \text{ and } r \notin D^-(C)\}.$$

In order to ensure connectivity of the solution, Backes et al. [1] add the following constraints, that we will refer to as the *subtour elimination constraints*:

$$y(C) - y(D^-(C)) \leq |C| - 1, \quad \forall C \in \mathcal{C}. \tag{6}$$

These constraints state that for each cycle $C \in \mathcal{C}$ whose node set is contained in the solution, at least one of the neighboring nodes outside of that cycle needs to belong to the solution as well. The model, that we will denote by $CYCLE_r$ reads as follows:

$$(CYCLE_r) \quad \max\left\{p^T y \mid \mathbf{y} \text{ satisfies (3)-(6)}, \mathbf{y} \in \{0,1\}^n\right\}.$$

A Flow-Based Model. Alternatively to the previous model, to ensure connectivity, we can use multi-commodity flows where the available arc capacities are defined as the minimum node capacities at each end of an arc. Finding a feasible solution now means allocating node capacities that will enable to send one unit of flow from the root to each of the nodes taken into the subnetwork. In this context, constraints (5) and (6) can be replaced by the following set of constraints that ensure that there is enough capacity on the nodes so that a unit of flow can be sent from the root to any other node $i \in V \setminus \{r\}$ with $y_i = 1$. These constraints state that (i) whenever an arc is part of a feasible solution of the RMWCS, both of its end nodes are included into the solution and (ii) the induced subgraph is connected:

$$\sum_{(i,j) \in \delta^-(S)} \min\{y_i, y_j\} \geq y_k, \quad \forall k \notin \{r\} \cup D^+(r), \ \forall S \subseteq V \setminus \{r\}, \ k \in S. \quad (7)$$

Constraints (7) represent just a compact way of writing $2^{|\delta^-(S)|}$ inequalities (see also [3] where these constraints have been proposed for a problem arising in the design of telecommunication networks). They can be separated in polynomial time by solving a maximum-flow problem in an auxiliary support graph. Observe finally that indegree constraints (5) are also implied by these constraints: For each node $i \notin r \cup D^+(r)$, we have $y(D^-(i)) \geq \sum_{(j,i) \in \delta^-(i)} \min\{y_j, y_i\} \geq y_i$. We can now define the B-RMWCS as

$$(CUT_m) \qquad \max\left\{p^T y \mid \mathbf{y} \text{ satisfies (3),(4),(7) and } \mathbf{y} \in \{0,1\}^n\right\}.$$

Formulation Based on Node Separators. The other way of modeling the connectivity of a solution using only node variables is to consider *node separators*. This idea has been recently used in Fügenschuh and Fügenschuh [8], Carvajal et al. [2] and Chen et al. [3] to model connectivity in the context of sheet metal design, forest planning, and telecommunication network design, respectively. The following inequalities will be called *node-separator constraints*:

$$y(N) \geq y_k, \quad \forall k \notin \{r\} \cup D^+(r), \ N \in \mathcal{N}(r,k). \quad (8)$$

These constraints ensure that for each node k taken into the solution, either k is a direct neighbor of r, or there has to be a path from r to k such that for each node i on this path, $y_i = 1$. Notice that whenever $\mathcal{N}(k, \ell) \neq \emptyset$, $D^-(k) \in \mathcal{N}(k, \ell)$ and in this case the in-degree inequalities (5) are contained in (8). Thus, we can formulate the B-RMWCS as

$$(CUT_r) \qquad \max\left\{p^T y \mid \mathbf{y} \text{ satisfies (3),(4),(8), } \mathbf{y} \in \{0,1\}^n\right\}.$$

2.3   Some More Useful Constraints

In case that the budget constraint (4) is imposed, the following family of cover inequalities can be used to cut off infeasible solutions.

Cover Inequalities. We say that a subset of nodes $V_C \subset V$ is a cover if the sum of node costs in V_C is greater than the allowed budget B. In that case, at least one node from V_C has to be left out in any feasible solution. A cover V_C is minimal if $C \setminus \{i\}$ for any $i \in V_C$ is not a cover anymore. Let $\mathcal{V}_C$ be a family of all minimal covers with respect to B. Then, the following *cover inequalities* are valid for the B-RMWCS:

$$\sum_{i \in V_C} y_i \leq |V_C| - 1, \quad \forall V_C \in \mathcal{V}_C \tag{9}$$

For further details on cover inequalities, see e.g. [10].

3 Polyhedral Results

In this section we compare the proposed MIP formulations with respect to their quality of LP bounds and we show that, under certain conditions, the newly introduced node-separator inequalities are facets of the rooted connected subgraph polytope.

3.1 Theoretical Comparison of MIP Models

Let $\mathcal{P}_{\mathrm{LP}}(.)$ denote the polytope of the LP-relaxations of the MIP models presented above and $v_{LP}(.)$ their optimal LP-values. We can show that:

Proposition 1. *We have* $\mathcal{P}_{\mathrm{LP}}(CUT_r) \subsetneq \mathcal{P}_{\mathrm{LP}}(CUT_m) \subsetneq \mathcal{P}_{\mathrm{LP}}(CYCLE_r)$, *and there exist instances for which the strict inequality holds.*

Proof. $\mathcal{P}_{\mathrm{LP}}(CUT_m) \subsetneq \mathcal{P}_{\mathrm{LP}}(CYCLE_r)$: Consider a feasible solution $\hat{y}$ of the LP relaxation of model CUT_m. We will show that each such solution is feasible for the model $CYCLE_r$. Let C be an arbitrary cycle from $\mathcal{C}$. Then, obviously, for any node $k \in C$, we have $\hat{y}_i(D^-(C)) \geq \sum_{(i,j) \in \delta^-(C)} \min\{\hat{y}_i, \hat{y}_j\} \geq \hat{y}_k$. Adding up this inequality with inequalities $1 \geq \hat{y}_i$, for each $i \in C \setminus \{k\}$, we obtain: $\hat{y}(D^-(C)) + |C| - 1 \geq \hat{y}(C)$ which is exactly the subtour elimination inequality associated to C. To see that the strict inequality holds, consider the directed graph shown in Figure 1(a).

$\mathcal{P}_{\mathrm{LP}}(CUT_r) \subsetneq \mathcal{P}_{\mathrm{LP}}(CUT_m)$: Consider a feasible solution $\hat{y}$ of the LP relaxation of the CUT_r model. Let $k \in V \setminus (\{r\} \cup D^+(r))$ be an arbitrary node such that $\hat{y}_k > 0$ and let $S \subset V \setminus \{r\}$ be a set such that $k \in S$. Then, we will show that $\sum_{(ij) \in \delta^-(S)} \{\hat{y}_i, \hat{y}_j\} \geq \hat{y}_k$, i.e., $\hat{y}$ satisfies (7). Let $N_1 = \{i \mid (i,j) \in \delta^-(S)\}$. Observe that $r \notin N_1$ and by definition, N_1 is a node separator for k, i.e., $N_1 \in \mathcal{N}(r,k)$. Let $N_2 = \{j \mid (i,j) \in \delta^-(S)\}$: (i) If $k \notin N_2$, then N_2 is a node separator for k ($N_2 \in \mathcal{N}(r,k)$). Consider the bipartite graph defined by $\delta^-(S)$. Each possible vertex cover $N' \subset N_1 \cup N_2$ on this graph, induces a node separator for k, i.e., $N' \in \mathcal{N}(r,k)$. There are $2^{|\delta^-(S)|}$ vertex covers in total, and constraints (8) associated to them imply constraint (7); (ii) if $k \in N_2$, then all vertex covers involving k trivially satisfy $\hat{y}(N') \geq \hat{y}_k$ for $k \in N'$. Together with the remaining vertex covers, inequality (7) is implied. An example shown in Figure 1(b) shows an instance for which the strict inequality holds. $\square$

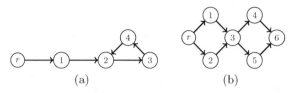

(a)   (b)

Fig. 1. Examples that prove the strength of the new formulations. (a) The LP-solution of $CYCLE_r$ sets $y_2 = y_3 = y_4 = 2/3$ and $y_1 = 0$, and this solution is infeasible for the model CUT_m. (b) The LP-solution of CUT_m satisfies $y_1 = \cdots = y_5 = 1/2$ and $y_6 = 1$. This solution is infeasible for CUT_r.

Proposition 2. *The (SA_r) model and the (CUT_r) model are equally strong, i.e., $v_{LP}(SA_r) = v_{LP}(CUT_r)$.*

Proof. We first show that $v_{LP}(SA_r) \geq v_{LP}(CUT_r)$: Let $(\hat{z}, \hat{y})$ be a feasible solution for the relaxation of the SA_r model. Let $k \in V \setminus \{r\}$ be a node such that $\hat{y}_k > 0$ and let $N \in \mathcal{N}(r, k)$. Because of in-degree constraints of the SA_r model, we have that $\sum_{i \in N} \hat{y}_i = \sum_{i \in N} \hat{z}(\delta^-(i))$. If N is removed from G, k cannot be reached from r. Let $S_r \subseteq V$, $r \in S_r$, be all the nodes i that can be reached from r after removing N, and let $S_k = V \setminus (N \cup S_r)$, $k \in S_k$. Because of inequalities (2), it holds that $\hat{z}(\delta^+(S_r)) \geq \hat{y}_k$. Moreover, observe that for each $(i, j) \in \delta^+(S_r)$ we have that $i \in S_r$ and $j \in N$, which means that $\sum_{i \in N} \hat{z}(\delta^-(i)) \geq \hat{z}(\delta^+(S_r))$. Therefore, $\sum_{i \in N} \hat{y}_i \geq \hat{y}_k$, which proves that any LP solution of the SA_r model can be projected into a feasible solution of the CUT_r with the same objective value.

To show that $v_{LP}(CUT_r) \geq v_{LP}(SA_r)$ consider a solution $\check{y} \in \mathcal{P}_{LP}(CUT_r)$. We will construct a solution $(\hat{y}, \hat{z}) \in \mathcal{P}_{LP}(SA_r)$ such that $\check{y} = \hat{y}$. On the graph G' (see Section 4.1, separation of separator inequalities) with arc capacities of (i_1, i_2) set to $\check{y}_i$ for each $i \in V \setminus \{r\}$ and to 1 otherwise, we are able to send $\check{y}_k$ units of flow from the root r to every (k_1, k_2) such that $\check{y}_k > 0$. Let f_{ij}^k denote the amount of flow of commodity k, sent along an arc $(i, j) \in A'$. Let **f** be the minimal feasible multi-commodity flow on G' (i.e., the effective capacities on G' used to route the flow cannot be reduced without violating the feasibility of this flow). We now define the values of $(\hat{y}, \hat{z})$ as follows:

$$\hat{z}_{ij} = \begin{cases} \max_{k \in V \setminus \{r\}} f_{i_2 j_1}^k, & i, j \in V \setminus \{r\} \\ \max_{k \in V \setminus \{r\}} f_{i, j_1}^k, & i = r, j \in V \setminus \{r\} \end{cases} , \forall (i, j) \in A, \quad \text{and}$$

let $\hat{y}_i = \hat{z}(\delta^-(i))$, for all $i \in V \setminus \{r\}$. Obviously, the constructed solution $(\hat{y}, \hat{z})$ is feasible for the (SA_r) model, and, due to the assumption that **f** is minimal feasible, it follows that $\check{y} = \hat{y}$, which concludes the proof. $\square$

Finally, regarding the strength of the three MIP models studied by Dilkina and Gomes [5], we notice that their single-commodity flow model is weaker than the multi-commodity model, which is equally strong as the cut-set model (SA_r) (see, e.g., [13]).

3.2   Facets of the RCS Polytope

In this section we consider the RMWCS with $R = \emptyset$, and let $\mathcal{P}$ denote the *rooted connected subgraph (RCS) polytope* defined in the natural space of $\mathbf{y}$ variables:

$$\mathcal{P} = \text{conv}\{\mathbf{y} \in \{0,1\}^n \mid \mathbf{y} \text{ satisfies (8)}\}.$$

In this section we establish under which conditions some of the presented inequalities are facets of the RCS polytope.

Lemma 1. *The RCS polytope is full-dimensional (i.e., $\dim(\mathcal{P}) = n$) if and only if there exists a directed path between r and any $i \in V$.*

Proof. We first generate a spanning arborescence T in G rooted at r. We will then apply a *tree pruning technique* in order to generate $n+1$ affine independent feasible RMWCS solutions. We start with the arborescence T in which case $\mathbf{y}$ consists of all ones. We iteratively remove one by one leaf from T, until we end up with a single root node (in which case $\mathbf{y}$ is a zero vector). Thereby, we generate a set of $n+1$ affinely independent solutions. Conversely, if $\mathcal{P}$ is full dimensional, then in order to create a feasible solution containing an arbitrary node $i \in V$, there has to be a directed path between r and i in G.   □

Lemma 2. *Inequality $y_i \geq 0$ for $i \in V$ is facet defining if and only if in the graph $G - i$, any node $j \in V \setminus \{i\}$ can be reached from r.*

Lemma 3. *Inequality $y_i \leq 1$ for $i \in V$ is facet defining if and only if every node in V can be reached from r and there either exists $(r, i) \in A$, or there exist two node disjoint paths between r and i in G.*

Given some $k \in V$ and $N \in \mathcal{N}(r, k)$, let us now consider the corresponding node separator inequalities: $y(N) \geq y_k$. Let $S_r \subset V$ denote the subset of nodes that can be reached from r in $G - N$, and let S_k be the remaining nodes, i.e., $S_k = V \setminus (N \cup S_r)$. Then, we have:

Proposition 3. *Given some $k \in V$ and $N \in \mathcal{N}(r, k)$, the associated node separator inequality $y(N) \geq y_k$ is facet defining if N is minimal, every node in V can be reached from r and every node in S_k can be reached from k.*

Proof. For a given $k \in V$ and $N \in \mathcal{N}(r, k)$, that satisfy the above properties we prove the statement using the indirect method. Let $F(k, N) = \{y \in \{0,1\}^n \mid \sum_{i \in N} y_i = y_k\}$. Consider a facet defining inequality of the form $\mathbf{a}^t \mathbf{y} \geq a_0$. We will show that if all points in $F(k, N)$ satisfy

$$\mathbf{a}^t \mathbf{y} = a_0 \tag{10}$$

then $\mathbf{a}^t \mathbf{y} \geq a_0$ is a positive multiple of (8). Observe first that the zero vector belongs to $F(k, N)$. By plugging it into (10), we get $a_0 = 0$. Consider now an arbitrary node $\ell \in S_r$. Consider a path P from r to ℓ in S_r, and its subpath Q obtained by deleting ℓ. Characteristic vectors of both of them belong to $F(k, N)$, and by subtracting them, we obtain $a_\ell = 0$, for all $\ell \in S_r$. Consider now an arbitrary $\ell \in S_k$. Let P be a path from r to ℓ that passes through exactly one node $i \in N$ and through k. We can find such a path for the following reasons: (i) A path from r to k over a single node $i \in N$ exists because N is minimal. (ii) A path from k to ℓ fully contained in S_k also exists by our assumption. Let Q be a subpath of P obtained by deleting ℓ. Characteristic vectors of P and Q belong to $F(k, N)$, and by subtracting them, we obtain $a_\ell = 0$, for all $\ell \in S_k$. Finally, consider an arbitrary $i \in N$ and a path P' from r to k passing through i and no other nodes from N. Characteristic vector of P' belongs to $F(k, n)$ and after plugging it into (10), we obtain $a_i + a_k = 0$, for all $i \in N$. Therefore, we have $a_i = -a_k = \alpha$, and (10) can be written as $\alpha(y(N) - y_k) = 0$, which concludes the proof.   □

4   Computational Results

In this section, we study the computational performance of Branch-and-Cut (B&C) algorithms for the models (SA_r) and (CUT_r) for both the RMWCS and the B-RMWCS.

4.1   Branch-and-Cut Algorithms

Constraint Separation. At each node of the branch-and-bound tree, constraints (2) of the (SA_r) formulation are separated by solving a max-flow problem (see Ljubić et al. [14] for further details). For the (CUT_r) model, inequalities (8) can be separated in polynomial time on an auxiliary support graph G' that splits all nodes except the root into arcs so that each $i \in V$ is replaced by an arc (i_1, i_2). All ingoing arcs into i are now connected to i_1, and all outgoing arcs from i are now connected from i_2. For a given node fractional solution $\tilde{y}$ and $k \in V \setminus (\{r\} \cup D^+(r))$ such that $\tilde{y}_k > 0$, to check whether there are violated inequalities of type (8) we calculate the maximum flow between r and (k_1, k_2) in G' whose arc capacities are defined as $\tilde{y}_i$ for splitted arcs and to zero, otherwise. For both cases, we also use *nested, back-flow* and *minimum cardinality* cuts in order to insert as many violated cuts as possible (see Koch and Martin [11], Ljubić et al. [14]). At each separation callback, we limit the number of inserted cuts to 25.

For the B-RMWCS, the cover inequalities (9) are separated by solving a knapsack problem (which is weakly NP-hard) for each fractional solution $\tilde{y}$:

$$(P_{CI}) \qquad \min\{\sum_{i \in V}(1 - \tilde{y}_i)a_i \mid \sum_{i \in V} c_i a_i > B, a_i \in \{0,1\}^n\};$$

if the optimal value of (P_{CI}) is less than one, the nodes $i \in V$ such that $a_i = 1$ are the nodes of a cover V_C for which the corresponding inequality (9) is violated. Finally, once the violated cover inequality is detected, we insert the following *extended cover inequality* in the MIP:

$$\sum_{i \in V_C \cup V^*(C)} y_i \leq |V_C| - 1, \quad \forall V_C \in \mathcal{V}_C \tag{11}$$

where $V^*(C) = \{i \in V \setminus V_C \mid c_i \geq \max_{j \in V_C} c_j\}$. We solve the knapsack problem P_{CI} within the B&C using CPLEX. Only at the root node of the branch-and-bound tree the problem P_{CI} is solved to optimality; in the remaining nodes it is solved until reaching a 0.01% gap.

Primal Heuristic. At a given node of the branch-and-bound tree, we use the information of the current LP solution $\tilde{y}$ in order to construct feasible primal solutions for the (B-)RMWCS. The procedure, which is equivalent for both (SA_r) and (CUT_r), consists of a (restricted) breadth-first search (BFS) that starts from the root node r and constructs a connected component. A node is incorporated into this component if its weight $\tilde{p}_v := p_v \tilde{y}_v$ is non-negative and its cost c_v added to the cost of the current component does not violate the budget B.

MIP Initialization. As described in §4.2, part of our benchmark set consists of 4-grid graphs. In this case, all 4-cycles are easily enumerated by embedding the grid into the plane and iterating over all faces except for the outer face. Let $\mathcal{C}_4$ be the set of all 4-cycles C such that $r \notin C \cup D^-(C)$ and let $A[C]$ be the set of arcs associated to it. Therefore, in case of 4-grids, the (SA_r) model is initialized with the following *4-cycle inequalities*:

$$z(A[C]) \leq y(C \setminus i), \quad \forall i \in C, \, \forall C \in \mathcal{C}_4. \tag{12}$$

The corresponding 4-cycle inequalities for the (CUT_r) model are:

$$y(D^-(C)) \geq y_i, \quad \forall i \in C, \, \forall C \in \mathcal{C}_4. \tag{13}$$

Additionally, indegree constraints (1) (or (5)) and $z_{ij} + z_{ji} \leq y_i \, \forall e : \{i_{\neq r}, j\} \in E$ are added to the MIP.

Implementation. The B&C algorithms were implemented using CPLEX^{TM}12.3 and Concert Technology. All CPLEX parameters were set to their default values, except that: (i) CPLEX cuts, CPLEX heuristics, and CPLEX preprocessing were turned off, and (ii) higher branching priorities were given to **y** variables in the case of the (SA_r) model. All the experiments were performed on a Intel Core2 Quad 2.33 GHz machine with 3.25 GB RAM, where each run was performed on a single processor. We denote as "Basic" the B&C implementation for which neither the separation of CI nor the addition of 4-cycle inequalities, (12) or (13), is considered.

4.2   Benchmark Instances

Wildlife Corridor Design Instances. We have considered three real instances provided in [5] that are instances of the corridor design problem for grizzly bears in the Rocky mountains, labeled as CD-40×40-sq (242 nodes, 469 edges), CD-10×10-sq (3299 nodes, 6509 edges) and CD-25-hex (12889 nodes, 38065 edges). In all of them, three reserves are given and the root is chosen as one of them.

We have also considered 4-grid instances generated using the instance generator of Dilkina and Gomes [5]. The description of the parameters used for setting up the instances and the generator itself are available online at [6]. These instances are labeled as CD-O-C-T (see [6] for further details). In our experiments we have generated instances with $n + 1 = O^2$, where $O \in \{10, 15, 20\}$. We also generated both, correlated and uncorrelated instances ($C = \{U, W\}$). Weights and costs are independently and uniformly taken from $\{1, \ldots, 10\}$. We also considered $T = \{2fR, R\}$ and, in addition to the root, we consider two more terminals. For each combination of these parameters we have generated 20 instances.

These instances were used for both the RMWCS and the B-RMWCS. For the B-RMWCS, for a given instance I with set of terminals R, let $\hat{C}_{min}$ be the cost of the minimum Steiner Tree on R with arc costs $\hat{c}_{ij} = c_j$. Values of the available budget B are defined using slacks over $\hat{C}_{min}$ (see also [5]). For example, a 10% of budget slack corresponds to $B = 1.10 \times \hat{C}_{min}$. For the RMWCS, we redefine weights as $w'_v = p_v - c_v$, which can be done because p_v and c_v have comparable units. That way, w'_v somehow represents the net-profit of including node v into the solution. For the RMWCS we set $R = \emptyset$ and we take as root node the reserve node with the smallest index.

Network Design Instances. These Euclidean instances with a topology similar to street networks are generated as proposed in Johnson et al. [9]: First, n nodes are randomly located in a unit Euclidean square. A link between two nodes i and j is established if the Euclidean distance d_{ij} between them is no more than $\alpha/\sqrt{n}$, for a fixed $\alpha > 0$. For a given n and a given α, weights and costs are independently and uniformly taken from $\{1, \ldots, 10\}$.

We generated instances using $n = \{500, 750, 1000\}$ and $\alpha = \{0.6, 1.0\}$; in case that for a given distribution of n nodes in the plane the value of α is not enough for defining a connected graph, it is increased by 0.01 until connecting all components. For each combination of n and α, 20 instances are generated. We take as root the node with index 0 and when considering a set of terminals, these correponds to those nodes with labels 1 and 2.

4.3   Analyzing the Computational Performance

Results for the B-RMWCS. Table 1 shows a comparison of (SA_r) and (CUT_r) models (including 4-cycle and CI) on the set of corridor design instances. The first three rows correspond to the real instances provided by [5], so for each of them we report statistics over a set of 18 problems (obtained for different budget slacks taken from $\{10, 15, \ldots, 95\}$). For the remaining rows, since we create 20 instances for each parameter setting, the reported values correspond to statistics over $18 \times 20 = 360$ instances. In columns $T_{av}(s)$ and $T_{med}(s)$ we report the average and median running times (in seconds), respectively, of those instances solved to optimality, in columns Gap we show the gaps (as percentages) of those instances that were not solved to optimality within 1800 seconds. Columns #(2) and #(8) show the number of connectivity cuts of the (SA_r) and (CUT_r) model, respectively. Column #NOpt shows the number of instances that are not solved

Table 1. Computational performance on B-RMWCS (+C4+CI) instances from [5]

	SA_r				CUT_r				
Instance	T_{av}(s)	T_{med}(s)	Gap	#(2)	#NOpt	T_{av}(s)	T_{med}(s)	Gap	#(8)	#NOpt
CD-40×40-sq	5.28	4.45	0.00	388	0	4.28	3.27	0.00	90	0
CD-10×10-sq	619.58	332.40	0.07	1262	10	1389.07	1441.68	1.39	871	14
CD-25-hex	–	–	5.17	11524	18	–	–	4.81	2958	18
CD-10-U-2fR	1.67	1.12	–	527	0	2.71	1.82	–	360	0
CD-10-W-2fR	1.80	1.00	–	535	0	2.22	1.50	–	389	0
CD-10-U-3R	0.91	0.71	–	362	0	0.63	0.38	–	157	0
CD-10-W-3R	3.08	0.50	–	389	0	0.82	0.42	–	190	0
CD-15-U-2fR	12.47	7.71	–	1085	0	26.33	13.78	–	883	0
CD-15-W-2fR	12.40	8.08	–	1222	0	26.61	10.98	–	1071	0
CD-15-U-3R	4.56	2.98	–	814	0	7.84	2.81	–	513	0
CD-15-W-3R	4.86	2.88	–	809	0	7.34	3.24	–	539	0

to optimality within 1800 seconds. We observe that for all 4-grid instances, except for the CD-10×10-sq graph for which a more detailed analysis is given below, both approaches are able to solve all instances in more or less reasonable times, although the (SA_r) model is slightly better than the (CUT_r) model. On the other hand, the number of inserted violated cuts of the (CUT_r) model is in all of the cases significantly smaller than the corresponding number for the (SA_r) model. The efficacy of the (SA_r) model can be explained by the sparsity of 4-grid graphs. On the contrary, for the only more dense instance of this group, namely CD-25-hex, which is a 6-grid with 12889 nodes and 38065 edges, the (CUT_r) model performs better than the (SA_r) model. More precisely, the avg. gap and its standard deviation for the (SA_r) model are 5.17% and 1.11%, resp., while for the (CUT_r) model these values are 4.81% and 0.81%, resp.'

To analyze the effects of special inequalities, namely 4-cycle and CI, we compare three approaches: Basic, Basic plus 4-cycle inequalities (denoted by "+C4") and Basic plus 4-cycle and CI (denoted by "+C4+CI"). In Figure 2 we present the box-plots of the gaps attained within 1800 seconds when solving real instance CD-10×10-sq for budget slacks taken from $\{10, 15, \ldots, 95\}$. The values marked with an asterisk and × correspond to the mean and maximum running time, respectively. Below the bottom of each box the number of instances solved to optimality is indicated, and next to "#Cuts:" we report the average number of detected cuts of type (2) and (8), respectively.

The box-plots indicate that for the Basic setting the (CUT_r) model significantly outperforms the (SA_r) model on this instance, in terms of the quality of the solutions (smaller gaps), the stability of the approach (smaller dispersion), and the number of instances solved to optimality. This is mainly due to the fact that in the (CUT_r) model there are less variables, so the optimization becomes easier and more stable. However, when including 4-cycle inequalities, although both approaches perform better, (SA_r) now outperforms (CUT_r). The average number of inserted cuts of type (2) decreases from 5989 to 1264 when 4-cycle inequalities are added, while for the (CUT_r) model this reduction is more attenuated (only 18%). This means that for this instance constraints (12) are empirically more effective than (13) in reducing too frequent calls of the maximum flow procedure. When adding the separation of CI ("+CI") we observe that

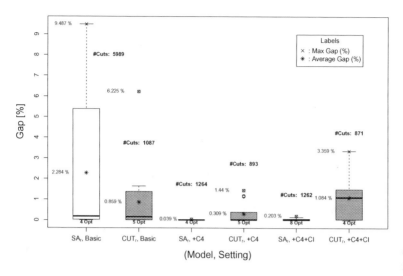

Fig. 2. Box-plots of the gaps [%] reached within 1800 sec for the CD-10×10-sq instance considering (SA_r) and (CUT_r) and three different settings of the B&C (Budget slack [%] taken from $\{10, 15, \ldots, 95\}$)

these constraints are more beneficial for the (SA_r) model than for the (CUT_r) model - the latter one even slows down with addition of these cuts. This can be explained by some numerical instability that can appear when dealing with the separation of CI. We conclude that the advantage of the (CUT_r) model of having less variables vanishes when more sophisticated ideas are considered.

For the Network Design instances (whose complete results are not reported due to space limitation), the graph density plays a role in the performance of the two models. For instance, for $n \in \{500, 750\}$ and $\alpha = 0.6$, the (SA_r) model solves 536 instances out of 760 within the time limit, while the (CUT_r) model solves 443. However, when $\alpha = 1.0$, the (SA_r) approach solves 483 while the (CUT_r) approach solves 502. In both cases, the average running times of the (CUT_r) model needed to prove optimality are smaller than those of the (SA_r) model.

Results for the RMWCS. For the RMWCS we have considered the same corridor design instances and, in addition, the network design instances with a weight transformation as described in § 4.2. In Table 2, equivalent to Table 1, we report the results obtained for the corridor design instances. In this case, time limit is set to 3600 seconds. We observe that the (CUT_r) model outperforms the (SA_r) model on real instances, and on random lattices it is the other way around, although the differences are less visible.

The results on the network design instances are reported in Table 3. For a given n and α equal to 0.6 and 1.0, respectively, column #nodes shows $n+1$ and column #edges shows the average number of edges for a set of 20 instances created using this setting. All instances of this group were solved to optimality,

Table 2. Computational performance on instances from [5] when solving the RMWCS

Instance	SA_r				CUT_r			
	Time(sec)	Gap(%)	#(2)	#NOpt	Time(sec)	Gap(%)	#(8)	#NOpt
CD-40×40-sq	0.70	–	254	0	0.16	–	10	0
CD-10×10-sq	316.11	–	3998	0	88.70	–	60	0
CD-25-hex	3600.00	1.99	20304	1	2611.13	–	14756	0
CD-10-U-2fR	0.15	–	231	0	0.14	–	34	0
CD-10-W-2fR	0.14	–	239	0	0.18	–	40	0
CD-10-U-3R	0.13	–	226	0	0.13	–	28	0
CD-10-W-3R	0.15	–	241	0	0.12	–	26	0
CD-15-U-2fR	1.28	–	720	0	11.59	–	99	0
CD-15-W-2fR	1.35	–	755	0	3.66	–	94	0
CD-15-U-3R	1.24	–	763	0	2.02	–	73	0
CD-15-W-3R	1.45	–	809	0	2.26	–	78	0
CD-20-U-2fR	7.67	–	1618	0	166.32	–	223	0
CD-20-W-2fR	7.41	–	1615	0	74.46	–	234	0
CD-20-U-3R	7.57	–	1667	0	16.90	–	133	0
CD-20-W-3R	8.39	–	1765	0	86.18	–	195	0

Table 3. Computational performance on the RMWCS network design instances

		SA_r		CUT_r	
#nodes	#edges	Time(sec)	#(2)	Time(sec)	#(8)
500	2535	11.42	1218	2.29	22.8
500	6484	3.50	211	0.84	<10
750	3845	57.07	2541	5.67	25.8
750	9944	7.69	287	1.71	<10
1000	5180	97.41	3188	15.59	36.3
1000	13397	10.16	302	2.77	<10

therefore in Table 3 we only report the average running times and the average number of detected connectivity cuts. For these instances, the (CUT_r) approach clearly outperforms the (SA_r) approach; for these instances, the ratio between the number of edges and the number of nodes is, depending on the value of α, around 5 or 13, in contrast to the corridor design instances, where this ratio is close to two. This characteristic implies a practical difficulty for the (SA_r) model due to the increase of the number of variables. Besides, for this group of instances, 4-cycle constraints and CI cannot be used in the initialization.

Conclusion. The obtained computational results let us conclude that both models (CUT_r) and (SA_r) perform very well in practice, and that their performance is complementary. Using the (CUT_r) model (i.e., having less variables) pays off for denser graphs with many zero-weight nodes for both, B-RMWCS and RMWCS.

Acknowledgement. We are thankful to Bistra Dilkina from the Department of Computer Science, Cornell University, who helped in the understanding and interpretation of the Wildlife Corridor Design instances considered in this paper. This research is partially conducted during the research stay of Ivana Ljubić at the TU Dortmund, supported by the APART Fellowship of the Austrian Academy of Sciences. This support is greatly acknowledged. Eduardo Álvarez-Miranda thanks the Institute of Advanced Studies of the Università di Bologna from where he is a PhD Fellow.

References

[1] Backes, C., Rurainski, A., Klau, G., Müller, O., Stöckel, D., Gerasch, A., Küntzer, J., Maisel, D., Ludwig, N., Hein, M., Keller, A., Burtscher, H., Kaufmann, M., Meese, E., Lenhof, H.: An integer linear programming approach for finding deregulated subgraphs in regulatory networks. Nucleic Acids Research 1, 1–13 (2011)

[2] Carvajal, R., Constantino, M., Goycoolea, M., Vielma, J.P., Weintraub, A.: Imposing connectivity constraints in forest planning models (2011) (submitted)

[3] Chen, S., Ljubić, I., Raghavan, S.: The generalized regenerator location problem (2012) (submitted)

[4] Conrad, J.M., Gomes, C.P., van Hoeve, W.-J., Sabharwal, A., Suter, J.F.: Wildlife corridors as a connected subgraph problem. Journal of Environmental Economics and Management 63(1), 1–18 (2012)

[5] Dilkina, B., Gomes, C.: Solving connected subgraph problems in wildlife conservation. In: Lodi, A., Milano, M., Toth, P. (eds.) CPAIOR 2010. LNCS, vol. 6140, pp. 102–116. Springer, Heidelberg (2010)

[6] Dilkina, B., Gomes, C.: Synthetic corridor problem generator (2012), http://www.cs.cornell.edu/~bistra/connectedsubgraph.htm

[7] Dittrich, M., Klau, G., Rosenwald, A., Dandekar, T., Müller, T.: Identifying functional modules in protein-protein interaction networks: an integrated exact approach. Bioinformatics 24, i223–i231 (2008)

[8] Fügenschuh, A., Fügenschuh, M.: Integer linear programming models for topology optimization in sheet metal design. Mathematical Methods of Operations Research 68(2), 313–331 (2008)

[9] Johnson, D.S., Minkoff, M., Phillips, S.: The prize-collecting Steiner tree problem: Theory and practice. In: Proc. 11th ACM-SIAM Symp. Discrete Algorithms, SODA 2000, San Francisco, USA, January 9-11, pp. 760–769 (2000)

[10] Kaparis, K., Letchford, A.N.: Separation algorithms for 0-1 knapsack polytopes. Mathematical Programming 124(1-2), 69–91 (2010)

[11] Koch, T., Martin, A.: Solving Steiner tree problems in graphs to optimality. Networks 32, 207–232 (1998)

[12] Lee, H., Dooly, D.: Decomposition algorithms for the maximum-weight connected graph problem. Naval Research Logistics 45, 817–837 (1998)

[13] Ljubić, I.: Exact and Memetic Algorithms for Two Network Design Problems. PhD thesis, Vienna University of Technology (2004)

[14] Ljubić, I., Weiskircher, R., Pferschy, U., Klau, G., Mutzel, P., Fischetti, M.: An algorithmic framework for the exact solution of the prize-collecting Steiner tree problem. Mathematical Programming, Series B 105, 427–449 (2006)

[15] Moss, A., Rabani, Y.: Approximation algorithms for constrained node weighted Steiner tree problems. SIAM J. Comput. 37(2), 460–481 (2007)

An Empirical Evaluation of Portfolios Approaches for Solving CSPs

Roberto Amadini, Maurizio Gabbrielli, and Jacopo Mauro

Department of Computer Science and Engineering/Lab. Focus INRIA,
University of Bologna, Italy
{amadini,gabbri,jmauro}@cs.unibo.it

Abstract. Recent research in areas such as SAT solving and Integer Linear Programming has shown that the performances of a single arbitrarily efficient solver can be significantly outperformed by a portfolio of possibly slower on-average solvers. We report an empirical evaluation and comparison of portfolio approaches applied to Constraint Satisfaction Problems (CSPs). We compared models developed on top of off-the-shelf machine learning algorithms with respect to approaches used in the SAT field and adapted for CSPs, considering different portfolio sizes and using as evaluation metrics the number of solved problems and the time taken to solve them. Results indicate that the best SAT approaches have top performances also in the CSP field and are slightly more competitive than simple models built on top of classification algorithms.

1 Introduction

The past decade has witnessed a significant increase in the number of constraint solving systems deployed for solving *Constraint Satisfaction Problems* (CSP). It is well recognized within the field of constraint programming that different solvers are better at solving different problem instances, even within the same problem class [3]. It has also been shown in other areas, such as satisfiability testing [18] and integer linear programming [9], that the best on-average solver can be out performed by a portfolio of possibly slower on-average solvers. This selection process is usually performed by using *Machine Learning* (ML) techniques based on feature data extracted from the instances that need to be solved. Thus in general a *Portfolio Approach* [3] is a methodology that exploits the significant variety in performances observed between different algorithms and combines them in a portfolio to create a globally better solver. Portfolio approaches in particular have been extensively studied and used in the SAT solving field. On the other hand, to the best of our knowledge in the CSP field there exists only one solver that uses a portfolio approach, namely CPHydra [13]. This solver uses a rather small portfolio (just 3 solvers) and seems rather limited when compared to modern SAT portfolio approaches.

In this work we tried to investigate to what extent a portfolio approach can increase the performances of a CSP solver and which could be the best portfolio approaches, among the several existing, for CSPs. We considered 22 versions

C. Gomes and M. Sellmann (Eds.): CPAIOR 2013, LNCS 7874, pp. 316–324, 2013.

of 6 well known CSP solvers and using these 22 solvers we implemented two classes of CSP portfolio solvers, building portfolios of up to 16 solvers: in the first class we used relatively simple, off-the-shelf machine learning classification algorithms in order to define solver selectors; in the second class we tried to adapt the best, advanced, and complex approaches of SAT solving to CSP. A third portfolio solver that we considered was CPHydra, mentioned above. We then performed an empirical evaluation and comparison of these three different portfolio approaches. We hope that our results, described in the remaining of this paper, may lead to new insights, to a confirmation of the quality of some approaches and also to some empirical data supporting the creation of better and faster CSP solvers.

It is worth noticing that adapting portfolios techniques from other fields is not trivial: for instance, since portfolio approaches usually exploit features extracted from the various instances of the problems, a good features selection may be responsible of the quality and the performances of an approach. Moreover, differently from the SAT world, in the CSP field there is no a standard language to express CSP instances, there are fewer solvers, and sometimes only few features and constraints are supported. To overcome these limitations we tried to collect a dataset of CSP instances as extensive as possible. We used this dataset to evaluate the performances of the three different CSP portfolio approaches.

2   Preliminaries

In this section we describe CPHydra and the SAT specific portfolio approaches that we have adapted to CSP.

CPHydra. To our knowledge CPHydra [13] is the only CSP solver which uses a portfolio approach. This solver uses a k-nearest neighbor algorithm in order to compute a schedule of the portfolio constituent solvers which maximizes the chances of solving an instance within a time-out of 1800 seconds. CPHydra was able to win the 2008 International CSP Solver Competition.

SAT Solver Selector (3S). 3S [6] is a SAT solver that conjugates a fixed-time static solver schedule with the dynamic selection of one long-running component solver. It first executes for 10% of its time short runs of solvers. The schedule of solvers, obtained by solving an optimization problem similar to the one tackled by CPHydra, is computed offline (i.e. during the learning phase on training data). Then, at run time, if a given instance is not yet solved after the short runs a designated solver is executed for the remaining time. This solver is chosen among the ones that are able to solve the majority of the most k-similar instances in the training dataset. 3S was the best-performing dynamic portfolio at the International SAT Competition 2011.

SATzilla. SATzilla [18] is a SAT solver that relies on runtime prediction models to select the solver that (hopefully) has the fastest running time on a given problem instance. In the International SAT Competition 2009, SATzilla won all three major tracks of the competition. More recently a new powerful version of SATzilla has been proposed [17]. Instead of using regression-based runtime predictions, the newer version uses a weighted random forest approach provided with an explicit cost-sensitive loss function punishing misclassifications in direct proportion to their impact on portfolio performance. This last version consistently outperforms the previous versions of SATzilla and the other competitors of the SAT Challenge 2012 in the Sequential Portfolio Track.

ISAC. In [10] the Instance-Specific Algorithm Configuration tool ISAC [7] has been used as solver selector. Given a highly parametrized solver for a SAT instance, the aim of ISAC is to optimally tune the solver parameters on the basis of the given instance features. It can be easily seen as a generalization of an algorithm selector since it could be used to cluster the instances and when a new instance is encountered it selects the solver that solved the largest number of instances belonging to the nearest cluster.

3   Solvers, Features and Dataset

In this section we introduce the three main ingredients of our portfolios: the CSP solvers that we use; the features, extracted from the CSP instances, which are used in the machine learning algorithms; the dataset used to perform the tests.

Solvers. We decided to build our portfolios by using some of the solvers of the International CSP Solver Competition. We were able to use 5 solvers of this competition, namely AbsCon (2 versions), BPSolver, Choco (2 versions), Mistral and Sat4j. Moreover, by using a specific plug-in described in [11], we were able to use also 15 different versions of the constraint solver Gecode (these different versions were obtained by tuning the search parameters and the variable selection criteria of the solver). Thus we had the possibility of using, in our portfolio, up to 22 specific solvers which were all able to process CSP instances defined in the XCSP format [14].

Features. In order to train the classifiers, we extrapolated a set of 44 features from each XCSP instance. An extensive description of the features can be retrieved in [8]. We used the 36 features of CPHydra [13] plus some features derived from the *variable graph* and *variable-constraint graph* of the XCSP instances. Whilst the majority of these features are syntactical, some of them are computed by collecting data from short runs of the Mistral solver.

Dataset. We tried to perform our experiments on a set of instances as realistic and large as possible. Hence, we constructed a comprehensive dataset of CSPs based on the instances gathered from the 2008 International CSP Solver Competition that are publicly available and already in a XCSP normalized format. Moreover, we added to the dataset the instances from the MiniZinc suite benchmark. These instances written in FlatZinc [12] were first compiled to XCSP (by using a FlatZinc to XCSP converter provided by the MiniZinc suite) and then normalized following the CSP competition conventions. Unfortunately, since FlatZinc is more expressive than XCSP not all the instances could be successfully converted. The final benchmark was built by considering 7163 CSP

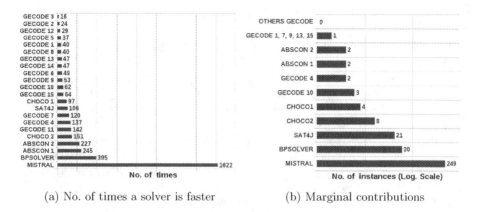

(a) No. of times a solver is faster   (b) Marginal contributions

Fig. 1. Solver statistics

instances taken from the Constraint Competition, 2419 CSP instances obtained by the conversion of the MiniZinc instances and then discarding all the instances solved by Mistral during the first 2 seconds computation of the dynamic features. We obtained a dataset containing 4547 instances (3554 from the Constraint Competition and 993 from MiniZinc). For all the instances in the dataset we run all the 22 version of the solvers collecting their results and computation times with a time limit of 1800 seconds (which is the same threshold used in the Constraint Competition). Among the dataset instances, 797 could not be solved by any solver in our portfolio within the time cap. Figure 1a indicates the relative speed of the different solvers by showing, for each solver, the number of instances on which the considered solver is the fastest one. Considering this metric, Mistral is by far the best solver, since it is faster than the others for 1622 instances (36% of the instances of the dataset). In Figure 1b following [17] we show instead the marginal contributions of each solver, that is how many times a solver is able to solve instances that no other solver can solve. Even in this case Mistral is by far the best solver, almost one order of magnitude better than the second one. It is worth noticing that there are also 8 versions of Gecode that do not give a marginal contribution.

4   Methodology

In order to evaluate and compare different portfolio approaches we tested every approach using a 5-repeated 5-fold cross-validation [2]. The dataset was randomly partitioned in 5 disjoint sets called folds. Each of these folds was treated in turn as the test set, considering the union of the 4 remaining folds as training data. In order to avoid a possible overfitting problem (i.e. a portfolio approach that adapts too well on the training data rather than learning and exploiting the generalized pattern) the random generation of the folds was repeated 5 times, thus obtaining 25 sets of instances used to test the portfolio approaches. Every test set was therefore constituted by approximately 909 instances and the portfolio approach for a single fold was built by taking into account (approximately) 3638 training instances. For every instance of every test set we computed the solving strategy proposed by the portfolio approach and we simulated it by using a time cap of 1800 seconds, checking if the solving strategy was able to solve the instance and the time required. To evaluate the performances of the portfolio approach we measured the average solving time (AST) and the percentage of solved instances (PSI) of the portfolio solver, computed on all the 22735 instances of the 25 test sets. In order to present a more realistic scenario, we have considered in the simulation also the time taken to compute the instance features. All the portfolio approaches were tested with portfolios of different sizes. Since we realized that some solvers had a very low marginal contribution we considered portfolios consisting of up to a maximum of 16 solvers. For every size $n = 2, \ldots, 16$ the portfolio composition was computed by using a local search algorithm that maximized the number of instances solved by one of the solvers in the portfolio. Possible ties were broken by minimizing the average solving time for the instances of the dataset by the solvers in the portfolio.

For the approaches that used off-the-shelf machine learning classification algorithms we used a training set to train a classifier in order to select the best solver among those in the portfolio. For the instances that were not solved by any solver we added a new label *no solver* that could be predicted. For every instance of the test set we simulated the execution of the solver selected by the model. In case the predicted solver was labeled *no solver* or it finished unexpectedly before the time cap the execution of a *backup solver* was simulated for the remaining time. To decide the backup solver, we simulated an election scenario by considering CSPs as voters who have to elect a representative among the 22 candidates solvers. Each CSP could express one or more preferences according to its favorite solver. The election outcomes clearly sustained Mistral as the backup solver since it was the *Condorcet winner*, i.e. the candidate preferred by more voters when compared with every other candidate.

To train the models we used the WEKA tool [5] which implements some of the most well known and widely used classification algorithms. In particular we used a k-nearest neighbors algorithm (IBk), decision trees based algorithms (RandomForest, J48, DecisionStump), bayesian networks (NaiveBayes), rule based algorithms (PART, OneR), support vector machines (SMO), and meta classifiers (AdaBoostM1, LogitBoost). For all the classification algorithms we tried different parameters in order to increase their accuracy. This task was performed following the best practices when they were available or manually trying different parameters starting from the default ones of WEKA. The above approaches based on a ML classification algorithm have been compared against the other approaches described in Section 2.

In order to reproduce the CPHydra approach, we computed the scheduling that it would have produced for every instance of the test set and simulated this schedule. Since this approach does not scale very well w.r.t. the size of the portfolio we were able to simulate this approach only for small portfolios (i.e.

containing less than 9 solvers). To compute the PSI and AST we did not take into account the time needed to compute the schedule; therefore the results of CPHydra can be considered only an upper bound of its real performances.

We simulated the SATzilla approach by developing a MATLAB implementation of the cost-sensitive classification model described in [17], with the only exception that ties during solvers comparison are broken by selecting the solver that in general solves the largest number of instances. We employed Mistral as a backup solver in case the solver selected by SATzilla ended prematurely.

To simulate the 3S approach we did not use the original code to compute the static schedule since it is not publicly available. To compute the schedule of solvers we used instead the mixed integer programming solver Gurobi [4] to solve the problem described in [6]. However, in order to reduce the search space, instead of using the column generation method as used by the developers of 3S, we imposed an additional constraint requiring every solver to be run for an integer number of seconds. If the instance was not solved in this time window the solver that solved the majority of the most k-similar instances was used for the remaining time (possible ties were broken by minimizing the average solving time) and, in case of failures, Mistral was used as a backup solver.

Thanks to the code kindly provided by Yuri Malitsky, we were able to adapt ISAC cluster-based techniques to create a solver selector using the *"Pure Solver Portfolio"* approach as done for SAT problems in [10]. Also in this case Mistral was used as a backup solver in case of failures of the selected solver. All the code developed to conduct the experiments is available at http://www.cs.unibo.it/~amadini/cpaior_2013.zip.

5   Results and Assessments

This section presents the experimental results of our work.

In Fig. 2 for brevity we just show the comparison between the approaches of SATzilla, ISAC, 3S, CPHydra and the best approach using off-the-shelf classifiers which was the one using Random Forest as solver selector (please see [1] for a more extensive comparisons) setting as baselines the performances of Mistral with a time cap of 1800 seconds and of the Virtual Best Solver (VBS), i.e. an oracle that for every instance always chooses the best solver. As already stated, due to the computational cost of computing the schedule of solvers, for CPHydra we report the results obtained using just less than 9 solvers.

It is possible to notice that the best approaches used in SAT, namely 3S and SATzilla, have peak performances. 3S is able to solve usually few more instances than SATzilla (3S have a peak PSI of 78.15% against the 78.1% peak performance of SATzilla) while SATzilla is usually faster (the AST of SATzilla with a portfolio of size 6 was 466.82 seconds against the 470.30 seconds of 3S). Even though conceptually 3S and SATzilla are really different they have surprisingly close performances. This is confirmed also from a statistical point by using the

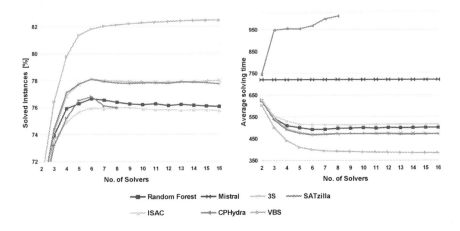

Fig. 2. Performances of portfolio approaches

Student's paired t-test with a p-value threshold of 0.05. 3S and SATzilla are instead statistically better than all the other tested approaches for portfolios of size greater than 3 (3S is able to close 26% of the gap of Random Forest w.r.t. the VBS). Moreover, the decay of performances due to the increase of the portfolio size is less pronounced that what usually happens when a classifier is used as a solver selector. As in the classification based approaches, the peak performance was reached with a relatively small portfolio (6 solvers) and the peak performances of both 3S and SATzilla are statistically significant w.r.t their performances with different portfolios sizes. The performances of ISAC are slightly worse than those of Random Forest: the maximum PSI reached was 75.99% while the Random Forest approach obtained 76.65%.

As far as CPHydra is concerned we saw that it solved the maximum number of instances with a portfolio of size 6 reaching a PSI of 76.81% that was slightly better than the peak performance obtained by Random Forest, even though not in a statistically significant way. After reaching the maximal number of solved instances CPHydra performances are decreasing and in a real scenario they would be rather poor since computing the optimal solvers schedule can consume a lot of time. From Figure 2 it is possible to note that CPHydra differs from other approaches because it is not developed to minimize the average solving time. There is no heuristic to decide which solver needs to be run first in order to minimize the solving time. For this reason, CPHydra is the only approach, among those we have considered, where the PSI and AST values have a positive correlation. Indeed, the Pearson correlation coefficient between PSI and AST values is 0.921, which means that PSI and AST are almost in linear relationship. Conversely for the other best performing approaches the correlation coefficient was always below -0.985 meaning that minimizing the average solving time was like requiring to maximize the number of instances solved and vice versa.

6   Conclusions

In this work we have implemented different portfolio approaches for solving Constraint Satisfaction Problems (CSPs). These approaches have been obtained both by using machine learning techniques and adapting to CSPs other algorithms proposed in the literature, mainly in the SAT solving field. We have evaluated and compared the different approaches by considering a dataset consisting of 4547 instances taken from two different kind of constraint competitions and a selection of 22 versions of different solvers. The portfolio approaches were evaluated on the basis of the number of problems solved and the time taken to solve them. The experimental results show that the approaches that won the last two SAT competitions, namely SATzilla and 3S, are the best ones among those considered in this paper, both for the instances solved and the time needed to solve them. However approaches using off-the-shelf classifiers as solver selector are not that far from the best performances and can potentially be used in scenarios were the time needed to build the model to make the predictions matters. Another interesting empirical fact is that, for all but one the portfolio approaches considered here, there was a strong anti-correlation between the average solving time and the number of solved instances.

We are aware of the fact that our results are not as exhaustive as those existing in the SAT field. However we believe that we made a first step towards a clarification of the importance of the portfolio approaches for solving CSPs. As a future work we plan to extend the number of portfolio approaches by considering also the dynamic schedule approach of 3S [6], the regression based approach of the previous version of SATzilla and other approaches which are not based on feature extraction like [15]. Moreover we are also interested in studying the impact of instance-specific algorithm configuration tools like ISAC or HYDRA [16] in the CSP field by allowing the automatic tuning of search and other solver parameters in order to boost the solver performances.

References

1. Amadini, R., Gabbrielli, M., Mauro, J.: An Empirical Evaluation of Portfolios Approaches for solving CSPs. ArXiv e-prints (December 2012),
   http://arxiv.org/pdf/1212.0692v1
2. Arlot, S., Celisse, A.: A survey of cross-validation procedures for model selection (July 2009)
3. Gomes, C.P., Selman, B.: Algorithm portfolios. Artif. Intell. 126(1-2), 43–62 (2001)
4. Inc. Gurobi Optimization. Gurobi Optimizer Reference Manual (2012)
5. Hall, M., Frank, E., Holmes, G., Pfahringer, B., Reutemann, P., Witten, I.H.: The WEKA data mining software: an update. SIGKDD Explor. Newsl. 11(1), 10–18 (2009)
6. Kadioglu, S., Malitsky, Y., Sabharwal, A., Samulowitz, H., Sellmann, M.: Algorithm selection and scheduling. In: Lee, J. (ed.) CP 2011. LNCS, vol. 6876, pp. 454–469. Springer, Heidelberg (2011)
7. Kadioglu, S., Malitsky, Y., Sellmann, M., Tierney, K.: ISAC - Instance-Specific Algorithm Configuration. In: ECAI, pp. 751–756 (2010)

8. Kiziltan, Z., Mandrioli, L., Mauro, J., O'Sullivan, B.: A classification-based approach to managing a solver portfolio for CSPs. In: AICS (2011)
9. Leyton-Brown, K., Nudelman, E., Shoham, Y.: Learning the Empirical Hardness of Optimization Problems: The Case of Combinatorial Auctions. In: Van Hentenryck, P. (ed.) CP 2002. LNCS, vol. 2470, pp. 556–572. Springer, Heidelberg (2002)
10. Malitsky, Y., Sellmann, M.: Instance-Specific Algorithm Configuration as a Method for Non-Model-Based Portfolio Generation. In: Beldiceanu, N., Jussien, N., Pinson, É. (eds.) CPAIOR 2012. LNCS, vol. 7298, pp. 244–259. Springer, Heidelberg (2012)
11. Morara, M., Mauro, J., Gabbrielli, M.: Solving XCSP problems by using Gecode. In: CILC, pp. 401–405 (2011)
12. Nethercote, N., Stuckey, P.J., Becket, R., Brand, S., Duck, G.J., Tack, G.: MiniZinc: Towards a standard CP modelling language. In: Bessière, C. (ed.) CP 2007. LNCS, vol. 4741, pp. 529–543. Springer, Heidelberg (2007)
13. O'Mahony, E., Hebrard, E., Holland, A., Nugent, C., O'Sullivan, B.: Using case-based reasoning in an algorithm portfolio for constraint solving. In: AICS 2008 (2009)
14. Roussel, O., Lecoutre, C.: XML Representation of Constraint Networks: Format XCSP 2.1. CoRR, abs/0902.2362 (2009)
15. Silverthorn, B., Miikkulainen, R.: Latent class models for algorithm portfolio methods. In: Fox, M., Poole, D. (eds.) AAAI. AAAI Press (2010)
16. Xu, L., Hoos, H., Leyton-Brown, K.: Hydra: Automatically Configuring Algorithms for Portfolio-Based Selection. In: AAAI (2010)
17. Xu, L., Hutter, F., Hoos, H., Leyton-Brown, K.: Evaluating Component Solver Contributions to Portfolio-Based Algorithm Selectors. In: Cimatti, A., Sebastiani, R. (eds.) SAT 2012. LNCS, vol. 7317, pp. 228–241. Springer, Heidelberg (2012)
18. Xu, L., Hutter, F., Hoos, H.H., Leyton-Brown, K.: SATzilla-07: The Design and Analysis of an Algorithm Portfolio for SAT. In: Bessière, C. (ed.) CP 2007. LNCS, vol. 4741, pp. 712–727. Springer, Heidelberg (2007)

Mixed Integer Programming
vs. Logic-Based Benders Decomposition
for Planning and Scheduling[*]

André Ciré, Elvin Coban, and John N. Hooker

Carnegie Mellon University, USA
{acire,ecoban,jh38}@andrew.cmu.edu

Abstract. A recent paper by Heinz and Beck (CPAIOR 2012) found that mixed integer software has become competitive with or superior to logic-based Benders decomposition for the solution of facility assignment and scheduling problems. Their implementation of Benders differs, however, from that described in the literature they cite and therefore results in much slower performance than previously reported. We find that when correctly implemented, the Benders method remains 2 to 3 orders of magnitude faster than the latest commercial mixed integer software on larger instances, thus reversing the conclusion of the earlier paper.

1 Introduction

Logic-based Benders decomposition (LBBD) [12, 21, 22] is a generalization of classical Benders decomposition [3] that accommodates an arbitrary optimization problem as the subproblem. LBBD is particularly attractive for planning and scheduling problems, in which the master problem can use MIP to allocate jobs to resources and the subproblem can use CP to schedule jobs on each resource. Implementations of this method have obtained computational results superior to those of state-of-the-art MIP and CP solvers, sometimes by several orders of magnitude [4–6, 8–10, 13, 15–17, 19, 23–27].

A recent study by Heinz and Beck [11] finds, however, that MIP software has improved to the point that it is competitive with or superior to LBBD on a class of planning and scheduling problems on which LBBD previously excelled [19]. MIP software has in fact improved significantly since the earlier results were published (2007). Yet the computation times reported in [11] for LBBD are much longer than those obtained in earlier studies, including [19].

A partial explanation for this discrepancy is that [11] incorrectly implements the Benders cuts described in the references it cites [16, 18, 19]. In addition, it solves the CP subproblems with a significantly slower method than the state of the art. We therefore re-implemented LBBD, for the purpose of reproducing previous results and comparing them with recent MIP software. We found that LBBD remains superior to state-of-the-art MIP on this class of problems, as

---

[*] Partial support from NSF grant CMMI-1130012 and AFOSR grant FA-95501110180.

C. Gomes and M. Sellmann (Eds.): CPAIOR 2013, LNCS 7874, pp. 325–331, 2013.

it is 2 to 3 orders of magnitude faster than CPLEX 12.4 on larger instances. These results have subsequently been confirmed by Beck and Ku in unpublished work [2].

2   Previous Work

Table 1 shows computational results for several methods tested by Heinz and Beck [11] as well as previous LBBD results from [19]. The results are for the "c" instances used in [13, 19] and available online [14]. There are n jobs to be assigned to m facilities and then scheduled on each facility, using cumulative scheduling. The objective is to minimize total cost of assigning jobs to facilities, although other objectives have been used [16, 17, 19].

The results from [11] shown in the table are implemented in SCIP on an Intel Xeon E5420 2.5 GHz machine. CIP(CP) and CIP(MIP) refer to CP/MIP hybrids. The LBBD results from [19] were obtained from an implemention in OPL Studio.

The discrepancy in LBBD results may be seen by examining the boldface figures in Table 1. The difference is actually greater than shown, because the SCIP results are shifted geometric means[1] while the earlier LBBD results are averages. There is an even larger discrepancy with results obtained in [28] by the integrated solver SIMPL for similar instances.

One possible explanation for the discrepancy is that [11] uses relatively weak nogood cuts as Benders cuts. When a set J of jobs assigned to facility i is found to have no feasible schedule, the Benders cut $\sum_{j \in J}(1 - x_{ij}) \geq 1$ is added to the master problem, where 0-1 variable $x_{ij} = 1$ when job j is assigned to facility i. The cut prevents this set same of jobs (or any superset) from being assigned to facility i in subsequent iterations. However, earlier studies use strengthened cuts that are obtained heuristically by re-solving the scheduling problem on facility i for subsets of J, so as to find a smaller set J' of jobs that have no feasible schedule. This results in the stronger cut $\sum_{j \in J'}(1 - x_{ij}) \geq 1$. Previous work suggests that the strengthening procedure can bring significant improvement in performance, and this may explain part of the discrepancy. We test this hypothesis in the next section.

3   Computational Results

Table 2 shows our results for the "c" instances. The MIP results are obtained by CPLEX 12.4.01. LBBD is implemented by solving the master problem with CPLEX 12.4.01 and the subproblems with IBM CP Optimizer 12.4.01 using extended filtering, DFS search, and default variable and value selection. All tests are run on an Intel Xeon E5345 2.33 GHz (64 bits) in single core mode with 8 GB RAM.

---

[1] The shifted geometric mean of $v_1, \ldots, v_n$ is $(\Pi_i(v_i + s))^{1/n} - s$ for shift s. Following [11], we use $s = 10$ seconds. The shifted geometric mean of a set of distinct values is smaller than the average.

Table 1. Computational results for planning and scheduling "c" instances as reported by [11] and [19], showing number of instances solved to optimality (out of 5) and solution time. Boldface figures show contrasting reuslts for LBBD.

Size		Reported in [11]								LBBD in [19]	
		MIP (SCIP)		LBBD		CIP(CP)		CIP(MIP)			
m	n	Solved	Sec[1]	Solved	Sec[1]	Solved	Sec[1]	Solved	Sec[1]	Solved	Sec[2]
2	10	5	1	**5**	**1**	5	0	5	2	**5**	**0**
	12	5	1	**5**	**1**	5	0	5	4	**5**	**0**
	14	5	1	**5**	**3**	5	1	5	5	**5**	**0**
	16	5	11	**5**	**17**	5	19	5	30	**5**	**2**
	18	4	162+	**5**	**89**	5	18	5	77	**5**	**9**
	20	3	401+	**3**	**158+**	5	3	5	139	**5**	**111**
	22	2	1442+	**2**	**703+**	2	1325+	4	2550+	**<5**	**1805+**
	24	2	2197+	0	-	3	707+	3	1180+		
	26	3	2977+	1	5193+	1	5440+	2	3261+		
	28	2	2503+	3	441+	3	160+	2	2598+		
	30	1	5429+	1	2972+	0	-	1	4180+		
	32	0	-	1	5680+	3	282+	1	6123+		
3	10	5	0	**5**	**1**	5	0	5	1	**5**	**0**
	12	5	1	**5**	**1**	5	1	5	5	**5**	**0**
	14	5	1	**5**	**1**	5	2	5	11	**5**	**0**
	16	5	14	**5**	**10**	5	22	5	60	**5**	**1**
	18	5	429	**5**	**21**	4	139+	4	296+	**5**	**3**
	20	4	1124+	**5**	**6**	5	35	5	253	**5**	**3**
	22	2	6014+	**5**	**149**	2	1352+	4	505+	**5**	**8**
	24	2	3253+	1	**2324+**	1	3165+	5	1001	**5**	**19**
	26	0	-	4	1351+	3	727+	1	4467+		
	28	2	1829+	0	-	2	1261+	2	3057+		
	30	0	-	0	-	0	-	2	5435+		
	32	0	-	0	-	1	6918+	1	6639+		
4	10	5	0	**5**	**1**	5	1	5	1	**5**	**0**
	12	5	1	**5**	**1**	5	1	5	4	**5**	**0**
	14	5	1	**5**	**1**	5	2	5	12	**5**	**0**
	16	5	2	**5**	**10**	5	1	5	11	**5**	**0**
	18	5	38	**5**	**21**	5	4	5	67	**5**	**1**
	20	5	309	**5**	**6**	5	27	5	106	**5**	**1**
	22	4	324	**5**	**149**	5	46	5	544	**5**	**3**
	24	0	-	1	**2324+**	2	1446+	2	4184+	**5**	**39**
	26	0	-	4	**1351+**	1	4070+	2	5530+	**5**	**29**
	28	1	5034+	0	-	1	2804+	2	3885+		
	30	0	-	0	-	1	2105+	0	-		
	32	0	-	0	-	0	-	0	-		

[1]Shifted geometric mean   [2]Average
+Computation terminated after 7200 sec for instances not solved to optimality.

The new LBBD results are somewhat better than the old ones in [19], presumably due to faster MIP and CP components. More importantly, LBBD remains significantly faster than the most recent CPLEX solver, particularly when there are 3 or 4 facilities. Decomposition is obviously more helpful when there are more facilities, because more decoupling is possible in the subproblem.

The "c" instances used in [13, 19] put LBBD at a disadvantage in two ways: they use relatively few facilities, and the facilities differ greatly (by a factor of m) in speed. The master problem assigns most jobs to the fastest facilities, resulting in harder scheduling problems for CP. We therefore ran MIP and LBBD on problem set "e," also used in [13, 19]. These instances have more facilities,

Table 2. Computational results for planning and scheduling "c" instances, showing the number of problem instances solved (out of 5) and computation time. The results are obtained for (a) the CPLEX MIP solver, (b) logic-based Benders decomposition implemented with simple nogood cuts, and (c) LBBD implemented with the strengthened cuts used in previous studies.

Size		MIP (CPLEX)			LBBD: Weak cuts			LBBD: Strong cuts		
m	n	Solved	Sec[1]	Sec[2]	Solved	Sec[1]	Sec[2]	Solved	Sec[1]	Sec[2]
2	10	5	0.1	0.1	5	0.1	0.1	5	0.1	0.1
	12	5	0.2	0.2	5	0.1	0.1	5	0.0	0.0
	14	5	0.1	0.1	5	0.1	0.1	5	0.0	0.0
	16	5	8.9	28	5	0.2	0.2	5	0.3	0.3
	18	5	65	388	5	0.5	0.5	5	0.6	0.7
	20	4	221+	1902	5	1.9	2.0	5	6.4	8.0
	22	3	1055+	3849+	5	38	617	5	67	955
	24	2	1000+	4351+	4	110+	1495+	5	267	1948
	26	1	6055+	6365+	5	94	327	5	299	1948
	28	1	1230+	4396+	5	509	1004	5	606	1133
	30	0	-	-	2	1916+	5391+	5	336	5401
	32	1	3932+	5828+	2	703+	4325+	2	704+	4325+
3	10	5	0.0	0.0	5	0.1	0.1	5	0.1	0.1
	12	5	0.1	0.1	5	0.4	0.5	5	0.1	0.1
	14	5	0.3	0.3	5	0.3	0.3	5	0.2	0.2
	16	5	8.6	13	5	2.5	2.7	5	0.8	0.8
	18	5	204	548	5	6.2	7.8	5	1.4	1.4
	20	4	326+	1712+	5	1.1	1.2	5	0.5	0.5
	22	3	1458+	3679+	5	6.2	7.5	5	2.4	2.6
	24	2	1941+	4438+	5	8.8	15	5	5.1	5.7
	26	0	-	-	5	119	191	5	62	98
	28	2	4035+	5254+	5	89	271	5	85	209
	30	0	-	-	4	612+	2356+	5	498	1856
	32	0	-	-	2	3091+	4666+	2	3350+	4751+
4	10	5	0.0	0.0	5	0.0	0.0	5	0.0	0.0
	12	5	0.1	0.1	5	0.1	0.1	5	0.1	0.1
	14	5	0.3	0.3	5	0.9	1.0	5	0.3	0.3
	16	5	1.0	1.0	5	0.4	0.4	5	0.1	0.1
	18	5	15	36	5	1.6	1.7	5	0.4	0.4
	20	5	109	522	5	1.1	1.1	5	0.3	0.3
	22	5	206	811	5	4.5	8.2	5	1.0	1.1
	24	1	5918+	6300+	5	16	23	5	6.8	9.1
	26	0	-	-	5	16	19	5	7.0	7.4
	28	1	3184+	5783+	5	19	36	5	11	11
	30	0	-	-	5	89	430	5	34	61
	32	0	-	-	5	286	679	5	206	478

[1] Shifted geometric mean   [2] Average
+Computation terminated after 7200 sec for instances not solved to optimality.

with speeds differing by no more than a factor of 1.5. The results appear in Table 3. The LBBD advantage is even greater on these instances, easily solving all 20 of the instances with 35 jobs or more, while MIP solved only 5 of them within a two-hour time limit.

The results show that strengthened cuts can make a significant difference when there are 3 or more facilities, but not enough to explain most of the discrepancy described above in the LBBD results. This suggests that the discrepancy is primarily due to SCIP's relatively slow solution of CP subproblems.

Table 3. Computational results for planning and scheduling "e" instances

| Size | | MIP (CPLEX) | | | LBBD: Weak cuts | | | LBBD: Strong cuts | | |
m	n	Solved	Sec[1]	Sec[2]	Solved	Sec[1]	Sec[2]	Solved	Sec[1]	Sec[2]
2	10	5	0.1	0.1	5	0.1	0.1	5	0.1	0.1
2	12	5	0.3	0.3	5	0.3	0.3	5	0.1	0.1
3	15	5	0.9	0.9	5	0.4	0.4	5	0.2	0.2
4	20	5	20	46	5	6.2	14	5	1.6	1.9
5	25	5	25	73	5	1.0	1.0	5	0.7	0.7
6	30	5	113	543	5	1.3	1.3	5	0.4	0.4
7	35	2	2956+	5122+	5	25	36	5	2.5	2.7
8	40	2	909+	4357+	4	130+	1527+	5	18	80
9	45	0	–	–	5	512	1050	5	29	35
10	50	1	6968+	6983+	5	22	45	5	5.0	5.4

[1]Shifted geometric mean   [2]Average
+Computation terminated after 7200 sec for instances not solved to optimality.

4 Conclusions

We conclude that despite improvements in MIP software, logic-based Benders decomposition results in significantly faster solution of the problem class studied here, generally with speedups of at least 100 to 1000 on larger instances. It is also superior to the other MIP/CP hybrid methods reported in [11] and shown in Table 1. The advantage of LBBD increases as the number of facilities increases, because it is a decomposition method that is naturally more effective when the problem decomposes into smaller subproblems.

Heinz and Beck [11] remark that even when MIP is slower than LBBD at proving optimality, it is more effective at finding good feasible solutions. This because LBBD does not find a feasible solution until it terminates with a proof of optimality.

In response, we first point out that this is a technical peculiarity of the minimum cost problem solved here, in which the Benders subproblem is a feasibility problem. It is not the case for other objective functions—such as makespan, number of late jobs, and total tardiness—for which the subproblem is an optimization problem that finds feasible solutions throughout the solution process [16, 17, 19].

More importantly, LBBD is proposed as an exact method, rather than a heuristic method for finding feasible solutions. The tests show that LBBD clearly excels at finding optimal solutions. If the goal is to find good feasible solutions, one can use primal heuristics. In fact, this is precisely the approach taken by the MIP solvers tested by [11]. LBBD can likewise find good feasible solutions if it is permitted to use a primal heuristic. However, in the tests reported in [11], only the MIP and MIP/CP hybrid methods are allowed to use primal heuristics.

Heinz and Beck also remark that unlike MIP, the decomposition approach has difficulty with constraints that couple the scheduling subproblems, such as precedence constraints. Decomposition in fact requires that the subproblems decouple and would presumably be less effective if the subproblems were linked. However, this is balanced by a weakness of the MIP model, which uses time-indexed variables. As the time horizon becomes longer, the number of integer variables increases, and the MIP problem becomes disproportionately harder. This is demonstrated in [7] even for a single-facility problem, in which case a LBBD method that decomposes the time horizon into short segments is far superior to MIP.

The more general issue is whether it is reasonable to pursue such special-purpose methods as LBBD when MIP software constantly improves. LBBD seems worthy of pursuit, for four reasons. One reason is that LBBD technology developed before 2007 is still far superior to the latest MIP solvers, at least on the problem class studied here. A second reason is that LBBD can also improve. Logic-based Benders cuts are more effective when they exploit information from the inference dual of the subproblem [21, 20]. This information is not available from the CP solver in pre-2007 implementations. An implementation that fully integrates the master and subproblem may be signficantly more effective.

A third reason is that LBBD is best conceived as an *enhancement* of MIP rather than as a *competitor* of MIP. After all, LBBD relies on MIP to solve the master problem, and conceivably the subproblems as well. MIP technology will doubtless continue to improve, but these improvements can be incorporated into LBBD. The relevant issue is whether the effectiveness of MIP can be increased *still more* by decomposing the problem and applying MIP to some or all of its components. So far, the answer appears to be yes.

Finally, LBBD need not be a "special-purpose" method, but can be part of an integrated approach to optimization. For example, LBBD has been implemented within the general-purpose solver SIMPL [1], which obtained the results in [28] mentioned earlier. SIMPL treats both the Benders method and MIP as special cases of a restrict-infer-and-relax algorithm. Rather than rely solely on engineering improvements in MIP solvers, it seems best to invest similar effort in integrated solvers, such as SCIP and SIMPL.

References

1. Aron, I., Hooker, J.N., Yunes, T.H.: SIMPL: A system for integrating optimization techniques. In: Régin, J.-C., Rueher, M. (eds.) CPAIOR 2004. LNCS, vol. 3011, pp. 21–36. Springer, Heidelberg (2004)
2. Beck, J.C., Ku, W.-Y.: Personal communication (September 14, 2012)
3. Benders, J.F.: Partitioning procedures for solving mixed-variables programming problems. Numerische Mathematik 4, 238–252 (1962)
4. Benini, L., Bertozzi, D., Guerri, A., Milano, M.: Allocation and scheduling for MPSoCs via decomposition and no-good generation. In: van Beek, P. (ed.) CP 2005. LNCS, vol. 3709, pp. 107–121. Springer, Heidelberg (2005)
5. Cambazard, H., Hladik, P.-E., Déplanche, A.-M., Jussien, N., Trinquet, Y.: Decomposition and learning for a hard real time task allocation problem. In: Wallace, M. (ed.) CP 2004. LNCS, vol. 3258, pp. 153–167. Springer, Heidelberg (2004)
6. Chu, Y., Xia, Q.: Generating Benders cuts for a class of integer programming problems. In: Régin, J.-C., Rueher, M. (eds.) CPAIOR 2004. LNCS, vol. 3011, pp. 127–141. Springer, Heidelberg (2004)
7. Coban, E., Hooker, J.N.: Single-facility scheduling over long time horizons by logic-based benders decomposition. In: Lodi, A., Milano, M., Toth, P. (eds.) CPAIOR 2010. LNCS, vol. 6140, pp. 87–91. Springer, Heidelberg (2010)

8. Corréa, A.I., Langevin, A., Rousseau, L.M.: Dispatching and conflict-free routing of automated guided vehicles: A hybrid approach combining constraint programming and mixed integer programming. In: Régin, J.-C., Rueher, M. (eds.) CPAIOR 2004. LNCS, vol. 3011, pp. 370–378. Springer, Heidelberg (2004)
9. Harjunkoski, I., Grossmann, I.E.: A decomposition approach for the scheduling of a steel plant production. Computers and Chemical Engineering 25, 1647–1660 (2001)
10. Harjunkoski, I., Grossmann, I.E.: Decomposition techniques for multistage scheduling problems using mixed-integer and constraint programming methods. Computers and Chemical Engineering 26, 1533–1552 (2002)
11. Heinz, S., Beck, J.C.: Reconsidering mixed integer programming and MIP-based hybrids for scheduling. In: Beldiceanu, N., Jussien, N., Pinson, É. (eds.) CPAIOR 2012. LNCS, vol. 7298, pp. 211–227. Springer, Heidelberg (2012)
12. Hooker, J.N.: Logic-Based Methods for Optimization: Combining Optimization and Constraint Satisfaction. Wiley, New York (2000)
13. Hooker, J.N.: A hybrid method for planning and scheduling. In: Wallace, M. (ed.) CP 2004. LNCS, vol. 3258, pp. 305–316. Springer, Heidelberg (2004)
14. Hooker, J.N.: Planning and scheduling problem instances (with documentation), website (2004), web.tepper.cmu.edu/jnh/instances.htm
15. Hooker, J.N.: A hybrid method for planning and scheduling. Constraints 10, 385–401 (2005)
16. Hooker, J.N.: Planning and scheduling to minimize tardiness. In: van Beek, P. (ed.) CP 2005. LNCS, vol. 3709, pp. 314–327. Springer, Heidelberg (2005)
17. Hooker, J.N.: An integrated method for planning and scheduling to minimize tardiness. Constraints 11, 139–157 (2006)
18. Hooker, J.N.: Integrated Methods for Optimization. Springer (2007)
19. Hooker, J.N.: Planning and scheduling by logic-based Benders decomposition. Operations Research 55, 588–602 (2007)
20. Hooker, J.N.: Integrated Methods for Optimization, 2nd edn. Springer (2012)
21. Hooker, J.N., Ottosson, G.: Logic-based Benders decomposition. Mathematical Programming 96, 33–60 (2003)
22. Hooker, J.N., Yan, H.: Logic circuit verification by Benders decomposition. In: Saraswat, V., Van Hentenryck, P. (eds.) Principles and Practice of Constraint Programming: The Newport Papers, pp. 267–288. MIT Press, Cambridge (1995)
23. Jain, V., Grossmann, I.E.: Algorithms for hybrid MILP/CP models for a class of optimization problems. INFORMS Journal on Computing 13, 258–276 (2001)
24. Maravelias, C.T., Grossmann, I.E.: Using MILP and CP for the scheduling of batch chemical processes. In: Régin, J.-C., Rueher, M. (eds.) CPAIOR 2004. LNCS, vol. 3011, pp. 1–20. Springer, Heidelberg (2004)
25. Terekhov, D., Beck, J.C., Brown, K.N.: Solving a stochastic queueing design and control problem with constraint programming. In: Proceedings of the 22nd National Conference on Artificial Intelligence (AAAI 2005), pp. 261–266 (2005)
26. Thorsteinsson, E.S.: Branch-and-check: A hybrid framework integrating mixed integer programming and constraint logic programming. In: Walsh, T. (ed.) CP 2001. LNCS, vol. 2239, pp. 16–30. Springer, Heidelberg (2001)
27. Timpe, C.: Solving planning and scheduling problems with combined integer and constraint programming. OR Spectrum 24, 431–448 (2002)
28. Yunes, T.H., Aron, I., Hooker, J.N.: An integrated solver for optimization problems. Operations Research 58, 342–356 (2010)

A Branch-and-Cut Algorithm
for Solving the Team Orienteering Problem[*]

Duc-Cuong Dang[1,3], Racha El-Hajj[1,2,**], and Aziz Moukrim[1]

[1] Université de Technologie de Compiègne, Département Génie Informatique
Laboratoire Heudiasyc, UMR 7253 CNRS, 60205 Compiègne, France
[2] Université Libanaise, Faculté de Génie, Département Contrôle Industriel
Campus Hadath, Beyrouth, Liban
[3] School of Computer Science, University of Nottingham
Jubilee Campus, Wollaton Road, Nottingham NG8 1BB, United Kingdom
{duc-cuong.dang,racha.el-hajj,aziz.moukrim}@hds.utc.fr

Abstract. This paper describes a branch-and-cut algorithm to solve the Team Orienteering Problem (TOP). TOP is a variant of the Vehicle Routing Problem (VRP) in which the aim is to maximize the total amount of collected profits from visiting customers while not exceeding the predefined travel time limit of each vehicle. In contrast to the exact solving methods in the literature, our algorithm is based on a linear formulation with a polynomial number of binary variables. The algorithm features a new set of useful dominance properties and valid inequalities. The set includes symmetric breaking inequalities, boundaries on profits, generalized subtour eliminations and clique cuts from graphs of incompatibilities. Experiments conducted on the standard benchmark for TOP clearly show that our branch-and-cut is competitive with the other methods in the literature and allows us to close 29 open instances.

Keywords: branch-and-cut, dominance property, incompatibility, clique cut.

Introduction

The Team Orienteering Problem (TOP) [4] is a widely studied Vehicle Routing Problem (VRP) which can be described as follows: a fleet of vehicles is available to visit customers from a potential set and each vehicle is associated with a predefined travel time limit and two particular depots, the so-called departure and arrival. Each customer is associated with an amount of profit that can be collected at most once by the fleet of vehicles. The aim of TOP is to select customers and organize an itinerary of visits so as to maximize the total amount of collected profits. The applications of TOP include athlete recruiting [4], technician routing [1, 9] and tourist trip planning [10].

[*] This work is partially supported by the Regional Council of Picardie and the European Regional Development Fund (ERDF), under PRIMA project.

[**] Corresponding author.

To the best of our knowledge, three exact methods have been proposed to solve TOP. Butt and Ryan [3] described a set covering formulation and developed a column generation algorithm for solving TOP. In Boussier et al. [2], the authors proposed a branch-and-price algorithm and a dynamic programming approach to deal with the pricing problem. More recently, a pseudo-polynomial linear model for TOP was introduced by Poggi de Aragão et al. [8] and a branch-cut and price algorithm was proposed. These methods are able to solve a large part of the standard benchmark [4] for TOP, however many other instances remain open. Furthermore, a recent effort [6] showed that it is hardly possible to improve the already-known solutions for TOP by heuristics.

In this paper, we propose a branch-and-cut algorithm for the exact solution of TOP based on a set of dominance properties and valid inequalities. This set includes symmetric breaking, generalized subtour eliminations, boundaries on profits/numbers of customers based on dynamic programming, as well as clique cuts based on the graphs of incompatibilities. Experiments conducted on the standard benchmark [4] for TOP clearly show the competitiveness of the approach, especially on the number of instances being solved to optimality. The algorithm also allows us to close 29 open instances.

1   Compact Formulation

TOP is modeled with a complete graph $G = (V, E)$. $V = \{1, \ldots, n\} \cup \{d, a\}$ is the set of vertices representing customers and depots. $E = \{(i, j) | i, j \in V\}$ is the set of arcs. Vertices d and a are respectively the departure and the arrival depot for the vehicles. We use V^-, V^d and V^a to denote the sets of customers only, customers with departure depot and customers with arrival depot respectively. Each vertex i is associated with a profit P_i ($P_d = P_a = 0$) and a travel cost C_{ij} is associated with each arc $(i, j) \in E$ ($C_{id} = C_{ai} = \infty, \forall i \in V^-$). The travel costs are assumed to satisfy the triangle inequality. A fleet F is composed of m identical vehicles and available to visit customers without exceeding a travel cost limit L for each vehicle. The problem can be then formulated in Mixed Integer Programming (MIP) using a polynomial number of decision variables x_{ijr} and y_{ir}: $x_{ijr} = 1$ if arc (i, j) is used by vehicle r to serve customer i then customer j and 0 otherwise; $y_{ir} = 1$ if client i is served by vehicle r and 0 otherwise.

$$\max \sum_{i \in V^-} \sum_{r \in F} y_{ir} P_i \tag{1}$$

$$\sum_{r \in F} y_{ir} \leq 1 \qquad \forall i \in V^- \tag{2}$$

$$\sum_{j \in V^a} x_{djr} = \sum_{j \in V^d} x_{jar} = 1 \qquad \forall r \in F \tag{3}$$

$$\sum_{i \in V^a \setminus \{k\}} x_{kir} = \sum_{j \in V^d \setminus \{k\}} x_{jkr} = y_{kr} \ \forall k \in V^-, \forall r \in F \tag{4}$$

$$\sum_{i \in V^d} \sum_{j \in V^a \setminus \{i\}} C_{ij} x_{ijr} \leq L \qquad \forall r \in F \tag{5}$$

$$\sum_{(i,j)\in U\times U} x_{ijr} \leq |U| - 1 \quad \forall U \subseteq V^-, |U| \geq 2, \forall r \in F \tag{6}$$

$$x_{ijr} \in \{0,1\} \quad \forall i \in V, \forall j \in V, \forall r \in F \tag{7}$$

$$y_{ir} \in \{0,1\} \quad \forall i \in V^-, \forall r \in F$$

The objective function (1) is to maximize the sum of collected profits. Constraints (2) guarantee that each customer is visited at most once. The connectivity of each tour is ensured by constraints (3) and (4). Constraints (5) describe the travel length restriction. Constraints (6) ensure that subtours are forbidden. Finally, the integral requirement on variables is guaranteed by constraints (7).

2   Generalized Subtour Eliminations

The number of constraints (6) in the formulation is exponential. In practice, these constraints are removed from the formulation and only added when needed. Moreover, we replace these constraints with stronger ones known as the Generalized Subtour Elimination Constraints (GSEC) [7] as follows.

For a given subset S of vertices, we define $\delta(S)$ as the set of arcs connecting vertices of S with vertices of V\S. We also define $\gamma(S)$ as the set of arcs interconnecting vertices of S. The following constraints ensure that each customer served by vehicle r is connected to the depots.

Property 2.1. *GSEC:*

$$\sum_{(u,v)\in\delta(S)} x_{uvr} \geq 2y_{ir}, \forall S \subset V, \{d,a\} \subseteq S, \forall i \in V \setminus S, \forall r \in F \tag{8}$$

Property 2.2. *Equivalent to the GSEC:*

$$\sum_{(u,v)\in\gamma(S)} x_{uvr} \leq \sum_{i\in S} y_{ir} - y_{jr}, \forall S \subset V, \{d,a\} \subseteq S, \forall j \in V \setminus S, \forall r \in F \tag{9}$$

$$\sum_{(u,v)\in\gamma(U)} x_{uvr} \leq \sum_{i\in U} y_{ir} - y_{jr}, \forall U \subseteq V^-, \forall j \in U, \forall r \in F \tag{10}$$

3   Dominance Properties

Given an instance X of TOP with m vehicles, we use X_I to denote the same instance for which profit of each customer is changed to 1. We also define X^g as the instance X where the number of vehicles is reduced to g ($g \leq m$). On the other hand, we use $LB(X)$ (resp. $UB(X)$) to denote a lower (resp. an upper) bound of instance X. The following properties hold for the formulation of TOP.

Property 3.1. *Symmetric breaking on profits, (without loss of generality) we focus on solutions in which profits of routes are sorted in a particular order:*

$$\sum_{i\in V^-} y_{i(r+1)}P_i - \sum_{i\in V^-} y_{ir}P_i \leq 0, \forall r \in F \setminus \{m\} \tag{11}$$

Property 3.2. *Boundaries on profits:*

$$\sum_{r \in H} \sum_{i \in V^-} y_{ir} P_i \leq UB(X^{|H|}), \forall H \subset F \tag{12}$$

$$\sum_{r \in H} \sum_{i \in V^-} y_{ir} P_i + UB(X^{m-|H|}) \geq LB(X), \forall H \subseteq F \tag{13}$$

Property 3.3. *Boundaries on numbers of customers:*

$$\sum_{r \in H} \sum_{i \in V^-} y_{ir} \leq UB(X_I^{|H|}), \forall H \subset F \tag{14}$$

$$\sum_{i \in V^-} y_{ir} \geq LB(\bar{X}_I^1), \ \forall r \in F \tag{15}$$

Values of LB in (13) are calculated using an efficient heuristic for TOP, such as the one used in [6]. Similarly to dynamic programming, values of UB in (12), (13) and (14) are computed as follows. We start our resolution by computing these values for the smallest instance ($|H| = 1$) using a stopping condition, i.e. computational time or number of branch-and-bound nodes, then we use the obtained values to solve larger instances ($|H| \leq m$). In (15), we use $LB(\bar{X}_I^1)$ to denote a lower bound obtained from solving the derived MIP model of X_I^1 for which the objective function is reversed to minimization and the constraints (12) and (13) with $|H| = 1$ are added.

4   Incompatibilities and Clique Cuts

Let S be a small subset of vertices of V^- (or arcs of E), we use $MinLength(S)$ to denote the length of the shortest path from d to a containing all vertices (or all arcs) of S. The graph of incompatibilities between customers is defined as: $G_{V^-}^{Inc} = (V^-, E_{V^-}^{Inc})$ with $E_{V^-}^{Inc} = \{[i,j] | i \in V^-, j \in V^-, MinLength(\{i,j\}) > L\}$. The graph of incompatibilities between arcs is defined as: $G_E^{Inc} = (E, E_E^{Inc})$ with $E_E^{Inc} = \{[i,j] | i = (u,v) \in E, j = (w,s) \in E, MinLength(\{(u,v), (w,s)\}) > L\}$. The following inequalities hold for the formulation of TOP.

Property 4.1. *(Clique) Let K (resp. Q) be a clique of $G_{V^-}^{Inc}$ (resp. G_E^{Inc}):*

$$\sum_{i \in K} y_{ir} \leq 1, \forall r \in F \tag{16}$$

$$\sum_{(u,v) \in Q} x_{uvr} \leq 1, \forall r \in F \tag{17}$$

The two graphs $G_{V^-}^{Inc}$ and G_E^{Inc} can be computed beforehand and archived for each instance. Maximal cliques are preferred in inequalities (16) and (17) since they provide tighter formulation. In practice a greedy decomposition of the two graphs into maximal cliques is used and the details are given in the next section.

5   Branch-and-Cut Algorithm

Branching Rule: Since the objective function is to maximize the collected profits, the selection of correct customers appears to be crucial for TOP. Therefore, our branching rule is prioritized on y_{ir} first then x_{ijr} [2, 8].

Presolving Steps: By definition, a customer is said to be *inaccessible* if the travel cost of the tour containing only that customer excesses the cost limit. An *inaccessible* arc can be similarly defined. So in order to make a proper linear formulation, we first eliminate all *inaccessible* customers and arcs. Additionally, during a limited computational time at the beginning (e.g. limited to 5% of the total solving time), the values required for inequalities (12)-(15) are computed using the method mentioned in Section 3. Then a greedy decomposition of $G_{V^-}^{Inc}$ into maximal cliques is generated using [5] and the associated inequalities (16) are added to the formulation.

Complete Algorithm: The MIP solver is initialized with a feasible solution generated from an heuristic of [6]. This initialization accelerates the resolution by eliminating portions of the search space composed of solutions with lower profits. In the first iteration, the linear model containing constraints (2)-(5), (7), (11)-(15) and (16) is solved and a solution is obtained. The solution is then checked for subtours. If the solution does not contain any subtour, then it is optimal and the resolution is terminated. Otherwise, the associated constraints (8), (9) and (10) are added into the linear model. Additionally, based on the sets of vertices and arcs from the solution, we extract the associated subgraphs from $G_{V^-}^{Inc}$ and G_E^{Inc}, then generate their greedy clique decompositions in order to add the corresponding constraints (16) and (17) to the linear model. In the next iteration, the same solving process is repeated with the new model.

6   Numerical Results

Our approach was tested on the benchmark of TOP instances proposed by Chao et al. [4]. It comprises 387 instances divided into 7 sets. The numbers of customers and vehicles are up to 100 and 4 respectively. Our algorithm is coded in C++. Experiments were conducted on an AMD Opteron 2.60 GHz and CPLEX 12.4 was used as MIP solver. We used the same 2h limit of solving time as in [2, 8].

In order to evaluate the usefulness of the proposed components, we activated them one by one so that the complete algorithm is obtained in the last activation. Table 1 shows the number of instances being solved to optimality for each activation. The average computational times on the subset of instances being solved by all configurations are also given. We notice that each component contributes to the improvement of the number of instances being solved, as well as to the reduction of the computational times. Table 2 shows a small comparison between our results and those of Boussier et al. [2]. Each instance is chosen so that only one of the two methods is able to prove the optimality. Columns *instance*, n, m, and L indicate respectively name of the instance, number of customers, number of vehicles and the cost limit. The main columns report the results of the branch-and-price method of Boussier et al. [2] in B-P and our results in B-C. Columns LB, UB and CPU report respectively the lower bound, upper bound and solving time in seconds for each instance and method. To summarize, our algorithm is able to prove the optimality of all the instances in the sets 1, 2, 3, and 6 and most of the instances in the other three sets. In total, we are able to solve 278 instances. Compared to [2] which solved 270 instances, our approach is clearly competitive. Moreover, it allows us to close 29 open instances.

Table 1. Performance of our branch-and-cut

Accumulated components	Solved instances	CPU (in seconds)
Standard model	177/387	198.9
+ Generalized subtour eliminations	233/387	61.75
+ Symmetric breaking	243/387	14.32
+ Boundaries on profit/customers	265/387	13.16
+ Clique cuts	278/387	3.24

Table 2. Comparison between the branch-and-price [2] and ours

Instance	n	m	L	B-P [2]			B-C		
				UB	LB	CPU	UB	LB	CPU
p1.2.p	30	2	37.5	250	–	2926	250	250	8.85
p1.2.q	30	2	40	–	–	–	265	265	10.96
p1.2.r	30	2	42.5	–	–	–	280	280	10.15
p3.2.l	31	2	35	605	–	4737	590	590	31.33
p3.2.m	31	2	37.5	–	–	–	620	620	58.41
p3.2.n	31	2	40	–	–	–	660	660	26.72
p3.2.o	31	2	42.5	–	–	–	690	690	34.64
p3.2.p	31	2	45	–	–	–	720	720	39.39
p3.2.q	31	2	47.5	–	–	–	760	760	16.55
p3.2.r	31	2	50	–	–	–	790	790	14.25
p3.2.s	31	2	52.5	–	–	–	800	800	0.11
p3.3.s	31	3	35	738.91	–	416	720	720	188.04
p3.3.t	31	3	36.7	763.69	–	4181	760	760	93.71
p4.2.h	98	2	60	–	–	–	835	835	2783.76
p4.2.i	98	2	65	–	–	–	918	918	1511.7
p4.2.t	98	2	120	–	–	–	1306	1306	1.29
p4.3.g	81	3	36.7	656.38	653	52	665	653	–
p4.3.h	90	3	40	735.38	729	801	761	729	–
p4.3.i	94	3	43.3	813.63	809	4920	830	809	–
p4.4.i	68	4	32.5	665.4	657	23	660	657	–
p4.4.j	76	4	35	741.47	732	141	784	732	–
p4.4.k	83	4	37.5	831.95	821	558	860	821	–

Table 2. (*Continued*)

Instance n m L				B-P [2]			B-C		
				UB	LB	CPU	UB	LB	CPU
p5.2.l	64	2	30	—	—	—	800	800	0.49
p5.2.m	64	2	32.5	—	—	—	860	860	0.97
p5.2.p	64	2	40	—	—	—	1150	1150	0.79
p5.2.t	64	2	50	—	—	—	1400	1400	3162.41
p5.2.x	64	2	60	—	—	—	1610	1610	38.81
p5.2.z	64	2	65	—	—	—	1680	1680	0.82
p5.3.l	64	3	20	605	595	33	615	595	—
p5.3.m	64	3	21.7	650	650	2	660	650	—
p5.3.n	64	3	23.3	755	755	42	765	755	—
p5.4.l	44	4	15	430	430	1	445	430	—
p5.4.m	52	4	16.2	555	555	0	560	555	—
p5.4.n	60	4	17.5	620	620	0	640	620	—
p5.4.o	60	4	18.8	690	690	1	720	690	—
p5.4.p	64	4	20	790	765	729	820	765	—
p5.4.q	64	4	21.2	860	860	1	880	860	—
p5.4.v	64	4	27.5	1320	1320	446	1340	1320	—
p6.2.j	62	2	30	—	—	—	948	948	0.46
p6.2.k	62	2	32.5	—	—	—	1032	1032	137.85
p6.2.l	62	2	35	—	—	—	1116	1116	13.92
p6.2.m	62	2	37.5	—	—	—	1188	1188	5.48
p6.2.n	62	2	40	—	—	—	1260	1260	1.03
p6.3.m	62	3	25	1104	—	33	1080	1080	574.45
p7.2.g	87	2	70	—	—	—	459	459	520.18
p7.3.h	59	3	53.3	429	425	8	436	425	—
p7.3.i	70	3	60	496.98	487	3407	535	487	—
p7.4.j	51	4	50	462	462	1	481	462	—
p7.4.k	61	4	55	524.61	520	73	586	520	—
p7.4.l	70	4	60	593.63	590	778	667	590	—

Conclusion and Future Work

In this article, we presented a branch-and-cut algorithm to solve TOP. Several cuts that strengthen the classical linear formulation were proposed. They include symmetric breaking, generalized subtour eliminations, boundaries on profits/numbers of customers and clique cuts. Experiments conducted on the standard benchmark show that our algorithm has the ability to solve a large number and a variety of instances. The algorithm permits to close several new instances. The obtained results clearly show the competitiveness and the robustness of our method on the classical linear TOP formulation. For future work, we plan to extend the approach to solve other combinatorial optimization problems.

References

[1] Bouly, H., Moukrim, A., Chanteur, D., Simon, L.: Un algorithme de destruction/construction itératif pour la résolution d'un problème de tournées de véhicules spécifique. In: MOSIM 2008 (2008)
[2] Boussier, S., Feillet, D., Gendreau, M.: An exact algorithm for team orienteering problems. 4OR 5(3), 211–230 (2007)
[3] Butt, S.E., Ryan, D.M.: An optimal solution procedure for the multiple tour maximum collection problem using column generation. Computers & Operations Research 26, 427–441 (1999)
[4] Chao, I.-M., Golden, B., Wasil, E.: The team orienteering problem. European Journal of Operational Research 88, 464–474 (1996)
[5] Dang, D.-C., Moukrim, A.: Subgraph extraction and metaheuristics for the maximum clique problem. Journal of Heuristics 18, 767–794 (2012)
[6] Dang, D.-C., Guibadj, R.-N., Moukrim, A.: A PSO-inspired algorithm for the team orienteering problem. European Journal of Operational Research (in press, accepted manuscript, available online March 2013)
[7] Fischetti, M., Salazar González, J.J., Toth, P.: Solving the orienteering problem through branch-and-cut. INFORMS Journal on Computing 10, 133–148 (1998)
[8] Poggi de Aragão, M., Viana, H., Uchoa, E.: The team orienteering problem: Formulations and branch-cut and price. In: ATMOS, pp. 142–155 (2010)
[9] Tang, H., Miller-Hooks, E.: A tabu search heuristic for the team orienteering problem. Computer & Operations Research 32, 1379–1407 (2005)
[10] Vansteenwegen, P., Souffriau, W., Vanden Berghe, G., Van Oudheusden, D.: Metaheuristics for tourist trip planning. In: Metaheuristics in the Service Industry. Lecture Notes in Economics and Mathematical Systems, vol. 624, pp. 15–31. Springer, Heidelberg (2009)

A Lagrangian Relaxation
Based Forward-Backward Improvement Heuristic
for Maximising the Net Present Value
of Resource-Constrained Projects

Hanyu Gu, Andreas Schutt, and Peter J. Stuckey

National ICT Australia, Department of Computing and Information Systems,
The University of Melbourne, Victoria 3010, Australia
{hanyu.gu,andreas.schutt,peter.stuckey}@nicta.com.au

Abstract. In this paper we propose a forward-backward improvement heuristic for the variant of resource-constrained project scheduling problem aiming to maximise the net present value of a project. It relies on the Lagrangian relaxation method to generate an initial set of schedules which are then improved by the iterative forward/backward scheduling technique. It greatly improves the performance of the Lagrangian relaxation based heuristics in the literature and is a strong competitor to the best meta-heuristics. We also embed this heuristic into a state-of-the-art CP solver. Experimentation carried out on a comprehensive set of test data indicates we compare favorably with the state of the art.

1 Introduction

We study the Resource-constrained Project Scheduling Problem with Discounted Cashflow (RCPSPDc) denoted as $m, 1|cpm, \delta_n, c_j|npv$ by [8] or $PS|prec|\sum C_j^F \beta^{C_j}$ by [1]. Specifically, given a set of activities J with precedence relationship $(i,j) \in L$, $i \in J$, $j \in J$, we need to decide the activity start time s_j, $j \in J$ within the project deadline T so that the net present value (NPV) of the project is maximised while the capacity of each renewable resource R_k, $k \in R$ is not violated. Each activity j has an associated cash-flow c_j and requires r_{jk} unit of resource $k \in R$ for a continuous period of time p_j. The net present value is calculated as the sum of the discounted cash flow of each activity defined as $c_j e^{-\alpha(s_j+p_j)}$ where α is the discount rate. A conceptual model can be formulated as

$$NPV = \text{maximise} \sum_{j \in J} c_j e^{-\alpha(s_j+p_j)} \tag{1}$$

$$\text{subject to} \quad s_i + p_i \leq s_j \qquad \forall (i,j) \in L \tag{2}$$

$$\sum_{j \in S(t)} r_{jk} \leq R_k \qquad k \in R, t = 0, \cdots, T-1 \tag{3}$$

$$0 \leq s_j \leq T - p_j \qquad j \in J \tag{4}$$

where $S(t)$ is the set of activities running in period $[t, t+1)$.

The RCPSPDc belongs to the class of NP-hard problems, and has been intensively investigated since it was first introduced in [17]. The reader is referred to [7]

C. Gomes and M. Sellmann (Eds.): CPAIOR 2013, LNCS 7874, pp. 340–346, 2013.
© Springer-Verlag Berlin Heidelberg 2013

for an extensive literature overview of solution approaches for RCPSPDC. Significant progress has been made in recent years for both complete and incomplete methods. The lazy clause generation approach to RCPSPDC [19] provides the state of the art complete method and outperforms the traditional branch-and-bound based methods [9,24,15]. For larger problems the evolutionary population based scatter search algorithm [23] achieved the bests results in comparison with other meta-heuristics such as genetic algorithms [11] and tabu search [25].

In spite of our success on the application of the Lagrangian Relaxation based Heuristic (LRH) [4] for very large problems (1400-10000 activities) and its reported superiority on smaller instances in [10] (up to 120 activities), our experiments with the set of test instances in [23] clearly shows that LRH has difficulty in finding feasible solutions on a significant percentage of instances. Careful analysis suggests that the test instances used in [10] have a much looser deadline and smaller duality gap compared with those of [23]. Since the Lagrangian relaxation solution may not be close to the optimal solution for the hardest cases, it is not surprising that the simple forward list scheduling heuristic failed.

We present in this paper a Lagrangian Relaxation based Forward-Backward Improvement heuristic (LR-FBI). The key improvements over LRH include: *(i)* the Lagrangian relaxation solution is perturbed to search more neighbours; *(ii)* the deadline infeasible solution is improved by the iterative forward/backward scheduling technique commonly used by meta-heuristics [12].

We compare LR-FBI with the state-of-the-art meta-heuristics [23] and CP solver [19] on a comprehensive set of test data. Our results show that LR-FBI is highly competitive especially for larger instances. We embed LR-FBI in the state-of-the-art lazy clause generation solution to further improve the performance.

2   Lagrangian Relaxation Based Forward-Backward Improvement Heuristic

We relax the resource constraints (3) as in [14,4] by introducing multipliers λ_{kt}, $k \in R$, $t = 0, \cdots, T$, and get the Lagrangian Relaxation Problem (LRP)

$$Z_{LR}(\lambda) = \text{maximise } LRP_\lambda(s), \quad \text{s.t.} \quad (2), (4) \tag{5}$$

where $LRP_\lambda(s) = \sum_{j \in J} rc_j(s_j) + \sum_{k \in R} \sum_t \lambda_{kt} R_k$ with

$$rc_j(s_j) = c_j e^{-\alpha(s_j + p_j)} - \sum_{t=s_j}^{s_j + p_j - 1} \sum_{k \in R} \lambda_{kt} r_{jk} \tag{6}$$

The multipliers λ are iteratively updated to minimise $Z_{LR}(\lambda)$ which is an upper bound of NPV. We omit here the technical details of the Lagrangian relaxation method for RCPSPDC which can be found in [4].

The solution s to $Z_{LR}(\lambda)$ is normally not feasible with respect to the resource constraints. Previously [4] we used a simple heuristic to construct a feasible solution from s, but this often fails for problems with tight deadline.

Algorithm 1. FBI(s)

---

1 $best\_NPV = -\infty$; generate keys $K(s)$;
2 **for** $x \in K(s)$ **do**
3      $right = true$; $s' = $ SGS_left(x) % decode x to schedule s';
4   **while** $makespan(s') > T$ **do**
5   **if** $right$ **then**
6              $s'' = $ SGS_right($s' + p$) % rightmost schedule using activity end times;
7   **else** $s'' = $ SGS_left(s') % leftmost schedule using activity start times;
8   **if** $makespan(s'') \geq makespan(s')$ **then** **return** $best\_NPV$;
9          $right = \neg right$; $s' := s''$;
10     $s' = $ shift(s'); **if** $NPV(s') > best\_NPV$ **then** $best\_NPV = NPV(s')$;
11 **return** $best\_NPV$;

For LR-FBI we try to find a feasible schedule similar to s using FBI(s) detailed in Algorithm 1. Firstly a set of keys $K(s)$ is created for s. The key is a vector $x \in \mathcal{R}^{|J|}$ which is decoded into a schedule by a Schedule Generation Scheme (SGS) [6]. The iterative forward/backward scheduling technique is used to reduce the makespan of a deadline infeasible schedule. Finally, the NPV of the schedule is further improved by shifting activities (shift) as in [10].

To calculate keys $K(s)$, rather than use a Linear Programming relaxation of the original problem [18,3,5], we use the computationally more efficient α-point idea of [14] which is based on a single LRP solution. The jth key element of the mth key is defined as $x_j^m = s_j + \alpha_j^m \times p_j$, $\alpha_j^m \in [0,1]$. We have two different strategies to create $K(s)$. Best-$\alpha(k)$ generates k uniformly distributed keys with $\alpha_j^m = m/k$, $m = 0, \cdots, k - 1$. Random-$\alpha(k)$ generates k random keys where each α_j^m is randomly chosen with a uniform distribution.

SGS_left(x) (SGS_right(x)) [2] greedily schedules activities one by one as early (late) as possible respecting the (reverse) precedence constraints and resource constraints, in the order where i is scheduled before (after) j if $x_i < x_j$. The resulting schedules are left(right)-justified [22]. SGS can be implemented in both serial and parallel modes [6].

3   Constraint Programming Hybrid Approach

LR-FBI can quickly find high quality solutions, but might converge to a local optima. Therefore we also investigate the possibility to further improve the solution quality using CP technology. The state-of-the-art complete method for RCPSPDC [19] is a constraint solver based on lazy clause generation [16]. Compared to the conceptual model on page 340, each resource constraint (3) is modeled by the global constraint cumulative($s, p, r_{.k}, R_k$) ($k \in R$) where the start times variables s_i ($i \in J$) are finite domain variables with an initial domain of $\{0, 1, \ldots, T\}$. As filtering algorithms for cumulative, the explanation-based version of the Time-Tabling [21] and Time-Tabling-Edge-Finder [20] are used.

In addition, the CP model uses the constraint $\texttt{max\_npv\_prop}(s, p, c, L, NPV)$, recently proposed in [19], in order to compute a tight upper bound on NPV function and filter impossible values from the start times domains. This constraint considers the subproblem of RCPSPDC in that (only) the resource constraints are relaxed, with the current bounds of the start times variables. Since this subproblem is polynomial solvable in time, the corresponding propagator computes its maximal NPV value, which is a valid upper bound for the original RCPSPDC, and, then, uses this NPV value for tightening the bounds on the NPV function and filtering the start times.

In [19], we compared different search strategies. Here, we consider the best-performing one VSIDS and combine this strategy with a Luby restart policy [13] having a restart base of 100.

The hybrid solution (CP-LR) runs after LR-FBI. Each solution s found by LR-FBI is stored in a set S. This immediately gives a much stronger lower bound for NPV. A two phase search strategy is then used. In the first phase the variable selected is the one with the largest average reduced cost defined as $\tilde{v}_j = \sum_{s \in S} rc_j(s_j)/|S|$. For value selection we maintain a reduced time window for each activity with window left (right) end defined as $w_j^l = \min_{s \in S} s_j$ ($w_j^r = \max_{s \in S} s_j$). The minimum feasible value in this reduced window will be chosen first. If no feasible value exists we rerun LR-FBI using the current bounds on the start times from the CP search to expand this window, using a limit of at most 5 iterations of LR, and adding any new solutions found to S. Our earlier work on RCPSP [21] shows that pure VSIDS search is quite robust, but can be improved using some more problem specific heuristics first. The same thing applies here. VSIDS is important for robustness, but the first phase helps find good solutions earlier, and set up VSIDS to be most productive. After one third of the time is used we swap to the second search phase which is pure VSIDS search.

4 Experiments

We carried out extensive experiments on the benchmark set available at www.projectmanagement.ugent.be/npv.html. The benchmark set consists of 17280 RCPSPDC instances which are split in 4 problem sizes, *i.e.*, 25, 50, 75 and 100 activities. A more detailed specification of these instances can be found in [23]. The NPV and CPU time for each instance is also available for the scatter search method in [23] which terminates when a maximal number of schedules are generated using a computer with a Dual Core processor 2.8GHz.

The parameter settings of LR can be found in [4]. We implemented our CP based approach using the LCG solver Chuffed. All tests were run on a computing cluster of which each node has two 2.8GHz AMD 6-Core CPU. We used a time limit of 5 minutes. The time limit for search in CP-LR is also 5 minutes.

We illustrate the effects of our improvements on the LRH in [10] in Table 1. We report the percentage of instances which have feasible schedule found (Fea%), the number of instances on which the best NPV is achieved (Best) and the average relative deviation (Dev) on instances for which all methods find a

Table 1. Comparison of feasibility results on 100-activities instances

	Fea%	Dev_V	Dev	Best
LRH	72.9	203.3	76.1	836
S-Best-α(1)	77.3	203.0	74.7	1005
P-Best-α(1)	91.9	207.6	79.5	450
P-Best-α(10)	95.8	197.1	69.4	1919
P-Random-α(2000)	99.5	197.1	69.4	2304

Table 2. Comparison of Scatter search, CP and LR

	Scatter(5000)			CP-Vsids			P-Random-α(2000)			CP-LR		
size	Fea%	Dev	Best	Fea%	Dev	Best	Fea%	Dev	Best	Fea%	Dev	Best
25	100.0	73.1	2507	100	72.8	3663	99.8	81.6	795	100.0	72.3	3708
50	99.9	91.6	1556	98.4	104.8	1124	99.9	82.4	937	100.0	78.6	2345
75	99.7	106.9	1196	90.3	-	-	99.8	98.3	1836	99.98	97.4	3054
100	99.6	100.2	1612	76.8	-	-	99.5	95.3	1524	99.7	93.6	2641

Table 3. Comparison with best scatter search results on size 100

	Fea%	Dev	Best	Ave(s)	Max(s)
Scatter(50000)	99.6	89.9	2003	26.2	139.8
P-Random-α(2000)	99.5	86.5	1283	167	607
CP-LR	99.8	85.4	2240	300	300

feasible schedule. Dev is defined as [23] $abs((Ub - Lb)/Ub)$. Dev_V is calculated with the upper bound in [23], while Dev uses the LR upper bound. The prefix S-(P-) stands for the serial (parallel) SGS. It can be seen that LRH has serious problems with feasibility. Parallel SGS is superior to serial SGS in terms of feasibility. The use of α-point further improves both feasibility and NPV. Since LR can produce much stronger upper bounds than [23] we only use Dev for the remaining tests.

We compare the reported results for scatter search [23] with at most 5000 schedules, with CP [19], LR-FBI and our hybrid CP approach (CP-LR) in Table 2. Scatter search is very fast (average computation time is 4.2s for size 100) and almost always finds a feasible solution. CP performs very well on the smallest instances but does not scale well. LR-FBI is highly competitive when problem sizes increase, generally finding better solutions, but requires more time than scatter search (82s on average for size 100). Running the hybrid CP-LR substantially improves on LR-FBI on a large number of instances. Clearly the hybrid is much more robust than a pure CP approach.

We also compare with the best results of scatter search with 50000 schedules in Table 3, showing also average and maximum solving time. Scatter search reduces the deviation by 10% with significant increase of solution time. The time limit for LR-FBI is set to 10 minutes. The LR-FBI has better deviation than that of scatter search. The hybrid method CP-LR improves the LR-FBI results further, resulting in the best known results on these instances.

5 Conclusion

We developed a new Lagrangian relaxation based heuristic for the RCPSPDC problem and achieved highly competitive results on a comprehensive set of test data. We also investigated the integration of our heuristic into a CP solver and obtained promising results. We have built an effective hybrid of local search and complete search, by using local search information not only for bounding but to direct the initial search phase. This hybrid is interesting since it runs the local search *on demand* when it can no longer provide useful guidance to the complete search. Our future research will focus on more efficient hybridisation of LR, CP and meta-heuristics for the RCPSPDC problem.

Acknowledgements. NICTA is funded by the Australian Government as represented by the Department of Broadband, Communications and the Digital Economy and the Australian Research Council through the ICT Centre of Excellence program.

References

1. Brucker, P., Drexl, A., Möhring, R., Neumann, K., Pesch, E.: Resource-constrained project scheduling: Notation, classification, models, and methods. European Journal of Operational Research 112(1), 3–41 (1999)
2. Debels, D., Vanhoucke, M.: A decomposition-based genetic algorithm for the resource-constrained project-scheduling problem. Operations Research 55(3), 457–469 (2007)
3. Gu, H.Y.: Computation of approximate alpha-points for large scale single machine scheduling problem. Computers & OR 35(10), 3262–3275 (2008)
4. Gu, H., Stuckey, P.J., Wallace, M.G.: Maximising the net present value of large resource-constrained projects. In: Milano, M. (ed.) CP 2012. LNCS, vol. 7514, pp. 767–781. Springer, Heidelberg (2012)
5. Gu, H., Xi, Y., Tao, J.: Randomized Lagrangian heuristic based on Nash equilibrium for large scale single machine scheduling problem. In: Proceedings of the 22nd IEEE International Symposium on Intelligent Control, pp. 464–468 (2007)
6. Hartmann, S., Kolisch, R.: Experimental evaluation of state-of-the-art heuristics for resource constrained project scheduling. European Journal of Operational Research 127, 394–407 (2000)
7. Hartmann, S., Briskorn, D.: A survey of variants and extensions of the resource-constrained project scheduling problem. European Journal of Operational Research 207(1), 1–14 (2010)
8. Herroelen, W.S., Demeulemeester, E.L., De Reyck, B.: A classification scheme for project scheduling. In: Weglarz, J. (ed.) Project Scheduling. International Series in Operations Research and Management Science, vol. 14, pp. 1–26. Kluwer Academic Publishers (1999)

9. Icmeli, O., Erengüç, S.S.: A branch and bound procedure for the resource constrained project scheduling problem with discounted cash flows. Management Science 42(10), 1395–1408 (1996)
10. Kimms, A.: Maximizing the net present value of a project under resource constraints using a Lagrangian relaxation based heuristic with tight upper bounds. Annals of Operations Research 102, 221–236 (2001)
11. Kolisch, R., Hartmann, S.: Experimental investigation of heuristics for resource-constrained project scheduling: An update. European Journal of Operational Research 174, 23–37 (2006)
12. Li, K., Willis, R.: An iterative scheduling technique for resource-constrained project scheduling. European Journal of Operational Research 56, 370–379 (1992)
13. Luby, M., Sinclair, A., Zuckerman, D.: Optimal speedup of Las Vegas algorithms. Inf. Proc. Let. 47(4), 173–180 (1993)
14. Möhring, R.H., Schulz, A.S., Stork, F., Uetz, M.: Solving project scheduling problems by minimum cut computations. Management Science 49(3), 330–350 (2003)
15. Neumann, K., Zimmermann, J.: Exact and truncated branch-and-bound procedures for resource-constrained project scheduling with discounted cash flows and general temporal constraints. Central European Journal of Operations Research 10(4), 357–380 (2002)
16. Ohrimenko, O., Stuckey, P.J., Codish, M.: Propagation via lazy clause generation. Constraints 14(3), 357–391 (2009)
17. Russell, A.H.: Cash flows in networks. Management Science 16(5), 357–373 (1970)
18. Savelsbergh, M., Uma, R., Wein, J.: An experimental study of LP-based approximation algorithms for scheduling problems. INFORMS J. on Computing 17, 123–136 (2005)
19. Schutt, A., Chu, G., Stuckey, P.J., Wallace, M.G.: Maximising the net present value for resource-constrained project scheduling. In: Beldiceanu, N., Jussien, N., Pinson, É. (eds.) CPAIOR 2012. LNCS, vol. 7298, pp. 362–378. Springer, Heidelberg (2012)
20. Schutt, A., Feydy, T., Stuckey, P.J.: Explaining time-table-edge-finding propagation for the cumulative resource constraint. In: Gomes, C., Sellmann, M. (eds.) CPAIOR 2013. LNCS, vol. 7874, pp. 234–250. Springer, Heidelberg (2013)
21. Schutt, A., Feydy, T., Stuckey, P.J., Wallace, M.G.: Explaining the cumulative propagator. Constraints 16(3), 250–282 (2011)
22. Sprecher, A., Kolisch, R., Drexl, A.: Semi-active, active, and non-delay schedules for the resource-constrained project scheduling problem. European Journal of Operational Research 80, 94–102 (1995)
23. Vanhoucke, M.: A scatter search heuristic for maximising the net present value of a resource constrained project with fixed activity cash flows. International Journal of Production Research 48(7), 1983–2001 (2010)
24. Vanhoucke, M., Demeulemeester, E.L., Herroelen, W.S.: On maximizing the net present value of a project under renewable resource constraints. Management Science 47, 1113–1121 (2001)
25. Zhu, D., Padman, R.: A metaheuristic scheduling procedure for resource-constrained projects with cash flows. Naval Research Logistics 46, 912–927 (1999)

Improving Strong Branching by Propagation

Gerald Gamrath

Zuse Institute Berlin, Takustr. 7, 14195 Berlin, Germany
gamrath@zib.de

Abstract. *Strong branching* is an important component of most variable selection rules in branch-and-bound based mixed-integer linear programming solvers. It predicts the dual bounds of potential child nodes by solving auxiliary LPs and thereby helps to keep the branch-and-bound tree small. In this paper, we describe how these dual bound predictions can be improved by including *domain propagation* into strong branching. Computational experiments on standard MIP instances indicate that this is beneficial in three aspects: It helps to reduce the average number of LP iterations per strong branching call, the number of branch-and-bound nodes, and the overall solving time.

1 Introduction

Since the invention of the linear programming (LP) based branch-and-bound method for solving mixed-integer linear programs (MIPs) in the 1960s [1,2], branching rules have been an important field of research in that context, being one of the core parts of the method (for surveys, see [3,4,5]). Their task is to split the current node's problem into two or more disjoint subproblems if the solution to the current LP relaxation does not fulfill the integrality restrictions, thereby excluding the LP solution from all subproblems while keeping at least one optimal solution.

The most common way to split the problem is to branch on trivial inequalities, which split the domain of a single variable into two parts (called *variable branching*). Alternatively, branching can be performed on general linear constraints (see [6,7,8,9,10]) or can create more than two subproblems, cf. [11,12]. In case of variable branching, the variable to actually branch on is typically chosen with the goal of improving the local dual bound of both created child nodes. This helps to tighten the global dual bound and prune nodes early (for recent research on alternative criteria, see, e.g., [13,14,15,16]). A very popular branching rule called *pseudo-cost branching* [17] uses history information about the change of the dual bound caused by previous branchings. More accurate, but also more expensive, is *strong branching* [18,19,4], which explicitly computes dual bounds of potential child nodes by solving an auxiliary LP with the branching bound change temporarily added. The *full strong branching rule* does this at every node for each integer variable with fractional LP value which empirically leads to very small branch-and-bound trees [5]. Modern branching rules typically combine these two approaches and use strong branching in the case of uninitialized or unreliable pseudo cost values (see [5,20]).

C. Gomes and M. Sellmann (Eds.): CPAIOR 2013, LNCS 7874, pp. 347–354, 2013.
© Springer-Verlag Berlin Heidelberg 2013

In practice, one can often observe a difference between the dual bound that strong branching computes for a node and the actual dual bound obtained later during node processing. This restrains the effectiveness of strong branching, which should predict the actual dual bound of the node and not just compute some valid dual bound. There are various reasons for the difference, most prominently *domain propagation* and global domain changes found in the meantime. The task of domain propagation (or *node preprocessing*) is to tighten the local domains of variables by inspecting the constraints and current domains of other variables at the local subproblem. It is the integral part of each constraint programming solver [21] and has also proven to improve MIP solvers significantly by tightening the LP relaxation, resulting in better dual bounds and detecting infeasibilities earlier [22,23,24].

While strong branching cannot do anything about the difference in the dual bounds caused by global domain changes, it should react upon the continuous improvement in domain propagation techniques. In this paper, we examine how strong branching can be improved by combining it with domain propagation in order to compute better dual bound predictions. This means that we perform the same domain propagation steps that are already performed at each node of the branch-and-bound tree also during strong branching, prior to solving the strong branching LP of a potential child node.

The general idea and an evaluation of the direct effects are presented in the next section. Based on that, we discuss additional improvements in Section 3 and provide benchmark results on a collection of MIPLIB [25,26,27] instances showing a reduction of both number of nodes and solving time when propagation is applied within a full strong branching rule.

2   Strong Branching with Domain Propagation

In the following, we regard mixed-integer linear programs of the form:

$$\min\{c^T x \mid Ax \geq b, x \geq 0, x_i \in \mathbb{Z} \ \forall i \in I\}. \tag{1}$$

The basic implementation of strong branching with domain propagation (*SBDP*) works as follows: Given the current problem P of form (1) and an integer variable $x_i, i \in I$ with fractional LP solution value $\hat{x}_i$, it computes dual bounds of the two potential child nodes that would be created by branching on x_i. Therefore, it creates two temporary subproblems P_d (the *down child*) and P_u (the *up child*) by adding to P the bound changes $x_i \leq \lfloor \hat{x}_i \rfloor$ and $x_i \geq \lceil \hat{x}_i \rceil$, respectively. After that, the variable domains of P_d are tightened by domain propagation. If propagation detects infeasibility, a dual bound of $+\infty$ is returned for P_d, otherwise the LP relaxation of P_d is solved and its optimal value provides the strong branching dual bound. The dual bound of P_u is computed analoguously.

The only difference to "standard" strong branching is that domain propagation is performed before solving the LP. Since this tightens the LP relaxation, the dual bounds obtained by solving the strong branching LP are always greater than or equal to the ones computed by standard strong branching.[1] The questions to be considered in this paper are: Is this worth the additional effort? In particular, how big is the propagation time and how does the number of LP iterations change? The simplex warmstart normally allows to solve the strong branching LPs with just a few iterations as there is only one bound changed, but additional changes performed by domain propagation might change this.

For answering these questions, we performed computational experiments using an implementation of SBDP based on the MIP solver SCIP 3.0 [22,28] with underlying LP solver SoPlex 1.7 [29]. They were performed on Intel Xeon E5420 2.5 GHz computers, with 6 MB cache and 16 GB RAM, running Linux (in 64 bit mode). A time limit of two hours per instance was imposed. We use full strong branching to measure the impact of our changes for each candidate variable at each node and concentrate on the branch-and-bound performance by providing the optimal objective value as objective cutoff and disabling primal heuristics and cutting plane separation as well as the components presolver[2] of SCIP. As test set, we used the MMM test set consisting of all instances from MIPLIB 3 [25], MIPLIB 2003 [26], and the benchmark set of MIPLIB 2010 [27]. We excluded all instances for which no significant amout of strong branching was performed (less than ten strong branching calls on single variables)—either because the instance was solved in presolving or at the root node prior to branching or because the time limit of two hours was hit. Additionally, we excluded the three infeasible instances from MIPLIB 2010 in order to be able to compute the additional gap closed by SBDP, which left us with a total number of 147 instances.

The experiments were then conducted as follows: After each standard strong branching call, we additionally performed a call of SBDP on the same variable, running the same domain propagation techniques as SCIP does on any node of the branch and bound tree (cf. [22]). We collected statistics about the differences, but did not use any of the information produced by SBDP within the branch-and-bound search. We chose this approach instead of running twice, one time with each variant, to exclude the difference in the branch-and-bound tree created by different branching methods and isolate the impact of the new method on each single strong branching call.

For analyzing the impact of SBDP, we divide the strong branching calls into three categories: *cutoff* if at least one of the two potential child nodes was detected to be infeasible, *better bound* if no infeasibility was detected and SBDP computed a better dual bound for at least one of the potential child nodes, and

---

[1] In this paper, we assume that the strong branching LPs are solved to optimality and no iteration limit is applied. This is also the case for the implementation of the full strong branching rule used in our computational experiments.

[2] The components presolver solves small independent subproblems in advance, excluding them from the main branch-and-bound search.

same bound if both strong branching variants computed the same (finite) bounds for both potential child nodes.

The results for each of these categories are presented in one line in Table 1, with an additional line that summarizes these results for all strong branching calls. Besides the number of strong branching calls (column calls), we show for both strong branching variants the number of potential subproblems detected infeasible (column cutoffs), the number of LP iterations for solving the LPs of the two subproblems (column LP iters), and the strong branching time in milliseconds (column time). Furthermore, we present the number of domain changes performed by SBDP (column dom. chgs.) and the percentage of the gap between primal bound and strong branching dual bound closed by using SBDP instead of standard strong branching (column gap closed). For each of the numbers listed, we compute the arithmetic mean over all strong branching calls for the single instances and average over the instances by taking a *shifted geometric mean*[3]. We use a shift of 100 for the number of strong branching calls, 10 for time, iteration number and domain changes, and 1 for the number of child nodes declared infeasible per call. Only for the gap closed, having only values between 0 and 100, do we average over the instances by arithmetic mean.

As expected, the *better bound* case—which happens only rarely—is typically caused by a high number of domain changes during propagation and leads to an increase in both the average number of LP iterations and time per strong branching call, thereby closing the gap by more than 20% on average. In the most common case, the *same bound* category, a smaller, but still relevant number of domains are changed by propagation. But instead of slowing down the simplex warm start, these bound changes even reduce the average number of LP iterations, e.g., by fixing variables that would otherwise need to be rendered feasible by some simplex pivots. Last, in the *cutoff* case, SBDP detects infeasibility of more potential child nodes—on average 1.11 of the two children regarded per call are declared infeasible compared to 0.92 otherwise. In about 15% of the cases, infeasibility is detected already during propagation, leading to a reduction of the average number of LP iterations and strong branching time. On average over all strong branching calls, SBDP can declare every twelfth instead of nearly every fourteenth strong branching child node infeasible and closes the gap by 2.66%. The average number of LP iterations is slightly decreased, while the time per strong branching call increases marginally. This demonstrates that the domain propagation time is relatively small compared to the total strong branching time; on average, it was less than 5%.

To summarize, SBDP exhibits benefits in all three categories. In the majority of strong branchings, where it yields no bound improvement, it reduces the number of LP iterations. In the remaining cases, significantly more child nodes can be cut off and about 20% additional gap is closed.

---

[3] For a definition and discussion of the shifted geometric mean, see Achterberg [22, Appendix A3].

Table 1. Impact of SBDP on the strong branching calls

category	calls	standard strong branching			strong branching with domain propagation				
		cutoffs	LP iters	time	dom. chgs.	cutoffs	gap closed	LP iters	time
better bound	376.02	–	44.00	19.5	38.85	–	20.73%	57.19	23.1
same bound	23801.96	–	82.33	39.7	23.74	–	–	78.99	40.6
cutoff	3342.63	0.92	56.81	27.3	35.70	1.11	8.50%	46.74	25.6
all	30469.42	0.14	81.03	40.2	26.26	0.17	2.66%	77.52	40.5

3   Further Improvements and Computational Results

In this section we describe further improvements motivated by the results of our first computational experiments and present the effect of SBDP on the overall performance when it is used within the full strong branching rule.

The first improvement treats the case of an *infeasible strong branching sub-problem*, which traditionally leads to simply tightening the domain of the candidate variable at the current node (or cutting off the current node if both subproblems are infeasible). While normally, strong branching methods always regard both subproblems, we interrupt a strong branching call when the first potential child is found infeasible, saving the effort we would spend for the second child node. As usual, the domain change of the other subproblem is then applied at the current node, causing a reoptimization of its LP, after which branching is started again, if needed.

In our computational experiments presented in Section 2, about 69% of the infeasible subproblems were up children. This is not surprising since problems are often modeled in a way such that changing a variable's lower bound—in particular, fixing a binary variable to one—has more impact than changing its upper bound (fixing a binary variable to zero). In order to profit from infeasible child nodes more often, we decided to investigate the potential up child first.

As in probing preprocessing (see [23]), we can often identify *valid local bounds* for some variables even if neither of the two potential child nodes is infeasible. If any variable's domain in the two potential child nodes was tightened to $[lb_d, ub_d]$ and $[lb_u, ub_u]$, respectively, we can change the domain of the variable in the local problem to $[\min\{lb_d, lb_u\}, \max\{ub_d, ub_u\}]$. For 94 of the 147 instances regarded in Section 2, this technique was able to identify tighter bounds, identifying on average 3.15 bounds that could have been tightened per strong branching call with both subproblems feasible. With this improvement, probing preprocessing is performed as a side product of SBDP.

Using these improvements, we performed computational experiments to compare the performance of SBDP against standard strong branching. We used the same computing environment as described in Section 2 and also the MMM test-set described there, this time without excluding any instances. Within SCIP, we exchanged the strong branching calls in the full strong branching rule for SBDP and again provided the optimum as cutoff bound, disabled primal heuristics, cutting plane separation, and the components presolver in order to focus on the branch-and-bound search and to reduce random performance changes (see [27]).

Table 2. Comparison of full strong branching with and without SBDP

test set	size	full strong branching			full strong with SBDP		
		solved	nodes	time	solved	nodes	time
MMM: complete	168	97	814	633.1	100	645	582.2
MMM: all optimal	94	94	321	86.3	94	253	78.9

The results are summarized in Table 2. We regard both the complete MMM test set (row MMM: complete) as well as the subset of instances that both variants solved to optimality (row MMM: all optimal), and present—besides the size of the sets—aggregated results for both sets. More specifically, we list the number of solved instances and the shifted geometric mean (with a shift of 10) of the number of processed branch-and-bound nodes and the solving time. The results are promising: with the improved strong branching method, SCIP is able to solve 100 out of the 168 instances of the MMM test set within the time limit of two hours, three instances more than with standard strong branching. For the subset of instances that both versions solved to optimality, the average number of nodes and the solution time are reduced by 21% and 9%, respectively. For detailed instance-wise results, we refer to [30].

4   Conclusions and Outlook

In this paper, we improved strong branching by applying domain propagation to compute more accurate dual bound predictions. First computational experiments on general MIP instances show that this comes with relatively small cost and, used in a full strong branching rule, can speed up the solution process while reducing the branch-and-bound tree size. For "structured" or more general problems classes like MINLP or CIP [22] where typically the LP misses more information which can be exploited by domain propagation, we expect an even larger improvement by the new method.

Our preliminary results show the potential of the approach. An integration into state-of-the-art branching rules like reliability branching [5] and a possible combination with other recent strong branching improvements like cloud branching [31] or nonchimerical branching [32] are fields for future research.

Already the improved full strong branching might prove useful when the branch-and-bound tree should be kept small, e.g., under tight memory restrictions or for massive parallel MIP solvers (see, e.g., [33,34]), where reducing the tree size has the added advantage of reducing the message passing overhead.

Acknowledgements. The author would like to thank Tobias Achterberg and Michael Winkler for fruitful discussions and Timo Berthold, Ambros Gleixner, and four anonymous reviewers for helpful comments on the paper.

References

1. Land, A.H., Doig, A.G.: An automatic method of solving discrete programming problems. Econometrica 28(3), 497–520 (1960)
2. Dakin, R.J.: A tree-search algorithm for mixed integer programming problems. The Computer Journal 8(3), 250–255 (1965)
3. Mitra, G.: Investigation of some branch and bound strategies for the solution of mixed integer linear programs. Mathematical Programming 4, 155–170 (1973)
4. Linderoth, J.T., Savelsbergh, M.W.P.: A computational study of search strategies in mixed-integer programming. INFORMS Journal on Computing 11(2), 173–187 (1999)
5. Achterberg, T., Koch, T., Martin, A.: Branching rules revisited. Operations Research Letters 33, 42–54 (2005)
6. Ryan, D.M., Foster, B.A.: An integer programming approach to scheduling. In: Wren, A. (ed.) Computer Scheduling of Public Transport Urban Passenger Vehicle and Crew Scheduling, pp. 269–280. North Holland, Amsterdam (1981)
7. Owen, J.H., Mehrotra, S.: Experimental results on using general disjunctions in branch-and-bound for general-integer linear programs. Computational Optimization and Applications 20, 159–170 (2001)
8. Mahajan, A., Ralphs, T.K.: Experiments with branching using general disjunctions. In: Chinneck, J.W., Kristjansson, B., Saltzman, M.J. (eds.) Operations Research and Cyber-Infrastructure. Operations Research/Computer Science Interfaces Series, vol. 47, pp. 101–118. Springer, US (2009)
9. Karamanov, M., Cornuéjols, G.: Branching on general disjunctions. Mathematical Programming 128, 403–436 (2011)
10. Cornuéjols, G., Liberti, L., Nannicini, G.: Improved strategies for branching on general disjunctions. Mathematical Programming 130, 225–247 (2011)
11. Borndörfer, R., Ferreira, C.E., Martin, A.: Decomposing matrices into blocks. SIAM J. Optim. 9(1), 236–269 (1998)
12. Lodi, A., Ralphs, T., Rossi, F., Smriglio, S.: Interdiction branching. Technical Report OR/09/10, DEIS, Università di Bologna (2009)
13. Patel, J., Chinneck, J.: Active-constraint variable ordering for faster feasibility of mixed integer linear programs. Mathematical Programming 110, 445–474 (2007)
14. Kılınç Karzan, F., Nemhauser, G.L., Savelsbergh, M.W.: Information-based branching schemes for binary linear mixed integer problems. Mathematical Programming Computation 1, 249–293 (2009)
15. Fischetti, M., Monaci, M.: Backdoor branching. In: Günlük, O., Woeginger, G.J. (eds.) IPCO 2011. LNCS, vol. 6655, pp. 183–191. Springer, Heidelberg (2011)
16. Gilpin, A., Sandholm, T.: Information-theoretic approaches to branching in search. Discrete Optimization 8(2), 147–159 (2011)
17. Benichou, M., Gauthier, J.M., Girodet, P., Hentges, G., Ribiere, G., Vincent, O.: Experiments in mixed-integer linear programming. Mathematical Programming 1, 76–94 (1971)
18. Applegate, D.L., Bixby, R.E., Chvátal, V., Cook, W.J.: On the solution of traveling salesman problems. Documenta Mathematica J.DMV Extra Volume ICM III, 645–656 (1998)
19. Applegate, D.L., Bixby, R.E., Chvátal, V., Cook, W.J.: The Traveling Salesman Problem: A Computational Study. Princeton Series in Applied Mathematics. Princeton University Press, Princeton (2007)

20. Achterberg, T., Berthold, T.: Hybrid branching. In: van Hoeve, W.-J., Hooker, J.N. (eds.) CPAIOR 2009. LNCS, vol. 5547, pp. 309–311. Springer, Heidelberg (2009)

21. Apt, K.R.: Principles of Constraint Programming. Cambridge University Press, Cambridge (2003)

22. Achterberg, T.: Constraint Integer Programming. PhD thesis, Technische Universität Berlin (2007)

23. Savelsbergh, M.W.P.: Preprocessing and probing techniques for mixed integer programming problems. ORSA Journal on Computing 6, 445–454 (1994)

24. Fügenschuh, A., Martin, A.: Computational integer programming and cutting planes. In: Aardal, K., Nemhauser, G.L., Weismantel, R. (eds.) Discrete Optimization. Handbooks in Operations Research and Management Science, vol. 12, pp. 69–122. Elsevier (2005)

25. Bixby, R.E., Ceria, S., McZeal, C.M., Savelsbergh, M.W.P.: An updated mixed integer programming library: MIPLIB 3.0. Optima (58), 12–15 (June 1998)

26. Achterberg, T., Koch, T., Martin, A.: MIPLIB 2003. Operations Research Letters 34(4), 1–12 (2006)

27. Koch, T., Achterberg, T., Andersen, E., Bastert, O., Berthold, T., Bixby, R.E., Danna, E., Gamrath, G., Gleixner, A.M., Heinz, S., Lodi, A., Mittelmann, H., Ralphs, T., Salvagnin, D., Steffy, D.E., Wolter, K.: MIPLIB 2010. Mathematical Programming Computation 3(2), 103–163 (2011)

28. Achterberg, T.: SCIP: Solving constraint integer programs. Mathematical Programming Computation 1(1), 1–41 (2009)

29. Wunderling, R.: Paralleler und objektorientierter Simplex-Algorithmus. PhD thesis, Technische Universität Berlin (1996)

30. Gamrath, G.: Improving strong branching by propagation. Technical Report 12-46, ZIB, Takustr. 7, 14195 Berlin (2012)

31. Berthold, T., Salvagnin, D.: Cloud branching. Technical Report 13-01, ZIB, Takustr. 7, 14195 Berlin (2013)

32. Fischetti, M., Monaci, M.: Branching on nonchimerical fractionalities. OR Letters 40(3), 159–164 (2012)

33. Shinano, Y., Achterberg, T., Berthold, T., Heinz, S., Koch, T.: ParaSCIP – a parallel extension of SCIP. In: Bischof, C., Hegering, H.G., Nagel, W.E., Wittum, G. (eds.) Competence in High Performance Computing 2010, pp. 135–148 (2012)

34. Shinano, Y., Berthold, T., Heinz, S., Koch, T., Winkler, M., Achterberg, T.: ParaSCIP – a parallel extension of SCIP. Technical Report ZR 11-10, Zuse Institute Berlin (2011)

Learning and Propagating Lagrangian Variable Bounds for Mixed-Integer Nonlinear Programming

Ambros M. Gleixner[1] and Stefan Weltge[2]

[1] Zuse Institute Berlin, Takustr. 7, 14195 Berlin, Germany
gleixner@zib.de
[2] Otto-von-Guericke-Universität Magdeburg,
Universitätsplatz 2, 39106 Magdeburg, Germany
weltge@ovgu.de

Abstract. Optimization-based bound tightening (OBBT) is a domain reduction technique commonly used in nonconvex mixed-integer nonlinear programming that solves a sequence of auxiliary linear programs. Each variable is minimized and maximized to obtain the tightest bounds valid for a global linear relaxation. This paper shows how the dual solutions of the auxiliary linear programs can be used to learn what we call *Lagrangian variable bound constraints*. These are linear inequalities that explain OBBT's domain reductions in terms of the bounds on other variables and the objective value of the incumbent solution. Within a spatial branch-and-bound algorithm, they can be learnt a priori (during OBBT at the root node) and propagated within the search tree at very low computational cost. Experiments with an implementation inside the MINLP solver SCIP show that this reduces the number of branch-and-bound nodes and speeds up solution times.

1 Introduction

Mixed-integer nonlinear programming studies the large class of mathematical programs specified by a nonlinear objective function, nonlinear constraints, and integrality requirements on some of the variables. It comprises the special cases of mixed-integer linear programming and nonlinear programming and provides a flexible modelling tool for a wide range of academic and industrial applications. For a detailed discussion, see, e.g., [1].

We consider *mixed-integer nonlinear programs* (MINLPs) of the form

$$\min\{\, c^\mathsf{T}x : x \in \mathcal{X}, x \in [\ell, u], x_j \in \mathbb{Z} \text{ for } j \in \mathcal{I} \,\}, \tag{1}$$

where $\mathcal{X} \subseteq \mathbb{R}^n$, ℓ and u are the vectors of lower bounds $\ell_j \in \mathbb{R} \cup \{-\infty\}$ and upper bounds $u_j \in \mathbb{R} \cup \{+\infty\}$, and $\mathcal{I} \subseteq \{1, \dots, n\}$ is the index set of integer variables. Without loss of generality, we assume a linear objective, since for a nonlinear objective function $f(x)$, we can append the constraint $f(x) \leqslant x_0$ and minimize x_0. The feasible region $\mathcal{X}$ is specified by a list of linear and nonlinear constraints $g_i(x) \leqslant 0$, where the g_i (and hence $\mathcal{X}$) may be nonconvex.

C. Gomes and M. Sellmann (Eds.): CPAIOR 2013, LNCS 7874, pp. 355–361, 2013.

Many complete algorithms for solving nonconvex MINLPs to (ε-)global optimality rely on spatial branch-and-bound combined with a convex relaxation. Domain reduction procedures have become a crucial element of state-of-the-art MINLP solvers because they not only reduce the size of the search space (as in mixed-integer or constraint programming), but specifically because smaller domains allow for tighter convex relaxations of the nonconvex constraints.

This paper is concerned with a specific domain reduction technique often referred to as *optimization-based bound tightening* (OBBT). Given a linear relaxation $\mathcal{R} \supseteq \mathcal{X}$, OBBT computes the tightest bounds valid for all relaxation solutions by in turn minimizing and maximizing each variable over $\mathcal{R}$,

$$\min / \max \{ x_k : x \in \mathcal{R}, x \in [\ell, u] \}. \tag{2}$$

Its first appearance in the literature we are aware of is an application to heat exchanger networks by Quesada and Grossmann [2] from 1993. Subsequently, it became a component of generic global optimization algorithms, see, e.g., [3,4,5].

An optimization algorithm may exclude suboptimal parts of the feasible region as long as at least one optimal solution remains. In OBBT, this can be exploited by adding an *objective cutoff* constraint $c^\mathsf{T} x \leqslant z$ to $\mathcal{R}$, where $z = c^\mathsf{T} \hat{x}$ is the objective value of the current incumbent solution $\hat{x}$. Zamora and Grossmann [6] have used this idea in a "branch-and-contract" algorithm, which employs OBBT aggressively at every node of the search tree.

Examples of MINLP solvers implementing OBBT are αBB [7,8], Couenne [9,10], GloMIQO [11,12], LaGO [13,14], and SCIP [15,16,17]. Since applying a full round of OBBT amounts to solving $2n$ linear programs—an expensive algorithm compared to the average amount of work performed at a branch-and-bound node—it is typically applied at the root node and within the search tree only with limited frequency or based on its success rate. For a recent theoretical study of an iterated version of OBBT see the paper by Caprara and Locatelli [18].

Contribution. Our paper presents a new idea for how to benefit from the potentially expensive solution of (2) beyond simply obtaining tighter bounds on variable x_k. To this end, we observe that the proof of optimality given by a dual solution of (2) can be used to learn globally valid inequalities whose propagation gives a local approximation of OBBT. These inequalities, which we call *Lagrangian variable bound constraints*, are redundant since they are obtained merely as an aggregation of the rows of the relaxation $\mathcal{R}$. Nevertheless, we demonstrate that propagating them during the tree search helps to speed up the solution process significantly.

In the remainder of the paper, Sec. 2 explains the derivation and propagation of Lagrangian variable bounds in detail. Section 3 presents computational results analyzing their effect on instances from MINLPLib and summarizes our findings.

2   Lagrangian Variable Bounds

Besides valid bounds for variable x_k, solving (2) yields dual multipliers for the constraints of $\mathcal{R}$ that prove that for no $x \in \mathcal{R}$—and by that for no feasible solution of (1)—variable x_k can lie outside these bounds. The following lemma uses basic LP duality to motivate our approach. For clarity, we restrict the presentation to upper bounds.

Lemma 1. *Let* $\mathcal{R} = \{x \in \mathbb{R}^n : a_i^\mathsf{T} x \leqslant b_i, i = 1, \ldots, m\} \supseteq \mathcal{X}$ *be given, where* $a_i \in \mathbb{R}^n$ *and* $b \in \mathbb{R}^m$. *Let* x^* *be an optimal solution of*

$$\max\{\, x_k : x \in \mathcal{R}, \, c^\mathsf{T} x \leqslant z, \, x \in [\ell, u] \,\}, \tag{3}$$

with $z \in \mathbb{R} \cup \{\infty\}$ *an upper bound on the optimal objective value of* (1). *Further, let* $\lambda_1, \ldots, \lambda_m, \mu \geqslant 0$ *be feasible dual multipliers with reduced costs*

$$r_j := \begin{cases} 1 - \sum_i \lambda_i a_{ij} - \mu c_j & \text{if } j = k, \\ -\sum_i \lambda_i a_{ij} - \mu c_j & \text{otherwise.} \end{cases} \tag{4}$$

Then

$$U(\ell, u, z) := \sum_{j:r_j<0} r_j \ell_j + \sum_{j:r_j>0} r_j u_j + \mu z + \lambda^\mathsf{T} b \tag{5}$$

is a valid upper bound for x_k. *If* $\lambda_1, \ldots, \lambda_m, \mu$ *are optimal multipliers then* $U(\ell, u, z) = x_k^*$, *otherwise* $U(\ell, u, z) > x_k^*$.

Proof. Multiplying the rows of (3) with their dual values and aggregating them gives the valid inequality $(\sum_i \lambda_i a_i + \mu c)^\mathsf{T} x \leqslant \lambda^\mathsf{T} b + \mu z$. Using (4), this becomes

$$x_k \leqslant \sum_j r_j x_j + \mu z + \lambda^\mathsf{T} b, \tag{6}$$

which for $x \in [\ell, u]$ is at most $U(\ell, u, z)$. Optimal multipliers are complementary slack with x^*, yielding the relation of $U(\ell, u, z)$ and the OBBT bound x_k^*.

We will refer to bounds of type (5) as well as their lower bound counterparts as *Lagrangian variable bounds* (LVBs). Figure 1 provides an illustrative example, which shows that LVBs can be learnt even when OBBT fails to tighten the bound.

Remark 1. If μ is nonzero then $U(\ell, u, z)$ depends on the primal bound; if some r_j, $j \neq k$, is nonzero, it depends on x_j. Hence, whenever an improving solution is found or $[\ell_j, u_j]$ is reduced, the LVB may tighten the bounds of x_k further.

Additionally, in stark contrast to OBBT, LVBs can be propagated very efficiently. This motivates the application of LVBs within a spatial branch-and-bound algorithm for nonconvex MINLP in the following scheme.

1. Learn LVB constraints while performing OBBT once during the root node.
2. Propagate them locally at the nodes of the search tree whenever bounds appearing on the right-hand side are reduced by branching or propagation.
3. Propagate them globally whenever an improving solution is found.

Already in [19] it has been observed that any dual feasible solution encountered during the solution process may be used to construct a one-row relaxation of the LP at hand and that this inequality can be used to tighten the bounds of each variable involved. Applied unconditionally, however, this idea appears too expensive. In this paper, we suggest to specifically select the dual solutions from OBBT and propagate LVBs only towards the left-hand side variable.

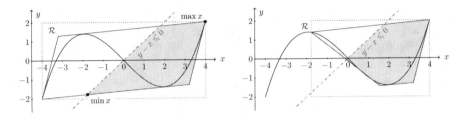

Fig. 1. Example $\min\{y - x : y = 0.1x^3 - 1.1x, x \in [-4, 4], y \in [-2, 2]\}$. On the left, the shaded region over which OBBT is performed is defined by the relaxation $\mathcal{R}$ and the dashed objective cutoff resulting from the zero solution. Minimizing x gives a lower bound of $-\frac{16}{9}$ and the LVB $x \geqslant -\frac{10}{9}z - \frac{16}{9}$. Maximizing x does not tighten its upper bound, still the LVB $x \leqslant \frac{10}{37}y + \frac{128}{37}$ can be learnt. In this two variable example, this is only the rightmost facet of $\mathcal{R}$, but in higher dimensions it may be nontrivial. On the right is the resulting, tighter relaxation.

Remark 2. The main purpose of the LVB constraints (6) is to identify bounds already implied by the relaxation and, by making them explicit, allow for improving the relaxations of nonconvex nonlinear constraints. Note that, unlike MIP cutting planes, they are not designed to cut off the LP optimum. Since they are redundant inequalities, it is not beneficial to add them to the LP relaxation.

3   Computational Results

Experimental Setup. The aim of our experiments was two-fold: first, to quantify how many *nontrivial LVBs* can be generated during OBBT, i.e., LVBs with $\mu \neq 0$ or $r_j \neq 0$ for some $j \neq k$; second, to evaluate the effect of propagating them during the solution process. Within the MINLP solver SCIP 3.0 [15,20,16,17] we have implemented an OBBT scheme that minimizes and maximizes each variable once subject to the LP relaxation after the first separation loop. We consider only nonbinary variables that appear in nonlinear constraints. By slightly relaxing the bounds on the variable that is currently minimized or maximized, we increase the chance to generate nontrivial LVBs when the bound is not tightened by OBBT. The generated LVB constraints are stored and propagated efficiently in a suitable topological order whenever their right-hand side improves.

As a test set, we used MINLPLib [21]. We excluded 18 instances which cannot be parsed or handled by SCIP 3.0.[1] Further 41 instances were linear after presolving or solved at the root node before OBBT was applied.[2] After removing two instances, for which SCIP 3.0 returned a wrong solution value due to numerical issues,[3] we were left with 211 instances. The experiments were conducted on a

---

[1] `blendgap`, `deb{6,7,8,9,10}`, `dosemin{2,3}d`, `prob10`, `var_con{5,10}`, `water{3, ful2,s,sbp,sym1,sym2}`, and `windfac`.

[2] `ex{1221,1222,1223a,1225}`, `feedtray2`, `gbd`, `hmittelman`, `lop97ic`, `lop97icx`, `mbtd`, `nvs{03,07,10}`, `pb*`, `prob{02,03}`, `qap`, `qapw`, `st_e{13,15,27}`, `st_miqp{1,2,3,4,5}`, `st_test{1,2,3,4,5,6,8}`, and `tln2`.

[3] `gear4` and `nvs22`.

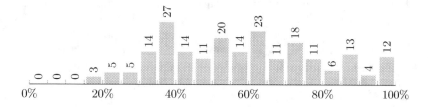

Fig. 2. Rate of generated LVBs per OBBT LP as distributed over 211 instances from MINLPLib for which OBBT was applied at the root node

cluster of 64bit Intel Xeon X5672 CPUs with 3.2 GHz, 12 MB cache, and 48 GB main memory. SCIP used CPLEX 12.4 [22] as LP solver, CppAD 20120101.3 [23], and Ipopt 3.10.2 [24,25].

Results. First, we measured the percentage of OBBT LPs solved that lead to a nontrivial LVB. The histogram in Fig. 2 shows the distribution of this success rate over the test set. For all instances, LVBs were generated from at least 15% of the OBBT LPs. For 132 out of 211 instances, the rate was above 50%.

Second, we compared SCIP with OBBT only and SCIP with OBBT and LVB propagation in a performance run with a time limit of one hour. To reduce distorting side effects from heuristic components of the tree search we deactivated primal heuristics in the tree, turned off conflict analysis, and used a simple first index branching rule with depth first node selection.

In this setting, two more instances could be solved with LVB propagation, while one instance solved before then hit the time limit of one hour. On 94 instances both solvers timed out; 109 instances were solved by both. Disregarding ten easy instances that were solved at the root by both variants, on the remaining 99 instances LVB propagation reduced the shifted geometric mean[4] of the number of branch-and-bound nodes by 14% and the solving time by 7%. Detailed results are given in [26].

For validation, we performed a control experiment using SCIP's default parameters as base setting. Here, for the instances solved by both solvers the number of nodes was reduced by 12% and the total solving time by 6%. Note that for a single propagation algorithm the achieved savings are substantial, in particular when considering its low computational overhead. Except for two easy instances, LVB propagation never took more than 2% of the total running time.

The fact that the solving time was reduced by less than the tree size is mostly explained by the longer processing time of the root. This general phenomenon is intensified by our experimental setup, since we applied a full round of OBBT without controlling the effort spent, e.g., by limiting LP iterations.

---

[4] The shifted geometric mean of values $x_1, \ldots, x_n \geqslant 0$ with shift $s > 0$ is defined as $\left(\prod_i (x_i + s)\right)^{1/n} - s$. We use a shift of five seconds and 100 nodes, respectively. This reduces the bias from outliers with large values as well as from very easy instances.

Conclusion. In this paper, we have introduced the notion of Lagrangian variable bound constraints, which are linear inequalities that can be learnt during OBBT and exploited during a spatial branch-and-bound algorithm. They can be propagated efficiently and give an approximation of reapplying OBBT where it may be overly expensive. Our experiments showed that on affected instances from MINLPLib this reduces the average number of branch-and-bound nodes by more than 10% and speeds up the solution process.

Future research should investigate whether LVB success correlates, for instance, with the tightness of the generating OBBT LP and how LVB propagation behaves in combination with a more sophisticated OBBT implementation.

Acknowledgments. The authors want to thank Timo Berthold, Pietro Belotti, and Domenico Salvagnin for the fruitful discussions and valuable comments on this paper.

References

1. Tawarmalani, M., Sahinidis, N.V.: Convexification and Global Optimization in Continuous and Mixed-Integer Nonlinear Programming: Theory, Algorithms, Software, and Applications. Kluwer Academic Publishers, Dordrecht Boston London (2002)
2. Quesada, I., Grossmann, I.E.: Global optimization algorithm for heat exchanger networks. Industrial & Engineering Chemistry Research 32(3), 487–499 (1993), doi:10.1021/ie00015a012
3. Quesada, I., Grossmann, I.E.: A global optimization algorithm for linear fractional and bilinear programs. Journal of Global Optimization 6, 39–76 (1995), doi:10.1007/BF01106605
4. Maranas, C.D., Floudas, C.A.: Global optimization in generalized geometric programming. Computers & Chemical Engineering 21(4), 351–369 (1997), doi:10.1016/S0098-1354(96)00282-7
5. Smith, E.M., Pantelides, C.C.: A symbolic reformulation/spatial branch-and-bound algorithm for the global optimisation of nonconvex MINLPs. Computers & Chemical Engineering 23, 457–478 (1999), doi:10.1016/S0098-1354(98)00286-5
6. Zamora, J.M., Grossmann, I.E.: A branch and contract algorithm for problems with concave univariate, bilinear and linear fractional terms. Journal of Global Optimization 14, 217–249 (1999), doi:10.1023/A:1008312714792
7. Adjiman, C.S., Androulakis, I.P., Floudas, C.A.: A global optimization method, αBB, for general twice-differentiable constrained NLPs—II. Implementation and computational results. Computers & Chemical Engineering 22(9), 1159–1179 (1998), doi:10.1016/S0098-1354(98)00218-X
8. Adjiman, C.S., Androulakis, I.P., Floudas, C.A.: Global optimization of mixed-integer nonlinear problems. AIChE Journal 46(9), 1769–1797 (2000), doi:10.1002/aic.690460908
9. Belotti, P., Lee, J., Liberti, L., Margot, F., Wächter, A.: Branching and bounds tightening techniques for non-convex MINLP. Optimization Methods & Software 24, 597–634 (2009), doi:10.1080/10556780903087124
10. COIN-OR: Couenne. Convex Over and Under ENvelopes for Nonlinear Estimation, http://www.coin-or.org/Couenne

11. Misener, R., Floudas, C.A.: GloMIQO: Global mixed-integer quadratic optimizer. Journal of Global Optimization, 1–48 (2012), doi:10.1007/s10898-012-9874-7
12. Computer-Aided Systems Laboratory, Princeton University: GloMIQO. Global Mixed-Integer Quadratic Optimizer, http://helios.princeton.edu/GloMIQO
13. Nowak, I., Vigerske, S.: LaGO: a (heuristic) branch and cut algorithm for nonconvex MINLPs. Central European Journal of Operations Research 16(2), 127–138 (2008), doi:10.1007/s10100-007-0051-x
14. COIN-OR: LaGO Lagrangian Global Optimizer, http://www.coin-or.org/LaGO
15. Achterberg, T.: Constraint Integer Programming. PhD thesis, TU Berlin (2007), http://opus4.kobv.de/opus4-zib/frontdoor/index/index/docId/1018
16. Vigerske, S.: Decomposition in Multistage Stochastic Programming and a Constraint Integer Programming Approach to MINLP. PhD thesis, HU Berlin (2012)
17. Zuse Institute Berlin, Department of Optimization: SCIP. Solving Constraint Integer Programs, http://scip.zib.de
18. Caprara, A., Locatelli, M.: Global optimization problems and domain reduction strategies. Mathematical Programming 125, 123–137 (2010), doi:10.1007/s10107-008-0263-4
19. Tawarmalani, M., Sahinidis, N.V.: Global optimization of mixed-integer nonlinear programs: A theoretical and computational study. Mathematical Programming 99, 563–591 (2004), doi:10.1007/s10107-003-0467-6
20. Berthold, T., Heinz, S., Vigerske, S.: Extending a CIP framework to solve MIQCPs. In: Lee, J., Leyffer, S. (eds.) Mixed41-Integer Nonlinear Optimization. The IMA volumes in Mathematics and its Applications, vol. 154, pp. 427–444. Springer (2012), doi:10.1007/978-1-4614-1927-3_15
21. Bussieck, M., Drud, A., Meeraus, A.: MINLPLib – a collection of test models for mixed-integer nonlinear programming. INFORMS J. on Comput. 15(1), 114–119 (2003)
22. IBM: ILOG CPLEX Optimizer, http://www-01.ibm.com/software/integration/optimization/cplex-optimizer/
23. COIN-OR: CppAD. A Package for Differentiation of C++ Algorithms. http://www.coin-or.org/CppAD
24. Wächter, A., Biegler, L.T.: On the implementation of a primal-dual interior point filter line search algorithm for large-scale nonlinear programming. Math. Prog. 106(1), 25–57 (2006)
25. COIN-OR: Ipopt. Interior point optimizer, http://www.coin-or.org/Ipopt
26. Gleixner, A.M., Weltge, S.: Learning and propagating Lagrangian variable bounds for mixed-integer nonlinear programming. ZIB-Report 13-04, Zuse Institute Berlin (2013), http://opus4.kobv.de/opus4-zib/frontdoor/index/index/docId/1763

Tight LP-Relaxations of Overlapping Global Cardinality Constraints⋆

Ioannis Mourtos

Department of Management Science and Technology, Athens University of Economics and Business, 95 Patision Avenue, 104 34 Athens, Greece
mourtos@aueb.gr

1 Motivation

The *global cardinality constraint (gcc)* [7], written as

$$cardinality(x, J; l, u), x_j \in D_j, j \in J,$$

states that each value d is received by at least l_d and by at most u_d of the variables $\{x_j : j \in J\}$, where $d \in D = \bigcup_{j \in J} D_j = \{0, \ldots, |D| - 1\}$; also, $0 \le l_d \le u_d$ and $u_d \ge 1$ for all $d \in D$. The gcc has several applications [2, 10], thus having been studied from the Constraint Programming community mainly for accomplishing various forms of consistency [4, 6–8] or for examining the tractability of a natural generalization [9].

From an Integer Programming perspective, two families of valid inequalities for a single gcc appear in [3, Section 7.10.1], while there is no polyhedral study of more than one gcc, and thus no theoretical results leading to tight LP-relaxations; notably, results of this kind exist for overlapping alldifferent constraints [1, 5]. Here, we study the polytope defined as the convex hull of vectors satisfying two gcc's, in the case where all variables share a common domain, thus initiating the study of overlapping gcc's. We establish a necessary and sufficient condition for this polytope to be full-dimensional and another such condition for the inequalities of the two known families to be facet-defining. These results hold trivially for the polytope associated with a single gcc; for that polytope, an example shows that further facets may exist, contrary to [3, Theorem 7.48]. Overall, the results presented here contribute to the construction of tight relaxations of (sets of) global constraints using just the variables appearing in these constraints.

2 The Dimension of the 2-gcc Polytope

Consider two gcc's, written as

$$cardinality(x, J_1; l, u), \; cardinality(x, J_2; l, u), \; x_j \in D, \forall j \in J_1 \cup J_2. \tag{1}$$

⋆ This research has been co-financed by the European Union (European Social Fund – ESF) and Greek national funds through the Operational Program "Education and Lifelong Learning" of the National Strategic Reference Framework (NSRF) - Research Funding Program: Thales. Investing in knowledge society through the European Social Fund.

C. Gomes and M. Sellmann (Eds.): CPAIOR 2013, LNCS 7874, pp. 362–368, 2013.

Let $C = \{1, 2\}, J_1 \cup J_2 = J, J_1 \cap J_2 = T$ and $J_1 \backslash T = I_1, J_2 \backslash T = I_2$, i.e., the set T indexes the variables common to both constraints, while I_c is the set of non-common variables of constraint $c \in C$. Notice that $T = \emptyset$ implies that the constraints are variable-wise disjoint while $I_1 = \emptyset$ or $I_2 = \emptyset$ yield that one constraint is 'dominated' by the other; thus we consider that none of I_1, I_2 or T is empty. The polytope examined here, namely P_I, is

$$P_I = conv\{x \in D^{|J|} : (1) \text{ is satisfied}\}.$$

A point of P_I satisfying (1) is hereafter called a *vertex* (with a slight abuse of notation). Assuming without loss of generality that $|J_1| \leq |J_2|$, it becomes easy to see that

$$P_I \neq \emptyset \text{ if and only if } \sum_{d \in D} l_d \leq |J_1| \leq |J_2| \leq \sum_{d \in D} u_d. \tag{2}$$

It is meaningful to study P_I as long as it contains more than one vertex.

Lemma 1. *For $P_I \neq \emptyset, P_I$ has more than one vertex if and only if $|D| \geq 2$ and $l_d < |J_2|$ for all $d \in D$.*

Proof. If $D = \{0\}$, P_I has a single vertex x^0 where $x_j^0 = 0$ for all $j \in J$; if $|D| \geq 2$ but, say, $l_0 = |J_2|, |J_1| \leq |J_2|$ and (2) yield $l_0 = |J_1|$, thus, x^0 is again the sole vertex of P_I.

If $|D| \geq 2$ and $l_d < |J_2|$ for all $d \in D$, one may construct a vertex $x^0 \in P_I$ by placing the smallest values in T and the largest ones in I_1 and I_2 and a (different) vertex $x^1 \in P_I$ by placing the largest values in T and the smallest ones in I_1 and I_2; it becomes easy to show that the two vertices being equal and $|D| \geq 2$ implies $l_d = |J_2|$ for some $d \in D$.

Let us add to our conventions that

$$l_d \geq |J_2| - \sum_{d' \in D \backslash \{d\}} u_d, d \in D \tag{3}$$

$$u_d \leq |J_1| - \sum_{d' \in D \backslash \{d\}} l_d, d \in D \tag{4}$$

Notice that this can be adopted without loss of generality; if, for example $l_d < |J_2| - \sum_{d' \in D \backslash \{d\}} u_d$, value d is bound to appear at least $l'_d = |J_2| - \sum_{d' \in D \backslash \{d\}} u_d$ times hence replacing l_d with l'_d yields two gcc's with an identical set of solutions.

Example 1. Let $I_1 = \{1, 2\}, T = \{3, 4\}$ and $I_2 = \{5, 6\}, D = \{0, 1\}, l = \{1, 1\}$ and $u = \{2, 6\}$. Value 0 occurring at most twice in J_1 imposes value 1 to occur at least twice thus $l_1 = 1$ can be replaced by $l'_1 = 2$.

Lemma 2. *If $\sum_{d \in D} l_d < |J_c| < \sum_{d \in D} u_d$ for some $c \in C$, then there exist $\{d_0, d_1\} \subseteq D$ having $l_{d_0} < u_{d_0}$ and $l_{d_1} < u_{d_1}$.*

Proof. $l_d = u_d$ for all $d \in D$ yields $\sum_{d \in D} l_d = |J_c|$ hence a contradiction. For $l_{d_0} < u_{d_0}, d_0 \in D$ and $l_d = u_d, d \in D \backslash \{d_0\}$, (3) and (4) yield

$$l_{d_0} \geq |J_2| - \sum_{d \in D \backslash \{d_0\}} u_d = |J_2| - \sum_{d \in D \backslash \{d_0\}} l_d \geq |J_1| - \sum_{d \in D \backslash \{d_0\}} l_d \geq u_{d_0},$$

which together with $l_{d_0} \leq u_{d_0}$ yields $l_{d_0} = u_{d_0}$, i.e., a contradiction.

It becomes convenient, and unambiguous, to say that a value $d \in D$ *appears in S* $(S \subseteq J)$ at a vertex $x \in P_I$ to denote that $x_j = d$ for some $j \in S$. Let $o(x; c, d)$ denote the number of occurrences of d in constraint c at vertex x, i.e., $o(x; c, d) = |\{j \in J_c : x_j = d\}|$. For two vertices $x^0, x^1 \in P_I$, $x^1 = x^0(j_1; d \to d')$ denotes that x^1 is derived from x^0 by only changing the value of variable x_{j_1} $(j_1 \in J)$ from d to d', i.e., $x^0_{j_1} = d \neq d'$ while $x^1_{j_1} = d'$ and $x^1_j = x^0_j$ for all $j \in J\backslash\{j_1\}$. Also, $x^1 = x^0(j_1 \leftrightarrow j_2)$ denotes that x^1 is derived from x^0 by only swapping the values of variables x_{j_1} and x_{j_2} $(\{j_1, j_2\} \subseteq J)$, i.e., $x^1_{j_1} = x^0_{j_2}, x^1_{j_2} = x^0_{j_1}$ and $x^1_j = x^0_j$ for all $j \in J\backslash\{j_1, j_2\}$. Last $x^1 = x^0(t_1 \leftrightarrow \{i_1, i_2\})$ denotes that x^1 is derived from x^0 by only swapping the value of variable x_t $(t \in T)$ with the *common* value of variables x_{i_1} $(i_1 \in I_1)$ and x_{i_2} $(i_2 \in I_2)$; that is, $x^0_{i_1} = x^0_{i_2} \neq x^0_t$ while $x^1_{i_1} = x^1_{i_2} = x^0_t, x^1_t = x^0_{i_1}$ and $x^1_j = x^0_j$ for all $j \in J\backslash\{i_1, i_2, t\}$.

Theorem 1. $\dim P_I = |J|$ *if and only if* $\sum_{d \in D} l_d < |J_1| \leq |J_2| < \sum_{d \in D} u_d$.

Proof. The 'only-if' part is direct under (3) and (4), since $|J_1| = \sum_{d \in D} l_d$ implies $\sum_{j \in J_1} x_j = \sum_{d \in D} l_d \cdot d$ while $|J_2| = \sum_{d \in D} u_d$ yields $\sum_{j \in J_2} x_j = \sum_{d \in D} u_d \cdot d$. To prove the 'if' part, we show that an equality $\alpha x = \alpha_0$ being satisfied by all $x \in P_I$ implies $\alpha_j = 0$ for all $j \in J$.

Let $c = 1$ and $i_0 \in I_1$. Since $\sum_{d \in D} l_d < |J_1| < \sum_{d \in D} u_d$, Lemma 2 becomes applicable hence let $d_0 \in D$ be a value having $l_{d_0} < u_{d_0}$. It is easy to obtain a vertex $x \in P_I$ such that $x_{i_0} = d_0, i_0 \in I_0$. Let us first show that, for any vertex $x \in P_I$ having $x_{i_0} = d_0$, we can derive a vertex $x^0 \in P_I$ such that $o(x^0; 1, d_0) > l_{d_0}$ and $x^0_{i_0} = d_0$. Assuming $o(x; 1, d_0) = l_{d_0}$ implies (since $\sum_{d \in D} l_d < |J_1|$) that there is $d_1 \neq d_0$ such that $o(x; 1, d_1) \geq l_{d_1} + 1 \geq 1$. If $x_{i_1} = d_1, i_1 \in I_1\backslash\{i_0\}$, derive the vertex $x^0 = x(i_1; d_1 \to d_0)$ and observe that $o(x^0; 1, d_0) = o(x; 1, d_0) + 1 > l_{d_0}$, while $o(x^0; 1, d_1) = o(x; 1, d_1) - 1 \geq l_{d_1}$. Otherwise (i.e., for $x_i \neq d_1$ for all $i \in I_1$), there is $t \in T$ such that $x_t = d_1$. Since all occurrences of d_1 in the first constraint are in T, $o(x; 2, d_1) \geq o(x; 1, d_1) > l_{d_1}$. Here there are two cases.

- For $o(x; 2, d_0) < u_{d_0}, x^0 = x(t; d_1 \to d_0)$ is a vertex of P_I because of $o(x^0; 1, d_0) = o(x; 1, d_0) + 1 > l_{d_0}$ (since $o(x; 1, 0) = l_{d_0}$) and $o(x^0; 2, d_1) = o(x; 2, d_1) - 1 \geq l_{d_1}$ (since $o(x; 2, d_1) > l_{d_1}$).
- For $o(x; 2, d_0) = u_{d_0}, u_{d_0} = o(x; 2, d_0) > o(x; 1, d_0) = l_{d_0}$ implies $x_{i_2} = d_0, i_2 \in I_2$; at the vertex $x^0 = x(t \leftrightarrow i_2)$, $o(x^0; 1, d_0) > l_{d_0}, o(x^0; 1, d_1) \geq l_{d_1}$ whereas $o(x^0; 2, d_0) = o(x; 2, d_0)$ and $o(x^0; 2, d_1) = o(x; 2, d_1)$.

None of the above value changes affects variable x_{i_0}, hence vertex $x^0 \in P_I$ satisfies $o(x^0; 1, d_0) > l_{d_0}$ and $x^0_{i_0} = d_0$. If there is $d \in D\backslash\{d_0\}$ such that $o(x^0; 1, d) < u_d$, derive the vertex $\bar{x} = x^0(i_0; d_0 \to d)$. Since, by hypothesis, $\alpha x = \alpha_0$ holds for all points of $P_I, \alpha x^0 = \alpha \bar{x}$ yields, after deleting identical terms,

$$\alpha_{i_0} d_0 = \alpha_{i_0} d \tag{5}$$

or $\alpha_{i_0} = 0$, since $d \neq d_0$.

Otherwise, $o(x^0; 1, d) = u_d$ for all $d \in D\backslash\{d_0\}$. However, $|J_1| < \sum_{d \in D} u_d$ implies $o(x^0; 1, d') < u_{d'}$ for a value d'; evidently, d' can only be d_0, i.e.,

$o(x^0; 1, d_0) < u_{d_0}$. Since Lemma 2 holds, there is another value $d_1 \neq d_0$ such that $l_{d_1} < u_{d_1}$; in addition, $o(x^0; 1, d_1) = u_{d_1} \geq 1$ implies that d_1 appears in J_1. We examine two cases.

Case 1. $x^0_{i_1} = d_1, i_1 \in I_1 \backslash \{i_0\}$

Derive the point $x' = x^0(i_1; d_1 \to d_0)$; since $o(x^0; 1, d_0) < u_{d_0}, o(x^0; 1, d_1) = u_{d_1}$ but $l_{d_1} < u_{d_1}$, $o(x'; 1, d_0) = o(x^0; 1, d_0) + 1 \leq u_{d_0}$ and $o(x'; 1, d_1) = o(x^0; 1, d_1) - 1 \geq l_{d_1}$ thus x' is a vertex of P_I. Hence, derive the vertex $\bar{x} = x'(i_0; d_0 \to d_1)$ and obtain (5) from $\alpha x' = \alpha \bar{x}$.

Case 2. $x^0_t = d_1, t \in T$

For $o(x^0; 2, d_0) < u_{d_0}$, after deriving vertices x' and $\bar{x}$ as in Case 1, but with t in the place of i_1, $\alpha x' = \alpha \bar{x}$ yields (5). For $o(x^0; 2, d_0) = u_{d_0}$, $o(x^0; 1, d_0) < o(x^0; 2, d_0)$ yields $x^0_{i_2} = d_0, i_2 \in I_2$; at $x' = x^0(t \leftrightarrow \{i_0, i_2\})$, the number of occurrences of both d_0 and d_1 remain as in x^0 at both predicates (i.e., x' is a vertex of P_I) thus $o(x'; 1, d_0) < u_{d_0}, o(x'; 1, d_1) = u_{d_1} (> l_{d_1})$ and $x'_{i_0} = x^0_t = d_1$. After deriving the vertex $\bar{x} = x'(i_0; d_1 \to d_0)$, $\alpha x' = \alpha \bar{x}$ yields (5).

Having shown that $\alpha_i = 0$ for all $i \in I_1 \cup I_2$, it remains to establish that $\alpha_t = 0$ for all $t \in T$. For that purpose, we construct a vertex x^0 as follows. Let $d(T) = \min\{d \in D : \sum_{d'=0}^{d} l_{d'} > |T|\}$; that is, all variables indexed by T can be assigned values $0, 1, \ldots, d(T) - 1$ so as to 'cover' the minimum necessary occurrences of each such value, although the same is not possible for values $0, 1, \ldots, d(T)$. Then, at vertex x^0, l_d variables indexed by T receive value d, for $d = 0, \ldots, d(T) - 1$, and the remaining $|T| - \sum_{d=0}^{d(T)-1} l_d$ variables receive value $d(T)$. Further, always at vertex x^0, $\sum_{d=0}^{d(T)} l_d - |T|$ variables indexed by each of the sets I_1 and I_2 also receive value $d(T)$ (thus completing the minimum necessary occurrences for that value in each constraint); another $l_{d(T)+1}$ variables indexed by each of I_1 and I_2 receive value $d(T) + 1$ and so on. Having assigned any value d at least l_d times in each constraint (for all $d \in D$), which is possible since $\sum_{d \in D} l_d < |J_1| \leq |J_2|$, all remaining variables may receive values in a way that assigns to T the smallest possible and to $I_1 \cup I_2$ the largest possible values. Then, $l_d < |J_1| \leq |J_2|$ for all $d \in D$ (Lemma 1) imply that some value d_0 appears in T whereas some other value $d_1 > d_0$ appears in both I_1 and I_2. After deriving the vertex $\bar{x} = x^0(t \leftrightarrow \{i_1, i_2\})$, $\alpha x^0 = \alpha \bar{x}$ yields, after deleting identical terms,

$$\alpha_t(d_1 - d_0) = (\alpha_{i_1} + \alpha_{i_2})(d_1 - d_0)$$

or $\alpha_t = \alpha_{i_1} + \alpha_{i_2}$, because of $d_1 \neq d_0$; since $\alpha_{i_1} = \alpha_{i_2} = 0$, $\alpha_t = 0$.

3 Facets of the 2-gcc Polytope

Notice that the sum of any subset of variables is minimized (maximized) when the smallest (largest) values occur as many times as possible. Although the definition of a gcc already yields that each value $d \in D$, occurs at most u_d times, the actual upper bound on the occurrences of a value depends also on the lower bounds of all other variables, e.g., consider value 1 in Example 1. For $c \in C$, let us recall the following definitions from [3, Section 7.10.1] (in which a more elaborate presentation can be found):

$$p_c(d) = \min \left\{ u_d, |J_c| - \sum_{i=0}^{d-1} p_c(i) - \sum_{i=d+1}^{|D|-1} l_i \right\}, \quad d = 0, \ldots, |D| - 1, \quad (6)$$

$$q_c(d) = \min \left\{ u_d, |J_c| - \sum_{i=d+1}^{|D|-1} q_c(i) - \sum_{i=0}^{d-1} l_i \right\}, \quad d = |D| - 1, \ldots, 0. \quad (7)$$

Considering now the sum of the variables indexed by S where $S \subseteq J_c$, we denote as $p_c(|S|, d)$ and $q_c(|S|, d)$ the number of occurrences of value d once the sum of $|S|$ variables is minimized or maximized, respectively. For $c \in C$,

$$p_c(|S|, d) = \min \left\{ p_c(d), |S| - \sum_{i=0}^{d-1} p_c(|S|, i) \right\}, d = 0, \ldots, |D| - 1, \quad (8)$$

$$q_c(|S|, d) = \min \left\{ q_c(d), |S| - \sum_{i=d+1}^{|D|-1} q_c(|S|, i) \right\}, d = |D| - 1, \ldots, 0. \quad (9)$$

Since $p_c(|S|, d)$ and $q_c(|S|, d)$ depend on both $|S|$ and $|J_c|$, let for conciseness $|J_1| = |J_2| = n$ and omit 'c' from the above definitions. The inequalities

$$\sum_{j \in S} x_j \geq \sum_{d \in D} p(|S|, d) \cdot d, S \subseteq J_c, c \in C, \quad (10)$$

$$\sum_{j \in S} x_j \leq \sum_{d \in D} q(|S|, d) \cdot d, S \subseteq J_c, c \in C, \quad (11)$$

are valid for P_I and separable in $O(n \log n)$ steps [3, Section 7.10.1]. Notice that, for $S = \{j\}$, these inequalities reduce to the trivial inequalities $0 \leq x_j \leq |D| - 1$. Also, since $p(n, d) = p(d)$ and $q(n, d) = q(d)$, for $|S| = n$ (10) and (11) become

$$\sum_{j \in J_c} x_j \geq \sum_{d \in D} p(d) \cdot d, c \in C, \quad (12)$$

$$\sum_{j \in J_c} x_j \leq \sum_{d \in D} q(d) \cdot d, c \in C. \quad (13)$$

We examine which of these inequalities cannot be facet-defining for P_I.

Lemma 3

(i) *(10) is redundant for $2 \leq |S| \leq p(0)$ or $n - p(|D| - 1) \leq |S| \leq n - 1$;*
(ii) *(11) is redundant for $2 \leq |S| \leq q(|D| - 1)$ or $n - q(0) \leq |S| \leq n - 1$.*

Proof. We only show (i), since the proof of (ii) is obtainable in an analogous manner. Notice that, for $2 \leq |S| \leq p(0)$, (8) implies $p(|S|, 0) = |S|$ but $p(|S|, d) = 0$ for $d \in D \backslash \{0\}$; thus, (10), written as

$$\sum_{j \in S} x_j \geq |S| \cdot 0,$$

equals the sum of inequalities $x_j \geq 0, j \in S$.

For $n - p(|D| - 1) \leq |S| \leq n - 1$ (no such S exists if $p(|D| - 1) = 0$), (8) yields $p(|S|, d) = p(d)$, for all $d \in D\{|D| - 1\}$, since $n - p(|D| - 1) \leq |S|$; hence

$$p(|S|, |D| - 1) = \min\{p(|D| - 1), |S| - \sum_{d \in D \backslash \{|D| - 1\}} p(|S|, d)\} =$$
$$= \min\{p(|D| - 1), |S| - \sum_{d \in D \backslash \{|D| - 1\}} p(d)\}. \quad (14)$$

Since $\sum_{d \in D} p(d) = n, \sum_{d \in D \setminus \{|D|-1\}} p(d) = n - p(|D| - 1)$ thus (14) becomes

$$p(|S|, |D| - 1) = \min\{p(|D| - 1), |S| - n + p(|D| - 1)\},$$

implying $p(|S|, |D| - 1) = |S| - n + p(|D| - 1)$, since $|S| \leq n - 1$. But then, (10), written as

$$\sum_{j \in S} x_j \geq (|S| - n + p(|D| - 1)) \cdot (|D| - 1) + \sum_{d \in D \setminus \{|D|-1\}} p(d) \cdot d =$$
$$= \sum_{d \in D} p(d) \cdot d - (n - |S|) \cdot (|D| - 1),$$

can be obtained by adding (12) and inequalities $-x_j \geq -(|D| - 1), j \in J_c \setminus S$.

Let $P^l(S) = \{x \in P_I : x$ satisfies (10) as equality$\}$ and $P^u(S) = \{x \in P_I : x$ satisfies (11) as equality$\}$ be the face defined by (10) and (11), respectively; $P^l(J_c)$ and $P^u(J_c)$ are defined accordingly. Let also $D^l(S) = \{d \in D : p(|S|, d) = p(d)\}$ and $D^u(S) = \{d \in D : q(|S|, d) = q(d)\}$.

Lemma 4

(i) $\sum_{d \in D^l(S)} p(d) + \sum_{d \in D \setminus D^l(S)} l_d \geq n$ *implies* $P^l(S) \subseteq P^l(J_c)$;
(ii) $\sum_{d \in D^u(S)} q(d) + \sum_{d \in D \setminus D^u(S)} l_d \geq n$ *implies* $P^u(S) \subseteq P^u(J_c)$.

Proof. We only show (i). At an arbitrary vertex $x \in P^l(S)$, $o(x; c; d) = p(d)$ for any $d \in D^l(S)$, by definition of $D^l(S)$. Also, $o(x; c; d) = l_d$ for any $d \in D \setminus D^l(S)$, since the opposite, i.e., $o(x; c; d') > l_{d'}$ for some $d' \in D \setminus D^l(S)$, yields

$$n = \sum_{d \in D^l(S)} o(x; c; d) + \sum_{d \in D \setminus D^l(S)} o(x; c; d) > \sum_{d \in D^l(S)} p(d) + \sum_{d \in D \setminus D^l(S)} l_d \geq n,$$

i.e., a contradiction. In addition, $p(d) = l_d$ for all $d \in D \setminus D^l(S)$, since $p(d') > l_{d'}, d' \in D \setminus D^l(S)$ yields (because of $\sum_{d \in D} p(d) = n$)

$$n = \sum_{d \in D} p(d) = \sum_{d \in D^l(S)} p(d) + \sum_{d \in D \setminus D^l(S)} p(d) > \sum_{d \in D^l(S)} p(d) + \sum_{d \in D \setminus D^l(S)} l_d \geq n$$

i.e., a contradiction. But then, at any vertex $x \in P^l(S)$, each value $d \in D$, appears $p(d)$ times thus x is also a vertex of $P^l(J_c)$. Since a face of P_I is the convex hull of vertices on it, it follows that $P^l(S) \subseteq P^l(J_c)$.

Please note that Lemmas 3 and 4 hold irrespectively of whether $|J_1| = |J_2|$, since the latter is not assumed within the corresponding proofs. They also generalize directly to the polytope associated with more than 2 overlapping gcc's, since their proofs rely on the properties of a sole gcc (the same holds for Lemma 2). This is no longer the case in the next lemma. Define $D' = \{d' \in D : l_{d'} < u_{d'}\}$.

Lemma 5. *Let* $|J_1| = |J_2|, S \supseteq I_c, c \in C$. *If all values appearing in S at any vertex of $P^l(S)$ are in $D \setminus D'$, then* $\dim P^l(S) \leq \dim P_I - 2$.

Proof. Let $S \supseteq I_1$. Any value $d \in D \backslash D'$ appears $l_d = u_d$ times in both J_1 and J_2 at any vertex of P_I and hence at any vertex of $P^l(S) \subseteq P_I$. The set $S' = (S \backslash I_1) \cup I_2$ satisfies $|S'| = |S|$ (because of $|J_1| = |J_2| = n$) and $S' \supseteq I_2$ (because of $S \supseteq I_1$); then, at any vertex $x \in P^l(S)$, since all values appearing in $S \supseteq I_1$ are in $D \backslash D'$, each value $d \in D \backslash D'$ appearing in I_1 appears also in I_2 (recall that $o(x, 1, d) = o(x, 2, d) = l_d$). But then, at any vertex $x \in P^l(S)$ the same values appear the same number of times in both S and S' (since $S \cap S' = S \cap T$), thus all vertices of $P^l(S)$ satisfy (10) as equality for both S and S'. Since these (two) equalities are linearly independent, $\dim P^l(S) \leq \dim P_I - 2$.

One can then show the following.

Theorem 2. *If P_I is full-dimensional then $P^l(S)$ $(P^u(S))$ is a facet of P_I if none of the conditions listed in Lemmas 3, 4 and 5 hold.*

Our results hold trivially for the case of a single gcc. For the same case, [3, Theorem 7.48] states that P_I is completely described by (10) and (11). The polytope of $cardinality(x, \{1, 2, 3, 4\}; [0, 1, 0], [2, 2, 2])$ has $\dim P_I = 4$ (by Theorem 1) thus any valid inequality satisfied as equality by 4 affinely independent vertices is facet-defining; this holds for the inequality $x_1 + x_2 - x_3 - x_4 \leq 3$ (which is different from (10) and (11)) and the vertices $\{(2, 1, 0, 0), (1, 2, 0, 0), (2, 2, 1, 0), (2, 2, 0, 1)\}$. Therefore, the facial structure of P_I admits further investigation.

References

1. Bergman, D., Hooker, J.N.: Graph Coloring Facets from All-Different Systems. In: Beldiceanu, N., Jussien, N., Pinson, É. (eds.) CPAIOR 2012. LNCS, vol. 7298, pp. 50–65. Springer, Heidelberg (2012)
2. Bulatov, A.A., Marx, D.: Constraint Satisfaction Problems and Global Cardinality Constraints. Communications of the ACM 53, 99–106 (2010)
3. Hooker, J.N.: Integrated Methods for Optimization. International Series in Operations Research & Management Science. Springer (2012)
4. Katriel, I., Thiel, S.: Complete Bound Consistency for the Global Cardinality Constraint. Constraints 10, 191–217 (2005)
5. Magos, D., Mourtos, I.: On the facial structure of the AllDifferent system. SIAM Journal on Discrete Mathematics 25, 130–158 (2011)
6. Quimper, C.G., Golynski, A., López-Ortiz, A., van Beek, P.: An Efficient Bounds Consistency Algorithm for the Global Cardinality Constraint. Constraints 10, 115–135 (2005)
7. Regin, J.C.: Generalized arc consistency for global cardinality constrain. In: Proceedings of AAAI 1996, pp. 209–215 (1996)
8. Regin, J.C.: Cost-Based Arc Consistency for Global Cardinality Constraints. Constraints 7, 387–405 (2002)
9. Samer, M., Szeider, S.: Tractable cases of the extended global cardinality constraint. Constraints 16, 1–24 (2011)
10. van Beek, P., Wilken, K.: Fast optimal instruction scheduling for single-issue processors with arbitrary latencies. In: Walsh, T. (ed.) CP 2001. LNCS, vol. 2239, pp. 625–639. Springer, Heidelberg (2001)

An Adaptive Model Restarts Heuristic

Nina Narodytska and Toby Walsh

NICTA and UNSW, Sydney, Australia
{Nina.Narodytska,Toby.Walsh}@nicta.com.au

Abstract. We propose an adaptive heuristic for model restarts that aligns symmetry breaking with the dynamic branching heuristic. Experiments show that this method performs very well compared to other symmetry breaking methods.

1 Introduction

Symmetry is an important but often problematic feature of constraint satisfaction problems. One way to deal with symmetry is to add constraints to eliminate symmetric solutions [1–7]. Posting static symmetry breaking constraints has both good and bad features. On the positive side, static constraints are easy to post, and a few simple constraints can eliminate most symmetry in a problem. On the negative side, static symmetry breaking constraints pick out particular solutions in each symmetry class, and this may conflict with the branching heuristic. An alternative is a dynamic approach that modifies the search method to ignore symmetric states [8–11]. Whilst this reduces the conflict with the branching heuristic, we may get less propagation. In particular there is no pruning of symmetric values deeper in the search tree. An effective method to tackle this tension is model restarts [12]. This restarts search frequently with new and different symmetry breaking constraints. The hope is that we will find symmetry breaking constraints that do not clash with the branching heuristic. The original model restarts method proposed a random choice of symmetry breaking constraints. We show here that we can improve performance with an adaptive heuristic that aligns symmetry breaking with the dynamic branching heuristic.

2 An Adaptive Model Restarts Heuristic

Our adaptive heuristic collects information about branching decisions in earlier restarts in order to build a heuristic friendly ordering of variables within the static symmetry breaking constraints. This ordering is based on variable scores. We describe three different techniques to obtain these scores. The first two reuse statistics collected by the branching heuristics. If we use the domain over weighted degree variable heuristic, then we can use the DOWD ratio to compute variable scores. $Score(X_k) = D(X_k)/(w \times deg(X_k))$ [13], where $D(X_k)$ is the domain size, $deg(X_k)$ is the number of constraints involving the variable, and w the sum of the counters associated with these constraints. We order variables in increasing order of their scores. We call this the DOWD-based heuristic. Similarly, we can use statistics associated with impact based branching heuristics to build variable scores [14]. $Score(X_k) = \sum_{v \in D(x)}(1 - impact(x, v))$ where

C. Gomes and M. Sellmann (Eds.): CPAIOR 2013, LNCS 7874, pp. 369–377, 2013.

$impact(x, v)$ is the impact of a branching decision measured by the reduction of the search space induced when the decision was posted. We now order variables in decreasing order of their scores. We call this the IMPACT-based heuristic. Our third approach is based on the branching levels of variables. This offers some robustness to the choice of branching heuristic. For example, it could be used with some other branching heuristic than DOWD or IMPACT. If a variable is instantiated at level i then it gets a Borda type score of $n - i$, where n is the number of variables, and 0 otherwise. We order variables in decreasing order of the average Borda score over the last restart. We call this the ADAPT heuristic.

We use these three scoring heuristics within model restarts [12]. Model restarts was proposed to use a random variable ordering within symmetry breaking constraints in each restart. Frequent restarting ensures we eventually select a good representative symmetric solution that is aligned with the dynamic branching heuristic. Instead of using randomization, our adaptive heuristics build a variable ordering for symmetry breaking in each restart that is aligned with the branching heuristic. This variable ordering is a permutation of the original variables, and hence itself can be seen as a variable symmetry. As noted in [15], applying a symmetry to a (sound/complete) set of symmetry breaking constraints generates a new (sound/complete) set of symmetry breaking constraints. Thus, we can safely use this permutation to reorder the variables in the symmetry breaking constraints.

3   Experimental Results

We carried out experiments with 3 sets of commonly used benchmarks. We used Choco 2.1.2 on an Intel Core 8 CPU, 2.7 Ghz, 4Gb RAM with 1000 sec timeout. We branch with DOWD or IMPACT heuristics [13, 14].[1]

The first set of benchmarks, DIMACS graph colouring problems was used in earlier studies of symmetry breaking for interchangeable values [4, 16]. Such problems are particulary suitable to a dynamic symmetry breaking labeling rule that avoids symmetric solutions (DYN) [10]. We compared four symmetry breaking methods, including DYN, the static symmetry breaking precedence constraint (PREC) [4, 16], model restarts and one modification of model restarts. We use the suffix ADAPT, DOWD and IMPACT to denote that variables are reordered in the symmetry breaking PRECEDENCE constraint based on the corresponding scores.

Model restarts constructs a random permutation of variables in the scope of the symmetry breaking PRECEDENCE constraint (MR). Our adapted method works in the following way. The search starts on the model without symmetry breaking constraints. Until the first restart, we collect statistic about the search tree. If we use the ADAPT heuristic, we store the information about variables that the solver branched on as described in Section 2. If we use DOWD or IMPACT heuristics then the solver accumulates statistics in weights and impact factors. On the first restart, we order variables based on their scores obtained from the heuristic. The scores are described in Section 2. Then we post the PRECEDENCE constraint and align variables in the scope of the constraint

---

[1] We would like to thank Charles Prud'homme for his help in implementing the model restarts technique.

Table 1. Graph coloring. The average **time**(sec) over ten runs with a random seed to initialize the branching heuristic. **s** is the number of runs that finished within the timeout. #vals is the number of values in a problem.

problem	#val	Domain over weighted degree branching						Impact based branching					
		DYN (s, time)	PREC (s, time)	MR+DOWD (s, time)	ADAPT (s, time)	MR_sh+DOWD (s, time)	ADAPT (s, time)	DYN (s, time)	PREC (s, time)	MR+IMPACT (s, time)	ADAPT (s, time)	MR_sh+IMPACT (s, time)	ADAPT (s, time)
SAT (Satisfiable instances)													
queen8 8	9	10 18.22	10 4.39	10 6.80	10 5.47	10 5.65	10 **3.50**	10 3.70	10 3.55	10 105.28	10 5.72	10 5.94	10 4.26
DSJC125.1	5	10 **0.22**	10 1.92	10 0.84	10 0.96	10 0.40	10 0.42	10 0.80	10 3.86	10 94.40	10 1.60	10 0.82	10 0.82
school1	14	10 1.41	10 4.53	10 9.34	10 8.76	10 1.75	10 1.79	10 0.26	10 7.30	9 **0.14**	10 0.41	9 0.19	10 0.30
school1	15	10 0.85	10 3.33	10 3.17	10 6.09	10 1.46	10 1.54	10 0.26	10 9.41	9 0.19	9 0.48	9 0.22	10 **0.38**
DSJC125.5	19	10 **13.30**	–	–	–	2 99.79	10 128.78	1 128.41	–	6 550.26	7 380.84	7 285.50	9 264.50
DSJC125.5	20	10 **0.25**	10 205.94	9 58.85	9 65.31	10 1.49	10 0.86	10 0.44	10 269.34	10 9.73	10 3.56	10 3.24	9 1.89
DSJC250.5	35	9 65.09	–	–	7 4.69	9 132.61	9 20.60	7 48.81	1 499.66	1 **0.22**	8 0.28	8 197.88	3 415.35 295.45
DSJC250.5	36	10 0.62	5 0.20	6 2.26	10 0.40	10 0.66	10 0.60	7 3.20	10 0.24	6 **0.18**	8 8.26	10 6.69	10 32.15 1.87
DSJC250.5	37	10 0.29	9 18.65	10 0.53	10 31.30	10 0.25	10 0.29	10 0.28	6 172.89	9 **0.18**	10 2.25	10 0.24	10 0.24 0.26
queen9 9	10	10 194.14	10 23.13	10 103.86	10 15.97	10 148.92	10 **11.60**	10 22.77	10 65.37	8 205.80	8 18.88	8 49.78	8 126.22 20.39
le450 15b	15	10 93.00	9 213.62	10 21.52	10 8.73	10 2.24	10 3.86	6 132.62	9 60.84	9 175.59	9 60.84	9 153.89	10 72.78 40.84
school1 nsh	14	10 0.32	10 **1.99**	10 1.76	10 3.51	10 0.87	10 1.07	10 0.32	3 261.06	3 **0.12**	3 1.40	3 0.14	3 0.13 0.71
school1 nsh	15	10 0.21	10 60.39	10 **0.18**	10 0.18	10 **0.16**	10 0.20	10 0.73	–	3 **0.16**	3 1.35	3 **0.16**	3 **0.16** 0.57
school1 nsh	16	10 0.22	10 95.76	10 0.17	10 0.18	10 0.18	10 0.19	9 1.10	4 –	5 **0.15**	4 2.07	4 **0.15**	10 67.12 0.76
queen10 10	11	–	2 190.35	1 **120.96**	10 0.86	1 979.08	5 516.73	3 496.60	3 500.89	–	–	3 517.91	1 414.93 836.35
queen10 10	12	10 8.81	10 1.29	10 1.17	10 0.86	10 **0.34**	10 0.40	10 0.41	10 1.15	10 1.27	10 1.00	10 0.53	10 1.66 0.47
TOTALS solved problems/total		15/16	13/16	14/16	13/16	15/16	16/16	16/16	11/16	15/16	15/16	16/16	16/16
UNSAT (Unsatisfiable instances)													
myciel5	5	10 **0.95**	10 2.98	10 14.93	10 11.27	10 6.84	10 6.53	10 1.52	10 2.39	10 16.45	10 12.05	10 9.30	10 6.73
R50 5gb	9	10 **0.28**	10 2.92	10 4.61	10 2.18	10 1.97	10 1.80	10 0.68	10 3.45	10 84.77	10 2.90	10 10.14	10 2.09
queen8 8	8	10 **5.16**	10 10.64	10 169.65	10 201.60	10 202.86	10 140.87	10 18.25	10 18.83	8 352.89	8 214.68	2 729.62	10 153.14
mulsol.i.1	47	–	–	–	3 41.62	2 24.42	2 56.00	2 75.09	–	–	–	–	**0.38**
mulsol.i.1	48	–	–	–	–	1 98.82	1 48.21	10 108.72	–	–	3 0.60	–	**0.50**
school1	13	10 1.11	10 2.75	10 3.65	10 7.27	10 0.91	10 1.85	10 **0.15**	10 2.58	–	3 0.59	3 –	3 0.35
fpsol2.i.2	29	10 **1.14**	–	10 15.02	10 7.13	10 1.32	10 80.80	8 1.32	–	–	3 0.67	–	3 0.35
myciel6g	6	10 **0.70**	10 1.36	10 29.35	10 9.22	10 6.83	10 6.49	10 7.51	10 2.01	10 52.29	5 9.83	6 89.98
DSJC125.5	12	10 **4.90**	–	10 41.13	10 –	10 90.04	10 65.71	3 477.69	–	–	5 19.82	10 27.88	6 14.86
DSJC250.5	14	10 **131.62**	–	–	7 67.18	3 820.79	3 749.73	–	–	–	1 225.70	–	3 480.15
le450 15b	13	10 150.46	–	10 0.52	10 0.63	10 **0.22**	10 **0.22**	9 70.15	–	10 52.29	10 6.65	–	10 1.40
school1 nsh	12	10 0.14	10 3.64	10 1.80	10 0.85	10 0.67	10 0.50	10 **0.11**	10 2.12	10 84.77	10 0.69	1 563.31	10 0.35
school1 nsh	13	10 **0.14**	10 133.17	10 1.41	10 0.92	10 0.63	10 0.53	10 0.18	10 514.70	–	10 2.46	–	10 0.88
4-FullIns 4	7	10 121.38	10 407.39	10 3.61	10 1.97	10 1.01	10 0.71	6 213.38	8 241.43	8 257.65	10 1.61	10 171.45	10 206.65 **0.51**
2-FullIns 5	5	10 0.95	10 5.12	10 3.33	10 0.57	10 0.36	10 **0.22**	10 5.90	10 2.20	10 3.91	10 0.82	10 1.60	10 3.15 0.41
4-Insertions 3	3	2 896.16 599.48	10 –	–	–	3 820.79	–	10 **56.72**	10 92.25	8 689.13	10 567.20	8 564.31	9 566.94 464.88 533.42
buck	10	10 2.09	10 13.84	10 46.62	10 0.19	10 **0.11**	10 0.14	10 63.85	10 19.25	10 63.85	8 564.31	9 566.94	10 16.49 541.38 0.21
homer	12	10 158.25	–	10 2.73	10 2.51	10 **0.48**	10 0.58	3 752.27	–	10 36.45	4 741.95	10 4.02	10 0.92
TOTALS solved problems/total		16/18	10/18	7/18	14/18	17/18	17/18	17/18	11/18	8/18	17/18	9/18	8/18 17/18

Table 2. Graph coloring. The branching heuristic is impact based branching. The average #backtracks over ten runs with a random seed to initialize the branching heuristic. #vals is the number of values in a problem.

problem	#val	Domain over weighted degree branching							Impact based branching						
		DYN	PREC	MR + DOWD	ADAPT	MR_sh + DOWD	ADAPT		DYN	PREC	MR + IMPACT	ADAPT	MR_sh + IMPACT	ADAPT
		bt	bt	bt	bt	bt	bt		bt	bt	bt	bt	bt	bt
SAT (Satisfiable instances)														
queen8 8	9	909599	109529	159069	115148	380833	229315	122541	99789	59922	387503	82926	130842	1.99e+06	78942
DSJC125.1	5	**504**	7087	994	928	2847	937	826	9902	400019	5977	6512	142994	3896	
school1	14	8037	5667	6397	6669	2673	5699	826	231	7325	**4**	**4**	**10**	93	
school1	15	3299	3472	1192	3904	1987	3452		39	10283	**10**	**10**	125	127	
DSJC125.5	19	**230634**		2.07e+06	12855	1.05e+06	2.42e+06		1.45e+06	1.89e+06	390780	14711	27984	3.31e+063.51e+063.06e+06	11432
DSJC125.5	20	**826**	2.29e+06	1e+06	1.05e+06	47883	2591	2256	2324	1.34e+06	**87**	**87**	914841	421366 2.03e+061.42e+06	
DSJC250.5	35	572804		1125	3846	493	138271	145	237564	763464	17	7277	25888	159019	4204
DSJC250.5	36	2693		**1**	**5**	**1**	23	15	15455		**87**	**87**	159019		20
DSJC250.5	37		85164	86	44	23			86		925	1379	24	44	
queen9 9	10	8.87e+06	575929	1.15e+06	763242	6.75e+03	426910		589342	1.21e+061.18e+062.61e+06	9	278081	1.18e+062.57e+06	679485	450170
le450 15b	15	1.79e+06	**11684**	81512	45835	13011	32670		367	297147 714565	**1**	317851	**1**		322694
school1 nsh	14	134	2.17e+06	376	1423	483	1274		2088		440	437	**0**	255628	440
school1 nsh	15	**0**	111659	**0**	**0**				4404	478604	**0**	999			305
school1 nsh	16	**0**	170079	**0**	**0**						3342				739
queen10 10	11		4.11e+06	**2.66e+06**		3.8e+07	1.89e+07		1.05e+07	6614		2481	1.04e+07 7.71e+061.66e+07		
queen10 10	12		10618	3688	2120	1168	1427	1432	3007		4640		2139	17143	2178
TOTALS		353154		7400							3149				
solved problems/total		15/16	13/16	11/16	14/16	13/16	15/16	16/16	16/16	11/16	15/16	14/16	15/16	16/16	16/16
UNSAT (Unsatisfiable instances)														
myciel5	5	**27100**	47759	1.1e+06	727665	518428	1.03e+06	423985	52560	30503	540891	459642	327131	500700	271010
R50 5gb	9	**653**	26311	150714	47840	10339	128823	10848	7068	40257	174693	1.24e+06	207598	633030 1.37e+06	13847
queen8 8	8	222329	337913	1.86e+07 4.87e+065.93e+06	1.93e+07	145618	6.64e+066.06e+06	494667	521777	425590	6.97e+06 4.15e+061.73e+07	4.15e+06	6.76e+064.01e+06	4.23e+06	
mulsol.i.1	47				83800				986138				101	101	101
mulsol.i.1	48			8022			371767	5910	1.39e+06	903		101	101	101	101
school1	13	5935	1645	1634	6040	1038	1.06e+06	530532	211			125	125	125	125
fpsol2.i.2	29	6241		14005		**1249**			13968			5036	557867	563975	563975
myciel6g	6	**7757**	8552	1915	140871	148121	142850		161052	21137	520652	232815	738818	267651	267651
DSJC125.5	12	**68793**	771308	516916	520442	1.29e+06 4.75e+06	933692 4.64e+06		5.93e+06		806651	1.26e+06			4.23e+06
DSJC250.5	14	**760284**						113	606155						
le450 15b	13	3.64e+06	3458	**108**	113	**108**	321	113	43	2011	8237	8237	3.87e+06		8237
school1 nsh	12	**13**	330336	521	136	236	133	133	150		115	115			115
school1 nsh	13	**28**		279	827	2506	156	156	1.83e+06 1.53e+06	582695	1293	1293	1.37e+06		1293
4-FullIns-4	7	2.96e+06	881513	2679	179	554	1019	173	118198	9011	632518	424	1.5e+06		424
2-FullIns 4	5	11263	31267	657	10679		173		15523		29506	1618	15523	31130	1833
4-Insertions 3	3	1.23e+06 6.22e+07			1.04e+06	1.04e+06	1.89e+07	334	4.75e+06 5.37e+06	1.02e+06	5.97e+064.35e+073.49e+07	3.56e+07	5e+07	4.42e+074.74e+074.43e+07	395
huck	10	361738 334880			322	214	214	1450	8.01e+06	339937	3.56e+07	395	974031	4.41e+07	3181
homer	12	1.15e+07			1279	813						2422			
TOTALS															
solved problems/total		16/18	10/18	7/18	15/18	8/18	17/18	17/18	17/18	11/18	8/18	17/18	9/18	8/18	17/18

with the obtained variable ordering. Statistics for the ADAPT heuristic are reset to zero after first and later restarts. Statistics for DOWD and IMPACT heuristics have built-in mechanisms to gradually forget pervious decisions. Between the first and the second restarts we again collect search statistics. On the second and later restarts, we remove the PRECEDENCE constraint that we posted on the previous restart from the model and post a new PRECEDENCE constraint where the variables in its scope are aligned with the ordering obtained from the heuristic. We continue this procedure until we find a solution or timeout. Note that our adaptive approach can be applied to all problems where model restarts can be applied as we replace a random ordering of variables with one derived from heuristics.

MR + DOWD and MR + IMPACT use the DOWD and IMPACT heuristics, whilst MR + ADAPT uses our adaptive version of model restarts.

We also consider limiting the cost of symmetry breaking. In MR_{sh}, MR_{sh} + ADAPT, MR_{sh} + DOWD and MR_{sh} + IMPACT‡ we shorten the PRECEDENCE constraint to the first $2m$ variables, where m is the number of values. The intuition behind this idea is based on an empirical observation that an instantiation of a relatively small number of variables in the scope of the PRECEDENCE constraint entails the constraint in most benchmarks. The value $2m$ was chosen based on statistical analysis of the benchmarks. We use a geometric restart policy with the base of 100 backtracks and a growth coefficient of 1.1. This ensure that restarts are rapid as in [12]. Tables 1–2 give average times and the number of backtracks for the DOWD and IMPACT branching heuristics over 10 runs. In addition, Table 1 shows the number of runs where a problem was solved. We also computed geometric means for these instances to reduce impact of outliers. However, as this gives the same picture of results and we have limited space, we do not include these results here. We removed instances solved by all methods in under 3 seconds and separated results for satisfiable and unsatisfiable.

Effect of the adaptive heuristic. By comparing PREC, MR and MR_{sh} with their adaptive counterparts, we see that our adaptive heuristic ADAPT dramatically improves performance on the majority of instances. For example, the adaptive heuristic helps solve 9 additional benchmarks if we compare MR and MR + ADAPT. The adaptive heuristic is especially useful on unsatisfiable instances. Note that many of these additionally solved benchmarks are easy once we remove conflict between the branching heuristic and static symmetry breaking. We observed that DOWD-based adaptive ordering also performs well. Unfortunately, the IMPACT ordering does not perform well on these benchmarks.

Effect of shortening. By comparing MR + ADAPT and MR_{sh} + ADAPT, as well as other models with their shortened counterparts, we see that shortening achieves much better performance. However, it slightly increases the number of backtracks in some cases. Shortening does not increase significantly the number of solved instances, or change substantially the search tree. However, it improves the efficiency of search. Overall, MR_{sh} + ADAPT gives the best performance over all benchmarks among all symmetry breaking methods using the DOWD and IMPACT branching heuristic.

Our second and third case studies consider classes of problems on which model restarts has been shown to outperform other static and dynamic symmetry breaking methods [12]. We ran experiments with the "signature" based static symmetry breaking constraints proposed for variable and value interchangeability in [11] and denoted here as GCC-based. We decomposed the GCC constraint into AMONG constraints so we can have access to the cardinality variables. Following [12] we only order partitions within the symmetry breaking constraints. We compute a score for all variables in each partition with respect to the used heuristic and sort the partitions according to these scores. Again, the main advantage of our approach is that instead of random ordering of partitions in model restarts we align them with branching heuristics.

We generated 20 problems of each size and averaged statistics over these problems. We report time to find an optimum solution and prove optimality. Note that all results are shown on instances that are solved by all techniques for at least 10 generated problems.

As in [4, 12], we tested on graph colouring and Concert Hall scheduling problems. In [12], the model restarts technique was shown to outperform other symmetry breaking methods on these benchmarks. Hence, we only compare our adaptive strategy with the simple static symmetry breaking constraints and the highly effective model restarts technique (GCC-based +MR). As previously, we biased the ordering of variables in the simple static symmetry breaking constraint to put large partitions first. Figure 1 (left part) shows the results for uniform and biased graph colouring problems with $q = 0.5$ using IMPACT branching heuristic. The results confirm that model restarts is better than static symmetry breaking. Our adaptive ordering of partitions significantly improves performance of model restarts. In particular, the ADAPT heuristic is more robust compared to the IMPACT heuristic.

For the Concert Hall Problem, we generated problems as in [4]. As it is important to put large partitions first, we assumed that any partition with size greater than 4 is a large partition (the maximum partition size is 8 in this setup). The number of halls is 12 or 14. Figure 1 (right part) shows the results for 14 halls. As can be seen from the graphs, using an adaptive heuristic to order partitions improves model restarts significantly using both DOWD or IMPACT branching heuristics. Moreover, ADAPT shows the best performance across all instances.

4   Other Related Work

Crawford *et al.* proposed a general method to break symmetry statically using lex-leader constraints [17]. Most static symmetry breaking constraints (including the PRECE-DENCE constraints used here) can be derived from such constraints. Efficient algorithms have been developed to propagate many static symmetry breaking constraints (e.g. [21–24]). Lex-leader constraints pick out the lexicographically smallest solution in each symmetry class. However, this may conflict with the branching heuristic. A number of dynamic methods have been proposed to deal with this conflict. For example, *SBDS* posts lex-leader constraints dynamically during search [8]. Another dynamic method for breaking symmetry is *SBDD* [9]. This checks if a node of the search tree is symmetric to some previously explored node. *GAPLex* is a hybrid method that combines together static and dynamic symmetry breaking [25]. However, it is limited to dynamically post-ing lex-leader constraints, and to searching with a fixed variable ordering (which can be a considerable burden). *Dynamic Lex* is another hybrid method that dynamically posts static symmetry breaking constraints during search which works with dynamic variable

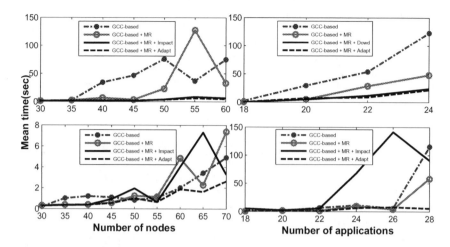

Fig. 1. Mean times for uniform (top left) and biased (bottom left) graph colouring benchmarks with q=0.5 using IMPACT based branching. Mean times for Concert Hall Problem benchmarks with 14 halls using DOWD (top right) and IMPACT based branching (bottom right).

ordering heuristics [26]. This method adds lex-leader constraints during search that are compatible with the current partial assignment. Hence the first solution found is not removed by symmetry breaking. However, unlike here, the method assumes that values are tried in a fixed order.

5   Conclusions

Static symmetry breaking constraints are often an easy and effective way to deal with symmetry in a constraint or optimisation problem. However, there can be a conflict between static symmetry breaking constraints and branching heuristics. To reduce this conflict, we propose a simple adaptive heuristic for model restarts. This orders variables within symmetry breaking constraints to align with the dynamic branching heuristic. Experimental results suggest that it is a very promising alternative between purely static and purely dynamic symmetry breaking methods. In particular, the results show that the proposed ADAPT heuristic works well across all benchmarks and two state-of-the-art branching heuristics. Our adaptive method thus appears to be more robust compared to the original model restarts algorithm.

References

1. Puget, J.F.: On the satisfiability of symmetrical constrained satisfaction problems. In: Komorowski, J., Raś, Z.W. (eds.) ISMIS 1993. LNCS (LNAI), vol. 689, pp. 350–361. Springer, Heidelberg (1993)
2. Shlyakhter, I.: Generating effective symmetry-breaking predicates for search problems. In: Proc. of Workshop on Theory and Applications of Satisfiability Testing, SAT 2001 (2001)

3. Flener, P., Frisch, A.M., Hnich, B., Kiziltan, Z., Miguel, I., Pearson, J., Walsh, T.: Breaking row and column symmetries in matrix models. In: Van Hentenryck, P. (ed.) CP 2002. LNCS, vol. 2470, pp. 462–477. Springer, Heidelberg (2002)

4. Law, Y., Lee, J.: Symmetry Breaking Constraints for Value Symmetries in Constraint Satisfaction. Constraints 11(2-3), 221–267 (2006)

5. Walsh, T.: Symmetry breaking using value precedence. In: Proc. of ECAI 2006, pp. 168–172 (2006)

6. Walsh, T.: General Symmetry Breaking Constraints. In: Benhamou, F. (ed.) CP 2006. LNCS, vol. 4204, pp. 650–664. Springer, Heidelberg (2006)

7. Walsh, T.: Breaking value symmetry. In: Proc. of the 23rd National Conf. on AI, pp. 1585–1588. AAAI (2008)

8. Gent, I., Smith, B.: Symmetry breaking in constraint programming. In: Proc. of ECAI 2000, pp. 599–603 (2000)

9. Fahle, T., Schamberger, S., Sellmann, M.: Symmetry breaking. In: Walsh, T. (ed.) CP 2001. LNCS, vol. 2239, pp. 93–107. Springer, Heidelberg (2001)

10. Hentenryck, P.V., Agren, M., Flener, P., Pearson, J.: Tractable symmetry breaking for CSPs with interchangeable values. In: Proc. of the 18th IJCAI (2003)

11. Flener, P., Pearson, J., Sellmann, M., Van Hentenryck, P.: Static and dynamic structural symmetry breaking. In: Benhamou, F. (ed.) CP 2006. LNCS, vol. 4204, pp. 695–699. Springer, Heidelberg (2006)

12. Heller, D., Panda, A., Sellmann, M., Yip, J.: Model restarts for structural symmetry breaking. In: Stuckey, P.J. (ed.) CP 2008. LNCS, vol. 5202, pp. 539–544. Springer, Heidelberg (2008)

13. ChocoTeam: Documentation. CHOCO is a java library for constraint satisfaction problems (CSP) and constraint programming (CP), http://choco.mines-nantes.fr/

14. Refalo, P.: Impact-based search strategies for constraint programming. In: Wallace, M. (ed.) CP 2004. LNCS, vol. 3258, pp. 557–571. Springer, Heidelberg (2004)

15. Katsirelos, G., Walsh, T.: Symmetries of symmetry breaking constraints. In: Proc. of ECAI 2010 (2010)

16. Walsh, T.: Breaking value symmetry. In: Bessière, C. (ed.) CP 2007. LNCS, vol. 4741, pp. 880–887. Springer, Heidelberg (2007)

17. Crawford, J., Luks, G., Ginsberg, M., Roy, A.: Symmetry breaking predicates for search problems. In: Proc. of the 5th Int. Conf. on Knowledge Representation and Reasoning (KR 1996), pp. 148–159 (1996)

18. Luby, M., Sinclair, A., Zuckerman, D.: Optimal speedup of Las Vegas algorithms. Information Processing Letters 47, 173–180 (1993)

19. Boussemart, F., Hemery, F., Lecoutre, C., Sais, L.: Boosting systematic search by weighting constraints. In: Proc. of the 16th ECAI 2004, pp. 146–150 (2004)

20. Gomes, C., Selman, B., Crato, N.: Heavy-tailed distributions in combinatorial search. In: Smolka, G. (ed.) CP 1997. LNCS, vol. 1330, pp. 121–135. Springer, Heidelberg (1997)

21. Frisch, A.M., Hnich, B., Kiziltan, Z., Miguel, I., Walsh, T.: Global constraints for lexicographic orderings. In: Van Hentenryck, P. (ed.) CP 2002. LNCS, vol. 2470, pp. 93–108. Springer, Heidelberg (2002)

22. Frisch, A., Hnich, B., Kiziltan, Z., Miguel, I., Walsh, T.: Propagation algorithms for lexicographic ordering constraints. Artificial Intelligence 170(10), 803–908 (2006)

23. Law, Y.C., Lee, J.H.M., Walsh, T., Yip, J.Y.K.: Breaking symmetry of interchangeable variables and values. In: Bessière, C. (ed.) CP 2007. LNCS, vol. 4741, pp. 423–437. Springer, Heidelberg (2007)

24. Katsirelos, G., Narodytska, N., Walsh, T.: Combining symmetry breaking and global constraints. In: Oddi, A., Fages, F., Rossi, F. (eds.) CSCLP 2008. LNCS, vol. 5655, pp. 84–98. Springer, Heidelberg (2009)
25. Jefferson, C., Kelsey, T., Linton, S., Petrie, K.: GAPLex: Generalised static symmetry breaking. In: Proc. of 6th Int. Workshop on Symmetry in Constraint Satisfaction Problems, SymCon 2006 (2006)
26. Puget, J.-F.: Symmetry breaking using stabilizers. In: Rossi, F. (ed.) CP 2003. LNCS, vol. 2833, pp. 585–599. Springer, Heidelberg (2003)

Constraint-Based Fitness Function for Search-Based Software Testing

Abdelilah Sakti, Yann-Gaël Guéhéneuc, and Gilles Pesant

Department of Computer and Software Engineering
École Polytechnique de Montréal, Québec, Canada
{abdelilah.sakti,yann-gael.gueheneuc,gilles.pesant}@polymtl.ca

Abstract. Search-based software testing is a powerful automated technique to generate test inputs for software. Its goal is to reach a branch or a statement in a program under test. One major limitation of this approach is an insufficiently informed fitness function to guide search toward a test target within nested predicates (constraints). To address this problem we propose fitness functions based on concepts well known to the constraint programming community, such as constrainedness and arity, to rank test candidates. Preliminary experiments promise efficiency and effectiveness for the new fitness functions.

1 Introduction

The main goal of software testing is to expose hidden errors by exercising a program on a set of test cases. A major challenge in this research field is automating the generation of test cases. The widely used approach is Search Based Software Testing (SBST). This approach translates the testing problem into an optimization problem, then it uses a meta-heuristic algorithm and a fitness function to generate test cases that meet a test target. SBST is very sensitive to the effectiveness and the efficiency of its fitness function. Despite the known limitations (nested structures, flags) [6] of the most widely used fitness function [11], alternatives have been little studied in the literature [2].

In the last decade, constraint programming has been proposed to replace SBST for software testing [3,4,7]. Collavizza et al. [3] have proposed a constraint based approach and implemented it in a tool to verify a program property. Gotlieb [4] and Pasareanu et al. [7] have proposed two different approaches based on constraint programming to generate test data. But constraint based testing suffers from its inability to manage dynamic aspects of a program and complex or unavailable source code. Recent research [9,2] has shown that integrating constraint programming techniques in SBST may make it more effective and more efficient. This short paper proposes a new approach based on constraint programming to enhance the performance of SBST. In particular we propose novel fitness functions based on branch "hardness". We statically analyze the test target to collect information about branches leading to it. Then we define a penalty value for each branch (constraint) according to its arity and its constrainedness.

C. Gomes and M. Sellmann (Eds.): CPAIOR 2013, LNCS 7874, pp. 378–385, 2013.
© Springer-Verlag Berlin Heidelberg 2013

Branch penalties are used to determine how close a test candidate is to reaching a test target.

The contributions of this paper are: (1) We introduce two new metrics to measure the hardness to satisfy a branch predicate in software test case generation; (2) We show how to use the new metrics to define two new fitness functions that measure how far a test candidate is to satisfying a test target; (3) We present the results of an empirical study on a large number of benchmarks that are randomly generated, the results of which indicate that our new fitness functions are significantly more effective and efficient than representative fitness functions from the literature, Approach-level [11] and Symbolically Enhanced [2] Fitness Functions.

The remainder of the paper is organized as follows: Section 2 introduces necessary definitions and presents a motivating example; Section 3 describes the new metrics and their use in two fitness functions; Section 4 presents an empirical study; Section 5 concludes with some future work.

2   Background and Motivation

To generate test data, SBST uses a meta-heuristic guided by a fitness function. The *approach-level* [11] and *branch-distance* [10] measures are largely used for computing the fitness function value [6]. These measures are respectively the number of branches leading to the test target that were not executed by the test candidate and the minimum required distance to satisfy the branch condition where the execution of the test candidate has left the target's path (that branch is called the *critical branch*).

Assume that our test target is reaching the statement at line 5 in the code fragment at Fig. 1. The branch distance of a test candidate (a, b, c) is equal to $|b - c|$, from the first branch. If this distance is equal to 0 then the first branch is satisfied and the branch distance is computed according to the second branch $(y > 0)$ which is equal to $\max(0, 1 - b)$, and so on. The approach level is an enhancement of the branch distance fitness that links each branch to a level. The level of a branch is equal to the number of branches

```
sample(int x,int y,int z){      1
   if(y==z)                      2
      if(y>0)                    3
         if(x==10)               4
            ...//Target          5
}                                6
```

Fig. 1. Code fragment

that separate it from the test target. In Fig. 1 the branch at line 4 has level 1 because there is only one branch to satisfy to reach the test target. The branch at line 3 has level 2 because there are two branches to satisfy, and so on. The approach level fitness function (f_{AL}) for a test candidate i and a branch b is $f_{AL}(i) = level(b) + \eta(i, b)$, where b is the critical branch of i. The second term η is a normalized branch distance. Among the many methods to compute it [1], we use $\frac{\delta}{\delta+1}$ where δ is the branch distance. This fitness function does not take into account non-executed branches. SBST is a dynamic approach: it executes the program under test and observes its behaviour. Therefore it only analyzes executed branches. Suppose that we have two test candidates $i_1 : (10, -30, 60)$ and $i_2 : (30, -20, -20)$ and we need to choose one of them. The approach level fitness function chooses i_2 ($f_{AL}(i_2) = 2 + \frac{21}{22} = 2.9545$) because it cannot see that i_1 ($f_{AL}(i_1) = 3 + \frac{90}{91} = 3.9890$) satisfies the branch at line 4. A careful static analysis would detect that i_2 needs more effort than i_1 since it needs to change every value whereas i_1 could reach the target with a single change of the second value. Furthermore the likelihood of randomly choosing the value 10 for x in a large domain is almost null.

Recently, Harman et al. [2] proposed the Symbolic Enhanced Fitness Function (f_{SE}) that complements SBST with a simple static analysis (symbolic execution): $f_{SE}(i) = \sum_{b \in P} \eta(i, b)$ whether branch b is executed or not. For our example their fitness function also chooses i_2 ($f_{SE}(i_2) = 0 + \frac{21}{22} + \frac{20}{21} = 1.9068$) over i_1 ($f_{SE}(i_1) = \frac{90}{91} + \frac{31}{32} + 0 = 1.9577$) because it does not prioritize branch "$(x == 10)$" — all branches are considered equal.

In summary: (normalized) branch distance is accurate to compare test candidates on the same branch but can be misleading when comparing on different branches; approach level corrects this by considering the critical branch's rank on the path to the target but ignores non-executed branches; symbolic enhanced, a hybrid approach that complements the dynamic approach with static analysis, includes all branches on the path but does not consider their rank or any other corrective adjustment. We propose branch hardness as such an adjustement, complementing branch distance.

3   A Fitness Function Based on Branch Hardness

We claim that test candidates that satisfy "hard" branches are more promising than those that only satisfy "easier" branches. Therefore we consider the difficulty to satisfy a constraint (i.e. a predicate used in a conditional statement or branch) as key information to measure the relevance of a test candidate. We propose a *branch-hardness fitness function* to guide a SBST approach toward test targets: it prioritizes branches according to how hard it is to satisfy them. As in the Symbolic Enhanced approach, we combine SBST with static analysis of the non-executed branches, but additionally apply a hardness measure to every branch.

Here the difficulty to satisfy a constraint c is linked to its arity and its *projection tightness* [8]. The latter is the ratio of the (approximate) number of solutions of a constraint to the size of its search space (i.e. the Cartesian product of the domains of the variables involved in c), in a way measuring the tightness of the projection of the constraint onto the individual variables. The lower the arity of the constraint, the less freedom we have to choose some of its variables in order to modify the test candidate. A projection tightness close to 0 will indicate high constrainedness and hardness to satisfy a constraint. Correspondingly we define parameters $\alpha(c) = \frac{1}{arity(c)}$ and $\beta(c) = 1 - projection\_tightness(c)$ that range between 0 and 1 and indicate hardness.

3.1   Branch-Hardness Fitness Functions

In this section we define two metrics Difficulty Coefficient and Difficulty Level to complete the branch distance and use them to express two fitness functions. These two metrics are designed to be used within meta-heuristic search to solve an evolutionary testing problem.

Difficulty Coefficient (DC). The DC is a real number greater than 1. It is a possible representation of the hardness of a branch. Each constraint has its own DC that is determined according to its arity and tightness. DC is calculated by the following formula,

$$DC(c) = B^2 \times \alpha(c) + B \times \beta(c) + 1,$$

where $B > 1$ is an amplification parameter. In this paper we use $B = 10$.

To express a fitness function based on DC we use DC as a penalty coefficient for breaking a constraint. The key idea behind this fitness function is determining a *standard-branch-distance*, then sorting candidates according to their total standard-branch-distance.

To compute a fitness value for a test candidate i on a set of branches (constraints) C we apply the following formula:

$$f_{DC}(i, C) = \sum_{c \in C} DC(c) \times \eta(i, c).$$

This fitness function penalizes a broken constraint in a relative manner, i.e., it determines the penalty according to the DC and the normalized distance branch. In this way we may prefer a candidate that satisfies a hard branch but breaks an easier one with a large normalized distance, over another candidate that breaks the former with a small normalized distance but satisfies the latter.

Now we come back to our example at Fig. 1 and apply this new fitness function. Assume that all domains are equal to $[-99, 100]$. We compute DC for each branch as follows:

1. $DC("y == z") = 10^2 \times 0.5 + 10 \times 0.995 + 1 = 60.95;$
2. $DC("y > 0") = 10^2 \times 1 + 10 \times 0.5 + 1 = 106;$
3. $DC("x == 10") = 10^2 \times 1 + 10 \times 0.995 + 1 = 110.95.$

Then we use these DC values to compute a fitness value for each test candidate: $f_{DC}(i_1, C) = 60.95 \times \frac{90}{91} + 106 \times \frac{31}{32} + 110.95 \times \frac{0}{1} = 162.9677; f_{DC}(i_2, C) = 60.95 \times \frac{0}{1} + 106 \times \frac{21}{22} + 110.95 \times \frac{20}{21} = 206.8485.$ Contrary to f_{AL} and f_{SE} this fitness function makes the adequate choice by choosing the test candidate i_1 instead of i_2. This type of decision may make f_{DC} more efficient than f_{AL} and f_{SE}.

Difficulty Level (DL). DL is a representation of a relative hardness level of a constraint in a set of constraints. DL determines a constant penalty of breaking a constraint in a set of constraints to satisfy. We make the satisfaction of a hard constraint more important than satisfying all constraints at a lesser level. To favor a constraint over all other easier constraints, its DL must to be greater than the sum of DL values for all the constraints with a lower DC score. Therefore, we define the DL as

$$DL(c, C) = \begin{cases} |C|, & \text{if } r = 0 \\ 2^{r-1} \times (|C| + 1), & \text{if } r > 0 \end{cases}$$

where r is the rank (starting at 0) of c in the constraints set C in ascending order of DC.

To get a fitness value based on DL for a test candidate i on a set of branches C we apply the following formula:

$$f_{DL}(i, C) = \sum_{c \in C} \ell(i, c) + \eta(i, c), \text{ where } \ell(i, c) = \begin{cases} 0, & \text{if } \eta(i, c) = 0 \\ DL(c, C), & \text{if } \eta(i, c) \neq 0 \end{cases}$$

This fitness function allows to compare test candidates on different levels. It absolutely favours test candidates that have a smaller penalty value, i.e, test candidates that satisfy hard constraints. Contrary to the fitness function based on DC this one always prefers a test candidate that satisfies a hard constraint and breaks all easier constraints over a test candidate that breaks the hard constraint, even though it satisfies all the easier constraints.

4   Empirical Study

Our aim in this empirical study is to analyze the impact of our proposed fitness functions on SBST in terms of effectiveness and efficiency: SBST is considered more efficient if the new fitness functions are able to reduce the number of evaluations; it is considered more effective if the new fitness functions are able to cover more targets. We compare our proposal to the state of the art, f_{AL}, f_{SE}, and also to a natural combination of the two that applies the branch level as a corrective coefficient to each term of f_{SE}, which we denote f_{SEL}.

To perform our empirical study, we select two widely used meta-heuristic algorithms: Simulated Annealing (SA) and Evolutionary Algorithm (EA). To implement them we use the open source library opt4j [5]. To define an optimization problem in opt4j, the user needs only to define a fitness function and a representation of a test candidate which, in our case, is a vector of integers. We defined a Java class that exports a fitness evaluator for each fitness function: f_{AL}; f_{SE}; f_{SEL}; f_{DC}; f_{DL}. We kept all the default parameters for both algorithms (EA and SA), with one exception. As a default the mutation and the neighbourhood operators make changes uniformly at random and so they do not take into consideration the current value of the changed vector component of a test candidate. Because f_{AL} and f_{SE} are highly dependent on branch distance and expect a change close to the current value, to make the comparison more fair

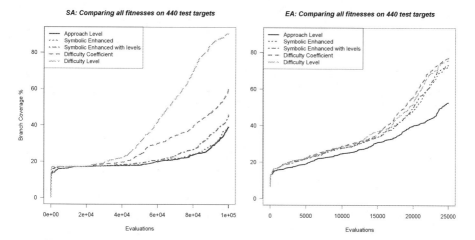

Fig. 2. SA on 440 test targets   **Fig. 3.** EA on 440 test targets

we defined new mutation and neighbourhood operators centered on the current value v: they select a new value uniformly at random in the range $[v-\triangle, v+\triangle]$, where $\triangle$ represents 1% of the domain $[-10^4, 10^4]$.

The study was performed on a well-studied benchmark, the triangle program [6], and on 440 synthetic test targets that were randomly generated. A synthetic test target is a simple program that contains a set of nested branches to be satisfied. To get realistic test targets, every test target is generated carefully, branch by branch. Every branch must: i) keep the test target feasible; ii) involve two variables (80%) or a variable and a constant (20%); iii) not be implied by the current test target.[1]

Each search for test data that meets a test target was performed 20 times for every combination of fitness function and meta-heuristic algorithm. If test data was not found after 25000 (respectively 100000) fitness evaluations for EA (respectively SA), the search was terminated. For all techniques, the 20 executions were performed using an identical set of 20 initial populations.

Experiments on the Triangle program show that all fitness functions perform in the same way on all branches except on the two branches that represent an isosceles and an equilateral triangle —they need at least two out of three parameters to be equal. On these two branches we observed that our fitness function f_{DL} outperforms the others in terms of number of evaluations necessary. Therefore f_{DL} makes SBST more efficient on Triangle, especially on these two branches.

Fig. 2 and 3 show the coverage achieved by the EA and the SA on the 440 test targets with respect to the number of evaluations. For each fitness function, a test target is considered as covered if at least one execution out of 20 succeeded to generate a test data. If more than one execution succeeded to do that, then the

---

[1] Benchmarks are available at
http://www.crt.umontreal.ca/~quosseca/fichiers/23-benchsCPAOR13.zip

median value of evaluations is used. These two plots confirm that the proposed fitness functions can enhance the effectiveness of SBST: the search based on f_{DL} or f_{DC} is able to cover test targets that are not covered by either f_{SE}, f_{AL} or their combination f_{SEL}. All fitness functions perform similarly on easy test targets but f_{DL} and f_{DC} outperform the rest on hard to reach test targets. The improved effectiveness is particularly clear on the SA results, where we observe that our fitness f_{DL} can reach 90% coverage whereas the best of f_{SE}, f_{AL} and f_{SEL} could not reach 45%.

In both Figures 2 and 3 we can observe that, from a certain point on, a given number of fitness evaluations allows our fitness functions to cover more targets than f_{SE} or f_{AL}. Therefore we may say that the proposed fitness functions improve the efficiency of SBST.

5   Conclusion

In this paper we defined two new metrics to measure the difficulty to satisfy a constraint in the context of test case generation for software testing. We used them to propose new fitness functions that evaluate the distance from a test candidate to a given test target. We presented results of an empirical study on a large number of benchmarks, indicating that the search algorithms EA and SA with our new fitness functions are significantly more effective and efficient than with the largely used fitness functions from the literature. Future work will focus on performing more experiments on real world programs to confirm whether the observed advantage of the proposed fitness functions represents a general trend.

References

1. Arcuri, A.: It does matter how you normalise the branch distance in search based software testing. In: ICST, pp. 205–214. IEEE Computer Society (2010)
2. Baars, A.I., Harman, M., Hassoun, Y., Lakhotia, K., McMinn, P., Tonella, P., Vos, T.E.J.: Symbolic search-based testing. In: Alexander, P., Pasareanu, C.S., Hosking, J.G. (eds.) ASE, pp. 53–62. IEEE (2011)
3. Collavizza, H., Rueher, M., Hentenryck, P.V.: Cpbpv: a constraint-programming framework for bounded program verification. Constraints 15, 238–264 (2010)
4. Gotlieb, A.: Euclide: A constraint-based testing framework for critical C programs. In: ICST, pp. 151–160. IEEE Computer Society (2009)
5. Lukasiewycz, M., Glaß, M., Reimann, F., Teich, J.: Opt4j: a modular framework for meta-heuristic optimization. In: Krasnogor, N., Lanzi, P.L. (eds.) GECCO, pp. 1723–1730. ACM (2011)
6. McMinn, P.: Search-based software test data generation: a survey. Software Testing Verification & Reliability 14, 105–156 (2004)
7. Pasareanu, C.S., Rungta, N.: Symbolic pathfinder: symbolic execution of java byte-code. In: Pecheur, C., Andrews, J., Nitto, E.D. (eds.) ASE, pp. 179–180. ACM (2010)
8. Pesant, G.: Counting solutions of CSPS: A structural approach. In: Kaelbling, L.P., Saffiotti, A. (eds.) IJCAI, pp. 260–265. Professional Book Center (2005)

9. Sakti, A., Guéhéneuc, Y.G., Pesant, G.: Boosting search based testing by using constraint based testing. In: Fraser, G., de Souza, J.T. (eds.) SSBSE 2012. LNCS, vol. 7515, pp. 213–227. Springer, Heidelberg (2012)
10. Tracey, N., Clark, J.A., Mander, K., McDermid, J.A.: An automated framework for structural test-data generation. In: ASE, pp. 285–288 (1998)
11. Wegener, J., Baresel, A., Sthamer, H.: Evolutionary test environment for automatic structural testing. Information and Software Technology 43(14), 841–854 (2001)

Coalition Formation
for Servicing Dynamic Motion Tasks

Udara Weerakoon and Vicki Allan

Computer Science Department, Utah State University,
4205, Old Main Hill, Logan, UT, 84322-4205, USA
u.w@aggiemail.usu.edu, vicki.allan@usu.edu

Abstract. When multiagent systems are used, design of an energy-efficient and autonomous data routing mechanism for wireless sensor networks is challenging. We see this challenge in the problem of coalition formation of agents (transmitters) for allocating dynamic-motion tasks (sensors) where the tasks have different service deadlines, and are in motion. The problem becomes harder when there are more tasks than agents, and when the data transmission is noisy. To address this, we design a novel and anytime decentralized heuristic algorithm to form coalitions. This algorithm can achieve at least $72 \pm 0.8\%$ and at most $102 \pm 2.2\%$ performance relative to the best known centralized coalition formation algorithm in such a sensor network.

Keywords: Coalition formation, sensor networks, data routing, simulation.

1 Introduction

Coalition formation for allocating tasks to agents in a multiagent system (MAS) is challenging when the tasks are *dynamic-motion tasks* [1]. Coalition formation happens when several agents come together, and make an agreement to cooperate, coordinate and communicate to service a number of tasks. A dynamic-motion task is a task where its workload changes over time, and where the task is in motion. For example, in a mobile Wireless Local Area Network (WLAN) wherein a number of *external mobile stations* (i.e. *mobile data sources*) need to be serviced, *the mobile servers* need to access data from those sources by collecting and transmitting the data, via a noisy communication channel, to a receiver before each data source has exhausted its memory. It has been proven that a multiagent based wireless network system can cope with the complexity of this problem [2].

In our model, agents are designed as *autonomous* and *self-adapting* mobile data transmission servers that move to the physical location of the dynamic-motion task to service it and transmit that data to the central receiver. Coalitions of these agents cooperate with each other to reduce the power required for the data transmission, and to increase the quality of this data transmission. For example, two agents service the sensor simultaneously in a way that both collect the data from it, and only one transmits data (utilizing modulation techniques) [3] to minimize the consumption of energy used for transmitting data. In another scenario, relay agents bridge the transmission channel to increase the quality of transmission.

C. Gomes and M. Sellmann (Eds.): CPAIOR 2013, LNCS 7874, pp. 386–393, 2013.

This paper moves the state of the art forward in the following ways. To our knowledge, this paper is the first to address the problem of allocating coalitions of agents to dynamic-motion tasks in a mobile WLAN. Furthermore, the paper devises an anytime decentralized heuristic search based algorithm to form coalitions and to allocate dynamic-motion tasks simultaneously. Finally, this paper shows the decentralized algorithm has a maximum of $160 \pm 1.1\%$ efficient solution compared to the random allocation and has $102 \pm 2.2\%$ best performance for the algorithm designed by Saad et al [2] in a crowded neighborhood.

2   Problem Statement

By forming coalitions, agents transmit data generated by a number of dynamic-motion tasks to a data receiver via a noisy communication link. Issues with regard to coalition formation can be formulated as following: *1)* Which agents should work together to service which tasks? *2)* What order should tasks be serviced? *3)* Which agents should serve as relays? and *4)* When should coalitions be reorganized and tasks reassigned?

The agent/task coalition consists of a number of tasks, attended by agents in a specific order. Like servers of the mobile WLAN, which poll several network nodes to transmit data packets, the agents travel among the physical locations of the tasks to service them (i.e. collect the task's data packets and transmit them to the receiver which is located at a static position in the environment). Agents of the coalition work either as a collector or as a relay. In the coalition of collector agents, only one collector needs to transmit modulated signals from the task serviced to the central data receiver. This allows the other collectors to save their transmission power, and to reduce the overhead of the coalition. Relay agents are additional coalition members. They are dedicated to bridge the transmission, but do no collection. In some cases, there are series of relays, the first of which receives data from the collectors of the coalition, and the last sends data to the receiver. Here, each relay maintains an equal distance to both its successor and predecessor. Probability $p_{(i,j)}$ of a data packet being transmitted from relay agent i to agent j is defined as follows [2,3]:

$$p_{(i,j)} = exp\left(-\frac{\sigma^2 \upsilon_0 d_{(i,j)}^\alpha}{\kappa \tilde{P}}\right) \tag{1}$$

where σ^2 is a variance of the Gaussian noise, υ_0 is the target level, $d_{(i,j)}$ is the distance from relay agent i to relay agent j, α is a path loss exponent, κ is a path lost constant, and $\tilde{P}$ is the maximum transmit power of the relay agent. The probability $p_{(i,j)}$ determines the strength of the communication link in an environment in which fading, like air resistance and destructive inference to the transmitted data packet, is affected.

In a mobile WLAN, a coalition theoretically needs to satisfy its primary constraint, the *utilization constraint*, where the total data generation rate must not exceed the total data transmission rate [3,2,4]. A coalition is a *feasible coalition* if it does not violate the utilization constraint. While infeasible coalitions may be necessary to give the best global outcome, our method, in an effort to divide the work fairly, attempts to produce feasible coalitions. Once the initial coalitions are determined, unallocated tasks are incorporated into proximate coalitions when possible. The utilization constraint is defined as follows:

$$\rho_c = \frac{\sum_{i \in T_C} \lambda_i}{\sum_{j \in C} \mu_j} \tag{2}$$

where ρ_c is the utilization factor of the coalition C, and T_C is the set of tasks allocated to coalition C.

In the coalition, agents service the tasks by visiting each of the tasks according to the order determined by the *dynamic routing mechanism*. With the dynamic routing mechanism, the decision of the agents as to the order in which the tasks are visited may depend on a certain amount of information available to the agents, such as the number of data packets stored by the task and the location of the task. When servicing the task, the collectors transmit either all (*exhausted strategy*) or a limited number (*gated strategy*) of data stored in the memory of the task.

Agents utilize the following seven functions for measuring utility $u(\cdot)$ of coalition C defined by Equation 3 based on the resources provided by the agents and the resources required by the allocated tasks.

1. $f_1(C)$ is the *gross profit* in terms of the amount of data collected from tasks allocated to the coalition
2. $f_2(C)$ is the *possible profit* in terms of the total amount of data generated by the tasks allocated to the coalition
3. $f_3(C)$ is the *net profit* in terms of the amount of data packets that have successfully been transmitted to the receiver by the coalition
4. $f_4(C)$ is the *total time* spent by the agents of a coalition in gathering and transmitting data packets
5. $f_5(C)$ is the *overhead* of the coalition which is the sum of the idle time and the traveling time of the agents
6. $f_6(C)$ is the *cost* of transmitting data and sending messages to form the coalition. The cost of transmitting data is proportional to the distance of transmission, and the cost of sending message is proportional to the distance divided by square root of two [5].

$$u(C) = \sum_{i=1}^{6} \alpha_i \cdot f_i(C) \tag{3}$$

where α_i is the coefficient of function $f_i(C)$, and α_i is determined empirically.

By defining the problem of coalition formation with spatial and temporal constraints (CFSTP), authors [1] show that CFSTP falls within the class of NP-hard problems. Likewise, we can also show that coalition formation for allocating tasks to agents in an environment where each task has a workload, a time deadline, and mobility is an NP-hard problem.

3   Methodology

This section presents a new method, termed Anytime Tabu and Variable neighborhood [6] heuristic algorithm (ATV), whereby agents form coalitions to service dynamic-motion tasks. ATV is composed of four stages: *1*) the generation of initial coalitions, *2*) the improvement of the coalitions, *3*) transmission of data, and *4*) the adaptation of the coalitions. To explain ATV, two new terms are introduced: *entity* and *communication radius*. An entity is either a task or an agent. An entity can recognize another entity that is proximate if the Euclidean distance between its location and the location of the other entity is less than a predefined distanced, which is termed the communication radius.

In the first stage (the generation of coalition formation), each agent within a set of agents (A_m) that are capable of communicating with each other cooperates to select a task group. A task group is a group of discrete tasks that are proximate. Tasks (T_m), known to the agents, are grouped into $\tilde{n}_m$ number of task groups by executing the well-known *k-mean clustering algorithm* [7]. To determine the value for $\tilde{n}_m$, Equation 4 is used. In Equation 4, T_m and A_m denote the set of known tasks and the set of known agents, respectively, and λ_i and μ_j respectively denote the data generation rate of task i and data transmission rate of agent j. At the end of first stage of the generation of initial coalitions, $|A_m|$ out of $\tilde{n}_m$ number of tasks groups are allocated to agents in A_m.

$$\tilde{n}_m = \left\lceil \frac{\sum_{i \in T_m} \lambda_i}{\left(\sum_{j \in A_m} \mu_j\right) / |A_m|} \right\rceil . \qquad (4)$$

To explain a new term *the neighboring coalition*, we introduce two attributes of the coalition: *centroid* and *radius*. The centroid of the coalition is the average x and y lo-cations (Cartesian coordinates) of tasks in the coalition. The radius of the coalition is defined as the distance to the furthest task of the coalition from the centroid. The neighboring coalition is the coalition that significantly overlaps with the considered coalition. The neighboring coalition is significantly overlapped when the total distance of two radii, minus the distance between the centers of the two centroids, is less than a predefined distance.

In the second stage, coalitions are improved by agents utilizing ATV search, which has three phases: *1*) the exploitation phase, *2*) the feasibility phase, and *3*) the explo-ration phase. Each of these three phases utilize two primitive operations: *1-move* and *swap*. The 1-move operation moves an entity from its own coalition to a neighboring coalition. There are four possibilities for a 1-move: *1*) move a task among coalitions, *2*) move a task from coalition to an unassigned task group, *3*) move a task from a task group to a coalition, or *4*) move an agent from coalition to task group. On the other hand, the swap operation exchanges two entities between coalitions or task groups. There are three possible swap options: *1*) swap tasks between coalitions, *2*) swap tasks among a coalition and a task group, and *3*) swap agents among coalitions.

In the first exploration phase of ATV search, the Variable Neighborhood Search (VNS) is used. VNS helps to move the search forward, as an agent may get stuck in poor local optima. In VNS search, an agent executes a limited number of 1-move or swap operations to exploit a better *solution* (the term solution is defined as the set of coalitions, consisting of agents and tasks) without considering the utilization constraint.

However, VNS does not use the exhaustive search of all feasible changes from the current solution. The stopping criterion for VNS is fixed at an empirically decided number of iterations without improvements. After finding the series of operations, the agent accepts them if VNS gives an improved solution from the current solution. To find the series of operations, the agent uses follwoing five heuristics:

1. *Move* - moves unallocated proximate tasks to the coalition
2. *Merge* - merges the coalition C_1 with a neighboring coalition C_2, when the new coalition C_3 is super additive: $u(C_3) \geq u(C_1) + u(C_2)$
3. *Share* - shares a batch of tasks that are proximate to the border, with the neighboring coalition
4. *Pass* - passes tasks to a neighboring coalition, when the utility of neighboring coalition is less than the empirically defined threshold
5. *Split* - splits the coalition with members, when the utility of the coalition is less than the empirically defined threshold

$$I(C) = \max \{\rho_c - 1, 0\}^2 . \tag{5}$$

In the second phase of ATV search, the feasiblility phase, agents make the coalition feasible. To evaluate coalition feasibility, *infeasibility* $I(\cdot)$ of a coalition is measured by Equation 5, where ρ_c is defined in Equation 2. This squared value where ρ_c is greater than one, helps agents to recognize coalitions with a small infeasibility rather than few coalitions with a great infeasibility. Thus, agents can make a small change according to Heuristic Share and Pass, e.g. move a single task with ease, and make the coalition feasible.

The third and last phase of ATV search is the exploration phase. Here, agents search through a sequence of solutions in which the next solution to be searched is derived from the current solution by executing a single 1-move or swap operation. To do so, agents utilize the tabu search. The tabu search is a meta-heuristic search that employs an adaptive memory to forbid the search to consider 1-move or swap operations, already considered, for a short period of time. Here, agents first find a 1-move of an entity to improve the utility of the solution. When agents cannot improve the solution by a 1-move, they search for possible swap operations. Once agents execute any of the operations, they run the *k-opt procedure* [8] to find the optimal path through the tasks of the coalition. The stopping criterion is fixed at an empirically decided number of iterations without improvements.

Finally, in the forth and last stage of ATV, agents deal with dynamism, where ATV is repeated in the significant motion pattern. In this technique, an agent constantly monitors the centroid and the radius of its own coalition for any significant deviation from the current values. The centroid and the radius are significantly changed by: *1)* arrival of a new task to the considered area of the coalition and *2)* the existing tasks move further apart.

4   Empirical Evaluation

To test the proposed ATV, simulations are set in the following WLAN configurations. A central receiver is placed at the center of a 4×4 km square area. The path loss parameters are set to $\alpha = 3$ and $\kappa = 1$, the target SNR is set to $\upsilon_0 = 10$ dB, and the noise variance $\sigma^2 = -120$ dBm [3]. All packets are designed to be 256 bits, which is the typical IP packet size. The agents have a transmission power of $\tilde{P} = 100$ mW. The data transmission capacity is 768 kbps and the communication radius is 0.8 km for each agent. Unless otherwise stated, the speed of each agent and task are 30 km/h and 0.6 km/h, respectively. The task generates data packets according to a Poisson process, and stores them in its 16 MB memory which is a queue. The task has a data generation rate of 128 kbps which could be mapped to video services. The simulation runs for 100 test instances. Each test instance requires 40 minutes to run. The results of ATV are here compared with three centralized algorithms: Saad [2], equal allocation of neighboring tasks, and random allocation. An additional comparison is also included with a version of Saad algorithm with periodic reconfiguration.

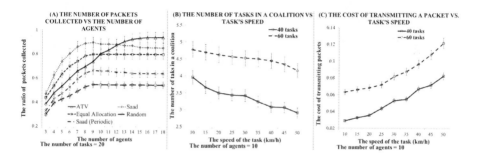

Fig. 1. Results for ATV

In Figure 1 (A), the ratio of successfully transmitted packets to generated packets is assessed. The ratio calculated is the number of data packets that have successfully been transmitted, divided by the total data packets generated by tasks. The ATV outperforms the Saad (periodic) algorithm. Agents in the ATV effectively and continually reconfigure themselves within a period of time much shorter than Saad. The ratio values increase due to the fact that when the number of agents increases, the distance that the agents need to travel within their coalition decreases. The increased number of agents also improves the probability of successful data transmission. In the crowded neighborhood, ATV is the most efficient and effective algorithm of this simulation.

Figure 1 (B) illustrates that when the velocity of the tasks is increased, the average number of tasks allocated to each coalition decreases. For this experiment, the speed of each agent is 60 km/h. Simulations are conducted with a constant number of ten agents. In one simulation, forty tasks are introduced. In the other simulation, sixty tasks are introduced. In each simulation, the velocity for the tasks is increased. As the velocity of the tasks increases, the necessity for forming new coalitions or dissolving existing coalitions increases significantly. These changes require agents to spend more time reconfiguring themselves. When the time spent for collecting data is low, the workload of data to be transmitted increases. The higher workload forces the number of tasks allocated to a coalition to drop gradually.

Figure 1 (C) shows variations in the cost incurred as a data packet is transmitted from a task to the central receiver. This cost has two components: *1*) the communication cost (the cost of forming coalitions and allocating tasks) and *2*) the actual transmission cost. The cost of sending a data packet or a communication message is measured by the distance of the transmission. This value is then divided by the number of successfully transmitted data packets where each data packet is a unit assigned a monetary value of one. As the possibilities of reconfiguration increase, there will always be an increase in the communication cost for successful data transmission.

5   Related Work

To route data from self-interested sensors (owned by different stake-holders) to the receiver, Rogers et al. [5] developed a communication protocol and a payment scheme using a multiagent design mechanism. Using only local information and the communication protocol, a sensor finds and selects another sensor that is willing to act as a hop to transmit its data to the receiver. However, Turget and Boloni [9] empirically tested sensor based multi-hop data routing via a simulation, and proved the lifetime of the sensor network can significantly be extended (500%) by eliminating the need for expensive multi-hop sensor routing. To eliminate the expensive sensor-based multi-hop data routing, the authors chose a sensor network with multiple mobile receivers. In contrast, Saad et al. [2] developed an agent based routing mechanism wherein mobile autonomous and self-adapting wireless transmitters (agents) visit the location of wireless and mobile sensors (tasks) to route data. These transmitters and sensors are controlled by the central commander. The issue of the Saad model is that the commander poses a possible threat of a single point of failure. In our paper, to overcome this threat, a novel any-time and decentralized heuristic search-based data routing mechanism has been devised. We refer reader to [10] for more details about approaches used in routing algorithms for wireless sensor networks. However, these approaches relate to network research, not multiagent research.

6   Conclusions

For the first time, the problem of coalition formation for allocating dynamic-motion tasks is presented from a multiagent perspective in a wireless network system. This paper shows this problem is NP-hard, and introduces a model to demonstrate the problem. In this model, data stored by dynamic-motion tasks is transmitted by agents to a central receiver via a noisy communication channel. A new anytime decentralized heuristic search based algorithm (ATV) is introduced to the model. The ATV assists agents to form coalitions, and to allocate tasks to those coalitions. The ATV utilizes the well-known Tabu search and variable neighborhood search. The simulation results show how the ATV allows agents to self-organize into coalitions. The use of the ATV improves the system performance as much as $160 \pm 1.1\%$ relative to the random allocation, and at least $72 \pm 0.8\%$ of performance relative to the best known centralized scheme by Saad et al. [2].

References

1. Ramchurn, S.D., Polukarov, M., Farinelli, A., Truong, C., Jennings, N.R.: Coalition formation with spatial and temporal constraints. In: AAMAS 2010, pp. 1181–1188 (2010)
2. Saad, W., Han, Z., Basar, T., Merouane, D., Hjorungnes, A.: Hedonic Coalition Formation for Distributed Tak Allocation Among Wireless Agents. IEEE Transactions on Mobile Computing, 1327–1334 (2011)
3. Proakis, J., Salehi, M.: Digital Communications, 5th edn. McGraw-Hill (2007)
4. Vinyals, M., Rodriguez-Aguilar, J.A., Cerquides, J.: A survey on sensor networks from a multiagent perspective. The Computer Journal, 455–470 (2011)
5. Rogers, A., David, E., Jennings, N.R.: Self-organized routing for wireless micro-sensor networks. IEEE Transactions on Systems, Man, and Cybernetics - Part A, 349–359 (2005)
6. Michalewicz, Z., Fogel, D.B.: How to Solve It: Modern Heuristics, 2nd edn. Springer (2010)
7. Aggarwal, N., Aggarwal, K.: An Improved K-means Clustering Algorithm for Data Mining, 2nd edn. LAP LAMBERT Academic Publishing (2012)
8. Helsgaun, K.: General k-opt submoves for the lin-kernighan tsp heuristic. Mathematical Programming Computation 1(2-3), 119–163 (2009)
9. Turgut, D., Bölöni, L.: Three heuristics for transmission scheduling in sensor networks with multiple mobile sinks. In: AAMAS Workshop on Agent Technology for Sensor Networks, pp. 1–8 (2008)
10. Akkaya, K., Younis, M.: A survey of routing protocols in wireless sensor networks. Ad Hoc Network 3(3), 325–349 (2005)

Computational Experience with Hypergraph-Based Methods for Automatic Decomposition in Discrete Optimization*

Jiadong Wang and Ted Ralphs

Department of Industrial and Systems Engineering, Lehigh University, PA 18015
{jiw508,ted}@lehigh.edu
coral.ie.lehigh.edu/~ted

Abstract. Branch-and-price (BAP) algorithms based on Dantzig-Wolfe decomposition have shown great success in solving mixed integer linear optimization problems (MILPs) with specific identifiable structure. Only recently has there been investigation into the development of a "generic" version of BAP for unstructured MILPs. One of the most important elements required for such a generic BAP algorithm is an automatic method of decomposition. In this paper, we report on preliminary experiments using hypergraph partitioning as a means of performing such automatic decomposition.

1 Introduction

We consider solution of a *mixed integer linear optimization problem* (MILP), which is to compute

$$z_{IP} = \min\{c^\top x \mid Ax \leq b, x \in \mathbb{Z}^r \times \mathbb{R}^{n-r}\}, \tag{1}$$

where $A \in \mathbb{Q}^{m \times n}$, $c \in \mathbb{Q}^n$, and $b \in \mathbb{Q}^m$. Although decomposition-based methods have the potential to generate strong dual bounds, to reduce symmetry, and to exploit special structure for such problems, they generally require that a decomposition of the constraints be either given as input or discovered as part of the solution process. In this paper, we focus on automatic detection of structure using hypergraph partitioning algorithms. Methods for automatic structure detection have been widely used in the context of certain linear algebraic computations, mainly for the purpose of efficient parallelization [1]. In solving MILPs, our goal is to exploit structure not only to allow the use of parallel computation but also to improve on the bounds yielded by solving the linear relaxation, eventually leading to improved solution times for certain classes of MILPs.

When detecting block structure automatically, it is important to have a measure of "quality" that is easy to compute, since there are usually multiple ways of decomposing a given matrix. When the goal is to parallelize a single matrix computation, the measures of quality are fairly straightforward, but in the MILP context, measuring the quality of a decomposition is much more difficult. Ultimately, our goal is to reduce the overall computation time as much as possible,

---

* NSF Grant CMMI-1130914.

C. Gomes and M. Sellmann (Eds.): CPAIOR 2013, LNCS 7874, pp. 394–402, 2013.

but solution time is difficult to predict and proxies for this measure must be used in practice. One such proxy is the bound improvement achieved in the root node, which also has the advantage of being unaffected by changes to other algorithmic strategies, such as branching. Unfortunately, the computation of this quantity is expensive, so one of our goals is to assess cheaper alternatives.

The work reported on herein follows lines of development similar to those of the study reported in [2]. The work in [2] was performed using GCG [3], a solver that implements an approach to generic decomposition similar to that used in the DIP framework [4] that is employed in this study. The present study differs by focusing on producing singly-bordered block-diagonal matrices, as opposed to the doubly-bordered structures favored in [2], which require both a different HP model and a slightly different solution technique. Furthermore, our measures of goodness take into account not only the distribution of nonzeros, but also the distribution of nonzeros in columns corresponding to integer variables. Finally, we also report on a number of issues that arise in tuning the HP software. A more detailed treatment of this material is available in [5].

2   Generic Branch-and-Price

A *decomposition* is a partition of the rows of $[A, b]$ into two sub-matrices $[A', b']$ and $[A'', b'']$. The decomposition is chosen such that the solution of the relaxation obtained by dropping the linear constraints represented by the sub-matrix $[A'', b'']$ is more tractable than that of the original problem. The principle underlying decomposition methods is to exploit the solvability of this relaxation in order to solve the original problem more efficiently. The methodology that is the basis of our experiments here is the branch-and-price (BAP) algorithm, which uses Dantzig-Wolfe (DW) Decomposition and column generation [6] to obtain bounds within a branch-and-bound framework (see [7] for details). A number of software frameworks have been developed to aid in the implementation of BAP algorithms, including BaPCod [8], DIP [4], GCG [3], BCP [9], and ABACUS [10]. However, it has proven difficult in general to develop a "generic" version of BAP, requiring no input from the user beyond the model itself. A generic variant of BAP can be obtained, however, by (1) using automatic methods of identifying block structure in the matrix to produce candidates for the decomposition, (2) solving the column generation subproblem using a generic MILP solver, and (3) branching on disjunctions derived from the original problem formulation. Identification of block structure involves determining a permutation of the rows and columns of A that reveals disjoint blocks of nonzero elements, as illustrated in Figure 1. When A' has this structure, the column generation subproblem decomposes naturally into independent (and much smaller) MILPs.

3   Hypergraph-Based Partitioning Methods

Previous work on identifying block structure in the context of MILP includes [11] and [12], but the methods proposed therein proved impractical. Multilevel algorithms for hypergraph partitioning provide a computationally practical heuristic alternative [13]. A *hypergraph* $\mathcal{H} = (V, E)$ is a generalization of a traditional graph, in which V is the set of nodes and E is a set of *hyperedges* or *nets*, which are subsets of nodes of arbitrary cardinality (in contrast to a traditional graph in which edges are subsets of cardinality two). Given a vector of weights $w \in \mathbb{R}^E$, the K-way HP problem is to partition the nodes of a hypergraph $\mathcal{H}$ into at most K subsets while minimizing the weight of the resulting *cut*, which is the sum of the weights of all edges having a nonempty intersection with more than one member of the partition. A secondary objective, usually expressed as a constraint, is for the cardinalities of the members of the partition to be approximately balanced. If the nodes have weights, we want the sum of the weights of the nodes in each member of the partition to be balanced. To use an HP algorithm to find (singly-bordered) block-diagonal structure in a matrix, we use a *row-net* model in which we identify each column of the matrix with a node in an associated hypergraph and identify each row of the matrix with a hyperedge consisting of the nodes associated with the columns in which the row has nonzero elements [14]. After mapping the matrix to a hypergraph in this way and finding a partition of the hypergraph, we can identify the structure of the original matrix as follows. The rows of each block consist of the set of hyperedges whose elements are completely contained in one of the partitions of the node set. The coupling rows consist of all hyperedges in the cut, i.e., hyperedges having nonempty intersection with more than one element of the partition. Thus, minimizing the size of the cut is the same as minimizing the number of coupling rows (when the edges have unit weights). The balance constraint can then be interpreted as ensuring that the blocks have approximately equal numbers of columns. Figure 1(a) shows the pattern of nonzeros in the constraint matrix of instance `a1c1s1` from MIPLIB 2003 [15], while Figures 1(b)–1(d) show the hidden block structure detected through the use of HP with different numbers of blocks (see Section 4.2 for details).

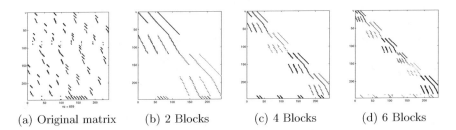

(a) Original matrix   (b) 2 Blocks   (c) 4 Blocks   (d) 6 Blocks

Fig. 1. Hidden block structure in MIPLIB instance `a1c1s1`

4 Computational Experiments

The experiments here were performed on compute nodes with dual eight core 0.8GHz AMD Opteron(tm) processors having 32G memory and 512KB cache per core. We used the hMETIS and PaToH hypergraph partitioning tools [16, 17] to detect the block structure and the DIP (Decomposition in Integer Programming) framework [4], an open-source package available from the COIN-OR repository, to implement the resulting generic BAP algorithm. We used the open-source solvers CLP [18] and CBC [19] to solve the restricted master problem and the subproblems, respectively. The test instances were selected from MIPLIB2003 and MIPLIB2010 [20] from among instances of medium size.

We want to emphasize that the results reported here are preliminary in nature and should be considered the first steps in what we hope will be a long line of future studies. While we do think that our results indicate promise, there is much work to be done and the tests are necessarily limited in scope. The small set of MILPs on which we experiment here should not be considered "representative." Our overarching goal is to motivate further study and further the development of a general framework and direction for that study.

4.1 Comparing PaToH and hMETIS

We first compared PaToH and hMETIS and obtained the following (subjective) insights, based on which we choose hMETIS for experiments in the remainder of this paper.

1. hMETIS produced more balanced solutions in general and the number of blocks produced was typically equal to the given maximum.
2. In cases for which PaToH produced fewer than the given maximum number of blocks, the bound tended to be better, since fewer blocks yield stronger bounds in general (but result in more difficult subproblems). When using PaToH, one might set the maximum number of blocks higher than what is desired in order to avoid the situation described above.
3. Generally, when the number of blocks produced was the same, hMETIS performed better than PaToH in terms of bound improvement at the root node.

4.2 Parameter Tuning

Before executing the partitioning procedure, there are two primary parameters that need to be set, both of which may affect the resulting bound improvement and overall decomposition quality. One is the maximum number of elements in the partition (number of blocks in the resulting partitioned matrix) and the other is the weights of the nodes and hyperedges. The experiments below illustrate the impacts of variation of these parameters on the decomposition.

Number of blocks. Figure 1 provides a visualization of block structure with different numbers of blocks for a single instance. It is difficult to tell from these figures which of these decompositions would be the most effective in our present setting and we have found this to be true in general. Preliminary experiments failed to find a strong correlation between the number of blocks and the bound achieved, though clearly the number of blocks must have some impact. Figure 2(a) shows the relationship between the optimality gap closed and the number of blocks on several instances. For the instance *go19*, the optimality gap closed decreases as the number of blocks increases, as perhaps would be expected. Not too surprisingly, there are no cases in which the bound increases reliably as the number of blocks increases. Interestingly, however, there are cases, such as the instance *swath*, for which the optimality gap increases and then decreases as the number of blocks increases.

To illustrate this phenomenon, we take computational time into consideration on some selected instances. Figures 2(b) and 2(c) show the relationship between the bound, the number of blocks, and the computation time required to obtain the bound for two instances. Generally, it can be observed that as the number of blocks increases, the time required decreases. The question remains how to choose the number of block a priori. Though this is a crucial question, we do not yet have a good answer. The most obvious strategy is to choose the number of blocks based on the number of available cores. In the remaining experiments, we fixed the number of blocks at three.

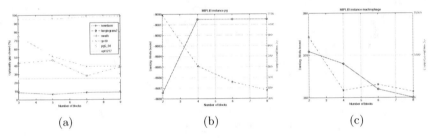

(a)   (b)   (c)

Fig. 2. Effect of number of blocks on gap, bound, and computation time

Weight of nodes and hyperedges. Our expectation is that decompositions in which the integer variables play a strong role in the individual blocks will yield strong bounds, since the DW bound is known to be equal to the LP bound when none of the blocks involve integer variables. One strategy, therefore, is to assign a higher weight to nodes corresponding to integer variables and to hyperedges that include a high number of such nodes. We compared the DW bound obtained with unit weighted and non-unit weighted nodes and hyperedges. In the non-unit weighted cases, we assigned a weight of two to integer variables and a weight of one to continuous variables. Weights of hyperedges were assigned to be the total number of nonzero elements in their corresponding rows. Out of 25 instances, six showed a decrease with weighting and the rest showed a bound at least as high as in the unit weight case (11 instances showed a strict increase). It thus appears that adjusting the weight of nodes and hyperedges intentionally should help in most cases and this agrees with our intuition.

4.3   Quality Measures for Decomposition

The core obstacle in automating all aspects of the BAP algorithm is to understand when BAP will work well based on easily discernible properties of the instance. As is traditional in machine learning, we refer to numeric quantities that may indicate the quality of a decomposition as *features* and propose five here ("integer elements" below are the nonzero elements of the matrix that are those in columns corresponding to the integer variables).

1. Ratio of the number of nonzero elements in the coupling rows to the total number of elements in the coupling rows (α).
2. Ratio of the number of integer elements in the coupling rows to the total number of nonzero elements in the coupling rows (β).
3. Ratio of the number of integer elements in the coupling rows to the total number of integer elements in the matrix (γ).
4. The average mean value of the ratio of the number of integer elements in a given block to the number of nonzero elements in that block (η).
5. The sample standard deviation of the measure η (θ).

Our initial conjecture was that lower values of α, β, γ, and $1 - \eta$ will lead to higher quality decompositions (as indicated by root bound), since it seems natural we would want more rows and nonzeros (both overall and in the columns corresponding to integer variables in particular) in the blocks. Motivated by the success of practical application in which BAP algorithms were used, we also conjectured that achieving balance in the number of rows and the number of integer variables in each block should yield good results. Here, we use θ as a measure of the balance among the blocks.

Figure 3(a) shows the relationship between the five potential measures and the DW optimality gap closed for an initial set of instances. Two out of the five measures (α, γ) appear to have the strongest correlation with optimality gap closed and DW bound improvement, although there are outliers and the relationship is not very strong. The standard deviation feature θ also does not clearly show potential. Based on the relationship between α, γ and bound improvement, we propose a prediction measure to estimate the quality of detected structure at the root node. It is a function of α and γ, expressed as

$$\Pi = 1 - \min(\alpha, \gamma),$$

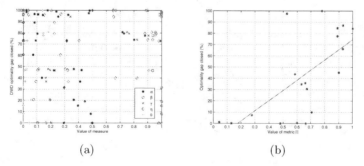

(a)   (b)

Fig. 3. The relationship between features, gap closed, and proposed metric

For this test set, the measure Π appears to have a positive correlation with optimality gap closed. Tests on 20 instances different from the ones used in Figure 3(a) used to derive the measure are shown in Figure 3(b) and illustrate the possible linear relationship obtained by linear regression between the metric and the optimality gap closed. While these results are promising, we repeat our earlier caveat that these conclusions should be considered preliminary.

5   Conclusions

We discussed the use of the row-net hypergraph model to detect block-diagonal structure of generic MILP problems automatically. The final goal is not to replace cutting plane methods, but rather to supplement them. By honing our ability to determine in a preprocessing step when decomposition might be effective, we hope to develop solvers capable of switching between cutting plane and column generation methods as appropriate. Moving forward, there are many obvious future research directions to pursue and we mention a few here.

1. This paper focuses on singly-bordered block-diagonal structure detection, but it may be advantageous to consider a more general framework allowing for doubly-bordered structure, as in [2].
2. There is much more work to be done in determining how to judge the quality of a given decomposition. Primarily, we need to take computation time into account as a component of goodness rather than simply considering the root bound, which can be easily manipulated. In this regard, there may be additional features that relate to decomposition quality that have not yet been explored. Even with an explicit metric, choosing the numeric threshold for determining whether the decomposition is good or not is difficult. Motivated by recent machine learning techniques, we may use support vector machine to classify the given decompositions according to their suitability for computation. Instances about which we have deep knowledge can be used for training.
3. In this study, we did not consider methods of choosing the number of blocks automatically based on properties of a specific instance. In many cases, the matrix may have a natural block structure to begin with and our methodology should be capable of detecting that and respecting this natural number of blocks. Otherwise, we may actually end up destroying existing structure with our automatic methods. We would also like to take into account the possibility of solving the subproblems in parallel by exploiting the block structure, in which case we should take into account the number of available cores/processors when choosing the number of blocks.
4. Finally, we have yet to experiment fully with how we can use the variance of node and edge weights to improve the quality of the decomposition. At the moment, we are using the partitioning software "out of the box," but it is our long-term goal to develop specialized methods for achieving decompositions that will be effective in the specific context discussed here.

Much work remains to be done and the potential for these methods to work in a completely generic fashion is still unclear. We are convinced that they do have a role to play and hope that this work will encourage more investigations.

References

[1] Catalyurek, U.V., Aykanat, C.: Hypergraph-partitioning-based decomposition for parallel sparse-matrix vector multiplication. IEEE Transactions on Parallel and Distributed Systems 10, 673–693 (1999)

[2] Bergner, M., Caprara, A., Ceselli, A., Furini, F., Lübbecke, M.E., Malaguti, E., Traversi, E.: Automatic Dantzig-Wolfe reformulation of mixed integer programs, http://www.optimization-online.org/DB_FILE/2012/09/3614.pdf

[3] Gamrath, G., Lübbecke, M.E.: Experiments with a generic Dantzig-Wolfe decomposition for integer programs. In: Festa, P. (ed.) SEA 2010. LNCS, vol. 6049, pp. 239–252. Springer, Heidelberg (2010)

[4] Ralphs, T.K., Galati, M.V.: DIP (2012), https://projects.coin-or.org/Dip

[5] Wang, J., Ralphs, T.K.: Computational experience with hypergraph-based methods for automatic decomposition in integer programming. Technical Report 12T-014, COR@L Laboratory, Lehigh University (2012), http://coral.ie.lehigh.edu/~ted/files/papers/CPAIOR12.pdf

[6] Barnhart, C., Johnson, E.L., Nemhauser, G.L., Savelsbergh, M.W.P., Vance, P.H.: Branch-and-price: Column generation for solving huge integer programs. Operations Research 46, 316–329 (1998)

[7] Galati, M.V.: Decomposition in Integer Programming. PhD thesis, Lehigh University (2009), http://coral.ie.lehigh.edu/ted/files/papers/MatthewGalatiDissertation09.pdf

[8] Vanderbeck, F.: BaPCod–a generic branch-and-price code (2005), http://wiki.bordeaux.inria.fr/realopt

[9] Ladányi, L.: BCP: Branch-cut-price framework (2012), https://projects.coin-or.org/Bcp

[10] Jünger, M., Thienel, S.: The ABACUS system for branch and cut and price algorithms in integer programming and combinatorial optimization. Software Practice and Experience 30, 1325–1352 (2001)

[11] Borndörfer, R., Ferreira, C.E., Martin, A.: Decomposing matrices into blocks. SIAM Journal on Optimization 9, 236–269 (1998)

[12] Ferris, M., Horn, J.: Partitioning mathematical programs for parallel solution. Mathematical Programming 80, 35–61 (1998)

[13] Catalyürek, U.V., Aykanat, C.: PaToH: A multilevel hypergraph partitioning tool, version 3.0. Technical Report 6533, Bilkent University, Department of Computer Engineering (1999)

[14] Aykanat, C., Pinar, A., Çatalyürek, Ü.V.: Permuting sparse rectangular matrices into block-diagonal form. SIAM Journal on Scientific Computing 25, 1860–1879 (2004)

[15] Achterberg, T., Koch, T., Martin, A.: The mixed integer programming library: MIPLIB 2003 (2003), http://miplib.zib.de/miplib2003

[16] Karypis, G., Kumar, V.: hMETIS 1.5: A hypergraph partitioning package. Technical report, Department of Computer Science, University of Minnesota (1998), http://www.cs.umn.edu/metis

[17] Çatalyürek, Ü.V., Aykanat, C.: PaToH: partitioning tool for hypergraphs (2012),
     http://bmi.osu.edu/~umit/software.html
[18] Forrest, J.J.: CLP: COIN-OR linear Programming Solver (2012),
     https://projects.coin-or.org/Clp
[19] Forrest, J.J.: CBC: COIN-OR branch-and-cut solver (2012),
     https://projects.coin-or.org/Cbc
[20] Koch, T., Achterberg, T., Andersen, E., Bastert, O., Berthold, T., Bixby, R.E.,
     Danna, E., Gamrath, G., Gleixner, A.M., Heinz, S., et al.: MIPLIB 2010. Math-
     ematical Programming Computation 3, 103–163 (2011)

Author Index